The Parental Brain

The Parental Brain

Mechanisms, Development, and Evolution

MICHAEL NUMAN

OXFORD
UNIVERSITY PRESS

OXFORD
UNIVERSITY PRESS

Oxford University Press is a department of the University of Oxford. It furthers the University's objective of excellence in research, scholarship, and education by publishing worldwide. Oxford is a registered trade mark of Oxford University Press in the UK and certain other countries.

Published in the United States of America by Oxford University Press
198 Madison Avenue, New York, NY 10016, United States of America.

Library of Congress Cataloging-in-Publication Data
Names: Numan, Michael, 1946- author.
Title: The parental brain : mechanisms, development, and evolution /
Michael Numan.
Description: New York : Oxford University Press, 2020. |
Includes bibliographical references and index.
Identifiers: LCCN 2019052894 (print) | LCCN 2019052895 (ebook) |
ISBN 9780190848675 (hardback) | ISBN 9780190848699 (epub) |
ISBN 9780190848705
Subjects: LCSH: Psychobiology. | Parental behavior in
animals—Physiological aspects. | Human behavior—Physiological aspects.
Classification: LCC QP360 .N87 2020 (print) | LCC QP360 (ebook) |
DDC 649/.1—dc23
LC record available at https://lccn.loc.gov/2019052894
LC ebook record available at https://lccn.loc.gov/2019052895

1 3 5 7 9 8 6 4 2

Printed by Integrated Books International, United States of America

Dedicated to my children: Bobbi, Suzanne, and Todd

Contents

Preface

The goal of *The Parental Brain: Mechanisms, Development, and Evolution* is to present a comprehensive, integrative, and multilevel analysis that examines how the brain regulates parental behavior in nonhuman animals and in humans, how these brain mechanisms develop, and how such development can go awry, leading to faulty parental behavior. Further, since maternal behavior is a defining characteristic of all mammals, the enduring mother–infant bond represents the most basic type of aid-giving social behavior. I will present evidence that the neural circuitry of the maternal brain has provided a foundation upon which natural selection could act to create other types of strong prosocial bonds in animals and humans when such enduring bonds have adaptive significance. A unique aspect of this book is the integration and comparison of animal and human research in order to create a complete understanding of the parental brain.

In 2003, Numan and Insel published *The Neurobiology of Parental Behavior*. To the best of my knowledge, that was the first modern authored book on the parental brain. Since the publication of that book, there has been tremendous progress in our understanding of the parental brain. The detailed neural circuitry, along with its neurochemical make-up, that underpins parental behavior in animals has expanded greatly, and the use of functional magnetic resonance imaging has provided an understanding of the human parental brain that was virtually nonexistent in 2003. Advances in genetics, epigenetics, and the effects of early life experiences on brain development have presented novel information relevant to the development of the parental brain and to the intergenerational continuity of normal and abnormal parental behavior. Finally, evolutionary perspectives on the parental brain, particularly with respect to alloparenting and cooperative breeding, have provided a framework for appreciating how the parental brain could have provided a foundation for other types of strong prosocial bonds. All of these advances, and many others, created the impetus for writing the current book.

This book will be a valuable resource for behavioral neuroscientists and neuroendocrinologists, social neuroscientists, developmental psychologists and psychobiologists, anthropologists, and evolutionary psychologists with an interest in parental behavior, mother–infant relationships, child development, and the evolution of prosocial behavior. Because it is a single-authored book, it presents a high degree of intellectual coherence and integration. The detailed analysis of the parental brain in both animals and humans will not only show that core subcortical neural circuits regulate parental motivation in all mammals but will also highlight the added complexity of the human parental brain due to the interactions between cortical and subcortical circuits.

I am a behavioral neuroscientist, and my research program has examined the neural mechanisms that control maternal behavior in rodents. This research, which was supported by the National Institutes of Health and the National Science Foundation, was undertaken while I was a Professor of Psychology at Boston College between 1975 and 2012. During that time, I was grateful for the support I received from James Russell while he was the chairperson of the department. I retired from Boston College in 2012 and moved to the Albuquerque, New Mexico region where I received an appointment as a Research Professor of Psychology (Letter of Academic Title) at the University of New Mexico (UNM). At UNM, I continue to engage in academic writing, and I authored the *Neurobiology of Social Behavior*, which was published by Elsevier in 2015. Writing that book greatly expanded my understanding of the neural control of a variety of social behaviors in animals and humans, and this knowledge contributed importantly to my ability to write this current volume.

To bring this book to completion, the contributions of my wife and colleague, Marilyn, were indispensable. She produced the artwork for all of the original figures and she read, commented on, and edited each chapter. My daughter, Suzanne Numan, performed the arduous task of compiling the reference list that appears at the end of this book, and for this I am very grateful. The aid I received from the staff of Oxford University Press, particularly Joan Bossert and Phil Velinov, and Sankari Balasubramanian at Newgen, is also greatly appreciated.

On a personal note, my first publication appeared in Moltz, Lubin, Leon, and Numan (1970). This current book will probably be published in 2020. I hope that my 50 years of scientific contributions have furthered our understanding of the nature of human nature.

Abbreviations Used for Major Brain Regions

ACC	Anterior cingulate cortex
ACN	Anterior commissural nucleus
AHN	Anterior hypothalamic nucleus
AI	Anterior insular cortex
Amyg	Amygdala
AOB	Accessory olfactory bulb
AON	Anterior olfactory nucleus
AVPV	Anteroventral periventricular nucleus of hypothalamus
BAOT	Bed nucleus of the accessory olfactory tract
BLA	Basolateral amygdala
BMA	Basomedial amygdala
BST	Bed nucleus of the stria terminalis
CeA	Central nucleus of the amygdala
CeAl	Lateral part of CeA
CeAm	Medial part of CeA
CoA	Cortical amygdala
CP	Caudate-Putamen (dorsal striatum)
dACC	Dorsal anterior cingulate cortex (area 24)
dBST	Dorsal BST
dlBST	Dorsolateral BST
dlPFC	Dorsolateral PFC
dmBST	Dorsomedial BST
dmPFC	Dorsomedial PFC (areas 9 and 10)
DR	Dorsal Raphe nucleus
GP	Globus pallidus
Hb	Habenular nuclear complex
Hipp	Hippocampus
IFG	Inferior frontal gyrus
IL	Infralimbic cortex
LA	Lateral amygdala
LC	Locus coeruleus
LDTg	Laterodorsal tegmental nucleus
LH	Lateral hypothalamus
LHb	Lateral habenula
LHbl	Lateral part of LHb
LHbm	Medial part of LHb
lPFC	Lateral PFC
LPOA	Lateral preoptic area

LS	Lateral septum
MeA	Medial amygdala
MeApd	Posterodorsal part of MeA
MOB	Main olfactory bulb
MOE	Main olfactory epithelium
mPFC	Medial PFC
MPN	Medial preoptic nucleus of MPOA
MPOA	Medial preoptic area
MPOAc	Central part of MPOA
MSN	Medium spiny neuron
NA	Nucleus accumbens
NAs	Shell region of NA
OFC	Orbitofrontal cortex
PAG	Periaqueductal gray
PFC	Prefrontal cortex
pgACC	Pregenual ACC (Area 32)
PIL	Posterior intralaminar complex of thalamus
PL	Prelimbic cortex
PM	Premammillary nucleus of the hypothalamus
PMv	Ventral part of PM
pOFC	Posterior OFC
PVN	Paraventricular nucleus of the hypothalamus
RMTg	Rostromedial tegmental nucleus
sgACC	Subgenual ACC (Area 25)
SON	Supraoptic nucleus of the hypothalamus
STG	Superior temporal gyrus
STS	Superior temporal sulcus
SVG	Subventricular zone
TPJ	Temporoparietal junction
vBST	Ventral BST
vlPAG	Ventrolateral PAG
vlPFC	Ventrolateral PFC
VMN	Ventromedial nucleus of the hypothalamus
VMNagg	VMN aggression-related neurons
VMNc	Central region of VMN
VMNdm	Dorsomedial region of VMN
VMNfear	VMN fear-related neurons
VMNvl	Ventrolateral region of VMN
vmPFC	Ventromedial PFC
VNN	Vomeronasal nerve
VNO	Vomeronasal organ
VP	Ventral pallidum
VTA	Ventral tegmental area

1
Introduction
The Parental Brain

The purpose of this book is to explain the mechanisms through which the brain regulates parental behaviors in nonhuman vertebrates (subsequently referred to as animals; my major focus will be on mammalian species) and in humans and how environmental and genetic factors influence how the parental brain develops. An additional emphasis will be to show that evolutionary forces (natural selection) likely utilized the parental brain neural circuitry as a neural foundation for other types of strong prosocial bonds. Research on the human parental brain, because it is mainly based on data derived from neuroimaging methods, is primarily correlational in nature, while the animal research is mostly experimental in nature and can therefore provide stronger information about cause–effect relationships. The integration of these two bodies of research will provide a firm body of data for understanding the organization and function of the parental brain and will highlight the similarities and differences between the animal and human parental brain. By understanding the normal and abnormal development of the parental brain, one can gain insights into the etiology of normal variations in parental style and into the etiology of abnormal parental responses and mental states that could lead to child abuse and neglect. I will describe research that shows that the manner in which a parent treats its offspring has significant effects on the socioemotional development of those offspring. Further, there is an intergenerational continuity in parental styles, and offspring who have been abused or neglected by their parents are more likely to abuse or neglect their own offspring. Since research has indicated that parental treatment of infants contributes to this intergenerational continuity, the way a parent treats its offspring can serve as a conduit through which faulty parental behavior and its associated socioemotional dysfunctions can be transmitted across generations. Clearly, environmental and/or pharmacological interventions that normalize parental responses will have broad beneficial effects on the psychological development of infants. Please note, however, that this book will not emphasize therapies for faulty parental behavior; it is a basic science book. That said, by understanding the normal and abnormal parental brain, a set of data will be provided that will be useful for applied researchers.

The book will be divided into three major topical themes, emphasizing the neural regulation of parental behavior (mechanisms), the intergenerational

The Parental Brain. Michael Numan, Oxford University Press (2020). © Oxford University Press.
DOI: 10.1093/oso/9780190848675.001.0001

continuity of parental behavior styles (development), and how the neural networks underlying parental behavior may have served as a foundation that was acted upon by natural selection to create broader forms of caregiving and prosocial behaviors (evolution). Most research on the parental brain in mammals, including humans, has been on maternal behavior, and that will therefore form the primary body of research in this book. However, paternal behavior and alloparental behavior (caretaking behaviors directed toward an immature conspecific other than one's own offspring) will also be discussed.

In the chapters that deal with mechanisms, I will describe research that shows that there is a remarkable similarity in the neural mechanisms controlling maternal, paternal, and alloparental behaviors, although the conditions that allow infant stimuli to gain access to these parental circuits differ between the three types of parental behaviors. This research will indicate that there are evolutionarily conserved core subcortical neural circuits that regulate parental behavior across mammalian species. In comparing the neural regulation of maternal behavior in animals with the neural correlates of maternal behavior in humans, I will show that there is excellent overlap in the subcortical circuitry implicated in animal and human maternal behavior. However, in humans, overlaying this core subcortical circuitry are cortical mechanisms that influence emotional empathy, mentalizing and cognitive empathy, and emotion regulation. These feeling states and mental states add complexity and variability to the human experience of motherhood. Importantly, such affective and cognitive cortical mechanisms influence maternal behavior by interacting with the core subcortical circuits that are present in both animals and humans. In the chapters dealing with mechanisms, the goal, for the most part, is to describe the normal function of brain mechanisms controlling parental behavior, and a major analysis of abnormal parental brain function will be presented in the chapters dealing with development. However, some analysis of faulty maternal brain function will be described in the section dealing with mechanisms when I describe the research on postpartum depression in human mothers.

The chapters on development will concentrate on the intergenerational continuity of maternal behavior. Research will be described that shows that female infants who have been abused or neglected by their mother are more likely, when compared to infants that have not been abused or neglected, to grow up to abuse and/or neglect their own offspring. Experimental animal research clearly shows that the way a mother treats her female offspring (an environmental effect) influences the development of maternal behavior in her offspring. However, not all offspring that have been abused and/or neglected grow up to become bad parents. Further, some parents who do not have a history of being maltreated as children do abuse and/or neglect their offspring. Therefore, additional factors, including genetics, influence the development of the maternal brain and behavior. In describing how maternal treatment effects influence the development of maternal behavior in the mother's female offspring, I will describe research derived from two hypotheses: (a) The way a mother treats her female offspring can influence the development of the neural circuits that specifically regulate maternal behavior and dysfunctions in these circuits could lead to the emergence of a neglectful or abusive maternal phenotype in the affected offspring, and (b) the way a mother treats her offspring can influence the development of neural circuits involved in emotion regulation, with dysfunctions in these circuits leading to enhanced anxiety and stress reactivity in the affected offspring. As adults, such offspring may not properly care for their own offspring, particularly under demanding environmental conditions. In describing this developmental research, and the underlying neural mechanisms, epigenetic modifications that influence gene expression and gene by environment (G × E) interactions will be seen to play important roles.

The last chapter in the book deals with evolutionary processes related to the parental brain and is based on the idea that maternal behavior is the most primordial caregiving system and is present in all mammals. I will present research that indicates that the neural networks of the parental brain may have provided the foundational neural circuitry for other types of strong social bonds outside the parent–infant relationship. For example, under socioecological conditions where it would be adaptive, natural selection may have utilized and appropriately modified the parental neural circuitry so that strong social bonds could be formed between mating partners in monogamous species. In addition, I will describe research that implicates the parental circuitry as providing a foundation for the hyper-cooperation found in human societies, where strong social bonds and aid-giving behaviors can be directed toward an individual other than one's spouse, infants, or other kin. The emergence of alloparental behavior and cooperative breeding in humans, which are unique among the great apes, will be shown as important factors in the evolution of such hyper-cooperation.

2
Parental Behavior
Descriptions, Terms, and Definitions

Parental Behavior in Vertebrates

Parental behavior can be defined as any behavior displayed by one member of a species toward an immature and developing organism of the same species (an immature conspecific) that increases the likelihood that the immature organism will survive (Numan & Insel, 2003). In egg-laying species (oviparous species), parental behavior includes caretaking activities directed toward the fertilized eggs, which can include protection of the eggs from predators and incubation of the eggs. In species that give birth to live young (viviparous species) and in oviparous species, various types of parental care can be directed toward the immature live-born young or toward the hatchlings, respectively (Smiseth, Kolliker, & Royle, 2012). Parental behavior occurs in a variety of vertebrates, including fish, amphibians, reptiles, birds, and mammals (Balshine, 2012; Dulac, O'Connell, & Wu, 2014). In fish, amphibians, and reptiles, most species do not provide postfertilization care of offspring. That is, they do not show parental behavior. However, under the proper ecological conditions where parental care is necessary for the survival of offspring, such behavior has evolved in certain species of fish (Goodwin, Balshine-Earn, & Reynolds, 1998; Teresa & Goncalves-de-Freitas, 2011), amphibians (Brown, Morales, & Summers, 2010), and reptiles (Chabert et al., 2015; Somaweera, Brien, & Shine, 2013). In contrast to fish, amphibians, and reptiles, parental care is necessary for the survival of offspring in birds and mammals, and it is an essential characteristic of these two classes of vertebrates (Balshine, 2012; Numan & Insel, 2003).

The type of mating and social system that a species exhibits influences which individuals show parental behavior. When comparing birds with mammals, approximately 90% of bird species have a monogamous mating system, while monogamy is rare in mammals, occurring in only about 10% of such species. Instead, approximately 90% of mammals exhibit either a polygynous (one male mates with several females) or promiscuous (both males and females mate with several different partners) mating system (Kleiman, 1977; Lukas & Clutton-Brock, 2013). In a monogamous mating system, one male

The Parental Brain. Michael Numan, Oxford University Press (2020). © Oxford University Press.
DOI: 10.1093/oso/9780190848675.001.0001

engages in sexual behavior with one female, and after the female has been impregnated, the mated pair stays together over an extended time and may form a strong affiliative pair bond (Numan, 2015; Numan & Young, 2016). Monogamy is frequently associated with a biparental care system where both the mother and the father care for the immature and developing offspring. It is not surprising, therefore, that most bird species show maternal and paternal behavior (Balshine, 2012; Kleiman, 1977). For mammals, in contrast, the typical parental care system is a uniparental maternal care system. In such non-monogamous mammalian species, the male and female leave each other after mating, and the impregnated female, upon giving birth, raises her offspring on her own (Lukas & Clutton-Brock, 2013). In all mammals, it is the female that lactates and is therefore absolutely necessary for the survival of young. This fact would have been the case during early human evolution, although in modern societies breastfeeding is not necessary for infant survival in humans. It is not surprising, therefore, that maternal behavior can be considered a defining characteristic of mammals.

When discussing monogamy, it is important to distinguish genetic monogamy from social monogamy (Numan, 2015; Phelps, Campbell, Zheng, & Ophir, 2010). A monogamous social system typically results in mating exclusivity within a pair and, therefore, genetic monogamy. However, in wild populations, extra-pair mating and fertilization have sometimes been observed ("infidelity"). Therefore, to be on the safe side, when I refer to *monogamy*, I am referring to social monogamy where the mated pair stays together for long periods of time after sexual activity has terminated.

For the remainder of this book, my main concern will be with parental behavior in mammals. There are two major reasons for this choice. First, most of the research on the hormonal and neural mechanisms of parenting have been conducted on mammals, and this research has primarily been focused on rodents, rabbits, sheep, and humans. Humans are mammals, and by comparing the neural mechanisms that make up the parental brain across the investigated mammalian species, I will attempt to define a core neural circuitry for parental behavior. Of course, there will be differences as well as similarities across species. The similarities will uncover a core neural circuitry that is evolutionarily conserved, as should be expected since maternal behavior is a characteristic of all mammals. Differences in regulatory mechanisms across mammals will highlight how the evolution of unique social systems has influenced and modified how the parental brain operates. When necessary, I will also discuss the neural regulation of parental behavior in nonmammalian vertebrate species whenever that research sheds light on the conserved neural mechanisms controlling parental behavior across vertebrates.

Parental Behavior in Mammals

Characteristics of Maternal Behavior in Select Mammalian Species That Exhibit Uniparental Maternal Care

Since uniparental maternal behavior is the dominant form of parental behavior in mammals, it will be worthwhile to describe some of the characteristics of the behavior in rats, rabbits, sheep, and rhesus monkeys (Numan & Insel, 2003; Rheingold, 1963), since in each of these species the mother is the infant's caregiver. These species represent a good cross section of the various characteristics of maternal behavior in mammals.

To begin, it needs to be emphasized that for most mammalian mothers, the hormonal and other physiological events associated with pregnancy and parturition act on the brain to trigger or turn on maternal responsiveness. The evidence for this statement will be presented in Chapter 3. Most nulliparous adult female mammals (those that have not borne young) do not show caretaking behaviors toward conspecific young, while at the time of parturition mammalian females are attracted to and care for conspecific infants. In other words, the physiological events of late pregnancy and parturition allow infant stimuli to gain access to the neural regulatory regions that make up the parental brain.

The nature of maternal behavior is affected by the level of maturity of the young at birth and the kind of social group within which the mother and her infant(s) reside (Gubernick, 1981; Numan, Fleming, & Levy, 2006). With respect to maturity, infants can be immature and immobile (altricial) at birth, precocial and mobile, or intermediate between these extremes (semi-altricial). The typical infant rat pup is helpless at birth (altricial). The mother typically gives birth to a large litter of such young in a secluded nest site (Calhoun, 1962) where she hovers over them to nurse them and keep them warm. The nest is typically constructed near the time of parturition and then maintained during the postpartum period. Within the nest, the mother also grooms the young by licking them, and this behavior not only cleans the pups but also facilitates urination and elimination. If pups become displaced from the nest or the mother moves her nest site to another location, then she shows retrieval behavior (transport behavior) during which she carries each pup in her mouth, one at a time, to transport them back to the nest or to a new nest, respectively. Rat pups mature quickly over the first 3 postpartum weeks and they are usually weaned and become independent by 4 weeks postpartum. Corresponding to this maturation, maternal behavior declines over the 4-week postpartum period. During the first postpartum week, the mother spends many hours per day in the nest with her pups, but this nesting time decreases as the postpartum period advances (Numan, 1994). Another important characteristic of maternal rats is the occurrence of maternal aggression, which is particularly evident during the early postpartum period. Maternal aggression is characterized by aggressive responses toward intruders at the nest site; such intruders can be potential predators or infanticidal conspecifics (Calhoun, 1962; Numan & Insel, 2003). In other words, the mother is protecting her young from potential threats to their safety. When describing maternal behavior in rats, one can refer to pup-directed behaviors (nursing, retrieval, grooming) and nonpup-directed behaviors (nest building, maternal aggression). Finally, rats and most other mammals that give birth to altricial offspring do not form selective attachments to their own young: If one *experimentally* cross-fosters or exchanges young between litters, the mother will care for young that are not her own (Numan & Insel, 2003; Numan & Young, 2016). Although this appears counter to evolutionary principles, note that under natural conditions altricial young are not capable of moving from one nest to another. Since confusion between own and alien young does not occur under these conditions, there is no need for mothers with altricial young to evolve selective social attachment mechanisms. Of course, the mother rat does need to learn and remember the location of her nest site, which will ensure that she is caring for her biological offspring.

The maternal behavior shown by rabbits is similar in some respects to that exhibited by rats (Gonzalez-Mariscal, Caba, Martinez-Gomez, Bautista, & Hudson, 2016). Under natural conditions, rabbit mothers give birth to relatively large litters of altricial young in a nest that is located within a burrow. Mothers do not form selective attachments to their young and will care for cross-fostered infants (Gonzalez-Mariscal & Gallegos, 2007). When the mother leaves the nest, she closes or hides the burrow entrance. What most distinguishes rabbits from rats is the extremely short duration of nursing behavior. The mother rabbit enters the nest to nurse her young for only 3 to 4 minutes per day! Therefore, the amount of mother–infant contact is extremely low. Maternal rabbits also do not show retrieval behavior, presumably because mothers do not move their nest sites and because the kits are unlikely to become displaced from a nest within a burrow (Gonzalez-Mariscal, Caba, Hoffman, & Melo, 2017).

In contrast to rats and rabbits, sheep give birth to precocial young that are relatively mature and mobile at birth. Maternal retrieval or transport is not necessary, since the lamb can follow its mother. The large herds that the lambs are born within are composed of genetically unrelated individuals, and because of seasonal breeding, many ewes give birth at around the same time. Under these social and ecological conditions, confusion between own and alien (genetically unrelated) young would be able to occur because mobile young could wander from one mother to another. Since it is usually disadvantageous, in terms of individual reproductive success, for a mother to care for unrelated young, it is not surprising that selective attachment mechanisms have evolved in maternal sheep

(Nowak, Keller, & Levy, 2011): Such a selective attachment mechanism between a mother and her lamb develops rapidly through an olfactory learning process within the first few hours postpartum (the mother learns and becomes attracted to the olfactory characteristics of her lamb), resulting in a mother that stays near her lamb, whom she will selectively nurse, groom, and protect, while at the same time rejecting (with head butts) any advances from alien lambs. Importantly, near the time of parturition, the pregnant ewe moves away from the main body of the herd, which allows the parturient female to bond to her particular offspring in relative isolation (Dwyer, 2008). After she learns her lamb's individual characteristics, she returns to the herd and cares only for her own lamb. Further, beginning at 12- to 24-hours postpartum the lamb learns the visual and auditory characteristics of its mother, which helps maintain the lamb's close proximity to its mother. This learning process in lambs is a sort of trial and error learning process based on being accepted and nursed by its mother as she emits low-pitched bleats, while being rejected by other ewes who head butt the unrelated lamb while emitting high-pitched bleats (Nowak, Keller, Val-Laillet, & Levy, 2007).

The maternal behavior of most primates, including humans, is adapted to infants that are semi-altricial at birth. Rhesus monkeys will serve as a typical example of maternal behavior in Old World monkeys (Numan, 1994). In this species, a singleton infant is born into a social group or troop composed of related and unrelated individuals (several adult males and females and other infants and juveniles). For the first month postpartum, the mother is in constant contact with her infant, whom she nurses, grooms, and protects (she prevents other members of the troop from gaining access to the infant). During this early period, the infant clings to its mother for transport. Subsequently, as the infant develops, it will begin to wander away from its mother, but the mother is always aware of the infant's location, and she will initiate contact and reunion when necessary (e.g., if the infant is threatened by another conspecific). Selective maternal attachments develop in most primates, but this development is not as rapid and temporally constrained as that which occurs in sheep, which coincides with the semi-altricial characteristics of primate infants.

A Critical Evaluation of the Mother-to-Infant Attachment Bond

As previously described in this chapter, rats and rabbits do not, while sheep and most primates do, form selective attachments to their offspring. I described the reasons why selective mother–infant bonds form in certain species while nonselective mother–infant bonds form in other species. When I use the term *mother–infant attachment* or *mother–infant bond*, I am referring to the enduring attraction that a mother develops toward her infant(s) across the postpartum period. More specifically, as I will show in Chapter 3, in those mammalian species for which experimental evidence is available, at the time of parturition, a mother will respond maternally to *any* conspecific infant. For most mammals, this onset of maternal responsiveness is caused by the hormonal and other physiological events associated with late pregnancy and parturition. During the mother's initial interaction with her infant, an enduring attraction or bond is formed that allows maternal behavior to continue throughout the remainder of the postpartum period in the absence of continued hormonal stimulation. When comparing rats with sheep, the mother–infant bond is strong in both species; what differs is the nature of the infant stimulus to which the mother becomes attached or attracted to. One can view the development of the bond between a mammalian mother and her infant(s) as a two-stage process (Numan & Young, 2016). The first stage involves a recognition process (in which certain infant stimuli gain access to parental brain motivational mechanisms), and the second stage involves a process that results in the persistent attraction to those infant stimuli across the postpartum period. Stage 2 is similar in all mammalian mothers: All mothers form an enduring attraction to their infants that persists at least until the young are weaned and become independent. Species do differ in the recognition stage. Females that give birth to altricial young exhibit a *nonselective* recognition process and become persistently attracted to a generic infant stimulus. That is, general infant stimuli continue to gain access to the brain's maternal motivational system—a mechanism that regulates the mother's enduring attraction to her infant(s)—so that maternal behavior occurs toward altricial conspecific young throughout the postpartum period. Since the nature of these species does not allow for confusion between own and unrelated young, this process still results in an enduring bond between a mother and her biological offspring under natural conditions. For the recognition stage in sheep and other species that give birth to precocial or semi-altricial young, learning mechanisms operate as the mother interacts with her offspring at parturition, and the mother becomes selectively attracted to (bonded to) the particular infant that she gave birth to. Subsequently, only those infant stimuli gain continued access to the brain's maternal motivational system. This *selective* recognition process prevents the mother from taking care of unrelated young, which would be likely in these species if a selective recognition process did not occur.

It is sometimes proposed that research on rats and other species that give birth to altricial young is not relevant to an understanding of the neurobiology of mother–infant bonds (Nowak et al., 2011). Such an argument is based on the assumption that an enduring mother–infant bond requires selectivity. But this assumption is not valid since all mammalian mothers form an enduring bond with their infants. If one were solely interested in how selectivity develops, then rats would not be a good model for studying that process. But if one were interested in the neural mechanisms underlying the enduring bond that allows a mother to be persistently attracted to infants across the postpartum period, then rats as well as sheep would serve as useful models for studying such mechanisms. Indeed, even the recognition process could be studied in rats to understand how generic infant stimuli gain access to the brain's maternal attraction mechanism across the postpartum period.

In an excellent review, Poindron, Levy, and Keller (2007) have clearly outlined the evidence for sheep that maternal motivation and the enduring attraction of a postpartum ewe for a lamb is a separate process from the development of maternal selectivity. As one example, the experimental induction of anosmia in ewes results in an ewe that cannot smell and therefore cannot learn the olfactory characteristics of its lamb. Such ewes will still show maternal behavior at parturition, and this maternal responsiveness endures during the postpartum period. These ewes, however, do not form a selective bond to their lamb and will care for any lamb that is presented to them. In other words, anosmic ewes act like mammalian mothers that give birth to altricial young: They form a nonselective bond to infants. Similar to the two-stage process that I have described, Poindron et al. distinguish between the mechanisms that regulate maternal selectivity and those that regulate maternal responsiveness.

By breaking down the neural mechanisms that regulate the mother–infant bond into a recognition stage and an attraction phase, one can explore conserved neural mechanisms across species even though the specific maternal behaviors shown by different species may differ (e.g., rats retrieve infants, but sheep and rabbits do not). Instead of concentrating on the neural mechanisms that regulate specific maternal responses, one can ask questions about the neural mechanisms that regulate the mother's attraction to infant stimuli and how maternal selectivity develops in certain species.

Monogamy and Paternal Behavior in Mammals

In terms of social/mating systems, Lukas and Clutton-Brock (2013) note that of the 2,545 mammalian species that could be classified, the females in about 70% of these species were solitary breeding females: After mating with a male, such females raise their young independently and without the help of others in their home ranges. In about 20% of the remaining species, females care for their young alone while residing in a social group with a mating system that is primarily polygynous or promiscuous. The remaining 10% of these species was classified as socially monogamous. Importantly, social monogamy does not imply biparental care: In about 50% of the socially monogamous species only maternal behavior is observed, while paternal behavior occurred along with maternal behavior in the others. Therefore, a good estimate is that paternal behavior occurs in only 5% of mammalian species, although if one were to examine different mammalian orders, paternal behavior would be more prevalent in some than in others. For example, paternal behavior has been observed in certain rodent, carnivore, and primate species.

Most evolutionary theorizing proposes that social monogamy in mammals evolved first for a male to maintain exclusive mating access to a single female under ecological conditions where males were unable to defend access to more than one female because of large female home ranges or territories (Kleiman, 1977; Lukas & Clutton-Brock, 2013). Paternal care is viewed as having evolved as a secondary adaptation in those cases where the occurrence of paternal care along with maternal care increased the reproductive success of both sexes (Stockley & Hobson, 2016). Although mammalian males do not lactate, paternal behavior can aid infant survival in other ways. Males can hover over infants to keep them warm, they can groom them and transport them, and provision them with food as they become less dependent on maternal milk.

Understanding the mechanisms underpinning paternal behavior is important because males do not undergo pregnancy and parturition. Therefore, under the proper ecological conditions, evolutionary forces can result in alternative mechanisms that allow males to form father–infant bonds. I will explore what we know about the regulation of paternal behavior in Chapters 7 and 8.

Cooperative Breeding and Alloparental Behavior

A cooperative breeding system occurs in only about 3% of mammalian species, and examples include certain vole species (voles are rodents), such as prairie voles, and certain species of New World monkeys, such as marmosets and tamarins (Cant, 2012; Emlen, 1995). In a cooperative breeding social system, some offspring, sometimes referred to as helpers, remain in their social group after they are weaned; they delay dispersal while they help their parent(s) rear subsequent offspring. Such alloparental helpers therefore show either allomaternal or allopaternal behavior, depending on their sex. A good definition of alloparental behavior would be parental behavior shown toward conspecific infants by individuals that are not the biological parents of the infants. Similar to paternal behavior, since alloparents in cooperative breeding species do not experience pregnancy and parturition, there must be alternate routes through which infant stimuli can gain access to the central neural circuitry that regulates parental motivation. I will discuss the mechanisms regulating alloparental behavior in Chapters 7 and 11.

A dominant hypothesis concerning the evolution of cooperative breeding is that it evolved from monogamous ancestors (Lukas & Clutton-Brock, 2013; Kramer & Russell, 2015). More specifically, given a monogamous social system, if it is advantageous for independent young to delay dispersal either because suitable breeding territories are scarce or because they are not strong enough to acquire and defend such territories, then it might be adaptive for such individuals to remain in their natal group and help their parent(s) raise additional offspring. Such a behavioral strategy might have evolved by kin selection because newly born infants are highly likely to be genetically related to the helpers (being their brothers or sisters).

A proper analysis of cooperative breeding and alloparenting is particularly important for an understanding of the human parental brain. In their analysis of traditional hunter-gatherer human societies, Hrdy (2009) and Kramer (2011) have provided strong evidence that such societies are cooperative breeding societies and that allomaternal behavior, in particular, is a dominant characteristic. What this suggests is that allomaternal behavior may have been crucial for infant survival during early human evolution. Interestingly, chimpanzees, the closest ape relative to humans, display a uniparental maternal care system. What might have driven early humans away for a uniparental maternal care system and toward a cooperative breeding system with high levels of allomaternal behavior? I will examine this question in Chapter 11, which deals with evolutionary processes relevant to the parental brain. Further, high levels of allomaternal motivation in women may have allowed them to rely less on the physiological events of pregnancy and parturition to activate maternal motivation. This view fits with our knowledge that nulliparous women can adopt infants and become excellent mothers.

Conclusions

In mammals, the typical parental care system is a uniparental maternal care system. This fact fits with the knowledge that the female mammal lactates and is therefore the main source of nurturance for the infant(s), and with the fact that the majority of mammalian mothers are socially solitary, raising their offspring without the help of others. For most female mammals, the physiological events associated with pregnancy and parturition act on the brain to promote maternal motivation or the attraction of the mother to her infants. Depending on the species, once maternal motivation is initiated, the enduring bond that forms between a mother and her infant(s) can be either nonselective or selective. Finally, since paternal behavior and alloparental behavior occur in certain species, there must be alternate mechanisms that do not require pregnancy and parturition that allow infant stimuli to gain access to the parental brain's neural circuitry. One theme of this book is that there is a common neural circuitry that underlies maternal, paternal, and alloparental behavior, but that different types of processes or mechanisms operate to allow infant stimuli to gain access to this circuitry in mothers, fathers, and alloparents.

3
Hormonal Control of Maternal Behavior in Nonhuman Mammals

Introduction

This chapter describes the hormonal mechanisms that regulate maternal behavior in those nonhuman mammalian species that are characterized by a uniparental maternal care system. Only a few species have received most of the experimental attention, and these are rats, mice (house mice), rabbits, and sheep. I will begin by reviewing hormonal mechanisms in rats, rabbits, and sheep. Mice will be discussed separately, because laboratory strains of this species exhibit an atypical hormonal regulation mechanism in comparison to the other species. After this analysis, I will review what is known about the hormonal regulation of maternal behavior in nonhuman primates.

Hormonal Regulation of Maternal Behavior in Rats, Rabbits, and Sheep

Introduction

In rats, rabbits, and sheep, adult nonpregnant nulliparous (virgin) females do not show maternal behavior toward conspecific neonates on their initial exposure to them (Gonzalez-Mariscal, Chirino, Beyer, & Rosenblatt, 2004; Levy, 2008; Numan & Insel, 2003; Rosenblatt, 1967; Wiesner & Sheard, 1933). In fact, such females seem to find infants aversive and will avoid them or may act aggressively toward them. In contrast, primiparous (giving birth for the first time) parturient rats, sheep, and rabbits will respond immediately with appropriate maternal responses toward *any* conspecific infant that is presented to them (Gonzalez-Mariscal & Gallegos, 2007; Levy, 2008; Numan & Insel, 2003). These data suggest that the physiological events associated with pregnancy and parturition activate maternal responsiveness in these species. Such a process makes sense since virgin females do not lactate, while the mammary glands of parturient females have been prepared for lactation (Tucker, 1994). Therefore, the occurrence of maternal behavior near the time of parturition ensures that a female will be able

The Parental Brain. Michael Numan, Oxford University Press (2020). © Oxford University Press.
DOI: 10.1093/oso/9780190848675.001.0001

to appropriately raise her offspring. Further, the lack of immediate maternal responsiveness in virgin females prevents such females from attempting to care for unrelated young that they might briefly encounter under natural conditions.

Hormonal Control of Maternal Behavior in Laboratory Rats

Hormones Are Not Absolutely Required for the Display of Maternal Behavior in Laboratory Rats

Although adult virgin female laboratory rats do not show maternal responsiveness on their initial exposure to young pups, if they are continuously housed with young pups in a single cage they will ultimately show maternal behavior (Rosenblatt, 1967; Wiesner & Sheard, 1933). The procedure occurs as follows. On each day adult virgin females are cohabited with freshly nourished young pups (2–5 days old) for 24 hours. The pups are provided by a group of "donor" lactating mothers. After about 5 to 12 days of exposure to pups, the virgin female will ultimately show maternal behavior. She will build a nest from wood shavings or paper strips, retrieve the young to the nest, and adopt a nursing posture over the pups, even though the virgin is unable to lactate, and the pups, therefore, cannot be fed. Because the virgin female is not lactating, she must be provided with fresh pups each day, even after the onset of maternal behavior, to prevent the pups from becoming severely malnourished. This pup-induced maternal behavior is usually referred to as sensitized maternal behavior or pup-stimulated maternal behavior. The number of days from first exposure to pups to the onset of maternal behavior is referred to as the female's sensitization latency. Pup-stimulated maternal behavior in virgin females appears to be primarily mediated through nonhormonal mechanisms since it occurs in females whose ovaries, adrenal glands, or pituitary gland have been removed (Rees, Panesar, Steiner, & Fleming, 2006; Rosenblatt, 1967). In other words, continuous sensory stimulation from neonates, over a period of days, ultimately results in pup stimuli gaining access to the neural mechanisms underlying maternal responsiveness without the need for hormonal mediation.

It should be obvious that pup-stimulated maternal behavior cannot be the mechanism that drives the onset of maternal behavior at parturition. If that were the mechanism, then the postpartum female would take several days to respond maternally to her pups, and these pups would therefore die of neglect. Given that experimentally induced pup-stimulated maternal behavior occurs over a period of days, one can view the role of the physiological events of late pregnancy and parturition as allowing pup stimuli to gain *immediate* access to the brain mechanisms mediating maternal behavior.

I am not aware of any experimental evidence with respect to whether feral virgin rats are capable of being sensitized. It is possible that the domestication process that has occurred in laboratory rats through selective breeding has tamed them, and this taming process may have allowed for the occurrence of sensitized maternal behavior, which might not occur in wild rats. Under natural conditions, although feral rats live in colonies, mothers care for their young in a secluded burrow and nest (Calhoun, 1962). Under these conditions, it is probably not even possible for a virgin to be exposed to another mother's pups for a period of days. If a virgin female were to enter a nest site while the mother was out foraging, because of the virgin's initial aversion to pups, she would either leave the nest site or attack the pups.

Fleming and Luebke (1981) found that sensitized maternal behavior in virgin laboratory rats actually occurs in a series of three stages. Females tend for avoid pups (move away from where they are located) for about 3 to 4 days. After this avoidance stage, the females tolerate being in the proximity of pups, and this allows the virgin the gain proximal sensory inputs from pups, although she does not show maternal behavior toward the pups during this phase. This tolerance stage may allow proximal pup stimuli to slowly begin to activate the neural mechanisms that mediate maternal behavior because after about 3 to 4 days of proximal contact the female begins to care for the pups: She builds a nest, retrieves the pups to the nest, and crouches over them in a nursing posture while also grooming them. Numan and Insel (2003) have therefore described pup-stimulated maternal behavior in virgins as a dual process of habituation and sensitization. Initially, the female habituates to the aversive qualities of infant stimuli and no longer avoids pups; after this avoidance period, proximal contact with pups sensitizes (stimulates) maternal neural circuits so that maternal behavior eventually occurs. Rosenblatt and Mayer (1995) have proposed that maternal behavior occurs when the tendency to approach and interact with pups is stronger than the tendency to avoid pups.

Several lessons can be learned from the sensitization process. First, the physiological events associated with late pregnancy and parturition may act on the brain to inhibit the neural system that mediates aversive responses to novel pup stimuli while at the same time activating neural circuits that promote maternal behavior so that pup stimuli can gain immediate access to this neural circuitry. In other words, maternal behavior is immediate at parturition because the avoidance and sensitization stage of pup-stimulated maternal behavior have been eliminated. Second, in the section on the maintenance (continuance) of maternal behavior throughout the postpartum period after its initiation at parturition, evidence will be presented for what has been referred to as the onset-maintenance dichotomy. While the immediate onset of maternal behavior at parturition requires hormonal stimulation, the continued maintenance of maternal behavior during the remaining postpartum period is relatively free from hormonal control. Pup-stimulated maternal behavior in virgins shows us that there are mechanisms that would allow for the expression of maternal behavior in the absence of hormonal mediation. Finally, pup-stimulated maternal behavior in virgins informs us that there can be alternate routes through which infant stimuli can gain access to the neural circuitry for parental behavior. Such knowledge serves as a starting point for understanding how paternal behavior and alloparental behavior can occur in individuals that are not exposed to the specific physiological events associated with pregnancy and parturition.

Differences Between Sensitized Maternal Behavior and Postpartum Maternal Behavior in Rats

While residing in their home cages, the maternal behavior of sensitized virgin rats has been compared to that of postpartum females in several studies (Fleming & Rosenblatt, 1974a; Reisbick, Rosenblatt, & Mayer, 1975). In these studies, once a virgin female becomes maternal, she is subsequently provided, on a daily basis, with freshly nourished test pups that advance in age throughout the testing period. This procedure allows investigators to study the course of maternal behavior in sensitized virgins, which can then be compared to maternal behavior of lactating females to pups of advancing age. The findings show a remarkable similarity in the pattern of maternal responsiveness in the two groups: Both sensitized virgins and lactating females showed high levels of retrieving, nest building, and time spent in the nest area with the pups for about the first 10 days of testing. As the pups advance beyond 10 days of age, these behaviors begin to decline and reach low levels by day 21. Lonstein, Wagner, and De Vries (1999) have detected slight differences in the maternal behavior of sensitized virgins and lactating female rats when tested on the fourth day of maternal responsiveness: Lactating females retrieved displaced pups back to the nest site more quickly than did sensitized virgins. The virgins took about 120 seconds to retrieve eight pups, while the postpartum females took about 30 seconds. This difference may have been due to the fact that lactating females in the Lonstein et al. study were tested with their own pups, which they had been caring for over the 4 postpartum days, while the virgins, of necessity, were tested with a novel group of freshly nourished pups. In addition, although both virgins and lactating females spent the same amount of time in the nest with their pups, the duration of particular types of nursing postures differed between the groups. This difference may have resulted from the fact that lactating females have developed nipples and can receive intense suckling stimulation, while this is not the case for virgins. I conclude from these findings that when tested in their home cages, maternal behavior is quite similar in both groups of animals, suggesting that a common neural circuitry underlies the behavior in the two groups.

Some investigators have asked whether sensitized virgins would behave like lactating females if the testing environment were made more challenging (Bridges, Zarrrow, Gandelman, & Denenberg, 1972; Stern & Mackinnon, 1976). These investigators examined whether females that were retrieving pups in their home cages would also retrieve pups that were placed in a novel T-maze extension attached to the home cage. In this situation, although both sensitized virgins and lactating females leave their home cages to enter the T-maze, lactating females are much more likely than the virgins to retrieve pups back into the home-cage nest. These investigators have suggested that the novel T-maze evokes fear-related states that suppress maternal retrieving in the sensitized virgins but not in the lactating females. One possibility is that maternal motivation may be higher in lactating females than in virgins, which allows them to overcome their fearfulness so that they can care for their young. It is also possible that maternal motivation is equal in the virgins and the lactating females, but that lactating females are less fearful of the novel environment than are the virgins. Finally, it is possible that maternal motivation is higher and fearfulness is lower in lactating females than in virgins, and that this combined effect of higher motivation and less fearfulness allows such females to take greater risks to care for and protect young infants. I will discuss these issues in more detail in the sections of this chapter that deal with maternal behavior in mice and also in the chapter on maternal aggression (Chapter 6 of this volume). One can conclude at this point, however, that the physiological events associated with late pregnancy, parturition, and the postpartum period appear to promote a higher level of maternal capability in postpartum female rats than that which exists in sensitized virgins.

Hormonal Induction of a Rapid Onset of Maternal Behavior at Parturition in Laboratory Rats

If the hormonal events associated with late pregnancy and parturition activate the immediate onset of maternal behavior in parturient rats, then it would be important to know what these hormonal events might be. Figure 3.1 schematically displays the blood plasma levels of five hormones, across the 22-day rat pregnancy, that have been shown to be important for the onset of maternal behavior (Bridges, 1984; Grattan, 2001; Levy, 2016; Lonstein, Pereira, Morrell, & Marler, 2015; Numan, 1994; Numan & Insel, 2003; Soares, 2004). The hormones shown are estradiol, progesterone, pituitary prolactin, and placental lactogens I and II. Estradiol and progesterone are steroid hormones produced primarily by the ovary in rats. Because of their lipid-like nature, steroid hormones passively cross the blood–brain barrier to gain access to central neural circuits. Prolactin, secreted by the anterior pituitary, and the placental lactogens, secreted into the blood from the pregnant female's placenta, are collectively referred to as

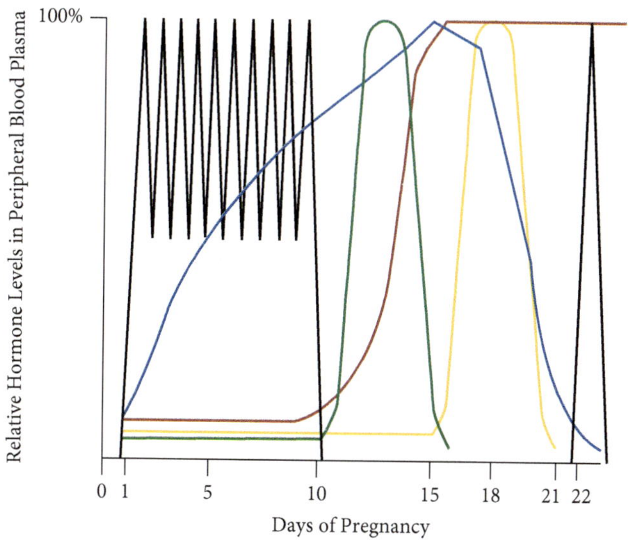

Figure 3.1. Relative blood plasma levels of several hormones across the 22-day pregnancy of rats. Pituitary prolactin is shown in black. Placental lactogen I is in green and placental lactogen II is in yellow. Estradiol and progesterone levels are shown in red and blue, respectively. Across the first 10 days of pregnancy, pituitary prolactin is released in two daily surges, represented by the 10 peaks shown. A surge in pituitary prolactin also occurs on day 22 of pregnancy.

Source: Based on data presented by Bridges (1984), Grattan (2001), and Soares (2004).

lactogens because they promote mammary gland development and lactogenesis. The lactogens are polypeptide hormones and peptides typically do not cross the blood–brain barrier. However, for the lactogens there is an active transport mechanism within the choroid plexus that allows these hormones to gain access to the cerebrospinal fluid, and from there, they can then diffuse into the brain (Bridges et al., 1996; Grattan, 2002; Numan, 1994). Additional mechanisms that allow lactogens to have direct access to the brain via the cerebral vasculature also appear to exist (Bridges & Grattan, 2019). Finally, estrogen receptors (which bind estradiol), progesterone receptors, and prolactin receptors (which bind prolactin and placental lactogens) are located in neurons within the brain (Bakowska & Morrell, 1997, 2003; Grattan et al., 2001; Numan et al., 1999; Pfaff & Keiner, 1973; Shughrue, Lane, & Merchanthaler, 1997). Therefore, the peripheral plasma levels of the hormones shown in Figure 3.1 should be able to influence the brain circuits that regulate maternal behavior.

In examining Figure 3.1, note that I am showing relative, not absolute, blood plasma levels for each hormone. In other words, the hormone levels across pregnancy are shown in relation to their maximal values exhibited during pregnancy (100%). A quick look at the figure might make one assume that peak estradiol levels during pregnancy are equal to peak progesterone levels. However, this is not the case, and peak levels of progesterone are much higher than estradiol, since if an absolute scale were being used, progesterone would be measured in nanograms while estradiol would be measured in picograms. With respect to the pattern of hormone secretion, the following is worth noting: (a) Plasma estradiol levels are low during the first part of pregnancy, but rise by day 15 and this relatively high level is then maintained through the day of parturition (usually day 22); (b) progesterone levels are high throughout the first part of pregnancy, and they peak by day 15, after which progesterone levels slowly decline, with an abrupt decline beginning on day 18; (c) prolactin is secreted from the anterior pituitary in daily surges during the first half of pregnancy and then again on day 22; (d) placental lactogens are secreted into the blood during the second half of pregnancy, with placental lactogen I being secreted over days 11 to 14 and placental lactogen II being secreted at high levels over days 17 to 21 of pregnancy. To summarize this pattern, during the second half of pregnancy the rat's brain would be exposed to high levels of estradiol, placental lactogens, and pituitary prolactin, with these high levels being superimposed on a dramatic decline in progesterone.

Based on the hormone patterns shown in Figure 3.1, if these hormones were important for the onset of maternal behavior in female rats, then perhaps the administration of these hormones to virgin females in a pattern that simulated the pattern that occurs naturally would be able to induce short latency maternal behavior. This approach was first taken by Moltz, Lubin, Leon, and Numan (1970) and was subsequently followed by the similar report of Zarrow, Gandelman, and Denenberg (1971). I will describe the early findings by Moltz et al. in some detail because they have stood the test of time.

Figure 3.2 shows the hormone injection regimen that was systemically administered (all injections were subcutaneous) to ovariectomized virgin female rats in an attempt to induce maternal behavior (Moltz et al., 1970). Estradiol was injected daily over days 1 to 11. Progesterone was administered over days 5 through 9, and prolactin was injected on days 9 and 10. Control females either received no hormone injections (vehicle injections were administered instead, with oil being substituted for the steroids and water for prolactin) or only two of the three critical hormones, with vehicle injections substituting for the third hormone. Young pups were presented to these females daily beginning on day 10 and on each subsequent day until full maternal behavior was displayed (nest building, retrieving of all pups to the nest, pup grooming, and the adoption of a nursing posture over the pups) or until 6 days elapsed (day 16 of testing). Females not responding maternally within this 6-day period were assigned

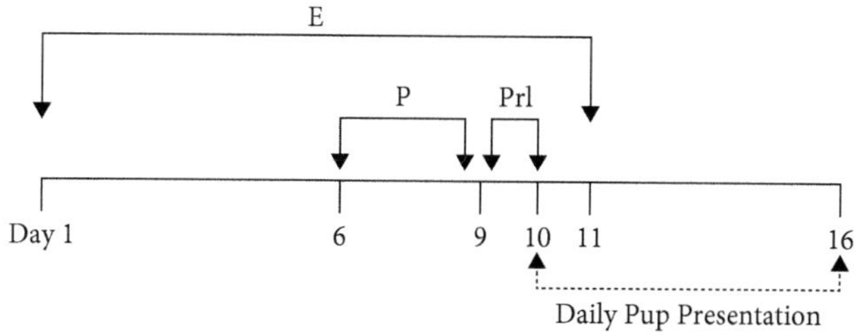

Figure 3.2. The hormone injection regimen utilized by Moltz, Lubin, Leon, and Numan (1970) that induced a short latency onset to maternal behavior in ovariectomized virgin female rats. Estradiol benzoate (E) was administered daily for 11 days. Progesterone (P) was administered twice daily on days 6 through 9. Prolactin (Prl) was injected on days 9 and 10 and pups were initially presented to the females on day 10 of the treatment schedule. All injections were administered subcutaneously.

onset latencies of 7 days for statistical purposes. As can be seen in Table 3.1, the females that were administered all three hormones had the shortest sensitization latencies, responding maternally after about 2 days. Females in the remaining groups had significantly longer latencies than did the females that received all three hormones. Of note, females that received estradiol and progesterone, without prolactin, did respond more quickly than did females in the remaining three groups, with the females in these latter three groups not differing significantly from one another. Moltz et al. noted that the relatively short 3-day latency shown by females that were administered only estradiol and progesterone

Table 3.1 Median Sensitization Latencies in Ovariectomized Virgin Female Rats Injected Systemically With Various Hormones

Groups	Median Latency to Onset of Maternal Behavior (Days)
E + P + Prl	2[a]
E + P	3[b]
E + Prl	4[c]
P + Prl	7[c]
No Hormones	7[c]

Notes: Groups with median sensitization latencies that do not share a letter differ significantly from one another. The data in this table were derived from the research of Moltz, Lubin, Leon, and Numan (1970). E = estradiol benzoate; P = progesterone; Prl = prolactin.

may have resulted from the well-known fact that estradiol is capable of stimulating the release of prolactin from the anterior pituitary (Neill & Nagy, 1994), and therefore endogenous prolactin was likely to have been high in this group. Moltz et al. proposed the following model to explain their results: Exposing the brain to high levels of progesterone followed by its abrupt decline increases the sensitivity of the brain to estradiol and prolactin, with these two key hormones then acting on the central neural circuits that underlie maternal behavior to stimulate a relatively fast onset of maternal behavior in response to continuous pup stimulation. Some support for this view can be seen in the data presented in Table 3.1. The estradiol–prolactin group had shorter sensitization latencies than the progesterone–prolactin and the nonhormone-treated groups, although these median differences did not reach statistical significance. In the estradiol–prolactin group, 5 of 10 females had onset latencies of 3 days or less, while none of the females in progesterone-prolactin group (n = 10), and only one female in the nonhormone-treated group (n = 11) had such short latencies. The overall analysis supports the view that increases in estradiol and prolactin are important stimulators of the onset of maternal behavior, and that this stimulatory effect is potentiated by progesterone withdrawal.

One can ask why the virgins injected with the three critical hormones did not display immediate maternal behavior, as is typically shown by parturient primiparous females. The most likely explanation is that the hormone regimen utilized by Moltz et al. (1970) did not replicate the duration, pattern, and amounts of each hormone that the pregnant and parturient rat is naturally exposed to. In addition, perhaps other hormones beyond the critical three are involved. Also, the time of pup presentation may be important; perhaps if pups had been presented on day 11 rather than day 10, shorter sensitization latencies would have been observed. It is also possible that the vaginal and cervical stimulation that occurs in normally parturient females provides and additional stimulatory factor (see the following discussion of the physiological regulation of the onset of maternal behavior in sheep). Finally, the stress of being handled and injected on each day may have had a disruptive effect on the immediate expression of maternal behavior in the virgins.

In subsequent studies, the results of Moltz et al. (1970) and Zarrow et al. (1971) have essentially been confirmed. Using physiological doses of estradiol and progesterone, Bridges (1984) found that the systemic administration of estradiol superimposed on a period of progesterone treatment followed by progesterone withdrawal could stimulate the onset of maternal behavior in ovariectomized virgin rats in as little as 24 hours of pup exposure. However, this stimulation of maternal behavior by estrogen and progesterone was prevented if estradiol-induced endogenous prolactin release was suppressed by the co-administration of bromocriptine, a drug that inhibits the release of prolactin from the pituitary

(Bridges & Ronsheim, 1990). Importantly, Bridges and Ronsheim showed that the inhibitory effects of bromocriptine could be reversed by the exogenous administration of prolactin to these steroid-treated rats. In further support of the model proposed by Moltz et al., Bridges and Russell (1981) have reported that if ovariectomized virgin female rats are treated with estradiol and progesterone, but progesterone is not withdrawn prior to testing for maternal behavior, then a short latency onset of maternal behavior was not observed. All of these behavioral results, along with the hormone pattern shown in Figure 3.1, support the proposal that estradiol and prolactin action on the brain that occurs after an abrupt decline in progesterone levels are critical components of the physiological events associated with pregnancy and parturition that stimulate the onset of maternal behavior in rats.

Research has also shown that placental lactogens can be administered instead of prolactin and they are as effective as prolactin in stimulating the onset of maternal behavior in virgin rats when they are combined with appropriate estradiol and progesterone treatment (Bridges et al., 1996, 1997). My only concern with these studies is that placental lactogen I or II was administered directly into the brain. I am not aware of any research showing that the systemic administration of placental lactogens can stimulate maternal behavior in rats, and such research would be needed to support the view that placental lactogens secreted into the general blood supply from the placenta can stimulate maternal behavior in steroid-primed rats.

S179D-PRL is a competitive prolactin receptor antagonist, and Bridges, Rigero, Byrnes, Yang, and Walker (2001) have shown that when this drug is injected directly into the lateral ventricle, from where it could eventually reach brain sites involved in the regulation of maternal behavior, it was capable of blocking the short latency onset of maternal behavior that is typically induced in ovariectomized virgin female rats that are treated with progesterone followed by estradiol. In other words, the facilitatory effects of estradiol-induced endogenous prolactin release would be prevented because endogenous prolactin would not gain effective access to prolactin receptors located in the brain. Since placental lactogens, like prolactin, are assumed to act on the prolactin receptor (Soares, 2004), it would certainly be interesting to determine whether the continuous central administration of S179D-PRL from midpregnancy through parturition would be able to block the natural onset of maternal behavior in parturient rats.

Rosenbatt's group (Rosenbatt & Siegel, 1975; Siegel & Rosenbatt, 1975b) have developed what is referred to as the pregnancy termination model to explore the endocrine basis of the onset of maternal behavior in rats. I want to spend some time describing this model because it has been used in several studies that have explored the brain mechanisms of maternal behavior. If primigravid (pregnant for the first time) rats are hysterectomized (H; removal of uterus, placentas, and fetuses) and ovariectomized (O) on day 15 of pregnancy (15HO rats) and are then presented with pups daily beginning 48 hours later, they exhibit sensitization latencies of about 3 days, which is significantly shorter than virgin sensitization latencies, which typically average about 7 days. What might account for this partial stimulation of maternal behavior in 15HO rats? A possible explanation can be arrived at by examining Figure 3.1. On day 15 of pregnancy, the brain of the pregnant female has been exposed to high levels of progesterone, estradiol, and lactogens, and the HO operation would cause a dramatic decline in these hormones. Perhaps the abrupt decline in progesterone allowed the preoperative exposure to estradiol and lactogens to partially stimulate maternal behavior. I would argue that the 15HO preparation represents a suboptimal hormonal stimulation paradigm that partially enhances the female's responsiveness to pup stimuli so that they show maternal behavior after only 3 days of pup exposure. It could be hypothesized that if estradiol levels were maintained after the 15HO procedure then perhaps the immediate expression of maternal behavior would be displayed. That is exactly what was found by Siegel and Rosenblatt: 15HO primigravid rats that receive 20 μg/kg of estradiol (E: administered subcutaneously) immediately postoperatively (15HO + E) show immediate maternal behavior (0 days sensitization latencies) when presented with pups 48 hours later. Since estradiol is assumed to have caused an increase in endogenous prolactin via its positive feedback effect, one can propose that maternal behavior occurred immediately because following HO + E the brain was exposed to high levels of estradiol and prolactin superimposed on an abrupt drop in progesterone. Perhaps this stimulatory effect of estradiol and prolactin was also potentiated by the preoperative exposure to estradiol and placental lactogens.

With respect to the importance of progesterone withdrawal, Numan (1978) has shown that maintaining progesterone levels in 15HO + E rats suppresses the facilitatory effects of estradiol on the onset of maternal behavior (also see Siegel & Rosenblatt, 1978): If 15HO + E rats are administered progesterone (P) subcutaneously on day 15 and again on the following day (15HO + E + P), when they are presented with pups 48 hours post HO they show sensitization latencies similar to that displayed by 15HO rats (about 3 days). In an important study, Numan et al. (1999) made use of the drug RU 486, which blocks the access of progesterone to progesterone receptors. In 15HO + E + P + RU 486 rats, maternal behavior was facilitated in a manner similar to that observed in 15HO + E rats. The results of all of these pregnancy termination studies are summarized in Figure 3.3.

One can conclude from these results that the decline in progesterone at the end of pregnancy potentiates the stimulatory effects of estradiol and lactogens on maternal responsiveness and that if progesterone does not decline it inhibits or antagonizes the effects of estradiol and lactogens (Numan & Insel, 2003).

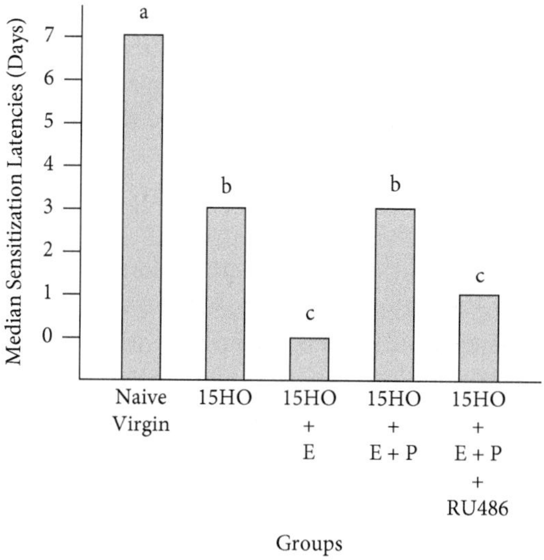

Figure 3.3. Median sensitization latencies to the onset to maternal behavior in female rats subjected to various treatments. Virgin female rats display long sensitization latencies, averaging about 7 days. The termination of pregnancy on day 15 in primigravid rats by hysterectomy (H) and ovariectomy (O) results in a moderate stimulation of the onset of maternal behavior when pups are presented to these females 48 hours after the HO procedure (15HO). The onset of maternal behavior is further enhanced if estradiol (E) is administered at the time of the hysterectomy and ovariectomy (15HO + E), but this effect is blocked by concurrent administration of progesterone (15HO + E + P). The inhibitory effects of progesterone are reversed by RU486, a progesterone receptor antagonist (15HO + E + P + RU486). Groups with different letters differ significantly from one another in their sensitization latencies to the onset of maternal behavior. All injections of E, P, and RU 486 were administered subcutaneously.

Hormones and the Induction of Maternal Behavior in Rabbits

Not only do virgin rabbits ignore rabbit pups (kits) on their initial exposure to them, but in contrast to rats, evidence shows that virgin rabbits do not show sensitized maternal behavior even after 15 days of constant exposure to young kits (Gonzalez-Mariscal, Lemus, & Aguilar-Roblero, 2015). The events associated with pregnancy and parturition appear to be essential for the display of maternal behavior in primiparous rabbits.

The hormonal pattern that occurs across the 30- to 32-day pregnancy in rabbits is similar in many respects to that which occurs in rats (Gonzalez-Mariscal, 2001; Levy, 2016; Numan & Insel, 2003). Progesterone and estradiol are present at high levels in blood plasma over days 21 to 27 of pregnancy, after which progesterone declines and estradiol, along with prolactin, exhibit substantial increases. Controversy exists with respect to whether placental lactogens are secreted in the blood of pregnant rabbits (Numan & Insel, 2003). To examine maternal behavior in the laboratory, a nest box and straw are placed in the pregnant female's cage. Nest building begins around day 28 of pregnancy, after the decline in plasma progesterone. Straw is carried into the nest box, and a nest is constructed. Upon giving birth and cleaning her kits in the nest box, the mother will subsequently nurse her young for only about 3 to 5 minutes each day and spend the rest of her time outside the nest box. The hormonal pattern occurring during pregnancy suggests, in a manner similar to rats, that the decline in progesterone and the rise in estradiol and prolactin trigger the onset of nest building and the subsequent nursing of young. Experimental evidence supports this reasoning.

In ovariectomized rabbits, long-term treatment with systemic injections of estradiol and progesterone, followed by progesterone withdrawal and continued estradiol treatment, induces nest-building behavior, and this is associated with increases in endogenous prolactin. Estradiol administration alone does not stimulate nest building (Gonzalez-Mariscal, Melo, Jimenez, Beyer, & Rosenblatt, 1996). When bromocriptine was administered to the rabbits treated with progesterone and estradiol, endogenous prolactin levels were suppressed, and nest building was not induced. These results support the view that declining progesterone and rising estradiol and prolactin levels in the blood stimulate maternal nest building in rabbits.

I am not aware of any studies that have attempted to induce pup-directed nursing behavior (couching over the young in a nest box) in virgin rabbits through the administration of exogenous hormones. However, experiments on pregnant rabbits have shown that prolactin is one of the hormones that is necessary to induce the onset of pup-directed nursing behavior (Gonzalez-Mariscal, Chirino, Flores-Alonso, Rosenblatt, & Beyer, 2004). Primigravid females that received daily systemic injections of bromocriptine from day 26 of pregnancy through parturition gave birth to and cleaned their young in the nest box, but they did not subsequently re-enter the nest box and adopt a nursing posture over their young. This inhibition of nursing behavior occurred on each of the 7 postpartum test days, although the bromocriptine injections were terminated on the day of parturition. This disruptive effect of bromocriptine was reversed if the rabbits were concurrently administered prolactin into the lateral ventricle of the brain (into the cerebrospinal fluid, from which it could diffuse to critical brain sites). Whether progesterone withdrawal and high levels of estradiol are necessary for prolactin to stimulate nursing behavior remains to be determined, but based on the research on rats, it appears likely that this is the case.

Clearly, more research is needed on the hormonal basis of pup-directed maternal behavior in rabbits, but the evidence available so far indicates significant similarities with the hormonal regulation of the onset of maternal behavior in rats.

Physiological Regulation of the Onset of Maternal Behavior in Sheep

Nonpregnant ewes will avoid lambs, will not allow them near their udder, and will reject them with head butts if the lamb is persistent; in contrast, at the time of parturition, prior to the development of maternal selectivity, a ewe will accept any lamb, licking it and allowing it access to the udder for nursing (Le Neindre, Poindron, & Delouis, 1979; Levy, 2008; Numan, 1994; Numan & Insel, 2003). There is no evidence that nonpregnant ewes can be sensitized to show maternal behavior through exposure to young lambs (Levy, Porter, Kendrick, Keverne, & Romeyer, 1996). This presumed lack of sensitization in sheep makes adaptive sense since under natural conditions nonpregnant ewes would be regularly exposed to unrelated lambs from other mothers in the herd. However, to my knowledge, no one has attempted to study whether maternal behavior could be induced in nonpregnant ewes after forced association with lambs over long periods of time.

In sheep, pregnancy lasts about 150 days, and starting at 2 to 4 days before parturition, progesterone levels decline, and this decline is followed by a rise in estradiol and prolactin (Levy, 2016; Nowak, Keller, & Levy, 2011). Given this hormone pattern, and knowing that estradiol stimulates the release of prolactin from the anterior pituitary, researchers have attempted to induce maternal behavior in nonpregnant ewes through the systemic administration of progesterone and estradiol (Poindron & Le Neindre, 1980). When nonpregnant, nonlactating ewes are treated concurrently with estradiol and progesterone, followed by the withdrawal of progesterone and the maintenance of estradiol administration, about 50% of the ewes will show maternal behavior toward a lamb in a 2-hour test. Interestingly, this stimulatory effect of steroid treatment only occurs in multiparous nonpregnant sheep (sheep that had given birth previously) and does not occur in nulliparous sheep (Le Neindre et al., 1979; Poindron & Le Neindre, 1980).

In an important study, Keverne, Levy, Poindron, and Lindsay (1983) injected ovariectomized multiparous nonpregnant, nonlactating ewes with progesterone followed by estradiol. Twenty-four hours after the estradiol injection a lamb was presented to each ewe. Half of the ewes received vaginocervical stimulation (VCS; with a vaginal probe) 5 minutes before lamb presentation. In a 1-hour test,

80% of the ewes that received P + E + VCS accepted the lamb and showed maternal behavior, while this was the case for only 20% of the P + E females that did not receive VCS. VCS in the absence of steroid treatment did not induce maternal responsiveness. In a subsequent study (Kendrick & Keverne, 1991), it was found that the P + E + VCS treatment was ineffective when applied to ovariectomized nulliparous females.

These results indicate that falling levels of progesterone and rising levels of estradiol, along with the cervical and vaginal stimulation that normally occurs at parturition, play a role in the induction of maternal behavior in sheep. However, additional factors appear to be necessary for the occurrence of immediate maternal responsiveness in primiparous mothers since P + E + VCS is ineffective in virgin ewes. The importance of naturally occurring VCS for the onset of maternal behavior in parturient ewes is further supported by the finding that disruption of neural feedback from the vaginal–cervical area near the time of parturition, through peridural anesthesia, disrupts the onset of maternal behavior in sheep (Krehbiel, Poindron, Levy, & Prudhomme, 1987). The central neural effects of VCS with respect to the onset of maternal behavior will be analyzed in Chapter 4.

Given the importance of VCS for the onset of maternal behavior in sheep, one can ask whether this stimulation is important for other species as well. In ovariectomized virgin female rats, steroids and prolactin can induce maternal behavior, but such behavior is rarely immediate and usually occurs after 1 to 2 days of pup stimulation. Perhaps the administration of VCS to these females would result in an immediate onset of maternal behavior in a manner similar to that observed by Kendrick and Keverne (1991) in sheep. Pregnancy termination studies in rats indicate that 15HO + E females do show immediate maternal behavior when pups are presented to them 48 hours later, but the hysterectomy surgery and its lingering effects may have provided sensory feedback to the central nervous system that mimicked the effects of the VCS that normally occurs at parturition. Finally, Mayer and Rosenblatt (1984) have noted that most primigravid female rats will show maternal behavior toward pups that are presented to them within 3.5 hours prior to parturition. Since this prepartum onset of maternal behavior was associated with the onset of uterine contractions, perhaps the neural feedback from these contractions, combined with the effects of progesterone withdrawal and rising estradiol and lactogens, resulted in immediate maternal responsiveness prior to the actual occurrence of parturition. Therefore, it is certainly possible that neural feedback from the uterus–cervix–vagina is involved in the immediate onset of maternal behavior in rats (cf. Yeo & Keverne, 1986).

With respect to the involvement of prolactin in the maternal behavior of sheep, Poindron, Orgeur, Le Neindre, Kann, and Raksanyi (1980) treated late pregnant ewes with bromocriptine. Although this treatment depressed plasma prolactin levels, it did not interfere with the onset of maternal behavior, suggesting that prolactin is not important for maternal behavior in this species. However, a cautious interpretation of this data is necessary. First, multiparous ewes were used as subjects. Perhaps previous maternal experience rendered the maternal behavior of these sheep less dependent on prolactin. Further, unlike the rabbit, and similar to the rat, sheep produce placental lactogens (Soares, 2004), and these hormones, which would not be suppressed by bromocriptine, could have acted on prolactin receptors in the brain to stimulate maternal behavior. In support of this logic, recall that bromocriptine injections during late pregnancy disrupt maternal behavior in rabbits. However, in pregnant (rather than virgin) rats, it is ineffective (Numan, Rosenblatt, & Komisaruk, 1977). Bromocriptine does have disruptive effects in nulliparous rats because such rats do not produce placental lactogens. An appropriate experiment to test the importance of pituitary prolactin in sheep would be to determine whether bromocriptine would disrupt the maternal behavior induced in nonpregnant ewes through P + E + VCS treatment.

Hormones and Maternal Behavior in Mice

Introduction

The maternal behavior of virgin females of most inbred and outbred strains of the laboratory house mouse contrasts sharply with that of virgin rats, rabbits, and sheep. Virgin female laboratory mice, when presented with foster pups in their home cages, show near immediate maternal behavior (Kuroda et al., 2011; Numan & Insel, 2003). Such females retrieve the pups, groom them, and crouch over them within 15 to 30 minutes, although in some studies 60 minutes of pup exposure is necessary to induce maternal behavior (Gandelman, 1973; Larsen, Kokay, & Grattan, 2008; Martin-Sanchez et al., 2015; Stolzenberg & Rissman, 2011; Tsuneoka et al., 2013; Wu, Autry, Bergan, Watabe-Uchida, & Dulac, 2014). Such near immediate maternal behavior in virgin mice has been referred to as spontaneous maternal behavior. Importantly, such short latency maternal behavior also occurs in ovariectomized mice (Gandelman & vom Saal, 1975; Stolzenberg & Rissman, 2011). Therefore, unlike rats, rabbits, and sheep, these results suggest that the physiological events associated with late pregnancy and parturition do not play a significant role in pup-directed maternal behavior in many laboratory strains of mice when these mice are tested with foster pups in their home cages.

The behavior of inbred and outbred strains of laboratory house mice, however, does not match that of virgin wild (feral) house mice that have been raised in a laboratory setting. When they are tested with foster pups in their home cages, virgin feral house mice do not show spontaneous maternal behavior, but instead attack and kill foster pups (Jakubowski & Terkel, 1982; McCarthy & vom Saal, 1985; Soroker & Terkel, 1988). These studies have also shown that feral mice will kill foster pups during late pregnancy, but that they will care for such unrelated pups during the postpartum period. The conclusion that I reach is that the physiological events of late pregnancy and parturition are essential for pup-directed maternal responses in wild mice, but are not absolutely necessary in many laboratory strains.

This comparison of the maternal behavior of virgin house mice that are feral with those that have been either inbred or outbred over many generations in a laboratory setting indicates that experimental genetic selection can alter the mechanisms that regulate maternal behavior, rendering laboratory strains less dependent on the physiological factors associated with pregnancy and parturition for the onset of maternal behavior. In laboratory strains tested in their home cages, pup stimuli have near immediate access to the brain mechanisms regulating pup-directed maternal behavior in naïve virgins. In support of a role for genetic factors in mouse maternal behavior, Perrigo et al. (1993) found that adult virgin female hybrid offspring derived from mating feral mice with CF-1 inbred mice showed maternal behavior toward foster pups during a 30-minute test. Irrespective of the mother's genotype (feral or CF-1), the hybrid virgin females showed the spontaneous maternal behavior typical of the inbred CF-1 strain. Chaflin et al. (2014) have provided more recent support for the proposal that experimental genetic selection has altered both maternal behavior and aggressive behavior in females of laboratory strains of mice, increasing their maternal responsiveness while decreasing their aggressiveness.

Why do I find these results to be so significant? The virgin female laboratory mouse's response to neonates is an atypical response that is not representative of the standard response shown by nulliparous females in species that are characterized by a uniparental maternal care system. This is clearly evident from the comparison between laboratory and feral mice. I think the response of laboratory virgin female mice gives us insight into the evolution of allomaternal behavior. That is, under natural conditions where it would be adaptive for allomaternal behavior to occur, natural selection, rather that experimental selection, could operate to affect the genotype of inexperienced virgin females so that they would be biased to care for, rather than to avoid or harm, conspecific offspring. Further, these feral versus laboratory mice findings provide insight into the evolution of paternal behavior because they allow us to understand that natural selection could, under the proper ecological conditions, alter the brain of fathers so that there could be mechanisms that would induce paternal behavior outside the boundaries of the specific physiological events associated with pregnancy and parturition.

Differences Between the Maternal Behavior of Virgin Laboratory Mice and Their Postpartum Counterparts

Although virgin female laboratory mice show near immediate maternal behavior when tested with foster pups in their home cages, their behavior differs significantly from postpartum dams when they are tested in more challenging settings. Similar to the findings reviewed for rats, virgin female mice that are showing maternal behavior in home cage tests do not initially retrieve pups that are placed in a novel T-maze extension attached to their home cage, while recently parturient primiparous females do (Numan & Insel, 2003; Stolzenberg & Mayer, 2019; Stolzenberg & Rissman, 2011). However, after 4 days (2 hours/day) of engaging in maternal behavior in their home cages with foster pups, such maternally experienced females, even if they are ovariectomized, will retrieve pups from the T-maze in a manner equivalent to that of postpartum females (Stolzenberg & Rissman, 2011). What these results suggest is that even in laboratory mice the physiological events of pregnancy and parturition result in a higher level of maternal capability in the postpartum female than that which is initially exhibited by the virgin female. Importantly, maternal experience can compensate for the absence of pregnancy/parturition factors, and after 4 days of such experience the maternal responsiveness of virgins in the T-maze is equivalent to that of the postpartum female.

In contrast to the findings of Stolzenberg and Risssman (2011), Stern and Mackinnon (1976) found that virgin rats that were displaying sensitized maternal behavior in their home cages for 4 days were still much less likely than postpartum females to retrieve pups from a T-maze. Therefore, maternal experience in virgin rats appears to be less effective in enhancing T-maze retrieval than it is in laboratory mice. Perhaps this difference is related to the lower level of maternal behavior that is initially exhibited by virgin rats when compared to virgin mice: In home-cage tests, mice are "spontaneously" maternal, while rats require many days of pup exposure before they show maternal behavior. Perhaps more than 4 days of maternal experience after the onset of pup-induced maternal behavior would be required for sensitized rats to retrieve pups during a T-maze test.

When I discussed the T-maze retrieval findings in virgin and postpartum rats earlier in this chapter, the question was asked as to whether a decrease in fearfulness or an increase in maternal motivation, or both, was involved in the finding that postpartum females are more likely to retrieve pups from a T-maze than are virgin females. Stolzenberg and Rissman (2011) examined aspects of this question in mice with respect to the effects of maternal experience on the maternal behavior of virgin mice. Virgin female mice were either naïve (no experience with pups and therefore no maternal experience) or received 4 days of maternal experience with pups in their home cages (2 hours/day). The mice that were exposed to pups and showed maternal behavior were separated from pups for 24 hours, and then both groups were tested for T-maze retrieval. As expected, the experienced virgin mice did, while the naïve virgins did not, retrieve pups in the T-maze. Twenty-four hours after the T-maze test, the mice were tested for fear-related behavior in an elevated plus maze (pups were not present). The elevated plus maze consists of two runways that cross each other, forming four arms. Two arms are called the closed arms because they have walls and are considered to be the safe or protected arms, while the other two arms have no walls are called the open arms and are considered to be less protected. The amount of time spent in the open arms is positively related to decreases in fearfulness. The maternally experienced virgin females spent the same amount of time in the open arms as did the naive virgins. Stolzenberg and Rissman concluded that the increased T-maze retrieval behavior in their pup-experienced virgin mice was related to an increase in maternal responsiveness or motivation rather than to a decrease in fearfulness.

A final study related to these issues in mice was conducted by Hauser and Gandelman (1985). Using an operant conditioning procedure, they examined the degree to which virgin and postpartum mice (Rockland-Swiss outbred strain) would press a lever to receive a pup as a reward. Each lever press deposited a pup in the female's home cage. Although both virgin and postpartum females were caring for young in their home cages, the postpartum females exhibited a much higher number of bar presses for a pup reward than did the virgins (approximately 150 versus 50 bar presses in a 45-minute session; after 20 bar presses, most pups were removed from the home cage and only 6 pups were allowed to remain in the nest). Subsequent experiments provided evidence that the endocrine changes associated with pregnancy termination were the cause of the higher bar press rate in the postpartum females. Given these results, it can be proposed that in laboratory mice, the endocrine events associated with pregnancy termination act to increase the level of maternal motivation above that shown by the "spontaneously" maternal virgin, so that pup stimuli become more attractive to the postpartum female than to the virgin.

An interesting finding of this study, when compared to the results of Stolzenberg and Rissman (2011), was that even after 20 days of home-cage maternal experience, the bar press rate of the virgins did not increase. These results of the Hauser and Gandelman (1985) study suggest that maternal experience in virgins did not increase maternal motivation as tested in the operant conditioning paradigm. What might account for this difference? A relevant event may be related to whether pups were present for the virgin to care for. In the Hauser and Gandelman study, pups remained in the cage with the virgins during the operant test, while in the Stolzenberg and Rissman study, the virgin mice were separated from pups for 24 hours before being tested in the T-maze. Perhaps it is essential to couple the absence of pups (pup deprivation) with maternal experience to increase maternal motivation in virgin mice.

In conclusion, although several strains of virgin laboratory mice are spontaneously maternal in home-cage tests, further analysis clearly indicates that the physiological events of late pregnancy and parturition and/or maternal experience can boost maternal motivation so that such mice become more interested in and attracted to pup-related stimuli and are willing to expend effort and overcome obstacles to gain access to pups.

A comparison of the results from laboratory rats and mice indicates the following. Inexperienced virgin rats initially avoid pups, but the events of pregnancy and parturition seem to have two effects in the parturient primiparous female that allow for the immediate onset of maternal behavior: They decrease the aversive qualities of pup stimuli that typically stimulate avoidance responses while they increase the attractive qualities of pup stimuli that activate approach tendencies and maternal responsiveness or motivation. In contrast, in naive laboratory mice, the aversive qualities of pup stimuli are low and the approach tendencies that exist in such mice allow for near immediate maternal behavior in home-cage tests. However, as in rats, the physiological events of late pregnancy further enhance the attractive qualities of pup stimuli, thereby increasing maternal responsiveness or motivation. Maternal experience also boosts maternal motivation in mice.

Prolactin and Maternal Behavior in Laboratory Mice

Null Mutation of the Prolactin Receptor Gene

In home-cage tests, the near immediate maternal behavior of nulliparous female laboratory mice, even when they are ovariectomized, suggests that this behavior is not under hormonal control and that pup stimuli have direct access to the neural circuits regulating maternal behavior. In apparent conflict with this proposal, Lucas, Ormandy, Binart, Bridges, and Kelly (1998) have reported that adult virgin females of a transgenic mouse line with a null mutation (knockout mutation) of the prolactin receptor (PrlR) gene show severe deficits in maternal behavior when tested with foster pups for 30 minutes per day for 6 days. While control virgins that retained the PrlR gene showed full maternal behavior during these tests, the females with the null mutation did not. Since these mutated mice do not produce the prolactin receptor protein (Ormandy et al., 1997), these results suggest that prolactin may be acting on its receptor, presumably in the brain, in the adult virgin female mouse to allow for the expression of near immediate maternal behavior. This result is somewhat difficult to understand, since in all mammalian species where prolactin has been shown to stimulate maternal behavior, its stimulatory actions are dependent on the co-administration of estradiol, yet ovariectomized laboratory virgin mice show spontaneous maternal behavior (see Stolzenberg & Rissman, 2011).

To begin an analysis of these results, we need to understand the transgenic methods (insertion of a new gene into an organism) used to produce mice with a null, or knockout, mutation (Numan, 2015). In transgenic mice, a native (normal or wild-type) gene is replaced with another gene through a procedure called homologous recombination. When the substitute gene is nonfunctional (does not regulate the synthesis of a functional protein), a null mutation of that gene is produced. In creating a transgenic mouse line with a null mutation of the PrlR gene, the researcher would first manufacture a double-stranded DNA sequence that mirrors the normal PrlR gene, except that changes are made to the gene to render it nonfunctional. Such a genetically engineered gene is then injected into the nucleus of pluripotent mouse embryonic stem cells, where it can replace the native gene. These effectively modified embryonic stem cells are then inserted into early mouse embryos and as the embryonic cells divide and differentiate, some of the cells in the developing mouse will contain the defective gene (not all cells contain the defective gene because the mutant stem cells were inserted into an embryo that already had some normal cells). For those mice in which the defective PrlR gene is incorporated into the gametes, adult males and females would be bred together to ultimately produce offspring that are homozygous for the PrlR gene knockout (PrlR –/–). Such offspring are referred to as containing whole body knockouts of the PrlR gene, since the nonfunctional gene would be present in all cells, including brain cells. Therefore, in those cells where normal mice would produce prolactin receptors, the mutated transgenic strain would not. It is important to note that this lack of prolactin receptor production in the knockout strain would occur throughout prenatal and postnatal development, as well as into adulthood.

In conflict with the findings of Lucas et al. (1998), Horseman et al. (1997) have reported that adult virgin female mice of a transgenic line that contained a null mutation of the prolactin gene show full maternal behavior within 30 minutes after foster pups were introduced into their home cages. In these females, prolactin is not produced and yet they showed spontaneous maternal behavior. How can we explain why the absence of prolactin does not disrupt virgin maternal behavior, while the absence of the prolactin receptor does? One possibility is that some ligand other than prolactin is capable of binding to and activating the prolactin receptor so that maternal behavior could be stimulated in the adult mouse even if prolactin were not present itself.

However, it is also possible that an absence of the developmental effects of lactogens influenced the findings of Lucas et al. (1998). Freemark, Driscoll,

Andrews, Kelly, and Royster (1996) have reported that although prolactin receptors are not present in the olfactory bulb of adult rats, they are expressed at very high levels in the olfactory bulb of fetal and neonatal rats. They suggest that stimulation of prolactin receptors in the olfactory bulb during fetal development might be involved in the differentiation and development of the full functional capacity of olfactory bulb neurons. Relevantly, prolactin receptors have also been detected in the olfactory bulb of the fetal mouse brain (Tzeng & Linzer, 1997). Therefore, in the absence of the prolactin gene and prolactin production by the fetus, maternal prolactin and placental lactogens during late pregnancy would likely be able to act on fetal olfactory bulb prolactin receptors to influence the development of the olfactory system in the fetus. Such a developmental effect would not be possible in PrlR –/– mice. Some pertinent data has recently been reported by Sairenji et al. (2017). They found that pups born to mouse mothers who secreted lower than normal levels of prolactin over the last few days of pregnancy showed deficits in their maternal behavior toward their own pups in adulthood. These maternal deficits in the offspring could be reversed by injecting their mothers with prolactin from day 15 of pregnancy through the day of parturition. Therefore, maternal prolactin secretion during pregnancy does appear to affect the development of the fetal brain, and these effects influence the development of the offspring's maternal behavior. Whether the developmental effects of prolactin observed in this study were the result of affecting the development of the fetal olfactory system remains to be determined.

As I will describe in detail in the Chapter 4, normal olfactory function is essential for the display of maternal behavior in both virgin and postpartum mice. With this understanding, it is possible that the lack of prolactin receptor production in the transgenic mouse line of Lucas et al. (1998) resulted in a developmentally induced deficit in normal olfaction, which subsequently caused a decrease in maternal behavior. It should be noted, however, that Lucas et al. did perform olfactory discrimination tests on their mice and they found no differences between the ability of normal and PrlR knockout mice to detect an extremely intense odor (isoamyl acetate; a banana-like odor). Because this odor was very intense, their test could not rule out the possibility that the PrlR knockout mice would have displayed olfactory deficits if a more sensitive olfactory discrimination task had been used. Further, their mice may have developed deficits in the ability to detect natural pup pheromones, and this possibility was not examined. Finally, more subtle deficits in development of the olfactory system, which do not produce anosmia, may have disrupted maternal behavior (cf. Wei, Meaney, Duman, & Kaffman, 2011).

It is hard to draw any firm conclusions from these diverse findings. It is certainly possible that a ligand other than prolactin binds to and activates the prolactin receptor in adult virgin mice to allow for spontaneous maternal behavior, even in the absence of estradiol. It is also possible that a deficit in the early development of the olfactory system is involved in the poor maternal behavior shown by adult virgin mice with a null mutation of the PrlR gene. Another possibility is that the absence of prolactin signaling in the fetus affects the proper development of other aspects of the neural circuits that underpin maternal behavior (see Sairenji et al., 2017). A resolution of these issues awaits the results of future research. It would certainly be important to determine whether the central administration of a prolactin receptor antagonist (S179D-PRL) to adult virgin mice would be capable of disrupting the spontaneous maternal behavior typically shown by such females. A further exploration of this question will be presented in the next chapter.

Prolactin, Neurogenesis, Olfaction, and Maternal Behavior in Adult Mice

It is now widely accepted that neurogenesis (the birth of new neurons) occurs within two important brain regions of the adult mammalian brain. Such neurogenesis occurs regularly in the subgranular zone in the dentate gyrus of the hippocampus and in the subventricular zone (SVZ) of the cells that line the lateral ventricle in the forebrain (Feierstein, 2012; Imayoshi et al., 2008; Leuner & Sabihi, 2016; Slattery & Hillerer, 2016). The new hippocampal neurons remain within the dentate gyrus, where they form glutamatergic granule cell neurons. In contrast, the newborn cells in the SVZ ultimately migrate to the olfactory bulb where they differentiate into inhibitory interneurons.

Because of the importance of olfaction for mouse maternal behavior, I will focus on the relevance of olfactory bulb neurogenesis for maternal behavior in the adult laboratory mouse. Shingo et al. (2003) were the first to report that the rise in prolactin that occurs during early pregnancy in mice stimulates neurogenesis within the SVZ of these pregnant mice. These newborn cells ultimately migrate to, and are integrated into, the olfactory bulb to form interneurons near the time of parturition (Larsen & Grattan, 2012).

In an ingenious study, Larsen and Grattan (2010) systemically administered low doses of bromocriptine, to suppress prolactin secretion, into primigravid laboratory mice over days 1 to 3 of pregnancy. Although prolactin during early pregnancy in mice maintains the secretion of progesterone from the ovaries and progesterone is necessary for the continuance of pregnancy, the low doses of bromocriptine used only truncated, but did not eliminate, prolactin secretion, and therefore pregnancy and parturition occurred normally. They found that this bromocriptine treatment blocked the normal increase in neurogenesis that occurs in the SVZ on day 7 of pregnancy and the normal increase in new neurons that are present in the olfactory bulb on day 2 postpartum. It would have been important to show that this bromocriptine-induced inhibition of olfactory bulb neurogenesis could have been reversed by the co-administration of prolactin, but this manipulation, unfortunately, was not performed.

The maternal behavior of additional groups of bromocriptine-treated and control mice was examined on day 2 postpartum. During these tests, the females were presented with foster pups rather than their own pups. In home-cage tests, the maternal behavior of bromocriptine-treated females was indistinguishable from control females that did not receive bromocriptine. However, when similarly treated females were placed in a novel clean cage with three foster pups, the control females showed full maternal behavior (retrieving all pups and adopting a nursing posture over them) within 5 minutes, while the bromocriptine-treated females did not retrieve the pups during the 60-minute observation period. The authors did not examine whether co-administration of prolactin could reverse this bromocriptine-induced deficit in maternal behavior.

Here is one possible interpretation of these results. The fact that all females, irrespective of treatment, cared for pups in their home cages certainly indicates that the bromocriptine-treated females were not anosmic (since olfaction is essential for maternal behavior in mice) and they were fully capable of displaying maternal behavior. However, the bromocriptine-induced inhibition of olfactory bulb neurogenesis may have affected the ability of the postpartum dam to make sharp discriminations between behaviorally relevant odors and other novel odors (Kopel, Schechtman, Groysman, & Mizrahi, 2012; Vinograd et al., 2017). The testing of females with foster pups in a novel cage, without the mother's own pheromones being present, may have been too much of an olfactory challenge for females with a compromised olfactory system. The bromocriptine-treated females may have therefore been too confused and distracted by novel odors, which, in turn, inhibited their ability to care for the foster pups within a 60-minute test period.

Other interpretations of these data are also possible and the reader is referred to Larsen and Grattan (2010), Feierstein (2012), and Feierstein et al. (2010). It should also be noted that the findings of Larsen and Grattan are correlational in nature. Although I made the assumption that bromocriptine's inhibition of prolactin-induced olfactory bulb neurogenesis was the underlying reason for the disruption of maternal behavior in the novel cage test situation, it is possible that the suppression of prolactin during early pregnancy had other effects that could have impacted maternal behavior.

Conclusions for Mice

The maternal behavior of female laboratory mice has clearly been influenced by inbreeding or selective breeding so that virgin females show near immediate maternal behavior in home-cage tests, indicating that the physiological events of late pregnancy and parturition are not absolutely required for prompt maternal responsiveness toward pups. However, the declining levels of progesterone superimposed on rising levels of lactogens and estradiol, by boosting maternal motivation, are probably essential for maternal responsiveness to occur under more challenging or stressful situations, such as retrieving pups in a novel environment such as a T-maze attachment to the home cage. Nevertheless, a sufficient amount of maternal experience can also boost maternal motivation and substitute for the physiological events of late pregnancy. The possibility that a ligand other than prolactin can activate the prolactin receptor to promote spontaneous maternal behavior in virgin mice in home-cage tests remains an open question that requires further research. I have also reviewed evidence in support of the possibility that prolactin may influence maternal behavior in mice by affecting olfactory bulb development during the fetal and neonatal period and/or by affecting olfactory bulb neurogenesis in adult mice during early pregnancy. Such effects could influence the ability of the female mouse to make the necessary olfactory discriminations required for effective maternal behavior.

The Maintenance of Maternal Behavior and the Onset-Maintenance Dichotomy in Rats, Rabbits, and Sheep

In the results that I have reviewed, it is clear that naïve virgin female rats, rabbits, and sheep do not show immediate maternal responsiveness when presented with infants of their species. However, at parturition, primiparous rats, rabbits, and sheep show immediate maternal behavior when presented with any young infant of their species. In rats and rabbits, rising plasma levels of estradiol and lactogenic hormones superimposed on a background of progesterone withdrawal, possibly also coupled with parturition-induced VCS, modify the brain's responsiveness to pup stimuli so that the primiparous mother is immediately responsive to general conspecific infant stimuli. Similar processes occur in sheep, although VCS is definitely coupled with the importance of declining progesterone and rising estradiol. The involvement of lactogens in the onset of maternal behavior in parturient sheep remains to be more fully examined.

After maternal behavior becomes established at parturition in these species, the question arises as to whether hormones are needed for the continuance of maternal behavior for the remainder of the postpartum period. Evidence indicates that hormones are not required, giving rise to the onset-maintenance dichotomy view of maternal behavior regulation, with hormones being essential for the onset, but not the maintenance of the behavior (Numan, 1994; Numan & Insel, 2003; Rosenblatt, Mayer, & Siegel, 1985). For rats, a variety of endocrine interventions when applied to postpartum primiparous mothers do not disrupt the continuance of maternal behavior after its onset at parturition. These ineffective manipulations include hypophysectomy (Erskine, Barfield, & Goldman, 1980a), ovariectomy (Grieb, Tierney, & Lonstein, 2017; Moltz & Weiner, 1966), adrenalectomy (Rees, Panesar, Steiner, & Fleming, 2004), treatment with drugs that are similar to bromocriptine and that block both prolactin release and lactation (Numan, Leon, & Moltz, 1972), and the administration of high doses of progesterone (Moltz, Levin, & Leon, 1969). Therefore, while estradiol and lactogens are essential for the onset of maternal behavior at parturition, they are not required for its continuance. (Note that prolactin is the only lactogen that is present in postpartum dams that have not mated during their postpartum estrus and are therefore lactating but are not pregnant; the inhibition of prolactin release in such lactating females inhibits lactation but does not disrupt maternal responsiveness.) The case of progesterone is also interesting. The near-term decline in progesterone levels is essential for the onset of maternal behavior at parturition in primiparous rats, and if this decline is prevented, the onset of maternal behavior is disrupted (Moltz et al., 1969, 1970; Numan, 1978; Numan et al., 1999). In contrast, the naturally high endogenous progesterone that is secreted by the ovary in rats beginning a few days after parturition and continuing throughout most of lactation (Smith & Neill, 1977) occurs alongside normal maternal behavior. Further, the administration of exogenous progesterone to lactating rats, which would result in supraphysiological progesterone levels, has no inhibitory effects. One potential explanation for why high progesterone can inhibit the onset of maternal behavior in rats but has no disruptive effects on its postpartum maintenance may be related to the expression of progesterone receptors in critical brain regions known to be essential for maternal behavior. Progesterone receptors are present at high levels in neurons of these brain regions near the time of parturition, but are expressed at very low levels on days 3 and 7 postpartum (Grieb et al., 2017; Numan et al., 1999). Therefore, progesterone may be ineffective in inhibiting maternal behavior during the postpartum period because it cannot effectively interact with those brain circuits essential for maternal behavior.

Although the research on rabbits and sheep is not as extensive as that in rats, researchers have reached similar conclusions: Although the onset of maternal behavior is hormone-dependent, its maintenance is not (Gonzalez-Mariscal, 2001; Nowak et al., 2011). The general conclusion reached by researchers studying the maternal behavior of rats, rabbits, and sheep is that the endocrine and other physiological events of late pregnancy alter brain mechanisms so that general infant stimuli elicit immediate maternal responsiveness. However, this physiological modulation of reactivity to infant stimuli is short-lasting: Infants must be present during the immediate postpartum period for maternal behavior to occur and become established. Maternal responsiveness will then continue during the remainder of the postpartum period without the continued requirement of hormonal mediation.

What is the evidence that there is a short time window after parturition, sometimes referred to as a sensitive period, when mother–infant interactions need to occur for maternal behavior to become established? For primiparous rats, rabbits, and sheep, if the newborn young are removed from the mother immediately after birth, precluding mother–infant interactions, and are then returned to the mother a few days later, the mother does not respond to the young, but instead avoids and/or rejects them; such females act like naïve virgins. However, if the young are removed from the mother after maternal behavior has become established, let's say on day 5 postpartum, and then returned a few days later, maternal behavior is normal (Gonzalez-Mariscal et al., 1998; Orpen & Fleming, 1987; Poindron, Levy, & Keller, 2007; Rosenblatt & Lehrman, 1963). These results suggest that hormones prepare the brain for immediate maternal responsive and if mother–infant interactions occur during this period of hormonal priming then maternal experience further modifies the brain so that (a) subsequent maternal behavior becomes relatively hormone-independent and that (b) an enduring bond or attachment occurs between the mother and her offspring that can persist following a period of mother–infant separation.

Figure 3.4 presents a summary of the research I have described in this chapter and in Chapter 2. Rats and rabbits are described separately from sheep for the following reasons: In rats and rabbits, once maternal behavior becomes established at parturition, subsequently during the nonhormonal maintenance phase of maternal behavior these species continue to respond to a generic infant stimulus and will care for any appropriately aged offspring of their species. In contrast, for sheep, mother–infant interactions at parturition have a dual effect. Maternal experience with a lamb at parturition not only results in a mother forming a strong bond with a lamb that persists postpartum without the need for hormonal mediation, but also results in maternal selectivity. As a result of interacting with a particular lamb, the mother learns its olfactory characteristics and then during the remaining postpartum period, she will only care for that lamb and will not respond maternally toward unfamiliar lambs.

In interesting research on the onset of maternal responsiveness in sheep and the subsequent development of maternal selectivity, Keverne et al. (1983) have reported that ewes that remained with their own young for the first 2 hours postpartum would reject alien young that are presented to them after this 2-hour postpartum interaction. However, if experimenter-applied VCS (to mimic parturition-related stimuli) is administered to postpartum ewes even after 24 hours of experience with their own lamb, such ewes will accept and form a new bond with an alien lamb while still maintaining the original bond with their own offspring (Kendrick, Levy, & Keverne,1991). These results are intriguing and

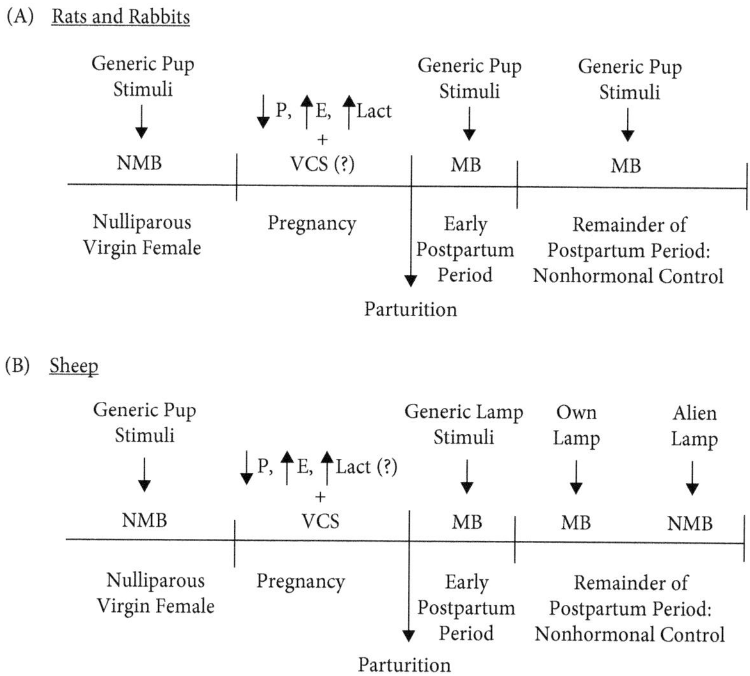

Figure 3.4. The onset-maintenance dichotomy with respect to the regulation of maternal behavior in (A) rats and rabbits and (B) sheep. For each species, the physiological events of late pregnancy and parturition allow for the immediate onset of maternal behavior, while the maintenance of maternal behavior throughout the remainder of the postpartum period is free from hormonal control. At parturition, each of these species will respond maternally to any conspecific infant. Rats and rabbits, throughout the remainder of the postpartum period, will act maternally toward any conspecific infant: Their maternal behavior is directed toward generic conspecific infant stimuli. In contrast, during the early postpartum period, maternal sheep learn the olfactory characteristics of the lamb to which they have been exposed, and thereafter, they will only care for that specific lamb, while rejecting the advances of unfamiliar lambs. E = estradiol; Lact = lactogens; MB = maternal behavior; NMB = no maternal behavior; P = progesterone; VCS = vaginocervical stimulation. An upward arrow indicates hormone increases and a downward arrow indicates hormone decreases. A question mark signifies that the involvement of a particular physiological factor has not been completely settled.

suggest that when the postpartum ewe's brain has been primed by late-pregnancy steroid hormones and perhaps lactogens, there is a postpartum sensitive period of about 24 hours that permits VCS to reactivate maternal responsiveness to a general lamb stimulus and then as a result of interacting with the new lamb an additional period of olfactory learning takes place that results in the mother forming a new selective bond. This process certainly makes sense since under natural conditions during each new birth cycle a parturient ewe forms a new selective bond with a newborn lamb. The neural mechanisms that underlie the ability of late pregnancy hormonal factors coupled with parturition-related VCS to activate maternal responsiveness in sheep to a generic lamb stimulus will be discussed in the next chapter. I would predict that if VCS were applied later during the postpartum period in sheep during the nonhormonal maintenance phase of maternal behavior and after the effects of parturient hormones have waned, then the ewe would not accept an alien lamb. Unfortunately, Kendrick and his colleagues did not test this possibility. My prediction, however, fits with the fact that in nonpregnant, nonlactating ewes, VCS can only induce maternal responsiveness if the ewe has been primed with progesterone and estradiol.

In rats, the continuance of maternal behavior during the postpartum period without the need for hormonal mediation is similar to a process referred to as maternal memory or the consolidation of maternal responsiveness (Bridges, 1975, 1977, 1978; Orpen & Fleming, 1987). Maternal memory refers to those findings that show that although first-time parturient (primiparous) rats require hormonal stimulation to initiate prompt maternal responsiveness, once a critical amount of maternal experience with young occurs at parturition, subsequent episodes of maternal behavior, even after very long periods of mother–infant separation, become relatively emancipated from hormonal control. In a standard maternal memory experiment, one group of primiparous mothers is permitted to interact with their young for 1 hour immediately following parturition, and then the pups are removed. In contrast, a second group of primiparous mothers is not allowed this 1 hour of maternal experience and the offspring are removed from the mother as soon as they are born. A retention test is then conducted 10 to 25 days later at which time all females are exposed to healthy young pups, obtained from donor lactating mothers, on a daily basis until sensitized maternal behavior occurs. The mothers that had the 1-hour maternal experience typically display sensitization latencies of between 1 and 2 days, while the mothers that lacked such experience act like virgins and exhibit sensitization latencies of about 7 days.

One might ask about the adaptive significance of this maternal memory process. One might conclude that this process, under natural conditions, might lead a maternally experienced nonpregnant and nonlactating female to attempt to care for unrelated pups from another mother. Such an event would not seem to be very adaptive. I do not think that such an event would occur under natural conditions. Most important, rats with previous maternal experience are not immediately maternally responsive; they require 1 to 2 days of continuous exposure to young before they show maternal behavior. Although this latency is much shorter than that exhibited by naïve virgins, it still represents an inhibitory process that would prevent a female from caring for unrelated young during a brief encounter. (Recall that under natural conditions, rats nest in secluded sites within burrows separated from other females. Although a female might enter a nest site during a mother's absence, the inhibitory mechanism that is present even in maternally experienced females should prevent maternal behavior.)

Therefore, I view both the maintenance of maternal behavior and maternal memory as representing a similar underlying process. Maternal experiences that occur during the early postpartum period of hormonal stimulation modify maternal brain circuits so that a strong and persisting bond occurs between a mother and her young. This process then allows maternal behavior to continue throughout the remainder of the postpartum period in the absence of hormonal stimulation. Maternal memory can simply be viewed as representing the residual effects of the strong mother–infant bond that forms during the postpartum period.

Although the maternal memory process occurs in rats, one would not predict that this process would occur in sheep. Ewes are constantly surrounded by unfamiliar lambs in their herd and are easily exposed to such lambs. If a process analogous to maternal memory in rats occurred in sheep, it would be very likely that an experienced nonpregnant and nonlactating ewe would attempt to care for unrelated young. There is no evidence for a maternal memory process in sheep that is similar to that in rats, since nonpregnant, nonlactating ewes with previous maternal experience will reject and avoid lambs (Poindron & Le Neindre, 1980). Does this mean that maternal experience does not influence maternal circuits in sheep after such sheep have been separated from their lambs for long periods? Recall that nonpregnant, nonlactating (estrous cycling) ewes can be induced to show maternal behavior with steroid treatment coupled with VCS but that this stimulatory effect only occurs in ewes that have had previous breeding experience, and does not occur in nulliparous virgin ewes (Kendrick & Keverne, 1991). Therefore, in addition to maternal experience affecting the nonhormonal postpartum maintenance of maternal behavior in sheep, it may also have other effects that influence the ease with which particular hormones and VCS can activate maternal responsiveness to lamb stimuli.

Hormones and Maternal Behavior in Nonhuman Primates

Do the endocrine and other physiological events which occur near parturition play a role in the onset of maternal behavior in nonhuman primates? A classic view is that the maternal behavior of nonhuman primates is much less dependent upon the physiological events of late pregnancy than that which has been demonstrated for most nonprimate mammals (for reviews, see Numan &

Insel, 2003; Saltzman & Maestripieri, 2011). However, because there is so little experimental evidence on this subject, I think it is unwarranted to make such a broad statement.

Under natural conditions, the majority of monkey and ape species live in either large multimale and multifemale groups with a promiscuous mating system or in one-male groups containing many females with a polygynous mating system (Saltzman & Maestripieri, 2011). In such groups, there are many lactating females with their infants living alongside other nonpregnant, nonlactating females. As Pryce (1996) has emphasized, it is hard to imagine the adaptiveness of a situation where all females in such primate groups would be equally attracted to infants irrespective of their physiological condition. Such a situation would result in intense competition to gain access to infants, which would disrupt the functioning of the group. Instead, it seems much more likely that the physiological events of late pregnancy function to stimulate a high level of maternal responsiveness to infant stimuli at parturition so that maternal behavior is directed primarily to one's own offspring. Subsequent mother–infant interactions may then solidify the mother–infant bond. Although all females may demonstrate a basal level of attraction to young infants, the natural mother, in contrast, is highly attracted to her own infant to whom she becomes strongly attached. It is not surprising, therefore, that the majority of monkey species and the great apes (orangutans, chimpanzees, gorillas) exhibit a uniparental maternal care system.

Subfamilies of Old World monkeys are divided into cercopithecine and colobine species (Hrdy, 2009). Cercopithecine species include the rhesus macaque and the savanna baboon, and the mothers in these species are typically highly protective of their young, and they do not allow other group members, including other females, to handle their young. Colobine monkeys, such as langurs, are more permissive, and nulliparous females in the mother's group that express interest in a young infant are allowed to carry it for brief periods of time (Hrdy, 2009; Numan, 1994). It should be noted that females in most colobine monkeys are typically genetically related to one another because the troop exhibits a matrilocal social organization (females born into the group remain there, while it is the postpubertal males that emigrate). Nevertheless, the infant's mother is still its primary caregiver and she can easily retrieve her infant back from any other female. Following Maestripieri (1994), I prefer to call the infant-directed behavior that is displayed by virgin langur monkeys as infant handling rather than full-fledged alloparental behavior because such behavior is limited in its duration and is not essential for infant survival. However, the occurrence of infant handling in langurs and other colobines indicates that there is a basal level of maternal motivation in nonpregnant, nonlactating nulliparous females of these species. It is likely that the interest in young infants that is demonstrated in these species of Old World monkeys may be affected by experiential factors that are analogous to the process of sensitization that I described for rodents. In many of these species, nulliparous females are surrounded by the infants of mothers in their group, and if these mothers are permissive and allow virgin females to approach and interact with their infants, then such stimulation may further arouse maternal responsiveness in virgin females (Numan, 1994; Numan & Insel, 2003).

In New World monkeys, existing in Central and South America, an entirely different picture emerges. Callitrichine species, primarily composed of marmosets and tamarins, are cooperative breeders and high levels of allomaternal behavior are exhibited by nulliparous nonbreeding females. In these cooperatively breeding groups, there is typically one dominant breeding female and one dominant breeding male, although instances of more than one breeding female and male have been reported (Diaz-Munoz, 2016; Digby, 1995). These New World monkeys are small in size (100–600 g) and the parturient mother typically gives birth to twins at 6-month intervals, and each pair of twins may weigh up to 20% of the mother's weight (Culot et al., 2011). These twins are carried not only by the mother but also by all other members of the group, which includes adult males and older siblings from previous births (Pryce, 1993; Saltzman & Maestripieri, 2011). In particular, most of the mother's older daughters delay dispersal from the natal group and do not reproduce themselves, but instead help their mother take care of their younger siblings. Such care not only involves carrying young infants for long periods of time, but also involves provisioning these younger siblings with food once weaning begins (Hrdy, 2009). Importantly, shared care of young offspring appears to be very important for infant survival, as research indicates that the new born infant survival rate is lower when helpers are either absent or low in number (Bardi, Petto, & Lee-Parritz, 2001). Hrdy (2009) considers marmoset and tamarins, along with humans, as the only primates that exhibit what she refers to as full-fledged cooperative breeding, where allomaternal care is an important component of the social group within which infants are reared.

In my review of maternal behavior under natural conditions in Old World monkeys and apes and in New World monkeys, one might predict that the physiological events associated with late pregnancy and parturition would be more important for the onset of maternal behavior at parturition in Old World monkeys and apes than they would be in New World monkeys since virgin females in the latter monkeys show high levels of allomaternal behavior. With respect to Old World monkeys and apes, very little evidence exists on this subject, and this research is typically correlational, rather than experimental, in nature. Basically, researchers have correlated prepartum changes in peripheral levels of steroids (estrone and progesterone) in either plasma, urine, or fecal samples, with the onset and quality of postpartum maternal behavior and with infant survival. Before I describe the few studies that exist with respect to Old World monkeys and apes, it should be noted that in many primate species progesterone does not exhibit a major decline before birth but estrogens (estrone and estradiol) do show increases as pregnancy progresses, and this steroid profile therefore typically increases the estrogen to progesterone ratio near the time of parturition (Levy, 2016; Numan, 1994).

For Japanese macaque monkeys observed in outdoor enclosures, Bardi, Shimizu, Barrett, Borgognini-Tarli, and Huffman (2003) reported a negative correlation between a rejecting maternal style (the mother interrupts or prevents infant contact) and prepartum concentrations of fecal estrone and the estrone/progesterone ratio. This finding coincides with the nonprimate mammalian literature where a rising estrogen to progesterone ratio promotes maternal behavior. In contrast, for gorillas observed in a zoo, Bahr, Martin, and Pryce (2001) could not detect any correlations between prepartum urinary estrone levels, progesterone levels, or the estrone/progesterone ratio and postpartum maternal behavior. Finally, for the hamadryas baboon observed in outdoor enclosures, French et al. (2004) reported that the display of poor postpartum maternal behavior (rejection of offspring) was correlated with higher levels of prepartum plasma and urinary estrone and estrone/progesterone ratios. The results of French et al. conflict with the nonprimate mammalian literature. It is very hard to draw any firm conclusions from these few findings because of their correlational nature and because of the conflicting results.

With respect to New World monkeys, important research has been performed on the common marmoset (*Callithrix jacchus*), some of it being experimental in nature, and these findings will be enlightening. Although the focus of this chapter has been on species that demonstrate a uniparental maternal care system, I decided to describe this research on cooperatively breeding marmosets because it represents the most significant research on the involvement of the physiological events of late pregnancy in the maternal behavior of a nonhuman primate.

Pryce and his colleagues have performed the seminal research on the common marmoset (Pryce, 1993; Pryce, Dobeli, & Martin, 1993). Overall, this research suggests that although allomaternal behavior occurs in virgin subadult and adult subordinate females in family groups of common marmosets tested in a laboratory setting, the endocrine events of late pregnancy enhance maternal responsiveness in primigravid females above the level shown by alloparents. It is interesting to note that this interpretation is quite similar to that which I described for laboratory female mice. Pryce and his colleagues used an operant conditioning procedure to measure maternal responsiveness (maternal motivation) in marmosets. The females were trained to press a bar to cause a 15-second visual presentation of a model or replica of an infant marmoset and the onset of this visual stimulus was coupled with the termination of an auditory playback stimulus of infant distress vocalizations. This procedure therefore involved both positive reinforcement (the visual presentation of the infant model) and negative reinforcement (the removal of an assumed aversive infant stimulus). Indeed, in marmosets, infant distress calls typically evoke infant carrying, which then calms the infant. The idea behind this protocol was that the operant bar press rate should increase as maternal motivation increases. In one study, adult virgin females that had previous allomaternal experience in their family groups learned to press the bar in this operant test, and they produced approximately 20 bar presses during a 20-minute test under a continuous reinforcement schedule. In a second study, plasma levels of estradiol and progesterone were measured across the 140-day pregnancy of *primigravid* marmosets with previous allomaternal experience, and this endocrine profile was related to the operant bar press rate across pregnancy. Progesterone levels rose throughout most of pregnancy and then declined slightly over the last 25 days of pregnancy. In contrast, estradiol rose consistently over the last 50 days of pregnancy and peaked near the time of parturition. The estradiol–progesterone ratio was highest over the last 10 days of pregnancy. Throughout most of pregnancy, the bar press rate of these primigravid females matched that of the virgin females, approximating 20 bar presses during each 20-minute session. In contrast, over the last 25 days of pregnancy, when progesterone was declining and estradiol was rising, there was a dramatic and significant increase in operant responding with a maximum rate of 70 bar presses during the 20-minute session. In a final important experiment, adult virgin common marmosets with previous allomaternal experience were treated systemically with steroids that mimicked the hormonal pattern that was known to occur over the last 25 days of pregnancy. A treatment schedule that resulted in declining progesterone, rising estradiol, and an increase in the estradiol–progesterone ratio, resulted in an operant bar press rate that matched that which was shown to occur at the end of pregnancy, and this bar press rate was significantly higher than that displayed by untreated virgin control females.

These results can clearly be interpreted as indicating that there is a certain level of maternal responsiveness in virgin female marmosets that allows for allomaternal behavior (note that such responsiveness might be further increased by allomaternal experience), but that maternal motivation is subsequently boosted by the endocrine events associated with late pregnancy, which include rising estradiol and declining progesterone. Further, it can be proposed that such a hormonal boost to maternal motivation near the time of parturition might be particularly important for the onset of maternal behavior in those primiparous females that lack previous alloparental experience (Pryce, 1993).

One last point needs to be emphasized. Pryce et al. (1993) clearly showed that adult virgin females with previous allomaternal experience could be easily trained in the previously described operant conditioning paradigm. However, Pryce et al. were unable to train nonpregnant, nonlactating *multiparous* females using their operant procedure. Instead, such females appeared highly aroused and agitated when presented with the infant stimuli. Note that these females had previously reared two or more sets of their own twins and were therefore the dominant breeding females in their family groups. These differences between adult virgin females and multiparous females will become important as I discuss research that conflicts with the findings of Pryce's group.

Saltzman and Abbott (2005) tested the responsiveness of pregnant *multiparous* common marmoset females to unrelated infants (5–10 days old) during early and late pregnancy. These females had successfully reared at least one previous litter of offspring. Prior to the tests for maternal responsiveness toward unrelated infants, all females had been housed in laboratory cages with their family group (an adult male pair mate and their older offspring). During the 15-minute tests with unrelated infants, other family members were removed from each female's cage. Saltzman and Abbott reported that while most females in early pregnancy approached and carried unrelated infants, those tested during late pregnancy did not. Measurement of plasma estradiol and progesterone levels indicated that estradiol and the estradiol/progesterone ratio were higher in late pregnancy than in early pregnancy, but absolute levels of progesterone were not significantly different at these two times. These findings contrast sharply with those of Pryce's group. The data from Pryce et al. (1993) indicated that maternal motivation increases at the end of pregnancy while Saltzman and Abbott report that maternal responsiveness is inhibited during late pregnancy. What can account for these differences? One obvious difference was that Saltzman and Abbott tested their females with live infants with whom the adult female could interact, while Pryce's group did not. However, I propose that the social situation and the breeding histories of the females were probably the most important determining factor: Pryce's group reported increased maternal responsiveness at the end of pregnancy in primigravid marmosets while Saltzman and Abbott found that maternal responsiveness was depressed in late-pregnant multiparous marmosets. The multiparous females tested by Saltzman and Abbott were the dominant females in their group, and marmosets depend upon other group helpers to raise their young. It would not be adaptive for a dominant female who is just about to give birth to react positively to unrelated infants in the family group who would then compete with the mother's own to-be-born infants for support from helpers. Therefore, this overall social situation may have acted, either directly or indirectly, on the maternal neural circuits of the late pregnant females to inhibit their maternal responsiveness to unfamiliar infants.

There is some support for this idea when New World monkey cooperative breeders are observed under natural free-ranging conditions and under laboratory conditions. While it is typical for marmosets and tamarins to display a singular breeding system, with one dominant female producing all the offspring while other group members serve as alloparental helpers, under certain conditions subordinate females do reproduce, and such groups therefore demonstrate a plural breeding system (Diaz-Munoz, 2016). However, subordinate females are much less successful in raising offspring to survival than are dominant females, and in many instances late-pregnant dominant females have been observed to kill the dependent young offspring of subordinate females (Culot et al., 2011; Digby, 1995; Hrdy, 2009; Saltzman, Liedl, Salper, Pick, & Abbott, 2008). Such young infants, of course, would compete for support from helpers with the late pregnant dominant female's infants that are about to be born.

If social competition for helpers depresses the maternal responsiveness of late pregnant dominant females and might also lead to infanticide of young infants that are not the dominant female's own offspring, what might be the underlying mechanisms that cause such a depression in maternal responsiveness? I can only speculate on some of the factors that might be involved. In the next chapter I will describe the neural circuits that regulate maternal behavior in female mammals that demonstrate a uniparental maternal care system. I will show that there are two circuits that influence maternal behavior, with one circuit inhibiting and another circuit promoting maternal motivation. Although cooperatively breeding virgin marmoset and tamarin females display allomaternal behavior, perhaps there are latent inhibitory circuits that can be activated in callitrichine females under certain social situations. Although the physiological events of late pregnancy may boost maternal motivation in primigravid females, as shown by Pryce's group, perhaps social competition for helpers in some way activates latent maternal inhibitory circuits in dominant multiparous females during late pregnancy, in some way overcoming the typical stimulatory effects of late pregnancy on maternal motivation. One possibility is that progesterone might be a factor that underlies this inhibition. Recall that in rats, sheep, and rabbits, progesterone withdrawal is important for the onset of maternal behavior at parturition, and if progesterone is maintained at high levels, it can depress the stimulatory effects of estradiol and lactogens. In marmosets, although progesterone declines slightly near term, it still remains at relatively high levels. For progesterone to act on the brain, it needs to interact with progesterone receptors that are present within certain neurons (Numan et al., 1999). Although progesterone may only decline slightly in late pregnant marmosets, perhaps under certain situations progesterone receptors decline in the brain, while in other situations they do not (cf. Thijssen, 2005). In particular, for late pregnant dominant females, prior to the birth of their own young and under socioecological conditions that promote interfemale competition for alloparental helpers, functional progesterone receptors might remain at high levels in certain brain regions, which would allow progesterone to inhibit maternal responsiveness. This analysis suggests that a full understanding of how hormones influence maternal behavior must take into account the presence or absence and functional activity of the hormones' receptors within the brain.

Another interesting possibility is that under conditions of obvious social competition for alloparental helpers, a late pregnant dominant female marmoset may require the VCS that occurs at parturition to display maternal behavior. Such females may require a particular hormone profile coupled with VCS to induce maternal behavior. That is, they may require a type of stimulation analogous to that described for sheep. This process would ensure that the female is caring for her own young.

Finally, it is also possible that the social situation may act directly on central neural mechanisms that influence maternal behavior and that social competition for helpers may directly stimulate circuits that inhibit maternal behavior and/or inhibit those neural circuits which stimulate maternal behavior. This proposal is supported by the finding of Pryce et al. (1993) that nonpregnant, nonlactating dominant multiparous females could not be trained in their operant paradigm. Further support comes from research reviewed by Hrdy (2016) and Bardi et al. (2001). Early *postpartum* marmosets and tamarins may actually abandon their *own* young under conditions where paternal and alloparental support from other group members is unreliable.

All of these hypotheses are interesting and they are not mutually exclusive, but the validation of any one of them will require further investigations.

General Conclusions

In female mammals that display a uniparental maternal care system, and particularly for rats, rabbits, and sheep, the physiological events of late pregnancy and parturition stimulate the onset of maternal behavior. However, it is also clear that maternal behavior is not rigidly tied to the physiological states of pregnancy and lactation. Sensitization processes occur in rats and may occur in some primates. Further, as a result of maternal experience with infants at parturition the maintenance of maternal behavior in rats, rabbits, and sheep becomes relatively emancipated from hormonal control.

Although I have emphasized the physiological factors of late pregnancy and parturition as triggers for the rapid onset of maternal behavior, it should be clear from the research I have reviewed that the physiological events throughout pregnancy set the stage for the stimulatory effects of late pregnancy and parturition. A prime example is the importance of the decline in progesterone from its high levels throughout most of pregnancy (progesterone withdrawal) in potentiating the effects of rising estradiol and lactogens.

In certain species, evolutionary forces have operated to boost the maternal responsiveness of virgin females that have not been exposed to the physiological events of late pregnancy and parturition. This enhanced maternal motivation in virgins is particularly evident in cooperatively breeding primates where allomaternal behavior occurs at high levels in virgin female marmosets and tamarins. Support for this view is also evident from research on the maternal behavior of virgin laboratory mice, where selective breeding and inbreeding has resulted in high levels of maternal responsiveness in such females. However, for both laboratory mice and for marmosets, the evidence suggests that the endocrine events associated with late pregnancy and parturition can boost maternal motivation to levels above that expressed by virgins, at least under certain situations.

The reader might be wondering why I have not discussed the role of oxytocin in the onset of maternal behavior in nonhuman mammals. As I will describe in detail in Chapter 4, oxytocin acts as both a hormone in the periphery and as a neurotransmitter/neuromodulator within the brain. Certain nuclei in the hypothalamus produce oxytocin, and the axons of these neurons project to both the posterior pituitary and to other nuclei within the brain. Hormonal oxytocin released from the posterior pituitary into the systemic blood supply has critical effects related to maternal physiology (Numan, 2015): It stimulates uterine contractions at the end of pregnancy, which then promotes parturition, and during the postpartum period, it is involved in the milk-ejection reflex (suckling stimulation activates the release of oxytocin from the posterior pituitary and then oxytocin acts on its receptors in the mammary glands to promote milk ejection into the suckling infant's mouth). However, oxytocin circulating in the blood as a hormone has poor penetration across the blood–brain barrier (Leng & Ludwig, 2016; Numan, 1994), and therefore it is unlikely to influence maternal circuits in the brain (cf. Yamamoto et al., 2019). Conversely, oxytocin released from axons within the brain, where it acts as a neurotransmitter/neuromodulator, has a very important role in maternal behavior. Therefore, I will describe this role of oxytocin in the next chapter, which deals with the brain mechanisms that regulate maternal behavior.

4

Brain Mechanisms Regulating Maternal Behavior in Nonhuman Mammals

Oxytocin and Olfaction

Introduction

In this chapter and the next, I will describe and evaluate the research that has defined the core neural circuits that regulate maternal behavior in nonhuman mammals. Most of this research has examined these essential neural circuits in laboratory rats and, with the advent of transgenic mouse lines, in laboratory mice. Additional research has been performed on sheep and rabbits, but scant research has been performed on nonhuman primates. The lack of research on nonhuman primates, however, has been compensated for by the growing research on the neural underpinnings of human maternal behavior, which has resulted from the application of functional magnetic resonance imaging technology to this issue. I will discuss the research on the human parental brain in Chapter 8.

In this chapter, my goal is to provide an introduction to the anatomy of the rodent brain, which will then be followed by an analysis of the general role of oxytocin (OT) neural systems in the maternal behavior of nonhuman mammals. The chapter will conclude with an evaluation of the sensory processes that regulate maternal behavior in these species, with an emphasis on olfaction.

Up to this point, I have used the terms *maternal motivation* or *maternal responsiveness* rather loosely. Here, I would like to define the processes underlying motivated behaviors more precisely. The concept of motivation has been defined in a variety of ways (Berridge, 2004; Hinde, 1970; Numan, 2015; Numan & Woodside, 2010; Pfaff, 1982). A straightforward definition of motivation is that it is an internal process that modifies the way an organism responds to the same external stimulus. With respect to maternal motivation in the typical female mammal, this definition would apply to the role of physiological events of late pregnancy and parturition, which modify brain circuits so that the manner in which a female responds to young changes, shifting the female away from avoiding or rejecting infants, as seen in the virgin state, and toward accepting infants and showing maternal behavior at parturition. The valence of the infant stimulus can be conceived as changing from negative to positive, not because

The Parental Brain. Michael Numan, Oxford University Press (2020). © Oxford University Press.
DOI: 10.1093/oso/9780190848675.001.0001

the stimulus itself has changed, but instead because of how the infant stimulus is processed by the brain (Numan, 2015).

Motivated behavior directed toward a pleasant or rewarding stimulus (a stimulus with a positive valence), as would be the case for the parturient female's response to infants, can further be divided into an initial appetitive phase and a terminal consummatory phase (Stolzenberg & Numan, 2011). The appetitive phase is the stage of attraction or the reward-seeking phase and is composed of those proactive voluntary behaviors that allow an organism to gain access to a desired goal object, while the consummatory phase is composed of those behaviors that occur once the desired goal has been obtained. While appetitive responses are voluntary and variable in nature, consummatory responses tend to be more reflexive and are elicited by proximal stimuli from the goal object.

In rodent mothers, retrieval behavior is usually defined as the major appetitive component of pup-directed maternal behavior, because it is a voluntary proactive pup-seeking response that allows a mother to be reunited with her young; nursing behavior, which is elicited by proximal stimuli from pups that include nuzzling and suckling stimuli, is considered the main consummatory pup-directed maternal response (Hansen, Harthon, Wallin, Lofberg, & Svensson, 1991; Numan & Stolzenberg, 2009; Stern, 1996). Of course, if the mother's pups are already located in the nest, then the return of the mother to the nest after a period of absence can also be considered an appetitive approach response, signifying the mother's attraction to her pups.

In addition to examining retrieval behavior or other natural appetitive responses, such as the return of the mother to her nest, the appetitive aspects of maternal behavior can be studied in other ways. For example, maternal rats will learn an operant bar press response to obtain pups as a reward, while nonmaternal rats will not learn such an operant response (Lee, Clancy, & Fleming, 2000). In other words, pups are rewarding stimuli that can reinforce operant responses that provide access to pups for maternal, but not for nonmaternal, rats. In Chapter 3 of the volume, I also reviewed the research that showed that hormone-primed female laboratory mice and marmosets will perform an operant response to gain access to infant stimuli at a much higher rate than their nonhormone-primed virgin counterparts, indicating that the appetitive aspects of maternal motivation are higher in such females after they have been primed by the hormonal events associated with late pregnancy. In another paradigm that examines the appetitive aspects of maternal behavior in rodents, it has been shown that lactating rats will learn a conditioned place preference when pups are used as a rewarding stimulus (Fleming, Korsmit, & Deller, 1994). In this procedure, over a series of training days, mothers are placed in a two-compartment cage, with pups present in one compartment, while the other compartment remains empty. Subsequently, during a test phase, the mother is placed in the two-compartment cage without any pups being present. During this test phase, maternal rats spend more time in the compartment that had previously contained pups, suggesting that they are searching for pups that had served as an attractive and rewarding stimulus. Importantly, nonmaternal virgin female rats do not learn such a conditioned place preference, indicating that pup stimuli are not attractive and rewarding to such females.

How might one examine the appetitive aspects of maternal behavior in rabbits and sheep? Operant procedures to test appetitive motivation have yet to be performed in these species. With respect to naturally occurring maternal responses, since these species do not retrieve young, some other measures would have to be used. In rabbits, the return of the mother to the nest box each day, to nurse her young for about 5 minutes, would probably be a good measure of appetitive maternal motivation, while the actual nursing behavior would be the consummatory response. To measure the appetitive aspects of maternal behavior in sheep, Perrin, Meurisse, and Levy (2007) have used a separation–reunion procedure. Postpartum ewes are separated from their lambs, with the lambs being placed in an adjoining pen. After a period of time, a gate is opened, and the time it takes for the mother to reunite with her lamb (latency to reunion) and the amount of time she spends near the lamb in the adjoining pen are measured.

In my review of the neural mechanisms controlling maternal motivation, I will discuss the neural control of both the appetitive and consummatory aspects of maternal behavior, but my emphasis will be on the appetitive aspects. One rationale for this approach relates to the importance of appetitive maternal motivation for adaptive maternal responses: If a mother is not attracted to her infants and does not perceive them to be rewarding stimuli, then this could lead to maternal neglect of offspring. Indeed, if a mother perceives infant stimuli to be aversive rather than attractive, she might either actively avoid her infants or even abuse (reject) them.

Such an emphasis on appetitive processes is not meant to minimize the importance of consummatory maternal responses. Berridge (2007) has indicated that rewarding stimuli have three important characteristics: (a) They are attractive and activate appetitive approach behaviors (such as retrieval or reunion with infants); (b) they have reinforcing properties so that once the reward is obtained, the occurrence of consummatory responses toward the rewarding stimulus (such as direct interaction and nursing of infants) strengthens or reinforces the particular appetitive responses that were used to obtain the desired goal (such as an operant bar press response to gain access to infants); and (c) at least in humans, the acquisition of the reward and the subsequent performance of consummatory responses result in pleasurable hedonic sensations (maternal love). Nevertheless, the available research on the neural basis of maternal behavior in nonhuman mammals has informed us more about the processes underlying appetitive maternal motivation, and although information is available on the neural circuits that regulate consummatory maternal responses, such as nursing behavior, less is known about how engaging in consummatory maternal responses can result in reinforcement. As I will propose, mechanisms analogous to such reinforcement processes may be involved in the establishment of the maintenance phase of maternal behavior and maternal memory. Finally, I will delay a discussion of the neural underpinnings of hedonic pleasurable emotions that give rise to feelings of maternal love in humans until Chapter 8.

A Tour of the Rodent Brain

Because so much research on maternal neural circuits has been performed on rodents, I want to give the reader a "lay of the land" and outline the names and the locations of many of the brain regions shown to be important for maternal behavior in these species. Figure 4.1 shows a sagittal section through the more medial parts of the rat brain. Areas in the telencephalon or cerebral hemispheres that have been shown to be involved in the regulation of rodent maternal behavior include the main olfactory bulb (MOB), the accessory olfactory bulb (AOB), medial prefrontal cortex (mPFC), nucleus accumbens (NA), ventral pallidum (VP), bed nucleus of the stria terminalis (BST), and lateral septum (LS). With respect to these regions, the following is worth emphasizing at this point: The nasal cavity of species, such as rodents, that depend highly on olfaction, contains two groups of peripheral olfactory sensory neurons, located in either the vomeronasal organ (VNO) or the main olfactory epithelium (MOE), and both of these groups of chemosensory neurons are involved in the detection of pheromones. Sensory neurons in the MOE project via the olfactory nerve to the MOB, while the VNO sensory neurons project via the vomeronasal nerve to the AOB. The NA and VP play important roles in the control of appetitive motivational processes relevant to maternal behavior, and the NA has strong projections to the VP, forming what can be referred to as the NA–VP circuit.

Caudal to the telencephalon lies the diencephalon, which includes the thalamus and the hypothalamus. In the very dorsal tip of the thalamus, in a region referred to as the epithalamus, lies the habenula, which is composed of a medial and lateral division. Within the hypothalamus, several regions have been implicated in maternal behavior control, and these include the medial preoptic area (MPOA), anterior hypothalamic nucleus (AHN), ventromedial nucleus (VMN), and the paraventricular nucleus of the hypothalamus (PVN). The PVN contains a major population of OT-producing neurons, some of which project to other brain regions, while additional OT neurons project to the posterior pituitary.

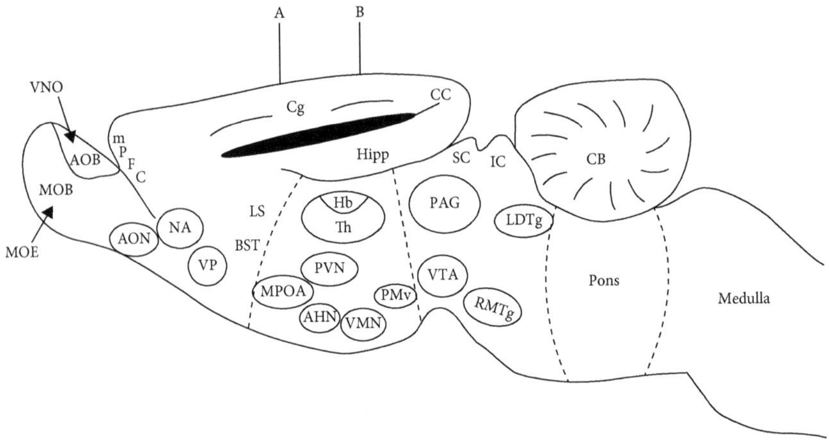

Figure 4.1. A sagittal section through the medial part of the rat brain. Many of the areas shown play significant roles in the regulation of maternal behavior in mammals. Within the telencephalon (cerebral hemispheres), the olfactory bulb is shown as containing the main olfactory bulb (MOB) and accessory olfactory bulb (AOB). The MOB receives peripheral olfactory input from the main olfactory epithelium (MOE) and the AOB receives peripheral olfactory input from the vomeronasal organ (VNO). Other important areas of the telencephalon are the medial prefrontal cortex (mPFC), cingulate cortex (Cg), hippocampus (Hipp), lateral septum (LS), nucleus accumbens (NA), ventral pallidum (VP), and bed nucleus of the stria terminalis (BST). The neural connections between the NA and VP (NA-VP circuit) are particularly important for maternal motivation. Moving posteriorly to the diencephalon, in the hypothalamus note the medial preoptic area (MPOA), a critical integrative region that regulates maternal behavior, and the paraventricular hypothalamic nucleus (PVN), a nucleus which contains oxytocin-producing neurons. Also shown are the anterior hypothalamic nucleus (AHN) and ventromedial nucleus of the hypothalamus (VMN). Dorsal to the hypothalamus in the diencephalon lies the thalamus (Th), with the habenular nucleus (Hb) shown in the dorsal part of the thalamus. Posterior to the diencephalon is the midbrain, and an important nucleus in this region is the ventral tegmental area (VTA), which contains dopamine neurons, some of which project to the NA. The midbrain also contains the periaqueductal gray (PAG), the rostromedial tegmental nucleus (RMTg), and the laterodorsal tegmental nucleus (LDTg). AON = anterior olfactory nucleus; CB = cerebellum; CC = corpus callosum; IC = inferior colliculus; PMv = ventral premammillary nucleus of the hypothalamus; SC = superior colliculus. Because this figure is a medial sagittal section, certain laterally located areas that are importantly involved in maternal behavior are not shown. The A and B lines are the approximate anterior-posterior locations of the frontal sections shown in Figures 4.2 and 4.3, respectively, which show these more laterally located regions.

The midbrain (MB) is located posterior to the diencephalon, and two important nuclei related to maternal behavior control are the periaqueductal gray (PAG) and the ventral tegmental area (VTA). As I will show, the VTA is involved in the appetitive aspects of maternal behavior, while the PAG influences nursing behavior. The VTA contains a population of dopamine neurons that project to the NA in the telencephalon. Just caudal to the VTA is the rostromedial tegmental nucleus, which contains GABAergic neurons that project to and inhibit VTA-dopamine neurons. Also note the location of the laterodorsal tegmental nucleus.

Because the midsagittal section in Figure 4.1 depicts mainly medially located regions, I want to also show some frontal sections through the rat brain, which will delineate some important regions involved in maternal behavior that are located more laterally in the brain. Figure 4.2 shows a frontal section through the rat brain at the level of the MPOA (indicated as a slice through point A in Figure 4.1). Note that the lateral preoptic area is situated lateral to the MPOA, and that the posterior part of the VP is lateral to the lateral preoptic area and ventral to the

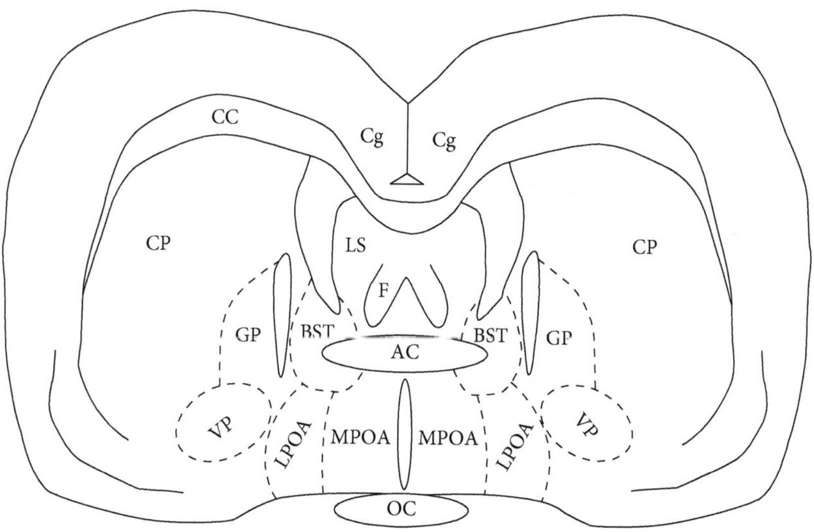

Figure 4.2. A frontal section through the rat brain at the level of the medial preoptic area (MPOA), indicated as a slice through point A in Figure 4.1. Note that the lateral preoptic area (LPOA) and posterior parts of the ventral pallidum (VP) lie lateral to the MPOA. AC = anterior commissure; BST = bed nucleus of the stria terminalis; CC = corpus callosum; Cg = cingulate cortex; CP = caudate-putamen; F = fornix; GP = globus pallidus; LS = lateral septum; OC = optic chiasm.

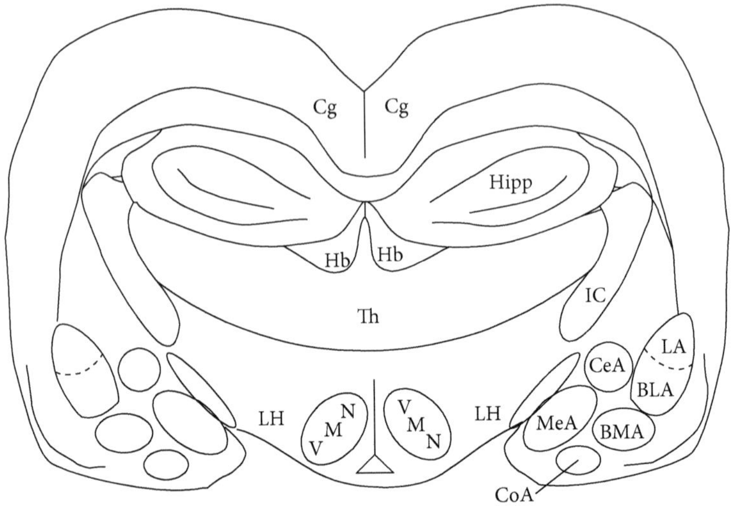

Figure 4.3. A frontal section through the rat brain at the level of the ventromedial nucleus of the hypothalamus (VMN), indicated as a slice through point B in Figure 4.1. Lateral to the VMN lies the lateral hypothalamus (LH). The habenula (Hb) lies in the dorsal part of the thalamus (Th). Within the telencephalon, important nuclei within the amygdala are the medial amygdala (MeA), the cortical amygdala (CoA), the central nucleus of the amygdala (CeA), the basomedial (BMA) and basolateral (BLA) amygdala. Cg = cingulate cortex; Hipp = hippocampus; IC = internal capsule; LA = lateral amygdala.

globus pallidus. Figure 4.3 shows a frontal section through the rat brain at the level of the VMN (indicated as a slice through point B in Figure 4.1). Note the locations of critical regions within the amygdala, located within the telencephalon lateral to the hypothalamus: Medial amygdala (MeA), cortical amygdala, central nucleus of the amygdala (CeA), basomedial amygdala (BMA) and basolateral amygdala. As I will show, basolateral amygdala/BMA projections to the NA–VP circuit are important for maternal behavior.

Oxytocin and Maternal Behavior

Introduction

Figure 4.4 shows a schematic view of OT neural systems in the mammalian brain. OT, a neuropeptide containing nine amino acids, is produced in the brain by two major nuclei in the hypothalamus, the PVN and the supraoptic nucleus (SON),

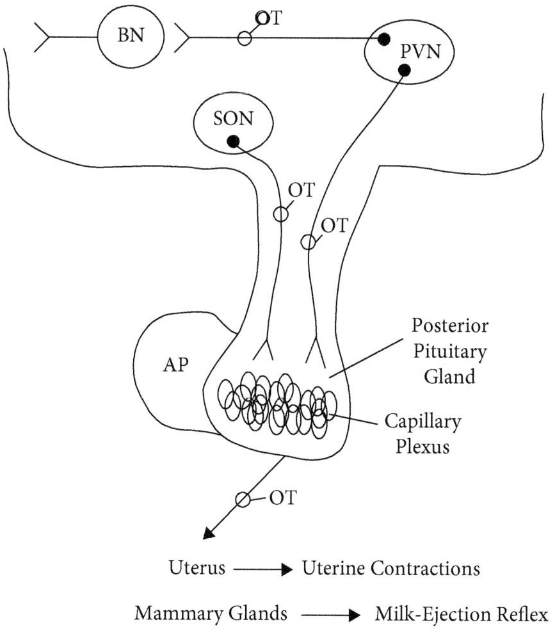

Figure 4.4. A schematic view of oxytocin (OT) neural systems. The brain's OT-producing neurons are primarily localized in the paraventricular nucleus (PVN) and supraoptic nucleus (SON) of the hypothalamus. OT neurons in each of these regions project to the neural lobe of the pituitary gland (the posterior pituitary gland), and action potentials in these neurons result in OT release into the general blood supply, where it acts as a hormone to affect the milk ejection reflex and uterine contractions. Importantly, OT neurons also project to other neurons in the brain (BN), where it acts as a neurotransmitter or neuromodulator. The primary location of OT neurons that project within the brain is the PVN. AP = anterior pituitary gland.

although additional OT-producing neurons are located in accessory nuclei between these two nuclei (Stoop, 2012). One group of OT-producing neurons is shown as projecting to the posterior pituitary gland. Action potentials in these neurons result in the release of OT into a capillary plexus in the posterior pituitary from where it is then released into the systemic blood supply. OT can then circulate in the blood to act on OT receptors (OTR) located in the uterus to promote uterine contractions that facilitate parturition and on receptors located in the mammary gland to stimulate milk ejection (Gimpl & Fahrenholz, 2001). In this context, OT is referred to as a neurohormone because it is released by neurons into the blood where it then acts on target cells distant from the site of its release. Figure 4.4 also shows that OT is released into the brain where it acts on other neurons to alter their activity. In this context, OT is acting as a neuromodulator/neurotransmitter because action potentials in these OT neurons cause the local release of OT onto other neurons in the brain. Although both the SON and PVN have prominent projections to the posterior pituitary, the primary (but not the only) source of OT that is released into the brain is derived from the PVN (Knobloch et al., 2012).

Since OT released into the systemic blood supply plays important roles in maternal physiology (uterine contractions during parturition and contractions of myoepithelial cells in the mammary glands during the milk-ejection reflex), it makes sense to predict that OT release into the brain might promote maternal responsiveness. As I will show, there is good evidence for this proposal. Most research on the role of OT in maternal behavior has been conducted on rats, sheep, and mice, with some research on rabbits and primates. In what follows, I will first describe the research on rats, sheep, and rabbits, and this will be followed by a separate analysis of the role of OT in the maternal behavior of mice. I will conclude with the research that has been conducted on primates.

OT Neural Pathways in the Brain of Rats, Sheep, and Rabbits

For rats, the axonal projections of PVN OT neurons within the brain, and the location of neuronal OTRs, have been described by Knobloch et al. (2012) and Grinevich, Knobloch-Bollmann, Eliava, Busnelli, and Chini (2016). Table 4.1 shows a summary of their findings. Overall, there is an excellent correspondence between the location of OT axon terminations derived from the PVN and the presence of OTRs. I would like to emphasize the presence of OT fibers and OTRs in the following brain regions: NA, amygdala, BST, anterior hypothalamus, VTA, and PAG. Note that the PVN also contains OTRs (Freund-Mercier, Stoeckel, & Klein, 1994). Such receptors can be affected by the release of OT from PVN cell bodies and dendrites, since OT is stored in dense-core vesicles located throughout all parts of PVN neurons (Bergquist & Ludwig, 2008; Stoop, 2012). Such OT release from PVN cell bodies and dendrites can also explain some of the mismatches shown in Table 4.1, where some brain nuclei contain OTRs but not OT axon terminals. In rats, a relevant example is the MPOA. The MPOA is located just anterior to the PVN; in fact, the most anterior part of the PVN actually lies just dorsal to the posterior MPOA (Swanson, 1998). (A part of this anterior PVN region has sometimes been referred to as the anterior commissural nucleus; Yoshihara, Numan, & Kuroda, 2018.) Therefore, it is highly likely that the release of OT from PVN cell bodies and dendrites can easily diffuse through the extracellular fluid to affect the functional activity of MPOA neurons by acting on MPOA OTRs. However, some mismatches are hard to explain by

Table 4.1 The Location of Oxytocin Receptors and Oxytocin-Containing Axon Terminals Originating from Neurons in the Paraventricular Hypothalamic Nucleus in the Rat Brain

Brain Region	**OTRs**	**OT Axon Terminals**
Telencephelon		
Amyg	yes	yes
AON	yes	yes
BST	yes	yes
Hipp	yes	yes
LS	yes	yes
mPFC	yes	yes
OB	yes	no
NA	yes	yes
VP	yes	no
Hypothalamus		
AHN	yes	yes
MPOA	yes	no
PVN	yes	no
VMN	yes	no
Midbrain		
PAG	yes	yes
VTA	yes	yes

Notes: AHN = anterior nucleus of the hypothalamus; Amyg = amygdala; AON = anterior olfactory nucleus; BST = bed nucleus of the stria terminalis; Hipp = hippocampus; LS = lateral septum; mPFC = medial prefrontal cortex; MPOA = medial preoptic area; NA = nucleus accumbens; OB = olfactory bulb; OT = oxytocin; OTR = oxytocin receptors; PAG = periaqueductal gray; PVN = paraventricular hypothalamic nucleus; VMN = ventromedial nucleus of hypothalamus; VP = ventral pallidum; VTA = ventral tegmental area.

Source: The data in this table were derived from research of Knobloch et al. (2012) and Grivevich, Knobloch-Bollman, Eliava, Busnelli, and Chini (2016).

this mechanism: The olfactory bulb is located far from the PVN, and it contains OTRs, but OT axon terminals have not been detected. It has been suggested that in some cases OT release from PVN neurons may enter the nearby ventricular system (located just medially to the PVN), and then OT may reach the olfactory bulb through the cerebrospinal fluid (CSF; Yu, Kaba, Okutani, Takahashi, Higuchi, & Seto, 1996).

In conclusion, for the rat brain, PVN OT neurons can reach diverse areas of the brain via long axonal projections, although in a few cases, OT may reach certain nuclei that contain OTRs either by local diffusion through the extracellular fluid or by transport to distant regions via the CSF.

Broad et al. (1999) and Jimenez, Young, Triano-Del Rio, LaPrairie, and Gonzalez-Mariscal (2015) have examined the distribution of OTRs in the sheep and rabbit brain, respectively (in the rabbit, only the forebrain was examined, excluding the olfactory bulbs). Many of the same nuclear regions that contain OTRs in rats were found to contain such receptors in sheep and rabbits: For sheep, these regions included the olfactory bulbs, mPFC, NA, LS, BST, amygdala, MPOA, PVN, VMN, PAG, and substantia nigra (located just lateral to the VTA in MB); for rabbits, these regions included the mPFC, LS, MPOA, and MeA.

It is important to note that the expression of OTRs in some brain nuclei is modified by the animal's physiological state. When parturient rats are compared to virgin rats, OTR mRNA expression is significantly higher in the olfactory bulbs, MeA, BST, and MPOA of parturient rats (Meddle, Bishop, Gkoumassi, van Leeuwen, & Douglas, 2007). These results indicate that the events associated with parturition increase the synthesis of OTRs in select brain regions. In other regions, the expression of OTR mRNA did not change with physiological state. Some of these physiological effects were probably due to the well-known fact that OTR expression is induced by estradiol, and therefore the estradiol peak that occurs near parturition (on a background of progesterone withdrawal) may have contributed to the stimulation of OTR synthesis (Gimpl & Fahrenholz, 2001; Numan, 2015). In support of this view, Kremarik, Freund-Mercier, and Stoeckel (1995), through the use of receptor autoradiography, reported that the systemic administration of estradiol to ovariectomized female rats increased the number of OT binding sites in MPOA, BST, MeA, and VMN (also see Champagne, Diorio, Sharma, & Meaney, 2001). Please note that some neural regions were insensitive to the administration of estradiol. For example, OT binding sites were detected in NA and in CeA and BMA amygdala nuclei, but estradiol administration did not further increase the number of these binding sites. The fact that some OTR-containing neurons are sensitive to the stimulatory effects of estradiol, which increase the number of OTRs, while other OTR-containing neurons are not is related to the fact that some neurons in certain brain regions contain intracellular estrogen receptors, while other cell groups do not.

Similar influences of physiological state on the expression of OTRs have also been reported for sheep (Broad et al., 1999) and rabbits (Jimenez et al., 2015). Therefore, estradiol and/or other events associated with parturition increase the expression of OTRs in select brain regions, and these brain regions, therefore, may be particularly involved in the immediate onset of maternal behavior that occurs at parturition in rats, sheep, and rabbits.

For rats and sheep, there is also excellent evidence, determined by microdialysis, that parturition (or vaginocervical stimulation) and suckling stimulation induce the release of OT into brain regions that are known to contain OTRs (Kendrick et al., 1997; Landgraf & Neumann, 2004; Landgraf, Neumann, & Pittman, 1991). Therefore, around the time of parturition, OT is released into critical OT-receptive brain regions, and I will show that such OT action, when coupled with effective steroid and lactogen priming, stimulates the onset of maternal behavior in the typical female mammal.

Finally, please note that there are major differences between mammalian species in the localization of OTRs in the brain, even when one compares closely related species (Freeman & Young, 2016; Numan & Young, 2016). The expression patterns that I have described, however, are good representations for rats, rabbits, and sheep. These species display a uniparental maternal care system and the rapid onset of maternal behavior in these species is hormone dependent.

Cellular Mechanisms of OT Action

Electrophysiological recordings performed in a variety brain regions that are known to contain OTRs have shown that OT primarily exerts excitatory effects on neurons and that these effects can be blocked by the local administration of selective OTR antagonists (OTA; Raggenbass, 2001). For example, locally applied OT increases the frequency of action potentials or increases excitatory postsynaptic currents in various nuclei in the amygdala (the lateral division of CeA [Huber, Veinante, & Stoop, 2005], BMA and MeA [Condes-Lara, Veinante, Rabal, & Freund-Mercier, 1994; Terenzi & Ingram, 2005]), in the anterior olfactory nucleus (AON; Oettl et al., 2016), and in the ventral midbrain (Tang et al., 2014). These results indicate that when OT acts on its neuronal receptors it enhances the neuronal excitability of certain neurons. How does it produce this effect? The OTR is a membrane bound receptor that is a member of the G protein–coupled receptor family. Research indicates that the OTR can be coupled to different types of G proteins that can lead to different functional effects. In this section, I want to describe just two of the known cellular effects of OT so that the reader will understand at least some of the mechanisms through which OT can act to increase neuronal excitability (Mitre et al., 2016; Stoop, 2012). Figure 4.5 shows how OT action on an OTR located on the membrane of a neuronal cell body can ultimately increase Na+ influx into the neuron, in this way increasing the neuron's excitability. In this example, when OT binds to its receptor it activates a stimulatory G protein (Gs) which then activates the enzyme adenylate cyclase. Adenylate cyclase catalyzes the conversion of adenosine triphosphate to cyclic adenosine monophosphate (cAMP). cAMP then acts

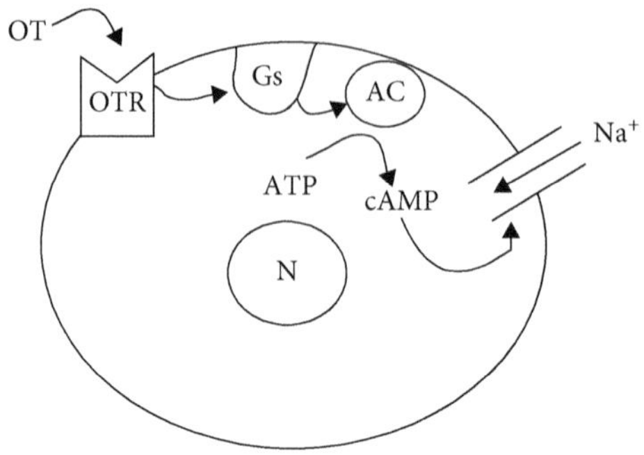

Figure 4.5. One (of many) cellular mechanism of action through which oxytocin (OT) can enhance neuronal excitability. On neuronal cell membranes, as well as on cell membranes in the uterus and mammary glands, OT binds to OT receptors (OTR). The OTR is a member of the G protein-coupled receptor family, and the OTR can be coupled to different types of G proteins each of which can result in different cellular effects when activated by OT. The example shown in this figure is a straightforward mechanism that shows how OT can enhance neuronal excitability. When OT binds to the OTR on the cell membrane of a neuronal cell body or dendrite, it activates a stimulatory G protein (Gs) which, in turn, activates the intracellular enzyme, adenylate cyclase (AC). AC catalyzes the conversion of adenosine triphosphate (ATP) to cyclic adenosine monophosphate (cAMP). In this example, cAMP acts directly on certain NA+ ion channels, causing them to open, which results in an influx of NA+ into the neuron. The positive inward current that results from the influx of NA+ causes the membrane potential of the neuron to depolarize from its resting state, which increases its neuronal excitability. N refers to the nucleus located within the neuronal cell body. See text for details.

directly on specific Na+ ion channels, causing them to open. In this way, sodium enters the cell to cause an inward positive current that results in a depolarization of the neuron (Raggenbass, 2001; Stoop, 2012). This effect of OT would be described as a neuromodulatory effect because OT is not directly stimulating action potentials. Instead, by causing a relatively small neuronal depolarization, it is bringing the neuron closer to its firing threshold so that when the affected neuron receives stimulatory inputs from other neurons, the OTR-containing neuron is more likely to produce an action potential or to produce a higher frequency of action potentials. (In many intracellular chemical cascades involving Gs, cAMP is known to activate protein kinase A, but in the case described here, protein kinase A is not involved in the stimulatory actions of OT on neuronal excitability.)

Another example of a neuromodulatory mechanism through which OT may act to increase the excitability of a particular neuron would be through an indirect mechanism that reduces inhibitory inputs to that neuron. More specifically, OT action on presynaptic OTRs located on the axon terminals of GABAergic inhibitory neurons has been proposed to exert presynaptic inhibition, in this way reducing GABA release on to a postsynaptic neuron. Such a disinhibitory process would then increase the responsiveness of the postsynaptic neuron to other excitatory inputs (Mitre et al., 2016). The specific cellular mechanisms through which OT may exert such presynaptic inhibition was not defined in the Mitre et al. (2016) study.

These mechanisms of OT action at the neuronal level represent only two examples; many other modes of OT action have been described (Dolen, 2015; Dolen, Darvishzadeh, Huang, & Malenka, 2013; Jurek & Neumann, 2018; Tang et al., 2014).

The Effects of PVN Lesions on Maternal Behavior in Rats and Rabbits

Given that the PVN is the main source of OT neural projections to other brain regions, if OT action within the brain were critical for maternal behavior one might expect that damage to the PVN would disrupt maternal behavior. In the first study to test this idea, Numan and Corodimas (1985) produced electrical lesions of the PVN on day 4 postpartum in primiparous female rats and found that such lesions did not disrupt nursing behavior or retrieval behavior, although pup weight gain was depressed in comparison to females with sham lesions. These results indicate that the PVN is not essential for the maintenance of maternal behavior in rats. The decreased pup weight gain in the offspring of the lesioned females could not be attributed to deficient nursing behavior and was likely due to an interference with the milk-ejection reflex. The fact that the pups did gain some weight was probably related to the fact that SON projections to the posterior pituitary were intact in the PVN-lesioned females.

Although the PVN is not essential for the maintenance of maternal behavior in postpartum rats, Insel and Harbaugh (1989) asked whether the PVN might be necessary for the onset of maternal behavior. They found that when electrical lesions of the PVN were performed on day 15 of pregnancy in primigravid rats, the subsequent onset of maternal behavior at parturition was disrupted. Their histological analysis of the brain indicated that those PVN lesions that included the more anterior parts of the PVN were particularly effective in disrupting the onset of maternal behavior. Further, replicating the findings of Numan and Corodimas (1985), Insel and Harbaugh found that when PVN lesions were performed on day 4 postpartum, ongoing maternal behavior was not disrupted.

These combined results suggest that an intact PVN, and presumably its central oxytocinergic neural projections, are required for the onset, but not the maintenance, of maternal behavior in rats. To emphasize the importance of central OT projections, an experiment by Herrenkohl and Rosenberg (1974) is informative. These investigators produced neural transections in the hypothalamus that blocked PVN and SON projections to the posterior pituitary, but left PVN intrabrain projections intact. These neural transections were performed during pregnancy in primigravid rats. Upon giving birth, the onset of nursing and retrieval behavior were normal, but the milk-ejection reflex was abolished, and the pups lost weight because they could not obtain milk from their mothers during suckling. Several conclusions can be derived from these results. First, hormonal OT released from the pituitary is essential for the milk-ejection reflex. Second, hormonal OT released into the blood is not essential for the onset of maternal behavior, and this makes sense because physiological levels of plasma OT cannot cross the blood–brain barrier (Leng & Ludwig, 2016). This second conclusion, therefore, focuses our attention on the central projections of PVN neurons for the onset of maternal behavior. Third, although hormonal OT released into the blood from the posterior pituitary promotes the uterine contractions involved in parturition, in rodents this effect is not essential for parturition to occur, and this has been confirmed by additional research (Takayanagi et al., 2005).

A few words of caution are required with respect to the interpretation that central PVN oxytocinergic neural projections appear to be required for the onset of maternal behavior in rats. First, electrical lesions not only destroy PVN neurons, but also destroy axons passing through the PVN (fibers of passage) but having their origins outside the PVN. Therefore, it could be that the disruptive effects of electrical PVN lesions on the onset of maternal behavior in rats were actually the result of damaging the neural connections of brain nuclei outside the PVN. Further, even if PVN cell bodies and their neural projections are involved in the onset of maternal behavior, it should be noted that certain populations of PVN neurons contain neurotransmitters/neuromodulators other than OT (Sawchenko & Swanson, 1982). Therefore, the disruption of maternal behavior after PVN lesions may have been the result of interfering with the neural projections of PVN neurons containing neurochemicals other than OT. Clearly, more specific experiments that selectively manipulate brain OT systems are needed to define the role of OT in maternal behavior, and some of this research will be described later in this chapter.

Dominguez, Aguilar-Roblero, and Gonzalez-Mariscal (2017) have reported on the effects of PVN lesions on the *maintenance* of maternal behavior in primiparous rabbits, and their results conflict with the findings from rats. The PVN was lesioned on day 7 of lactation with kainic acid. The use of kainic acid is important because it is an excitotoxic amino acid that destroys neuronal cell bodies while sparing fibers of passage. They found that such lesions had two different effects. In 50% of the lesioned females (4/8), the periodicity of nursing behavior was disrupted. Postpartum rabbits normally nurse their young once per day for about 5 minutes, but in those females with a disrupted periodicity, the pups were nursed more than once per day but the duration of each nursing bout was normal. In contrast to these females, nursing behavior was completely abolished in the remaining four females. Dominguez et al. did not describe any major differences between the PVN lesions that abolished nursing and those that disrupted nursing periodicity.

Why might the PVN, and potentially its central OT projections, be essential for the maintenance of nursing behavior in postpartum rabbits, but not in postpartum rats? One possibility may be related to the fact that rabbits, unlike rats, spend so little time in contact with their pups. Because of this very low duration of mother–infant interaction, perhaps rabbits require a surge of OT into the brain, induced by pup suckling, for maternal behavior to persist until the next nursing episode 24 hours later.

However, an important caveat with respect to the findings of Dominguez et al. (2017) is worth considering. Kainic acid is a very potent excitotoxic amino acid, and it is notorious for producing brain damage distant from its site of application (see Corodimas, Rosenblatt, Canfield, & Morrell, 1993). Since Dominguez et al. did not present a detailed analysis of neuronal damage at sites outside the PVN, it is possible that the disruption of nursing behavior might have resulted from damage to nearby sites, such as to neurons in the MPOA, and such distant effects might have been particularly relevant to those instances where nursing was completely abolished.

The Effects of Administration of OT or OTR Antagonists Into the Cerebrospinal Fluid on the Maternal Behavior of Rats and Sheep

In this section I will describe important research indicating that the injection of either OT or OTA into the CSF, which bypasses the blood–brain barrier and allows these agents to gain direct access to the brain, influence the onset of maternal behavior in rats and sheep. I will not describe the particular brain sites where OT acts to influence the onset of maternal behavior in this chapter. Instead, this research will be presented in the next chapter, where I will describe the specific central neural circuits that regulate maternal behavior in nonhuman mammals.

Pedersen and his colleagues (Pedersen & Prange, 1979; Pedersen, Ascher, Monroe, & Prange, 1982) were to first to show that intracerebroventricular (ICV) administration of OT stimulates the onset of maternal behavior in estradiol-primed ovariectomized virgin female rats. Ovariectomized virgins were injected systemically with estradiol and 48 hours later these rats received OT injections into the lateral ventricle. Pups were presented to these females immediately after the ICV injection, and 70% of these females displayed full maternal behavior toward pups within 1 hour after OT administration. In contrast, only 20% of control females that received saline injections instead of OT displayed maternal behavior. Data from additional control subjects showed that the administration of OT without estradiol priming was unable to stimulate this high level of maternal responsiveness, which suggests that estradiol-induced expression of OTRs within specific brain regions may be important for OT to exert its stimulatory effects on the onset of maternal behavior in rats.

In a subsequent study, Fahrbach, Morrell, and Pfaff (1984) replicated the findings of Pedersen's group. It is important to note that Pedersen's experiments and the study by Fahrbach et al. used a strain of Sprague–Dawley rats obtained from Zivic–Miller Laboratories. Virgin female rats from this strain have been found to be more responsive to pup stimulation and they have shorter sensitization latencies than those that have been observed in Sprague–Dawley rats obtained from other suppliers. For example, Fahrbach et al. found that their virgin females that were only treated with estradiol (without OT treatment) had median latencies to the onset of maternal behavior of 2 days. Such a facilitation of maternal behavior is not observed in estradiol-primed ovariectomized Sprague–Dawley rats obtained from Charles River Laboratories (Siegel & Rosenblatt, 1975a). Instead, these Charles River females first began to show full maternal behavior after about 4 days of pup exposure.

With these strain differences in mind, a study by Rubin, Menniti, and Bridges (1983) failed to detect a stimulatory effect of ICV OT on the onset of maternal behavior in steroid-primed Sprague–Dawley virgin rats obtained from Charles River Laboratories. The systemic steroid regimen used in this study, comprised of estradiol treatment superimposed on a background of progesterone withdrawal, was suboptimal, due to the low doses of the administered steroids, and did not induce a short-latency onset to maternal behavior but instead resulted in sensitization latencies of 3 to 5 days. ICV OT administration to these females did not further shorten these sensitization latencies.

One reason why ICV OT may have been ineffective in the Rubin et al. (1983) study is because the steroid hormone pretreatment was suboptimal. Recall from Chapter 3 that when an optimal steroid priming regimen is administered to virgin female rats, maternal behavior can be induced after about 1 to 2 days of pup exposure. In the particular studies described in Chapter 3, Zivic–Miller rats were not used as subjects. I would like to propose that OT would have been able to stimulate maternal behavior in these rats by acting on brain circuits only if those circuits had been optimally primed by progesterone withdrawal and rising estradiol and prolactin. I predict that if the Sprague–Dawley rats in the study by Rubin et al. were administered ICV OT after receiving an optimal steroid and prolactin treatment, then one would have been able to reduce sensitization latencies from about 1 to 2 days to zero days (females being maternally responsive to pups after less that 24 hours of pup exposure). Such an experiment, however, has yet to be performed.

Given my proposal, what accounts for the results with Zivic–Miller rats? It is very possible, because of the level of high maternal responsiveness in this strain, that estradiol pretreatment alone, which would also cause endogenous prolactin release, acted as an optimal treatment. The mechanism underlying the high level of maternal responsiveness in Zivic–Miller Sprague–Dawley rats will be described later in this chapter in the section on olfaction and maternal behavior in rodents, rabbits, and sheep.

Parturient primiparous rats are immediately maternally responsive, and rats that are hysterectomized and ovariectomized and treated with estradiol on day 15 of pregnancy show a near immediate onset of maternal behavior when presented with pups 48 hours later (see Chapter 3). It can be proposed that this immediate onset of maternal behavior is the result of the stimulatory effects of the physiological events associated with late pregnancy, which would include the release of endogenous OT into critical brain sites involved in maternal behavior. In support of this proposition, ICV administration of OTA to antagonize the effects of endogenous OT release into the brain disrupts the immediate onset of maternal behavior in parturient rats (van Leengoed, Kerker, & Swanson, 1987) and in pregnancy-terminated rats treated with estradiol (Fahrbach, Morrell, & Pfaff, 1985). Importantly, once maternal behavior has become established during the maintenance phase of postpartum maternal behavior, ICV injections of OTA do not disrupt ongoing maternal behavior (Fahrbach et al., 1985).

These results, taken together, strongly support the view that once the rat brain is properly primed with steroid hormones and lactogens, the endogenous release of OT, presumably derived from the PVN, acts at critical brain sites to stimulate the onset of immediate maternal behavior. Once maternal behavior becomes established, however, OT neural systems are no longer required for its maintenance. This analysis also conforms with the findings on the effects of PVN lesions on maternal behavior in rats.

Although OT neural systems are not required for the maintenance of maternal behavior in rats, research indicates that OT neural systems modulate the quantity and quality of postpartum maternal behavior. In postpartum rats, ICV administration of an OTA does not affect the amount of time a mother spends nursing her young, but such blockade of OTRs does decrease the amount of licking and grooming of the pups by the mother and also alters the nature of the specific nursing postures she displays (Champagne, Diorio, Sharma, & Meaney, 2001; Pedersen & Boccia, 2003). Therefore, although maternal behavior can continue without the action of endogenous OT in the brain, variations in the level of OT neural activity during the maintenance phase of maternal behavior can result in important variations in the quantity and quality of pup-directed maternal behaviors. There is a difference between maternal behavior being present and the detailed nature of the behaviors shown under conditions with or without OT action in the brain. It is also worth emphasizing that these statements about OT's role in the maintenance of maternal behavior have been obtained from the study of rats that were tested under standard laboratory conditions. It is certainly possible that OT's role in the ongoing maternal behavior of postpartum rats becomes more prominent under more challenging environmental conditions that mirror a more natural habitat. Research has shown that centrally released OT exerts anxiolytic effects in rodents. In a variety of stressful situations that evoke anxiety-related behaviors, OT is released both centrally and peripherally, and research indicates that such release helps the rodent cope with these stressful situations by reducing anxiety and fearfulness (Neumann & Landgraf, 2012; Neumann & Slattery, 2016; Ring et al., 2006). Activity in the elevated plus maze (EPM) is a validated test for anxiety-related behaviors, and Neumann, Torner, and Wigger (2000) have shown that late pregnant and lactating rats (peripartum rats) spend more time in the open arms of the EPM than their virgin counterparts, supporting the view that peripartum rats are less anxious than are virgins. Importantly, ICV infusion of OTA increased anxiety in peripartum rats, decreasing the amount of time they spent in the open arms. These results indicate that the combined actions of hormones and central OT during late pregnancy/parturition, along with OT action during lactation, not only stimulate the onset of maternal behavior in parturient females, but also lower the anxiety/fearfulness of peripartum females. This effect may allow a mother to effectively care for her offspring under challenging environmental conditions. This idea will be developed further later in this chapter and in Chapter 6.

For sheep, there is also good evidence that OT action on the brain is involved in the onset of maternal behavior. In multiparous nonpregnant ewes that were primed by the administration of exogenous steroids, ICV administration of OT stimulated maternal behavior toward a lamb in a 10- to 15-minute test, and this facilitation of maternal behavior was similar to that observed in steroid-primed ewes that received vaginocervical stimulation (VCS). This similarity conforms with the fact that VCS causes the release of endogenous OT into the brain (see the previous discussion of OT neural pathways in the brain of rats, sheep, and rabbits). The stimulatory effects of ICV OT were dependent upon prior steroid priming, and intravenous administration of OT was ineffective. In some experiments, systemic steroid priming only included estradiol

(Kendrick, Keverne, & Baldwin, 1987), while in others it included long-term treatment with progesterone superimposed on rising estradiol levels (Da Costa, Guevara-Guzman, Ohkura, Goode, & Kendrick, 1996). In an important study on primigravid ewes, the administration of a peridural anesthetic at the time of labor onset, a procedure that blocks afferent feedback from the vagina and cervix, disrupted the typical release of endogenous OT into the CSF that occurs at parturition and also blocked the onset of maternal behavior (Levy, Kendrick, Keverne, Piketty, & Poindron, 1992). Importantly, this inhibition of the onset of maternal behavior by peridural anesthesia was reversed by ICV administration of OT, while saline injections were ineffective.

In conclusion, these results on rats and sheep clearly indicate that the release of OT into the brain near the time of parturition, when acting in concert with the hormonal events that occur during late pregnancy, is an important trigger for the immediate onset of maternal responsiveness. Such OT release is induced, in part, by the VCS that occurs at parturition, and, in sheep, such VCS-induced OT release appears to be essential.

OT Neural Systems and Maternal Behavior in Mice

The research described in the previous two subsections supports the view that OT neural systems act in concert with the hormonal events of late pregnancy and parturition to facilitate the onset of maternal behavior in rats and sheep, and it is generally viewed that OT neural systems are essential for the immediate onset of maternal behavior in primiparous parturient females of most mammalian species that exhibit a uniparental maternal care system. The question that concerns us in this section is whether OT neural systems are also essential for maternal behavior in house mice (*Mus musculus*).

For feral mice, McCarthy (1990) has provided evidence that OT is involved in aspects of maternal responsiveness. Recall that feral virgin female mice typically exhibit infanticide when presented with young pups. McCarthy found that when OT was infused into the CSF (ICV injections) of feral virgin female mice and then pups were presented to them 20 minutes later, infanticide was inhibited but maternal behavior was not facilitated. This elimination of infanticide was shown in gonadally intact females and in ovariectomized females. Of course, parturient feral female mice not only do not show infanticide, but they also show prompt maternal behavior toward their own pups and to pups from another mother. What can be concluded from these results? Recall that the expression of OTRs in certain regions of the brain is increased by estradiol, while the expression of OTRs in other brain regions is independent of estradiol stimulation. It appears that in virgin feral females OT acts on estradiol-independent OTRs to inhibit infanticide. But for parturient feral females, it is likely that OT acts in concert with the hormonal events of late pregnancy and parturition to inhibit infanticide while also stimulating the immediate onset of maternal behavior. This dual effect is probably the result of OT action on both estradiol-independent and estradiol-dependent OTRs, respectively. Therefore, feral female mice seem to rely on OT neural systems to stimulate the onset of maternal behavior in a manner similar to that which occurs in other parturient female mammals.

What about laboratory strains of mice? As described in Chapter 3, virgin laboratory mice, whether intact or ovariectomized, display spontaneous maternal behavior when tested with foster pups in their home cages. Since this prompt maternal responsiveness in laboratory female mice is not dependent upon the physiological events associated with late pregnancy and parturition, it seems highly likely that OT neural systems are also not necessary for this behavior. Although the evidence with respect to this proposal is complex, overall it appears to be true, at least when the mice are tested in their home cages under low-stress conditions that do not arouse anxiety-related behaviors.

Some background information is necessary before I review the research on the involvement of oxytocinergic neural systems in the maternal behavior of laboratory mice. First, the location of OT neurons and axon terminals, as well as the location of OTRs, in the mouse brain closely matches that described in Table 4.1 for rats (Dolen et al., 2013; Mitre et al., 2016; Olazabal & Alsina-Llanes, 2016; Otero-Garcia, Agustin-Pavon, Lanuza, & Martinez-Garcia, 2016; Xiao, Priest, Nasenbeny, Lu, & Kosorovitskiy, 2017; Yoshida et al., 2009). Therefore, if OT neural systems are found to be less important for maternal behavior in laboratory mice when compared to rats, sheep, and other female mammals, it is not likely that this can be attributed to a lack of OT and OTR expression.

OT released into the systemic blood supply from the posterior pituitary acts on OTRs in the late pregnant uterus to facilitate uterine contractions, in this way promoting parturition. Although OT may not be essential for parturition to occur, it clearly facilitates parturition (Douglas & Meddle, 2008; Russell, Leng, & Douglas, 2003). For rats, Neumann, Douglas, Pittman, Russell, and Landgraf (1996) showed that during parturition, OT is released locally in the brain (somatodendritic release) onto OTRs within the SON to exert a positive feedback effect on SON projections to the posterior pituitary, in this way causing the release of bursts of OT into the systemic blood supply. When an OTA was directly infused into SON during parturition, parturition was prolonged. This prolongation was due to an increase in the interbirth interval. For laboratory mice, Douglas, Leng, and Russell (2002) injected mice systemically with either OTA (to block the effects of endogenous OT on the uterus) or saline during parturition. Labor was prolonged in the OTA injected females, although the pups were eventually born.

Finally, recall the research that has shown that centrally released OT exerts anxiolytic effects in rodents, which may allow a mother to effectively care for infants under challenging environmental conditions.

In laboratory mice, the primary way in which the role of OT in maternal behavior has been explored is through the use of transgenic mouse lines with either a null mutation of the OT gene (OT–/– mice) or the OTR gene (OXTR–/– mice). The maternal behavior of such mice has been compared to their normal counterparts (OT+/+; OXTR+/+; in some cases, heterozygotes+/– are used as controls). Before I describe this research, I want to present research on the behavior of virgin OT–/– mice in the EPM. It has been shown that such OT–/– adult female mice display more anxiety-related behavior in the EPM (less time in open arms) than do OT+/+ females (Amico, Mantella, Vollmer, & Li; 2004; Mantella, Vollmer, Li, & Amico, 2003). Further, these studies found that ICV administration of OT decreased anxiety in OT–/– females, while ICV injection of OTA increased anxiety in OT+/+ females. These results indicate that disruption of OT neural transmission in the brain of mice enhances anxiety and this effect needs to be taken into account when examining the effects of OT or OTR gene deletion procedures on the maternal behavior of mice.

Nishimori et al. (1996) examined the maternal behavior of primiparous OT–/– and OT+/– mice on the day of parturition. OT–/– females mated, became pregnant and delivered live young. There were no differences in retrieval behavior, nest-building behavior, and time spent in the nest caring for young between the two genotypes and maternal behavior appeared to be completely normal. Parturition also appeared normal. The only difference between the two groups was that the milk-ejection reflex was abolished in the OT–/– females, and their pups could not obtain milk during suckling. Significantly, abundant OT mRNA was present in PVN and SON in heterozygotes, but was completely absent in the OT–/– females. These results suggested that central OT neural systems are not essential for the onset of maternal behavior when parturient mice are tested in their home cages. This finding was replicated in a subsequent study when OT–/– females were compared to OT+/+ females (Takayanagi et al., 2005). In addition, this study also found normal maternal behavior in OT–/– virgin females, with such females demonstrating prompt maternal responsiveness when tested with foster pups in their home cages.

These results are comparable to the results I described with respect to the nonessential role of the hormonal events of late pregnancy for maternal behavior in laboratory mice. One can conclude that as a result of inbreeding and selective breeding, when tested in their home cages laboratory mice do not require steroid hormones, lactogens, and OT to show prompt maternal behavior.

A different picture emerges, however, when OT–/– virgin mice are tested under more challenging environmental conditions. Ragnauth et al. (2005)

examined the maternal behavior of virgin OT−/− and OT+/+ mice in a seminatural environment. Eighteen females of each genotype were placed in a large enclosure with various environmental challenges, such as food restriction. When a foster pup was introduced into this enclosure, all OT−/− virgin females exhibited infanticide, while 50% of OT+/+ females were maternal and infanticide only occurred in 17% of these females. Ragnauth et al. (2005) concluded that when virgin laboratory mice are tested under conditions meant to simulate a more natural environment, which includes group living and interindividual competition, the role of OT neural systems in maternal behavior becomes more evident (also see Pedersen, Vadlamude, Boccia, & Amico, 2006). One possibility is that this seminatural environment was a stressful condition that gave rise to an abnormally high level of anxiety- and fear-related behaviors in OT−/− females and that such behaviors, in turn, depressed maternal responsiveness and promoted infanticide. The anxiolytic properties of OT may be necessary for maternal behavior to occur under such stressful environmental conditions.

Figure 4.6 presents a preliminary analysis of the role of OT central neural circuits in the maternal behavior of a typical female mammal. OT acts along with the other physiological events of late pregnancy and parturition to stimulate the onset of maternal behavior. In laboratory mice, this effect of OT appears to be minimal, at least when the mice are tested under standard laboratory conditions. However, OT also has anxiolytic effects, and these effects probably allow a mother to show successful maternal behavior under challenging environmental conditions that simulate a more natural environment. This important, but indirect, role of OT in maternal behavior, seems to be retained in laboratory mice. This analysis does not exclude the possibility that in laboratory mice OT acts not only to decrease anxiety/fearfulness, but also to increase maternal responsiveness or motivation to enable the female to show adaptive maternal responses under challenging environmental situations. This possibility is also depicted in Figure 4.6.

In addition to OT, additional populations of PVN and SON neurons also produce the neuropeptide vasopressin, and mammalian vasopressin, because of its neurochemical make-up, is referred to as arginine vasopressin (AVP; Stoop, 2012). OT and AVP are very similar in chemical structure: They each contain nine amino acids; OT and AVP differ by only two out of the nine amino acids (Stoop, 2012). Importantly, AVP can act as a partial agonist at OTRs and OT can act as a partial agonist at AVP receptors (Smith, DiBenedictis, & Veenema, 2019). For example, in the uterus, both OT and AVP can cause uterine contractions, and in mice it has been shown that the uterotonic effects of AVP are due to its actions on OTRs (Kawamata et al., 2003). These findings have two important implications. First, parturition may be normal in OT−/− mice because endogenous AVP substitutes for OT. Second, the normal maternal behavior of

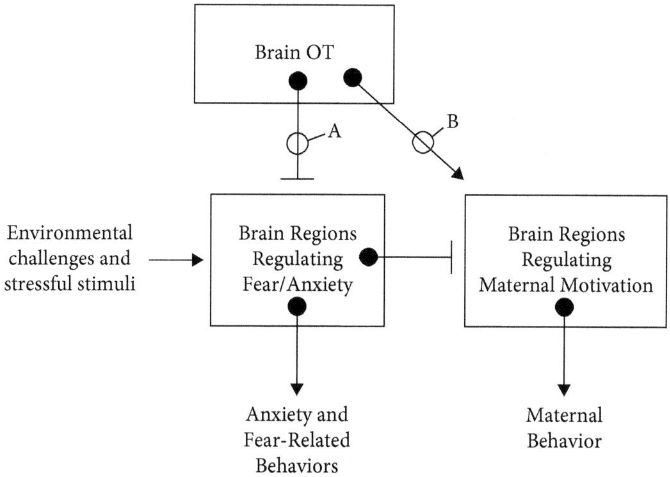

Figure 4.6. Two proposed mechanisms through which oxytocin (OT) action within the mammalian brain can increase maternal responsiveness under challenging environmental conditions. The diagram shows that stress-induced activation of anxiety- and fear-related neural systems may inhibit those neural systems that promote maternal motivation. Since OT exerts anxiolytic effects, it may indirectly enhance maternal motivation under stressful situations by depressing the activity of the brain systems that underpin fearfulness and anxiety (mechanism A). Since OT typically exerts excitatory neural effects, it would presumably inhibit these fear-related neural systems by activating inhibitory interneurons that suppress the output of these systems. In addition, OT may enhance maternal motivation by directly activating those neural systems that regulate maternal motivation (mechanism B). Axons ending in an arrow signify excitation and those ending in a bar indicate inhibition.

OT–/– mice in home cage tests may also be the result of AVP action on OTRs in the brain. To investigate this latter possibility, several research groups have explored the maternal behavior of transgenic mouse lines with a null mutation of the OTR gene (OXTR–/–). Nishimori's group was the first to examine maternal behavior in such mice (Takayanagi et al., 2005). On the morning of parturition, they found that OXTR–/– mice gave birth to young and spent as much time crouching over their pups in a nest as did control females (OXTR+/+). However, they noted that about three out of nine pups in the litters of OXTR–/– females were found scattered outside the nest area, while, on average, only about one of the nine pups were found outside the nest of control females. Each female was subsequently given a maternal behavior test on the day of parturition. To do this test (Hidema et al., 2016; Yoshihara, Numan, & Kuroda, 2018), each mother was briefly removed from her cage and placed in a novel cage. Three foster pups were then positioned outside the nest area in the home cage. Each mother was then returned to her home cage and retrieval and crouching behavior were observed over a 30-minute period. The OXTR−/− mice exhibited longer retrieval latencies and longer latencies to crouch over all three pups than did the control females. In particular, while OXTR+/+ postpartum mothers retrieved all three pups to the nest in about 3 minutes, it took about 13 minutes for the OXTR−/− mothers to complete retrieval of all three pups. My analysis of the data indicates that the longer latencies to crouch over the three pups in OXTR−/− mothers were primarily the result of the fact that it took these females longer to retrieve all the pups to the nest. Therefore, the primary deficit shown by OXTR−/− females was a mild retrieval deficit. Finally, similar mild retrieval deficits were also observed when the maternal behavior of virgin OXTR−/− females were tested, but all females eventually showed maternal behavior in a relatively short period of time (less than 20 minutes).

To indicate that the retrieval deficit was mild in OXTR−/− females, I want to compare their behavior to that of rats in which endogenous OT is blocked by ICV infusion of an OTA. During a 5-hour test period, when Fahrbach et al. (1985) administered OTA ICV to 15HO + E female rats, only 15% of OTA-treated females showed maternal behavior, while 90% of control females that were not administered OTA showed maternal behavior. Clearly, compared to rats, the involvement of OT and OTRs in mouse maternal behavior is very mild.

More recently, Kuroda's research group (Yoshihara, Numan, & Kuroda, 2018) has reported that OXTR−/− postpartum and virgin mice show absolutely no deficits in any aspect of maternal behavior when tested in their home cages. What might account for the different findings between Nishimori's group and Kuroda's group? An important methodological difference in the way maternal behavior was tested has been emphasized by Yoshihara et al. While Takayanagi et al (2005) removed each mother from her home cage prior to maternal behavior tests, Kuroda's group left the mothers undisturbed in their home cages and simply placed test pups outside the nest area while the mother was in her cage. Yoshihara et al. indicate that OXTR−/− females are highly responsive to stressful stimuli, presumably because of the lack of the anxiolytic effects of OT and that the removal and then the return of mothers to their home cages in the Takayanagi et al. study may have provided sufficient stress to cause the mild retrieval deficit observed in the OXTR−/− females.

Finally, Caldwell's research group has examined postpartum maternal behavior in a transgenic mouse line in which the deletion of the OTR was restricted to the forebrain (OXTRFB−/FB−; Macbeth, Stepp, Lee, Young, & Caldwell, 2010). The most important finding of this study was that for about 50% of the females in this transgenic line, the female's entire litter was found dead in the cage on postnatal day 1 (the day of parturition was designated as day 0; note that Takayanagi et al., 2005, studied their OXTR–/– mice on day 0). They referred to this high pup mortality as representing a significant incidence of pup abandonment in OXTRFB–/FB– parturient mice. Interestingly, for the remaining 50% of females in which the pups were not found dead, maternal behavior was relatively normal. At first glance, these results might suggest that OT action on OTRs in the forebrain plays a more important role in the onset of maternal behavior at parturition in mice than indicated by the studies that I have already reviewed. However, I would like to present an alternative interpretation. Perhaps a difficult parturition resulted in the pup deaths observed by Caldwell's group. Parturition may have been prolonged, and some of the pups may have been crushed during passage through the birth canal. Further, a difficult parturition may have affected maternal behavior due to the stress vulnerability of the OTR deficient mice. Recall that while OT action on uterine OTRs is not essential for parturition, it does facilitate parturition. Further, Yoshihara et al. (2018) have reported that parturition is indeed prolonged in OXTR–/– mice. It should also be considered that a disruption of the milk-ejection reflex may have exerted an additional contribution to the pup mortality that was observed 24 hours after birth. However, how can these proposals be used to explain the high pup mortality, since the uterus and mammary glands contain OTRs in the OXTRFB–/FB– transgenic line? In OXTRFB–/FB– mice, OTRs are deleted from the forebrain, which includes deletions from OT-containing neurons in SON and PVN (Lee, Caldwell, Macbeth, Tolu, & Young, 2008). Recall that the somatodendritic release of OT on to OTRs in the SON of parturient female rats exerts a positive feedback effect on its own release, causing bursts of systemic OT release from the posterior pituitary that facilitate parturition. Such OT bursts also facilitate the milk-ejection reflex (Neumann, Koehler, Landgraf, & Summy-Long, 1994). Therefore, parturition may have been difficult and less milk may have been ejected in response to suckling in a certain proportion of OXTRFB–/FB– mice. I think my proposal is worthy of future examination. For example, I would predict that the peripheral administration of OT to OXTRFB–/FB– mice might decrease the high postpartum pup mortality rate observed in these females. It should be noted that a high pup mortality by day 1 postpartum (24 hours after birth) was also observed in OT–/– mice (Nishimori et al., 1996), but this high mortality was not related to a lack of maternal responsiveness, but instead was due to a lack of the milk-ejection reflex. Indeed, in these OT–/– postpartum mice, systemic OT rescued the milk ejection reflex and increased pup survival.

These overall results indicate that central OT neural systems in laboratory mice are not critically involved in (a) the basic high level of maternal responsive shown by virgin females when they are tested under low stress conditions in their home cages and (b) are also not essential for the onset of maternal behavior at parturition in primiparous females tested under the same low stress conditions. This conclusion mirrors the results I described in Chapter 3 with respect to the lack of involvement of the other physiological events associated with late pregnancy and parturition for maternal behavior in laboratory mice. However, under challenging environmental circumstances that simulate a more natural environment, OT neural systems appear necessary for maternal behavior in laboratory mice, presumably related to the anxiolytic and fear-reducing properties of central OT; it is also likely that OT boosts maternal motivation under stressful conditions (see Figure 4.6). Therefore, by decreasing stress responsiveness and by boosting maternal motivation, the laboratory mouse will show adaptive maternal responses under challenging environmental conditions.

With respect to the research reviewed in Chapter 3, and related to the possible role of OT in boosting maternal motivation in laboratory mice, it would be interesting to examine whether OT–/– and OXTR–/– postpartum mice would perform an operant bar press response to receive pups as a reward at the same high level as do their control counterparts with an intact central OT system (see Hauser & Gandelman, 1985). Likewise, would maternal experience in virgin OT–/– and OXTR–/– mice enhance their ability to retrieve pups from a T-maze extension attached to their home cages in a manner similar to that observed for virgin laboratory mice with an intact central OT system (Stolzenberg & Rissman, 2011)?

These ideas and results gain in significance in the context of the knowledge that feral female mice require OT neural systems acting in concert with the other physiological events of late pregnancy and parturition to show proper maternal behavior. Selective breeding and inbreeding have clearly increased the basic attraction to pups and maternal responsiveness in laboratory strains of mice. As I indicated in Chapter 3, I think laboratory mice can serve as a useful model for understanding the neural underpinnings of naturally occurring alloparental behavior (see Chapters 7, 8, and 11 of this volume).

OT Neural Systems and Maternal Behavior in Nonhuman Primates

Very few studies have explored the role of OT in the maternal behavior of nonhuman primates and these studies, except for one experimental study, have been correlational in nature. Maestripieri, Hoffman, Anderson, Carter, and Higley (2009) measured blood OT levels in a free-ranging group of postpartum lactating rhesus monkey females. In their behavioral analysis, which occurred prior to the blood draws, they computed what they called a maternal warmth index, which was a combined measure of nursing and grooming behavior directed toward the mothers' infants. They recorded a strong positive correlation ($r = 0.84$) between the amount of time a mother nursed and groomed her infant and blood OT levels. These results, of course, are correlational in nature and therefore cannot provide evidence for a cause–effect relationship. Although high levels of peripheral OT may have also reflected the central release of OT in the brain (see the following discussion) with such central release promoting maternal responsiveness, it is just as likely, if not more so, that the opposite relationship accounted for their results. Increased infant suckling during nursing may have been the cause of the high plasma OT levels. Increases in maternal behavior may have been the cause of the increases in plasma OT rather than increases in OT being the cause of higher maternal warmth.

All of the remaining studies have been performed on the common marmoset (*Callithrix jacchus*). Marmosets are cooperative breeders and in addition to maternal behavior, paternal and alloparental behavior are observed in these social groups.

In a purely behavioral study, Sanchez, Ziegler, and Snowdon (2014) have provided evidence that postpartum marmoset mothers with dependent offspring (infants were 3 weeks of age and nursing) were more maternally responsive than either mothers with independent offspring (infants were 4 months of age) or former mothers without current offspring. To test maternal responsiveness, they used a two-chamber apparatus. The female was placed in one chamber, and the other chamber was referred to as the stimulus chamber. The stimulus chamber contained one of the following: an infant distress call that had been previously recorded from an isolated infant, a control sound, or no sound at all. Mothers with dependent offspring spent more time investigating the stimulus chamber that emitted the infant distress call, in comparison to control stimuli, than did mothers with independent offspring or females without current offspring. These results indicate that mothers with nursing infants are highly responsive to distal infant distress calls, and such responsiveness can be interpreted as implicating the involvement of peripartum hormonal and OT stimulation. These results also fit with findings of Pryce, Dobeli, and Martin (1993) described in Chapter 3. It is worth emphasizing that the two-chamber test used in this study can be considered a stressful condition since the females were separated from their family groups and tested while they were alone.

Finkenwirth, Martins, Dreschner, and Burkart (2016) examined parental behavior in marmoset family groups while also measuring urinary OT levels. Importantly, all females remained in their family groups during this study. For mothers, prepartum urinary OT levels were higher than those detected in other group members, and this was likely related to the oncoming parturition. During the first week postpartum, urinary OT rose to even higher levels in postpartum mothers and these levels were maintained until the time of weaning. Although these results suggest a relationship between the onset of maternal behavior in marmoset mothers and OT, the OT levels may also simply reflect the role of OT in parturition and milk ejection. Notably, however, urinary OT levels also rose during the first week postpartum for nonbreeding female helpers (and other helpers), although not to the high levels observed in mothers. This finding strengthens the case for the involvement of OT in the onset of parental motivation during the birth of young infants within the family group, since helpers do not experience parturition and do not lactate. Interestingly, however, during the early postpartum period, urinary OT levels did not correlate with the amount of infant carrying, but instead were positively correlated with the amount of infant licking. Therefore, it is also possible that newborn infants are automatically attractive to all marmosets in a family group and that the chemoreceptive stimulation associated with infant licking was the cause of the increase in urinary OT levels. A dual relationship is also possible, where chemoreceptive input activates OT release in the periphery and potentially in the brain, with such OT release in the brain further enhancing interest in young.

In reviewing these studies, it is clear that correlational measures cannot really answer questions of causation. Another concern with these studies is that OT was measured peripherally, either in blood plasma (for rhesus monkeys) or urine (for marmosets), so the question arises as to whether such peripheral measures are indicative of the central release of OT within the brain (Leng & Ludwig, 2016). Amico, Challinor, and Cameron (1990) measured blood plasma and CSF levels of OT in lactating rhesus monkeys, and they found that suckling caused increases in plasma OT but that variations in OT within CSF were independent of both suckling and plasma OT levels. They interpreted their results as indicating that peripheral measures of OT are not good indicators of the central release of OT within brain circuits relevant for maternal behavior. Similar results have been reported by Parker, Hoffman, Hyde, Cummings, and Maestripieri (2010). However, when OT is actually measured within brain nuclei (by microdialysis), rather than in the CSF, Neumann and Landgraf (2012) have made a convincing case for the coordinated release of OT both centrally and peripherally during parturition and suckling in mammals (also see Neumann & Landgraf, 1989). I conclude from these results that during the peripartum period OT is very likely to be released within critical neural regions of the primate brain in mothers and perhaps even in alloparents who are exposed to newborn infants and that peripheral measures of OT are probably reflective of such central release. Because OT is released in small amounts into select brain regions, the amounts that are released are probably too small to be measured in CSF.

The only experimental study on the role of OT in the maternal behavior of nonhuman primates that I am aware of was conducted by Taylor and French (2015). Eight adult female marmosets were studied; two of these were currently members of a breeding pair and had infants, while the remaining six females were in a breeding pair but did not have any infants (no experience with infants for at least 24 months). Each of these females was placed in the long arm of a T-maze, and the two other arms contained a stimulus box, one on each end. One box contained lifelike models of marmoset infants sitting on a branch along with a speaker that played infant distress and contact calls. The other arm simply contained a branch and emitted pure tones. Twenty minutes prior to being released into the T-maze each marmoset received intranasal infusions of either saline, OT, or AVP in a counterbalanced sequence. In comparison to saline infusions, the females that received intranasal AVP had shorter latencies to approach and look into the infant stimulus box. However, there were no treatment differences in time spent near the infant stimulus box, and all females spent more time near the infant box compared to the control box. Therefore, the primary effect was on the initial latency to approach and peek into the infant box. I want to emphasize that this latency effect was very mild. Females that received AVP approached and looked into the infant stimulus box within about 150 seconds, while it took the saline-treated females about 300 seconds to do so. Females treated with OT exhibited a latency of about 450 seconds. This relatively mild effect should be interpreted in the context of the fact that alloparental behavior occurs in marmosets and that all group members are usually attracted to infants (see the discussion in Chapter 3 on hormones and maternal behavior in nonhuman primates for some qualifications with respect to this statement). The significant effect observed after intranasal application of AVP may simply reflect a mild boost in maternal motivation, or sensitivity to infant vocalizations, from the basic high level that already exists in these cooperatively breeding females. Further, since each of these marmosets was separated from their family group during the T-maze test, the situation was probably stressful and the observed latency reduction may have required a boost in maternal motivation and, perhaps, a decrease in anxiety.

An analysis of these results requires several considerations. First, as you will recall, OT and AVP contain nine amino acids. The structure of AVP is conserved across mammals, but there are important differences in the exact structure of OT (French, Taylor, Mustoe, & Cavanaugh, 2016). For most mammals, leucine is the amino acid that is present in position 8 of the OT structure (Leu8-OT). But in marmosets and some other New World monkeys, proline is substituted for leucine at position 8 (Pro8-OT). Appropriately, Taylor and French (2015) administered Pro8-OT to their marmosets.

Another consideration is the mode of application of OT and AVP, which were administered intranasally. This procedure is commonly used as a noninvasive method to allow neuropeptides to gain access to the brain in primates and humans (Leng & Ludwig, 2016; Numan, 2015), although Leng and Ludwig have questioned its validity. However, Born et al. (2002) have shown that intranasal inhalation of AVP produces elevations of this peptide in human CSF within 10 minutes after its infusion, and Neumann, Maloumby, Beiderbeck, Lukas, and Landgraf (2013) have shown that intranasal infusion of Leu8-OT results in increases in OT within the amygdala of rats and mice within 30 to 60 minutes after its application. Finally, Chang, Barter, Ebitz, Watson, and Platt (2012) and Lee et al. (2018) reported that Leu8-OT increases in the CSF of rhesus monkeys within 35 to 60 minutes after intranasal application. Therefore, it is likely that the intranasal administration of AVP and OT to the marmosets in the Taylor and French (2015) study gained access to the brain.

Finally, why was AVP, but not Pro8-OT, effective in shortening the female marmosets' latencies to approach the infant stimulus box? As in other mammals, OT and AVP are produced within PVN and SON neurons of primates, and these neurons not only project to the posterior pituitary, but the PVN (primarily) also has projections within the brain (French et al., 2016). However, some interesting results for nonhuman primates have been obtained with respect to the locations of OTRs and AVP receptors within the brain. Like OT, AVP receptors are G protein–coupled receptors, and although several kinds of AVP receptors exist, the V1a receptor (V1aR) is the most common AVP receptor in the brain (Caldwell, Lee, Macbeth, & Young, 2008). Interestingly, in New Word monkeys (Freeman, Walum et al., 2014; Schorscher-Petcu, Dupre, & Tribollet, 2009), and in rhesus monkeys (Freeman, Inoue, Smith, Goodman, & Young, 2014), the distribution of V1aRs in the brain is much more widespread than that for OTRs. In particular, the distribution of V1aRs is very similar to the distribution of OTRs that I described for the rodent brain in Table 4.1, with some exceptions.

In the context of the above findings, Freeman, Inoue et al. (2014) bring up an interesting hypothesis with respect to the more widespread distribution of V1a receptors in the nonhuman primate brain. As I have already mentioned, OT is capable of binding to vasopressin receptors and vasopressin is capable of binding to OTRs, although each neuropeptide has a higher affinity for its own receptor. Freeman et al. raise the idea that OT regulation of social behavior, which would include maternal behavior, in nonhuman primates may occur through interactions with both OTRs and V1aRs. They suggest, based on the affinity of OT for each receptor, that OT may act on OTRs under conditions of low OT concentrations, and on V1aRs under conditions where OT is released into the brain at high concentrations. Since OT is presumably released into the primate brain at high levels during the peripartum period, it may act to boost maternal motivation by acting on V1aRs that are located in those brain regions that I will subsequently show to be critical for maternal responsiveness. When considering this hypothesis in the context of the results in the Taylor and French (2015) experiment, perhaps intranasal AVP activated V1a receptors to a greater degree that did intranasal Pro8-OT. Maybe higher doses of Pro8-OT, which would mimic the high levels of OT release that occur during the peripartum period, would have been necessary to detect an effect on the approach latencies, with such an effect being mediated by an action on V1a receptors.

The reader is referred to a recent review by Bayerl and Bosch (2019) that discusses the potentially complex role of vasopressin in mammalian maternal behavior.

Conclusions

My analysis of the role of OT in maternal behavior has been long and complex, but several important conclusions can be proposed. I consider the research on rats and sheep to be representative of how OT regulates maternal behavior in a typical female mammal that displays a uniparental maternal care system, although research on other species with this type of system needs to be performed. For the typical female mammal with a uniparental maternal care system, central OT systems, acting in concert with other peripartum physiological events, appear essential for the immediate onset of maternal behavior at parturition. During the postpartum period, after maternal behavior has become established, OT neural systems do not appear to be essential for ongoing maternal responsiveness under low stress conditions. However, under such low stress conditions, OT does modulate the quantity and quality of established maternal behavior, and one interpretation of this effect is that OT neural systems increase the mother's attentiveness to the needs of her infants. In contrast, under challenging environmental conditions that simulate a more natural environment, OT neural systems appear to be much more importantly involved not only in the onset of maternal behavior but also in the maintenance of competent and effective maternal responsiveness, and these effects may occur through OT's role in both boosting maternal motivation and decreasing anxiety and fear-related behaviors.

For laboratory mice, which represent an experimental model of allomaternal behavior, and for species, such as marmosets, where allomaternal behavior occurs under natural conditions, OT does not seem to be essential for the onset and maintenance of maternal behavior in mothers and allomothers under low stress conditions. However, under more challenging conditions, OT's positive role in potentiating both the onset and maintenance of maternal and allomaternal behavior becomes more evident.

Since typical mammals do not live under laboratory conditions but instead live under natural conditions that include many stressful and challenging events, it can be concluded that OT plays a very important role in all phases of mammalian parental behavior in all mammalian females, where it acts to promote adaptive and appropriate responses that allow mothers and allomothers to care for infants.

Finally, at least in the common marmoset, some of the maternal behavior–enhancing effects of OT may be mediated by its actions on V1a vasopressin receptors in the brain.

Olfaction and Maternal Behavior in Rodents, Rabbits, and Sheep

Introduction

To exhibit adaptive maternal responses, a mammalian mother must respond appropriately to sensory cues from her offspring. In rodents, behavioral studies have shown that both auditory and olfactory stimuli aid the mother in responding appropriately to young that have been displaced from the nest area. When young pups are displaced from the nest, they become hypothermic because their ability to regulate their body temperature has not yet developed, and in response to hypothermia they emit ultrasonic vocalizations (Okon, 1972). Smotherman, Bell, Starzec, Elias, and Zachman (1974) have shown that for lactating rodents, both ultrasounds and olfactory stimuli from displaced pups arouse mothers and activate searching behaviors. In rats, auditory cues provide the mother with directional information about the location of the pups, while in mice, olfactory cues are the primary source of such directional information (Smotherman et al., 1974). It seems obvious that under natural conditions, interference with a rodent mother's olfactory or auditory sensitivity would probably result in an increase in offspring mortality. However, I want to make the distinction between the ability of a mother to quickly detect displaced pups from a mother's actual motivation to engage in maternal behavior once the pups are actually found. For example, an anosmic or deaf mother may ultimately find a pup outside the nest as she engages in normal exploratory behavior. Would such a mother then care for the pup by retrieving it to the nest and nursing it?

For rats that are studied under laboratory conditions, Beach and Jaynes (1956) concluded that maternal behavior is under multisensory control. According to this concept, although many infant-related cues can be shown to influence maternal responses in rats, no single sensory modality is essential for maternal motivation or the ability to perform maternal behavior once a mother interacts with her infants. This multisensory view subsequently received support from a study by Herrenkohl and Rosenberg (1972). Different groups of primigravid rats were blinded, deafened, or rendered ansomic (via olfactory bulbectomy), and it was found that the subsequent postpartum maternal behavior (retrieving and nursing) of these females remained intact.

A more complete review of this multisensory concept of the control of maternal behavior in rats has been presented by Numan and Insel (2003). For the most part, this concept remains accurate today. I would like to provide one more example that highlights the point I am trying to make. During nursing behavior in rats, where infants provide important somatic sensory inputs to the mother, several different types of nursing postures are exhibited by the mother, and the duration of some of these postures is dependent upon suckling stimulation from the pups (Numan & Insel, 2003; Stern, 1996). Although removal of the mother's nipples (thelectomy) has been shown to abbreviate the duration of certain nursing postures, it clearly does not interfere with the mother's basic maternal motivation. Such females retrieve their young, even when they are placed in a T-maze extension attached to the home cage, and spend as much time hovering over their pups in the nest area, although the duration of certain nursing postures may be abbreviated (Numan & Numan, 1995; Stern & Mackinnon, 1976). Therefore, while elimination of any single sensory modality in rats may increase retrieval latencies (because it takes longer for a female to detect a displaced pup) or alter the duration of particular nursing postures, it appears that no one sensory modality is essential for maternal motivation and for the mother's attraction to the remaining infant cues that she can still detect.

Of all the sensory modalities that may influence maternal behavior in mammals, olfaction has received the most attention. In the following sections, I will describe this research in detail for the species that have been studied the most: rats, rabbits, mice, and sheep. This research will show that although olfaction is not essential for maternal motivation in rats, rabbits and sheep, a change in the ability of olfactory input from offspring to depress maternal responsiveness is important for the onset of maternal behavior in rats and rabbits. Further, although olfaction is not essential for maternal motivation in sheep, it is important for maternal selectivity—the maternal ewe's ability to distinguish her own from alien young. Finally, I will show that the concept of the multisensory control of maternal motivation cannot be extended to laboratory mice because olfaction is essential.

Olfaction and Maternal Behavior in Laboratory Rats and Rabbits

Pheromones are chemical odors released by one member of a species that influence the behavior and/or physiology of a conspecific. As indicated in Figure 4.1, there are two major chemosensory neural receptor systems within the nasal cavity of many mammalian species, which include rats, mice, rabbits, and sheep (Corona & Levy, 2015; Dulac & Wagner, 2006): the MOE, whose sensory neurons project to the MOB, and the VNO, whose sensory neurons project to the AOB. Research has shown that both of these chemosensory systems are involved in the detection of pheromones (Dulac & Wagner, 2006; Fraser & Shah, 2014).

In rats, various methods have been used to examine the involvement of these two pheromone-detection systems in maternal behavior. Total olfactory bulbectomy would eliminate both systems. Intranasal application of zinc sulfate (ZnSO4) has been used to selectively destroy the MOE, and VNO removal or cutting the vomeronasal nerves had been used to selectively eliminate the vomeronasal system. Consistent with the previously described research on the olfaction and maternal behavior in rodents, rabbits, and sheep, none of these interventions, when performed during pregnancy in primigravid rats, prevents the subsequent onset and maintenance of maternal behavior, although slightly longer retrieval latencies are observed when the main olfactory system is depressed because females take slightly longer to locate displaced pups (Fleming, Gavarth, & Sarker, 1992; Jirik-Babb, Manaker, Tucker, & Hofer, 1984; Numan & Insel, 2003; Numan & Numan, 1995). These results confirm the initial findings of Beach and Jaynes (1956) and Herrenkohl and Rosenberg (1974) that the detection of pheromones emitted by pups is not essential to the onset and maintenance of maternal motivation in laboratory rats.

Strikingly, however, research by Rosenblatt and Fleming (Fleming & Rosenblatt, 1974c; Fleming Vaccarino, Tambosso, & Chee, 1979) has shown that elimination of these pheromone detecting systems facilitates maternal behavior (shortens sensitization latencies) in naïve virgin estrous cycling female rats. Table 4.2 shows some of the results from the Fleming et al. (1979) study.

Table 4.2 Disruptions of the Olfactory System Facilitate the Onset of Maternal Behavior in Virgin Female Rats

Group	Average Latency to Onset of Maternal Behavior (Days)
pMOBX	6
VMNX	4[a]
pMOB + VMNX	2[b]
sham lesions	8

Notes: pMOB = partial lesions of the main olfactory bulb; VMNX = knife cuts severing the vomeronasal nerves.

Source: The data in this table were derived from the research of Fleming, Vaccarino, Tambosso, and Chee (1979).

[a]Significantly different from sham group.

[b]Significantly different from remaining groups.

Prior to pup exposure, virgin female rats received one of the following: (a) partial lesions of the MOB that spared the AOB (pMOBX); (b) knife cuts that severed the vomeronasal nerves (VNNX); (c) combined damage to the MOB and VNNs (pMOB + VNNX); or (d) sham lesions. Following daily exposure to foster pups, the virgins with disruption of both chemosensory systems exhibited the shortest sensitization latencies, showing maternal behavior after only about 1 to 2 days of pup exposure, which contrasted sharply with the 7- to 8-day sensitization latency shown by the sham females. An average sensitization latency of about 1 week is standard for untreated virgin female rats as described in Chapter 3. Please note that the relatively long sensitization latencies observed in the pMOB group would have been much shorter if the main olfactory system had been more extensively disrupted (Fleming & Rosenblatt, 1974b). These results indicate that inputs from both chemosensory systems inhibit maternal behavior in virgin rats.

Fleming and Rosenblatt (1974c) have proposed that virgin female rats find the novel odors/pheromones of pups aversive, and these odors therefore elicit avoidance or withdrawal responses. Recall from Chapter 3 that Fleming and Luebke (1981) divided pup-stimulated maternal behavior in virgins into a series of stages involving avoidance of pups, followed by tolerance of pup proximity, which then eventually leads to the onset of maternal behavior. Fleming et al. (1979) therefore concluded that interference of chemoreception mediated by the MOE and VNO abbreviates the avoidance phase, which then shortens the sensitization latencies of the affected females.

Given that the physiological events associated with late pregnancy and parturition stimulate the immediate onset of maternal behavior in primiparous rats, it can be proposed that one influence of these physiological events is to alter the processing of pup odors so that they no longer inhibit maternal behavior. One obvious possibility is these physiological events render the parturient female anosmic. This possibility is not supported by a variety of evidence. First, as I have already indicated, postpartum rats rely on pup odors to detect pups that have wandered outside the nest area, and other research shows that central neural structures that I will show to be important for maternal behavior respond to pup odors in lactating female rats (Hernandez-Gonzalez et al., 2005). Most important, toward the end of pregnancy, rather than being aversive, pup odors become highly attractive to primigravid female rats (Bauer, 1983; Kinsley & Bridges, 1990). Finally, treatment of naïve virgin female rats with a hormone regimen that is capable of stimulating short-latency maternal behavior increases the virgin female's attraction toward pup odors (Fleming, Cheung, Myhal, & Kessler, 1989).

These results, taken together, suggest that instead of rendering the puerperal female rat anosmic, the physiological events of late pregnancy and parturition alter the valence of pup-related olfactory stimuli, switching them from negative to positive; pup odors stimulate avoidance in virgins but are attractive to the parturient female. Note that this switch in the valence of pup odors so that they become attractive is not essential for maternal behavior and motivation, since anosmic females show maternal behavior. Therefore, this valence switch is primarily important for eliminating the occurrence of avoidance and rejection responses toward newborn pups in the parturient female. Of course, the attractive properties of pup odors, along with other pup stimuli, undoubtedly contribute to adaptive maternal responding, for example by allowing the postpartum female to quickly retrieve displaced pups back to the nest.

There are some interesting studies with respect to OT's role in stimulating the onset of maternal behavior in rats that involve an understanding of the role of olfactory input in depressing maternal behavior in virgin rats. Pedersen et al. (1982) and Fahrbach et al. (1984) were able to stimulate a very short latency onset of maternal behavior in estradiol-treated rats of the Zivic–Miller strain with ICV OT, while Rubin et al. (1983) were not able to replicate this finding using a rat strain obtained from Charles River laboratories. Earlier in this chapter, I argued that ICV OT was effective in the estradiol-treated Zivic–Miller rats because the baseline level of maternal responsiveness in these virgin females was very high even in the absence of OT treatment. Wamboldt and Insel (1987) have noted that at the time of the Pedersen and Fahrbach studies, the rats from Zivic–Miller laboratories were infected with pulmonary pathogens that might have rendered them hyposmic and that this might have caused the high baseline level of maternal responsiveness. Therefore, it can be proposed that estradiol treatment alone, in the absence of progesterone withdrawal, when coupled with decreased olfactory sensitivity, allowed ICV OT to be an effective treatment for stimulating a very short latency onset to maternal behavior in Zivic–Miller rats. Wamboldt and Insel tested this proposal on virgin female rats obtained from another breeder, Taconic Farms. In one aspect of this study, half of the females received intranasal application of $ZnSO4$ to destroy the MOE, while the remaining females received saline irrigation of the nasal cavity. All of these ovariectomized females received prior treatment with estradiol, and each group also received an ICV injection of OT prior to pup presentation. In a 3-hour test, about 85% of the $ZnSO4$ + OT females showed maternal behavior, while none of the females that received intranasal saline and ICV OT showed maternal behavior. In an additional control group, it was also found that only 10% of females that received $ZnSO4$ without ICV OT showed maternal behavior during the 3-hour test. One interpretation of these findings is that when combined with estradiol treatment, $ZnSO4$-induced damage of the MOE depressed the avoidance of pup odors while OT acted within the brain to promote approach responses and maternal motivation, with this combination of effects leading to a prompt maternal response.

In a related study, Yu, Kaba, Okutani, Takahashi, and Higuchi (1996) implanted bilateral cannulas into the MOBs of virgin rats (not of the Zivic–Miller strain). These females were ovariectomized and systemically injected with estradiol. Prior to pup presentation, OT or saline was injected directly into the MOB. In another group, the same amount of OT was injected into the lateral ventricle. In a 2-hour pup exposure test, 50% of the females that received OT into the MOB showed maternal behavior, while none of the females in the other two groups (saline MOB; OT ICV) did so. The authors suggested that OT injection into the MOB may have caused anosmia and in this way facilitated maternal behavior (see Yu, Kaba, Okutani, Takahashi, Higuchi, & Seto, 1996).

I would like to offer an alternative explanation for their results. Perhaps the injection of OT not only acted on the MOB, but also spread to other nearby brain regions to stimulate maternal behavior, and this multiple site of action effect was facilitated by the partial damage to the MOB caused by the bilateral cannula implants. Yu, Kaba, Okutani, Takahashi, and Higuchi (1996) suggested that their ICV OT treatment controlled for the possibility of spread of OT from the MOB to nearby regions. But to appropriately support this conclusion, they should have included a group that had both ICV OT and implants of control (saline) cannulas into the MOB. My interpretation certainly fits with the data from Wamboldt and Insel (1987).

The olfactory bulbs may be one of several sites where OT acts to stimulate the onset of maternal behavior and although selective action at this one site alone may not be sufficient to stimulate maternal behavior, its action there may be necessary for the normal onset of maternal behavior (see Yu, Kaba, Okutani, Takahashi, & Higuchi, 1996). As I will show in Chapter 5, OT acts at several brain sites to stimulate maternal behavior, and an action at each of these sites may be individually necessary, but not sufficient, for maternal behavior. The combined action of OT at multiple brain sites may be necessary to stimulate the onset of maternal behavior. The involvement of OT action on the olfactory bulbs, and nearby structures, in facilitating maternal behavior is attractive since OTRs are located in the OB (see Table 4.1), as well as in the nearby AON that lies just caudal to the OB (Oettl et al., 2016; see Figure 4.1 and Table 4.1). It is highly unlikely, however, that OT acts to cause anosmia, since late pregnant and parturient rats are attracted to pup odors. Perhaps OT action on the OB and AON is involved in the valence switch that shifts pup odors from being aversive to being attractive (cf. Oettl et al., 2016). But such attractive pup stimuli must still gain access to other brain regions involved in maternal motivation, and these regions as well may have to be modified by OT action for them to be receptive to inputs from pup odors and other pup stimuli.

Research on rabbits, although not as extensive as that in rats, also supports the view that chemosensory inputs from the main olfactory system and the vomeronasal system are involved in suppressing the maternal responsiveness of virgin female rabbits. Recall from Chapter 3 that virgin rabbits do not show sensitized maternal behavior even after 15 days of constant exposure to young kits. In contrast, following intranasal application of ZnSO4, 40% of virgin rabbits showed maternal behavior after 4 days of exposure to young kits, and this proportion rose to 70% after 14 days of exposure to the kits. The maternal behavior that was exhibited was similar to that shown by postpartum mothers: The maternal virgins entered the nest box and crouched over the young kits for about 3 minutes each day (Chirino, Beyer, & Gonzalez-Mariscal, 2007). Similar, although less dramatic results, were obtained by Gonzalez-Mariscal, Chirino, Beyer, and Rosenblatt (2004), where 40% of virgin rabbits whose AOB was removed displayed maternal behavior within 3 to 13 days of exposure to young. Of note, ovariectomy blocked these effects in both studies, suggesting that estradiol may be interacting with the chemosensory disruptions for maternal behavior to be facilitated. Similar to the rat studies, these results suggest that the physiological events of pregnancy and parturition eliminate the inhibitory effect of kit odors on maternal behavior. They also show that the detection of kit odors in not necessary for the performance of maternal behavior in domestic rabbits.

In Chapter 3, I indicated that postpartum rabbits, because they give birth to altricial young, will nurse their own and alien young (young from another mother). These findings were obtained from domesticated rabbits. Wild rabbits live in colonies, but they nest separately. There is some evidence that wild rabbit mothers will attack young kits from another rabbit colony and that they may discriminate young kits from their own and different colonies on the basis of olfaction (Myktowycz & Dudzinski, 1972). Therefore, although the valence of kit odors may change from negative to positive in postpartum wild rabbits, this change may only apply to a certain class of odors emitted by young kits from their colony, and the strange odors of young from another colony, presumably caused by different diets, may still retain a negative valence and elicit avoidance and/or rejection responses from the mother. This information shows us that the results obtained from studies on laboratory strains of rats, rabbits, mice, and even domesticated sheep may not be completely representative of the processes that occur in wild populations. This contrast has already been dramatically shown when I compared the maternal responsiveness of virgin laboratory mice with that of feral mice.

Olfaction and Maternal Behavior in Sheep

Recall that at parturition ewes will be maternally responsive to any lamb that is presented to them. However, after interacting with a particular lamb at parturition, the mother learns its olfactory characteristics and from that point forward she will show maternal behavior toward that particular lamb while rejecting the advances of novel or unfamiliar lambs. When analyzing the role of olfaction in the maternal behavior of sheep, several questions emerge: (a) Is olfaction necessary for the initial maternal responsiveness of the parturient mother to a lamb? (b) What are the olfactory mechanisms that underlie the development of maternal selectivity? and (c) Which olfactory system, the vomeronasal system or the main olfactory system is most important for maternal behavior in sheep?

The classic experiment that answered aspects of these questions was performed by Levy, Locatelli, Piketty, Tillet, and Poindron (1995). Primigravid sheep received one of the following treatments during pregnancy: intranasal application of ZnSO4 to destroy the MOE, knife cuts severing the vomeronasal nerves, or sham control procedures. One week prior to parturition, tests for anosmia were conducted by examining whether the ewes would avoid food that was contaminated with a foul odor. Upon giving birth, the maternal behavior of the ewes toward their own lamb was examined for about 1 hour, and instances of acceptance (allowing the lamb to suckle) and rejection (head butts) were recorded. The mothers were then left with their own lamb for 6 hours. Following this period, maternal selectivity tests were conducted, where the maternal behavior of the mother to her own and an alien lamb was examined.

Mothers that received ZnSO4 treatment were anosmic and did not avoid the contaminated food, while the remaining groups were able to detect the foul odor and did not attempt to eat the contaminated food. The sham females and the females receiving the vomeronasal nerve cuts exhibited normal maternal behavior at parturition and normal selectivity 6 hours postpartum, where they accepted their own lamb but rejected an unfamiliar lamb. Histological examination of the brains of the females with the vomeronasal nerve cuts confirmed that this procedure effectively disrupted the connection between the VNO and the AOB. These results indicate that the vomeronasal system is not essential for maternal responsiveness or for maternal selectivity. The females that received ZnSO4 presented a different picture. During the initial 1-hour test at parturition, approximately 60% of the ZnSO4-treated ewes would not allow their lamb to suckle. However, this simply represented a mild delay in the onset of maternal behavior because normal maternal responsiveness was observed in all of these ewes at 6-hours postpartum, except that maternal selectivity was abolished and these ewes accepted and were maternal toward both their own and alien lambs. My conclusion from these results is that the main olfactory system and the vomeronasal system are not essential for maternal responsiveness during the early postpartum period in primiparous sheep, but that the main olfactory system is essential for maternal selectivity. Because olfactory learning did not occur in the ZnSO4-treated ewes, maternal responsiveness was displayed toward any lamb.

What are the particular mechanisms that allow for olfactory learning and the development of maternal selectivity in the parturient ewe? The MOB receives a significant input from norepinephrine (NE) neurons, whose cell bodies are located in the locus coeruleus of the pons, and the vaginocervical stimulation that occurs during parturition causes the release of NE into the MOB (Kendrick et al., 1997; Levy, Gervais, Kindermann, Orgeur, & Piketty, 1990). Perhaps NE input to the MOB during parturition, coupled with the other hormonal and physiological events occurring at this time, allows for the formation of maternal selectivity. Evidence favoring this proposal has been presented by Levy et al. (1990) and by Pissonnier, Thiery, Fabre-Nys, Poindron, and Keverne (1985). Briefly, disruption of NE input to the MOB, which began during pregnancy and continued into the postpartum period, did not disrupt the onset of maternal behavior but did prevent the development of maternal selectivity in a large proportion of the affected mothers. Such ewes would care for their own and alien lambs. Importantly, the disruption of NE input to the MOB did not cause anosmia. One interpretation of these results is that the NE input to the MOB that occurs during parturition while a mother interacts with her lamb in some way narrows the types of lamb olfactory stimuli that can activate other central neural mechanisms that regulate maternal responsiveness so that only the "olfactory signature" of the particular lamb that she interacted with can gain access to these maternal behavior neural centers. That is, initially at parturition any lamb odor complex can activate acceptance, but after the mother interacts with a particular lamb, NE in some way is involved in allowing only that lamb odor complex to continue to gain access to those brain centers that regulate maternal responsiveness, while unfamiliar lamb stimuli subsequently activate brain regions that regulate avoidance and rejection responses.

Interestingly, parturition and vaginocervical stimulation also activate OT release into the MOB in sheep (Kendrick et al., 1997). In light of my previous discussion with respect to OT and the role of olfaction in the maternal behavior of rats, one can speculate that OT interacts with NE within the MOB to promote the development of maternal selectivity in sheep. However, there is currently no definitive proof for this proposal.

Figure 4.7 presents a hypothetical model of the mechanisms that might regulate maternal responsiveness and the development of maternal selectivity in postpartum ewes. Indicated in the figure are the locus coeruleus, the MOB, the PVN, which is the primary source of OT projections within the brain, and an undefined representation of the central neural brain systems that regulate rejection and avoidance behavior and those that regulate maternal motivation and acceptance of lambs (Chapter 5 of this volume will add specificity to these mechanisms). As a result of the action of hormonal events of late pregnancy and OT on central maternal motivation centers, the ewe responds maternally to any lamb, and this

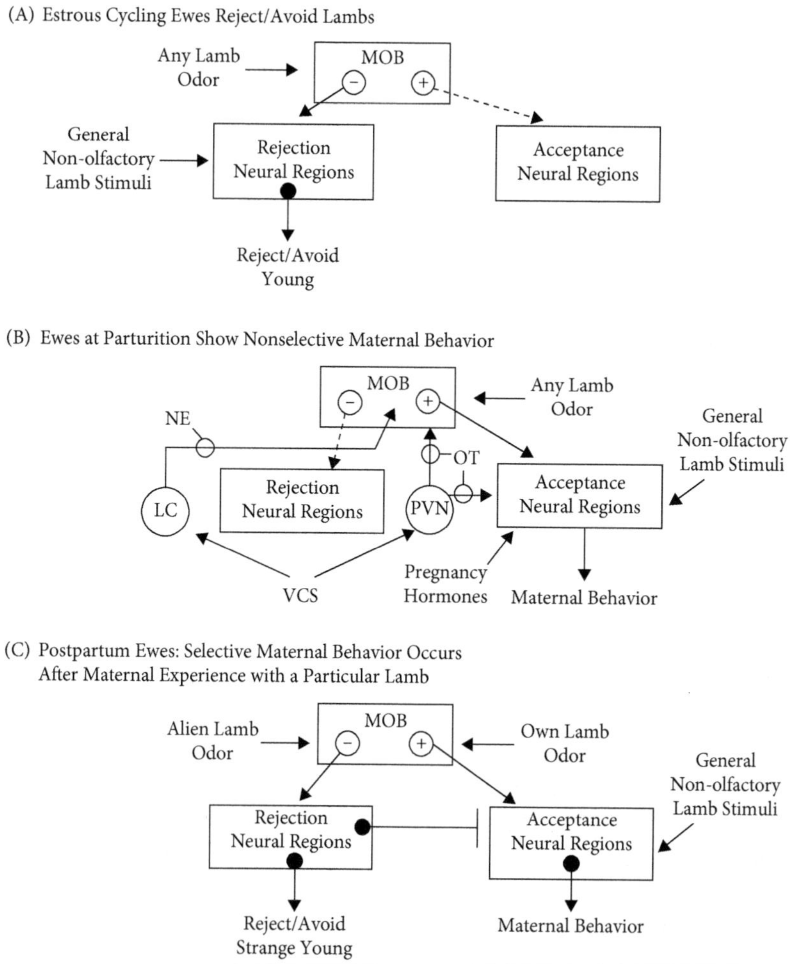

Figure 4.7. A hypothetical model of the mechanisms that might regulate maternal responsiveness and the development of maternal selectivity in postpartum ewes. In each part of this figure, undefined representations of the central neural systems that regulate rejection and avoidance of lamb-related stimuli and maternal motivation and acceptance of lambs are indicated. Lamb odors act at the level of the main olfactory bulb (MOB), while general (not specific for a particular lamb) nonolfactory lamb stimuli act at other neural sites. Axons ending in an arrow signify excitation and those ending in a bar indicate inhibition. (A) For estrous cycling ewes, which typically avoid and reject lambs, any lamb odor activates negatively valent MOB neurons, which, in turn, activate rejection/avoidance neural regions. These latter regions are also activated by general nonolfactory lamb stimuli. The projections of positively valent MOB neurons to maternal acceptance neural systems are not active (shown by a dashed line). The operation of this neural network results in estrous cycling ewes avoiding and rejecting all lambs. (B) At parturition,

is driven by both olfactory stimuli and general nonolfactory stimuli from the lamb that she gives birth to. Once she interacts with a particular lamb during the immediate postpartum period, the combination of her lamb's olfactory signature with NE and OT input to the MOB (undoubtedly, other factors are also involved) results in a reorganization of MOB neural circuits and their outputs to other brain regions, which include MOB projections to the medial and cortical amygdala. (See Figure 4.3, Figure 5.7 in the next chapter, and Keller, Perrin, Meurisse, Ferreira, and Levy, 2004.) This reorganization results in (a) "own" lamb odor maintaining the ability to activate maternal motivation centers in combination with general nonolfactory lamb stimuli and (b) alien lamb odors acquiring the ability to activate brain regions that regulate rejection behavior. These rejection regions not only promote rejection and avoidance responses directed toward the unfamiliar lamb, but they also inhibit the activity of maternal motivational systems, and such inhibition is hypothesized to prevent the postpartum ewe from responding in a maternal fashion to the general (present in all lambs) nonolfactory stimuli from the alien lamb.

This hypothetical model also explains why anosmic postpartum ewes show maternal responsiveness but not maternal selectivity. Such females show

---

Figure 4.7. Continued.
ewes will initially respond maternally toward any lamb. This is hypothesized to be due to the action of OT at the level of the MOB, and OT and pregnancy hormone action at the level of maternal acceptance neural systems. The source of OT is shown as originating from neurons in the paraventricular nucleus (PVN) of the hypothalamus. These effects allow any lamb odor to activate positively valent MOB neurons that project to brain regions that promote maternal motivation, and to allow general nonolfactory lamb stimuli to directly activate maternal acceptance regions. Negatively valent MOB neurons that project to rejection neural regions are inactive (shown as a dashed line). These combined effects result in a parturient ewe that will to accept any lamb at parturition. Vaginocervical stimulation (VCS) associated with the birth process also activates locus coeruleus (LC) norepinephrine (NE) neurons in the pons that project to the MOB. The action of NE at the level of the MOB, possibly combined with the effects of OT at this site, results in a reorganization of the MOB, which ultimately results in the development of maternal selectivity toward the lamb that the mother was exposed to at birth, as shown in plate C. (C) Once maternal selectivity develops, the specific lamb odor to which a mother was exposed to at parturition (own lamb odor) acquires the ability to activate positively valent MOB neurons that activate maternal acceptance neural regions. Alien lamb odors (unfamiliar lamb odors) now activate negatively valent MOB neurons that project to the rejection/avoidance neural regions. These rejection regions project to and inhibit maternal acceptance regions so that the mother does not respond maternally toward general nonolfactory lamb stimuli.

maternal behavior toward any lamb because the hormone- and OT-primed central maternal motivational systems respond to general nonolfactory lamb stimuli without being inhibited by the inactive rejections regions.

To complete this model, one can also propose that estrous cycling ewes reject all lambs because central maternal motivational systems have not been primed by the physiological events associated with late pregnancy and parturition and that olfactory and nonolfactory lamb stimuli activate central rejection regions. It is interesting to speculate that OT action on the MOB in the parturient ewe may allow any lamb odor to activate MOB circuits that project to regions involved in maternal motivation, while the subsequent action of NE causes an olfactory "imprint" so that subsequently only familiar lamb odors maintain the ability to access central maternal motivational systems.

Olfaction and Maternal Behavior in Mice

Laboratory mice when tested in their home cages do not depend on the physiological events of late pregnancy and parturition to display prompt maternal behavior. But is olfaction, either from the main olfactory system and/or the vomeronasal system, involved in mouse maternal behavior? In the first study that examined this issue, Gandelman, Zarrow, Denenberg, and Meyers (1971) found that olfactory bulbectomy eliminated pup-directed maternal behavior (retrieving and nursing of young) in both parturient and virgin female laboratory mice. While control parturient and virgin mice showed prompt maternal behavior toward pups, the bulbectomized females either ignored the pups or displayed infanticide. On the basis of these findings, Gandelman et al. concluded that the multisensory control of maternal behavior cannot be extended to laboratory mice and that the olfactory sense is essential for maternal behavior in such mice.

Since olfactory bulbectomy destroys both the MOB and the AOB, the question arises with respect to the importance of each system for maternal behavior in mice. The evidence clearly indicates that it is the main olfactory system that is involved in mouse maternal behavior. Irrigation of the nasal cavity with ZnSO4, which would destroy the MOE, disrupts the onset and maintenance of maternal behavior in parturient primiparous laboratory mice (Vandenbergh, 1973), while removal of the VNO does not disrupt maternal behavior in such females (Bean & Wysocki, 1989; Lepri, Wysocki, & Vandenbergh, 1985).

More recent studies have applied transgenic methods to delete genes that are involved in the ability of the VNO or the MOE to detect pheromones, and these studies have arrived at the same conclusion: Chemoreception by the MOE, but not the VNO, is essential for maternal responsiveness toward pups in laboratory female mice (Fraser & Shah, 2014; Wang & Storm, 2011; Wu, Autry, Bergan, Watabe-Uchida, & Dulac, 2014).

Feral mice, in contrast to laboratory strains of mice, rely on the physiological events of late pregnancy and parturition to initiate maternal behavior. Virgin feral mice, and even pregnant feral mice, display a high level of infanticide when presented with pups. Recent research indicates that the detection of pup pheromones by the VNO has depressive effects on the maternal behavior of virgin feral mice. The TRPC2 ion channel (transient receptor potential ion channel, subfamily C, member 2) is essential for pheromone detection by the vomeronasal organ, and this ion channel is not involved in chemoreception within the MOE (Dulac & Wagner, 2006). Genetic deletion of this ion channel does not disrupt the maternal behavior of female virgin and postpartum laboratory mice (Fraser & Shah, 2014; Wu et al., 2014), which conforms with the view that the vomeronasal system is not essential for maternal behavior in laboratory mice. Chalfin et al. (2014) examined the effects of genetic deletion of the TRPC2 ion channel on the maternal behavior of virgin feral and laboratory mice during a 15-minute test. In laboratory virgins, control and mutant females showed prompt maternal behavior toward pups. In contrast, feral control virgin mice that retained the critical ion channel either attacked (60%) or ignored the test pups, while only 30% *TrpC2–/–* females attacked pups, 30% were maternal, and 40% ignored the pups. These results indicate that in virgin feral females, but not in virgin female laboratory mice, detection of pup pheromones by the VNO is involved in promoting defensive and rejection responses toward young pups. With respect to postpartum maternal behavior in feral mice, since parturient female mice show maternal behavior toward their own or alien pups (as reviewed in Chapter 3 of this volume), one can conclude that the physiological events of late pregnancy and parturition act to preclude the inhibitory effects of pup pheromone detection by the vomeronasal system so that maternal behavior is initiated promptly at the time of parturition.

In analyzing all of these results, I can offer ideas about some of the factors that may contribute to the occurrence of spontaneous maternal behavior in many laboratory strains of virgin female mice. As a result of selective breeding or inbreeding, the brain of such females has been modified so that the detection of pup pheromones by the MOB promotes maternal behavior while any inhibitory effects on maternal behavior that result from the detection of pup odors by the VNO and the MOE have been eliminated. Such modifications undoubtedly interact with changes in the structure and function of other central neural systems involved in the regulation of maternal behavior (see Chapters 7 and 11 of this volume).

Conclusions

My discussion of the role of olfactory and other sensory inputs in the regulation of maternal behavior in mammals emphasized the concept of the multisensory control of maternal motivation. Such multisensory control was found to be accurate for rats, sheep, and rabbits. Laboratory mice were shown to be an important exception, where olfactory input from MOB was shown to be essential for maternal motivation.

I also emphasized (see the introduction to the previous section on olfaction and maternal behavior in rodents, rabbits, and sheep in this chapter), however, that even in those species in which maternal motivation (as distinguished from maternal selectivity, particularly in sheep) is under multisensory control, individual sensory systems do play important roles in competent maternal performance (as opposed to maternal motivation) under certain conditions. Recall that olfactory and auditory stimuli from pups aid the rodent mother in detecting pups that have become displaced from the nest. More specifically, the maternal physiological condition may render specific infant cues more salient to mothers in comparison to their virgin counterparts, and such increases in maternal attentiveness to specific infant cues are probably very important for infant survival under natural, rather than laboratory home cage, testing conditions (Banerjee & Liu, 2013; Elyada & Mizrahi, 2015; Krishnan, Lau, Ewall, Huang, & Shea, 2017; Marlin, Mitre, D'Amour, Chao, & Froemke, 2015; Mitre et al., 2016).

To emphasize this point, I want to describe the study by Marlin et al. (2015) in more detail. Although evidence indicates that the deafening of rodent mothers does not eliminate maternal motivation (Herrenkohl & Rosenberg, 1972), under more stringent testing conditions, OT action on the auditory cortex in female mice appears to help the recently parturient female mouse detect and quickly retrieve displaced pups. In Marlin et al.'s experiment, naïve virgin and postpartum female mice were individually placed in a novel cage with nest material and pups. Although most pups remained in the nest, one pup was removed and placed in the cage at a site distal from the nest. In a 2-minute retrieval test, postpartum females were much more likely to locate and retrieve the displaced pup. However, if OT was injected into the auditory cortex of naïve virgins, they quickly responded to the ultrasonic vocalizations of the displaced pup, quickly retrieving it back to the nest. Importantly, this study found that OTRs were located in the auditory cortex. Further, through electrophysiological recordings from single neurons in the auditory cortex, evidence was provided that OT may act to increase the responsiveness of these neurons to pup distress calls (also see Mitre et al., 2016). Finally, while OT action within the auditory cortex facilitated retrieval in naïve virgin laboratory mice under these challenging test conditions, the injection of an OTA into the auditory cortex did not interfere with the retrieval behavior of maternally experienced postpartum mice.

In my review of OT's role in maternal behavior in mammals, I stressed its primary involvement in the onset, but not the maintenance, of maternal behavior. In accordance with this view, the results of Marlin et al. (2015) suggest that OT action on the auditory cortex facilitates retrieval behavior in maternally inexperienced virgin laboratory mice by enhancing the salience of infant vocalizations but that this effect is no longer necessary once postpartum mice have acquired maternal experience. It would have been interesting to determine whether OTA injection in the auditory cortex would have disrupted the retrieval behavior of inexperienced parturient mice on their first exposure to pups in this particular situation.

Another important implication of this research deals with the site of OT action. In the next chapter, when I review the research on the central neural sites where OT has been shown to act to stimulate the onset of maternal motivation in mammals, I will emphasize OT action on subcortical neuronal circuits. The research of Marlin et al. (2015), however, shows that OT action on sensory neocortical systems can also modulate maternal attentiveness to infant cues without necessarily being essential for maternal motivation, per se.

General Conclusions

In this chapter, I described the general roles of OT in the maternal behavior of nonhuman mammals, and I also discussed the sensory regulation of maternal behavior, with an emphasis on olfaction. Chapter 3 discussed the hormonal mechanisms that stimulate the onset of maternal behavior in nonhuman mammals. Figure 4.1 outlined some of the brain regions that are particularly important for the regulation of maternal behavior. The next chapter deals with a detailed examination of the crucial central neural circuits that regulate maternal behavior and that are affected by hormones, OT, and sensory stimulation.

5
Central Neural Circuits Regulating Maternal Behavior in Nonhuman Mammals

Introduction

The goal of this chapter is to present a detailed analysis of the research on the central neural circuits that regulate the appetitive and consummatory aspects of maternal behavior in nonhuman mammalian species, as well as to describe those neural circuits that suppress maternal responsiveness. Much of this research has been conducted on rodents and has emphasized the crucial role of subcortical circuits in the regulation of maternal behavior. As I will demonstrate in Chapter 8, these subcortical circuits also underpin parental behavior in humans, but in humans these circuits interact with cortical control regions, which shape the complex nature of human parenting.

Research has clearly shown that the medial preoptic area (MPOA) represents a central integrative hub in the control of maternal behavior. I will begin by presenting the research on the MPOA, and, from there, I will describe the larger neural and neurochemical circuitry within which maternally relevant MPOA neurons are embedded and show how this overall circuitry influences various aspect of maternal behavior.

The Essential Role of the Medial Preoptic Area

Disruption of MPOA Function Depresses Maternal Behavior in Rats

The MPOA has been shown to be essential for the display of maternal behavior in all nonhuman mammalian mothers in which its involvement has been examined. Much of this research has been performed on laboratory rats, but significant research has also been conducted on other mammalian species. With respect to anatomy, the preoptic region, including the MPOA, is located in the most rostral part of the hypothalamus (see Figure 4.1), at the border between the

The Parental Brain. Michael Numan, Oxford University Press (2020). © Oxford University Press.
DOI: 10.1093/oso/9780190848675.001.0001

hypothalamus and the telencephalon (Simerly, 1995). Other neuroanatomists have proposed that the preoptic area is a telencephalic structure (Yoshihara et al., 2018). In either event, what is clear is that the preoptic area is a transitional region located at the crossroads between the telencephalon and diencephalon.

In rats, depression of MPOA neural activity disrupts retrieval behavior, nursing behavior, and nest building (Numan & Insel, 2003). A variety of methods have been used to inactivate the MPOA in rats, and these include electrical lesions (Jacobson, Terkel, Gorski, & Sawyer, 1980; Lee, Clancy, & Fleming, 2000; Numan, 1974), excitotoxic amino acid lesions, which destroy MPOA neurons while sparing fibers of passage (Kalinichev, Rosenblatt, & Morrell, 2000; Numan, Corodimas, Numan, Factor, & Piers, 1988), chemical inactivation with drugs that block voltage-gated sodium channels and therefore action potentials (Pereira & Morrell, 2009), and the direct injection of GABA receptor agonists into MPOA (Arrati, Carmona, Dominquez, Beyer, & Rosenblatt, 2006). GABA is an inhibitory neurotransmitter and the injection of GABA receptor agonists is presumed to inhibit MPOA output neurons essential for maternal behavior.

Given that MPOA neurons are essential for the display of maternal behavior in rats, research employing selective knife cuts has analyzed the trajectory taken by the axons of the critical MPOA neurons, and this research indicates that it is the lateral efferent output from the MPOA that needs to remain intact for maternal behavior to occur (Numan, 1974; Numan & Callahan, 1980; Numan & Corodimas, 1985; Numan, McSparren, & Numan, 1990; Terkel, Bridges, & Sawyer, 1979). In particular, knife cuts that sever the lateral MPOA connections, but not those that sever the dorsal, anterior or posterior MPOA connections, selectively and specifically disrupt maternal behavior (Numan & Callahan, 1980).

With respect to these lateral MPOA efferents originating from MPOA neurons, subsequent research has emphasized the importance of the dorsolateral MPOA connections over the ventrolateral MPOA connections, since the former, but not the latter, disrupt maternal behavior in postpartum rats (Numan et al., 1990; Terkel et al., 1979). The locations of the effective dorsolateral MPOA knife cuts and the ineffective ventrolateral MPOA knife cuts are shown in Figure 5.1. In examining this figure, note that the dorsolateral MPOA cuts would not only interfere with the dorsolateral connections of the MPOA, but would also interfere with some of the connections of the ventral part of the bed nucleus of the stria terminalis (vBST). Numan and Numan (1996) have suggested that neurons in the MPOA and adjoining vBST may form a common functional system important for maternal behavior in rats.

In examining the research findings of these various studies, it is important to point out that inactivation of the MPOA or its lateral projections during pregnancy disrupts the onset of maternal behavior at parturition, and when MPOA

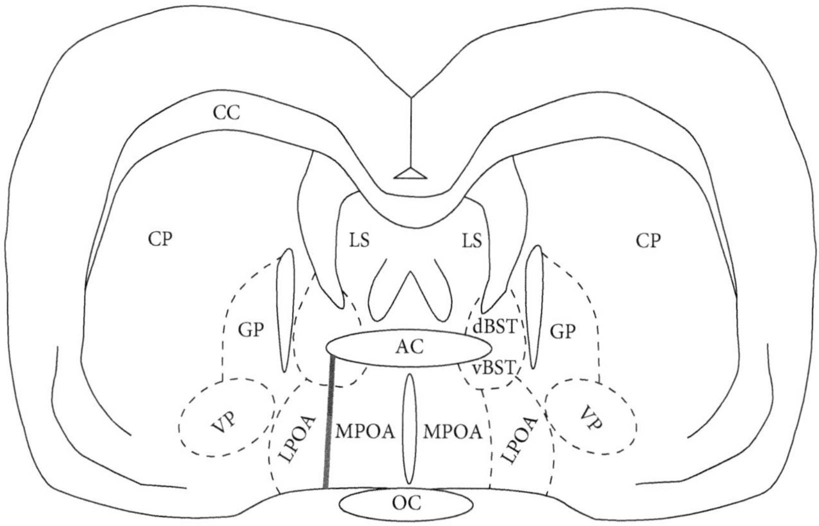

Figure 5.1. Bilateral knife cuts severing the lateral connections of the medial preoptic area (MPOA) disrupt maternal behavior in rats. Numan, McSparren, and Numan (1990) found that severing the dorsolateral connections of the MPOA and adjoining parts of the ventral bed nucleus of the stria terminalis (vBST; shown in red) disrupted maternal behavior, while severing the ventrolateral connections of the MPOA (shown in green) did not. For clarity, the knife cuts are shown on only one side of the brain, but bilateral knife cuts were actually performed on all animals. AC = anterior commissure; CC = corpus callosum; CP = caudate-putamen; dBST = dorsal bed nucleus of the stria terminalis; GP = globus pallidus; LPOA = lateral preoptic area; LS = lateral septum; OC = optic chiasm; VP = ventral pallidum.

activity is depressed during the postpartum period, after normal maternal behavior has become established, the subsequent maintenance of maternal behavior is also disrupted. MPOA lesions also disrupt sensitized maternal behavior in virgin rats. Therefore, in contrast to the paraventricular nucleus (PVN), MPOA activity is essential for both the onset and maintenance of maternal behavior in rats (Numan & Insel, 2003).

MPOA damage must be bilateral to severely disrupt maternal behavior in rats, since unilateral inactivation (with either unilateral knife cuts that sever the lateral MPOA connections or with unilateral MPOA excitotoxic amino acid lesions) is either ineffective or causes only mild and short-lasting deficits (Stack, Balakrishnan, Numan, & Numan, 2002). As I will show, this fact has been extremely important for research that has explored the larger neural circuitry within which the MPOA operates to regulate maternal responsiveness.

In Chapter 4, I divided maternal behavior into appetitive and consummatory components, with retrieving behavior representing the main appetitive (reward-seeking) component and nursing behavior representing the major consummatory component. Since inactivation of the MPOA disrupts both retrieving and nursing behavior, its neurons regulate both the appetitive and consummatory aspects of maternal behavior. As I will show, different populations of MPOA neurons, with different projections to distinct parts of the brain, appear to regulate these two aspects of maternal responsiveness. It should be noted that while retrieving behavior is totally abolished after MPOA damage, some, but not all, studies have observed some nursing behavior in females with MPOA damage, although the duration of such behavior is much lower than that observed in control females (cf. Numan & Callahan, 1980; Numan et al., 1988).

With respect to the abolition of retrieval behavior after MPOA inactivation, a question is whether this disruption simply represents an oral motor deficit. Research has clearly shown that this is not the case: While postpartum rats with MPOA inactivation no longer retrieve their pups, they will hoard or carry pieces of candy that approximate the weight and size of pups (Numan, 1990; Numan & Corodimas, 1985). Therefore, the disruption of retrieval behavior after MPOA damage can be described as a disruption of the appetitive aspects of maternal behavior, where pup stimuli are no longer recognized as rewarding stimuli and no longer elicit goal-directed appetitive maternal responses.

Additional research, using either operant learning procedures or the conditioned place preference (CPP) paradigm, add strong support for the involvement of MPOA neurons in the appetitive aspects of maternal motivation. Normal postpartum rats will learn to perform an operant bar press response to obtain pups as a reward, but rats with MPOA lesions will not learn and perform this operant response for pup reward but they will learn the operant response for a food reward (Lee et al., 2000). Therefore, rats with MPOA lesions do not show a generalized reduction in reward seeking behavior; the disruption is specific to the rewarding aspects of interacting with pups.

In the CPP paradigm, during the training stage of this procedure, postpartum rats learn to associate a particular chamber of a cage with pups. During the test phase, when pups are not present in the two-chambered cage, such rats will spend most of their time in the chamber that previously contained pups. It is as if the female is searching for her pups because they had served as a rewarding stimulus and the compartment in which they had previously been placed acquired the properties of a conditioned rewarding stimulus. Significantly, if MPOA activity is depressed with bupivacaine (a sodium channel blocker that prevents action potentials) during the test phase then the expression of this CPP is eliminated and the affected postpartum rats do not show a preference for the compartment that had previously been paired with pups (Pereira & Morrell, 2010).

Importantly, intra-MPOA injections of bupivacaine did not block the expression of a CPP when another rewarding stimulus was used during the training phase. Again, these results suggest that MPOA inactivation specifically blocks reward-seeking behavior when pups or pup-related stimuli are the rewarding stimulus, while other reward-related stimuli still remain attractive and activate nonmaternal appetitive responses.

Maternal Behavior in Rats is Associated With the Expression of Fos Proteins in MPOA/vBST

An important correlational method that has been used to analyze the involvement of the MPOA and other neural sites in the regulation of maternal behavior in rats and other species is the detection of *fos* gene activation within neurons, which can be measured by the production of various Fos proteins within those neurons during the display of maternal behavior. The *fos* family of genes, which are members of a class of genes referred to as immediate early genes, includes the *cfos* and *fos B* genes, and their protein products are referred to as cFos and Fos B (Morgan & Curran, 1991; Sheng & Greenberg, 1990). These proteins can be detected in neurons through the use of immunohistochemical methods. The detection of Fos proteins within neurons during the display of particular behaviors, such as maternal behavior, is used by behavioral neuroscientists to locate neurons that are active during those behaviors since the expression of Fos proteins is positively related to neuronal depolarization. More specifically, extracellular signals that act on neurons during the display of particular behaviors, such as neurotransmitters, neuromodulators, and hormones, produce intracellular signals within the affected neurons. Such intracellular signals function as transcription factors, which, in turn, activate *fos* genes, resulting in the production of the various Fos proteins. Interestingly, Fos proteins also serve as transcription factors to activate other genes, referred to as late-responding structural genes, that produce proteins that alter the structure and function of the Fos-expressing neurons.

With this brief analysis, one can appreciate that *fos* gene activation can be used as a neuroanatomical mapping procedure to locate those neurons within the brain that become active after an animal is exposed to certain external stimuli that activate specific behaviors. Since Fos proteins are nuclear proteins, their detection provides excellent anatomical resolution of individual cells. Fos production within neurons can also be used as a molecular neurobiological tool to determine the particular late responding structural genes that are activated by various Fos proteins. However, most research has used Fos protein expression as a marker to detect neuronal activation during maternal behavior; future research that explores those late responding genes that are activated by Fos proteins will ultimately provide information with respect to the molecular–genetic control of maternal behavior.

Many studies have shown the cFos and Fos B expression increases in the MPOA and the adjoining vBST during maternal behavior, suggesting that the MPOA/vBST region is activated during maternal behavior (Numan & Insel, 2003). I will describe some studies from my laboratory that have analyzed this issue. In our first study, we used the virgin sensitization paradigm (Numan & Numan, 1994). Virgin female rats were exposed continuously to foster pups, and some of the females became maternal (sensitized maternal behavior) during the test period, while others did not. Therefore, two groups of females were formed, maternal and nonmaternal, but the rats in each group were exposed to pups for the same number of days. The MPOA and vBST of the maternal females contained many more cFos-labeled cells than did the MPOA/vBST of the nonresponding females. This study is important because it shows that the retrieval of pups, along with adopting a nursing posture over them, is needed to detect high levels of MPOA-activated cells and that the mere exposure to pups does not have this effect.

Other research from my laboratory has studied Fos activation in the preoptic region of postpartum lactating rats. Stack and Numan (2000) separated primiparous lactating rats from their pups on day 5 postpartum. Forty-eight hours later, pups were returned to different groups of females for varying amounts of time (2–47 hours), and all females showed normal maternal behavior. Control females were not reunited with their pups. We found that the number of MPOA and vBST neurons that expressed cFos and Fos B in the females that were reunited with pups remained significantly above control levels through 47 hours of pup exposure. Some of these results are shown in Figure 5.2. Because MPOA/vBST neurons are essential for maternal behavior, it can be proposed that cFos and/or Fos B expression within these regions may be necessary to maintain their normal functional activity. More specifically, Fos activation of late-responding structural genes may result in the synthesis of proteins that maintain the structural and functional integrity of MPOA and vBST neurons that are essential for the maintenance of maternal behavior in postpartum rats. I would predict that if one were to selectively suppress the synthesis of Fos proteins in the MPOA region of such rats, then maternal behavior would be suppressed. Such an experiment has yet to be performed.

Stack, Balakrishnan, Numan, and Numan (2002) examined whether a knife cut that severs the lateral connections of MPOA/vBST would impact Fos expression in this region. We took advantage of the knowledge that unilateral knife cuts that sever the lateral connections of the MPOA/vBST do not disrupt maternal behavior, while bilateral cuts are disruptive. When postpartum rats received only

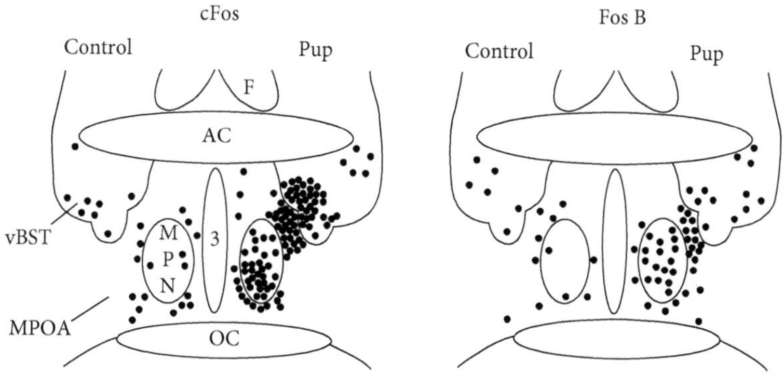

Figure 5.2. Frontal sections at the level of the medial preoptic area (MPOA) that depict the expression of cFos and Fos B in the MPOA and ventral bed nucleus of the stria terminalis (vBST) of postpartum female rats that showed maternal behavior toward pups (pup; shown on the right side of each frontal section) and control females that were not exposed to pups (control; shown on the left side of each frontal section). Each dot represents five cells that expressed either cFos or Fos B. AC = anterior commissure; F = fornix; MPN = medial preoptic nucleus within MPOA; OC = optic chiasm; 3 = third ventricle.

Source: Modified from Figure 4 in Stack and Numan (2002) with permission from the American Psychological Association.

unilateral cuts, they displayed normal maternal behavior. Significantly, when the brains of these females were subsequently immunohistochemically processed, it was found that the number of cFos and Fos B expressing neurons was significantly reduced on the side of the brain that contained the knife cut, while normal high levels of Fos proteins were expressed in the contralateral MPOA and vBST (the side of the brain without the knife cut). These results provide strong support for the view that Fos proteins may serve as a marker for those active MPOA and vBST neurons that contribute to brain circuits that regulate maternal behavior. Some of these findings are depicted in Figure 5.3.

In an important study, Mattson and Morrell (2005) showed that cFos expression in the MPOA is associated with the appetitive aspects of maternal behavior in postpartum rats. Using the CPP procedure, they found that during the test phase (when pups were not present), females that showed a preference for the chamber that previously contained pups demonstrated increased cFos expression in MPOA. These results suggest that certain MPOA neurons become active when postpartum female rats show reward-seeking responses that are aimed at achieving proximity to pups.

A line of reasoning that I am trying to develop is as follows: During the postpartum period, when maternal females interact with pups, Fos expression

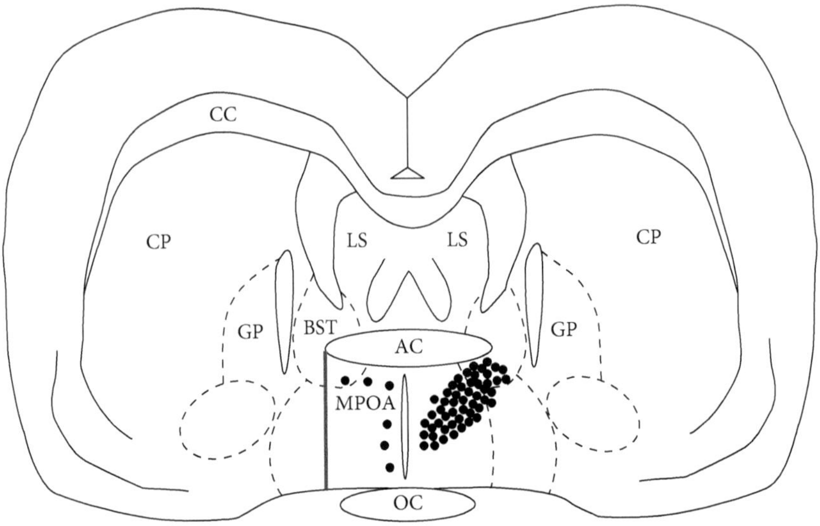

Figure 5.3. A frontal section through the rat brain schematically representing the effects of a unilateral knife cut severing the lateral connections of the medial preoptic area (MPOA) and adjoining ventral part of the bed nucleus of the stria terminalis (BST) on the expression of cFos within cells of the ipsilateral and contralateral MPOA and ventral BST of postpartum maternal rats. The knife cut (shown in red) substantially decreases cFos expression in MPOA and ventral BST on the same side of the brain (ipsilateral side) as the knife cut. The dots represent the general pattern of cFos expression. AC =anterior commissure; CC = corpus callosum; CP = caudate-putamen; GP = globus pallidus; LS = lateral septum; OC = optic chiasm.

Source: Based on the research findings reported in Stack, Balakrishnan, Numan, and Numan (2002).

increases in the MPOA and vBST and such expression, through the activation of late-responding genes and their associated proteins, maintains the functional integrity of MPOA/vBST neurons, which allows for the continuance of the appetitive and consummatory aspects of maternal behavior. When postpartum females are separated from their pups for 48 hours, Fos levels fall to basal levels. When such females are reunited with pups, they show maternal behavior immediately. Presumably this occurs because the MPOA/vBST had previously been modified by prior pup exposure. However, after the reemergence of maternal behavior, I would propose the maternal behavior-induced Fos expression is necessary for the behavior to continue. Here is one possibility: Maternal behavior-induced Fos expression results in Fos protein activation of the synthesis of a protein (or proteins), through its effects on late-responding genes, which is necessary for the proper function of MPOA neurons. Further, this protein is "used up" during the performance of maternal behavior, so that continued Fos expression during the display of maternal behavior is necessary for maternal behavior to continue. If Fos influences the structural and functional integrity of MPOA "maternal" neurons, then an interesting question arises. Why does the primiparous parturient female show immediate maternal behavior once her pups are born? Clearly, prior maternal behavior-induced Fos activation would not occur under this situation. One possibility is that the hormonal changes of late pregnancy activate Fos expression in MPOA neurons so that these neurons are properly prepared at the time of parturition. There is actually good evidence to support this view. Sheehan and Numan (2002) treated female rats with hormonal regimens that are known to be either effective or ineffective in inducing short-latency maternal behavior in female rats. Importantly, these females were not exposed to pups. Primigravid female rats were hysterectomized (H) and ovariectomized (O) during midpregnancy. One group of these females was systemically injected with estradiol (E) and the other group received control injections of oil. Forty-eight hours later, their brains were immunohistochemically processed to detect cFos-labeled neurons in the MPOA. The HO females that received E had about 400 Fos-expressing cells in the part of the MPOA that was examined, while a significantly lower number (80) of Fos-expressing MPOA neurons was detected in the oil-treated females. Subsequent behavioral experiments, on additional groups of females, showed that immediate maternal behavior occurred in the E-treated, but not the oil-treated, females. One conclusion that can be reached is that declining progesterone and rising E trigger Fos expression in the MPOA, and such expression prepares the MPOA for the onset of maternal responsiveness. The importance of declining progesterone (from high pregnancy levels) coupled with rising E is emphasized by the finding that when E was administered to virgin HO rats, cFos expression in MPOA was significantly lower than that which is observed in pregnant HO females that were treated with E (Sheehan & Numan, 2002).

The MPOA Is a Site Where Hormones and Oxytocin Act to Trigger the Onset of Maternal Behavior in Rats

Not only does inactivation of MPOA neurons disrupt maternal behavior in rats, but hormones and oxytocin (OT) also act on the MPOA to stimulate the onset of maternal behavior. With respect to E, several studies have shown that local application of this steroid to the MPOA/vBST region stimulates the onset of maternal behavior (Fahrbach & Pfaff, 1986; Matthews Felton, Linton, Rosenblatt, & Morrell, 1999a; Numan, Rosenblatt, & Komisaruk, 1977). I will describe the original study performed by Numan et al. (1977), which employed the pregnancy termination model. Crystalline E was implanted into the MPOA/vBST region immediately after hysterectomizing and ovariectomizing primigravid female rats on day 15 of pregnancy. The implants were left in place for the remainder of the study, and pups were presented to these females 48 hours after the HO procedure and E application to the MPOA. Eighty percent of these females showed full maternal behavior on their first day of exposure to pups, with the remainder of the females showing full maternal behavior by the next day. Control females that received implants of cholesterol into MPOA or implants of E into other hypothalamic regions did not show prompt maternal responsiveness, but instead showed sensitization latencies to the onset of maternal behavior of about 3 days, which is the standard latency shown by 15HO females that are not treated with E. This latency, of course, was significantly longer than that shown by the females that received E application to MPOA.

These results do not necessarily mean that the MPOA/vBST region is the only site where E acts to stimulate the onset of maternal behavior. On day 15 of pregnancy, endogenous plasma E has already risen to moderately high levels (Bridges, 1984), and E action at more than one brain site may have primed the brain to be responsive to the continued action of E on MPOA neurons. In this regard, it should be noted that Fahrbach and Pfaff (1986) found that E implants into the MPOA facilitated maternal behavior in ovariectomized virgin female rats, suggesting that the singular action of E on the MPOA is sufficient to stimulate maternal behavior in rats. Their results, however, should be interpreted with caution. First, they did not implant E into other brain regions, which would be necessary to show that E action of the MPOA was unique. Second, they used the Zivic–Miller strain, which most likely displayed olfactory deficits that may have substituted for the action of E at other neural sites.

Given that E acts in the MPOA/vBST to stimulate the onset of maternal behavior, what cellular effects of E might mediate this facilitation? There are two major types of estrogen receptors (ERs) to which E binds to exert its cellular effects, ER-alpha and ER-beta, and both of these receptors are expressed in MPOA neurons (Shughrue, Lane, & Merchenthaler, 1997). The classic model of E action is that E binds to ERs located in the nucleus of a cell, and then the E–ER complex binds to an estrogen response element located within the promoter region of specific genes, where the E–ER complex exerts transcriptional effects (Cornil, Ball, & Balthazart, 2015; Wilkenfeld, Lin, & Frigo, 2018). This mechanism is referred to as a genomic mechanism of action, since E would be activating particular genes with the result that the synthesis of particular proteins within the cell would be induced by E. Such E-induced protein synthesis would then alter the structure and function of the affected cell.

More recently, ERs have also been localized within the cell membrane of neurons where they could then stimulate intracellular effects by affecting biochemical cascades that ultimately activate cytoplasmic protein kinases (Cornil et al., 2015; Wilkenfeld et al., 2018). Although both of these types of E effects may occur within MPOA neurons to affect maternal responsiveness (Numan & Stolzenberg, 2009; Stolzenberg et al., 2009), at this juncture I would like to focus on the potential genomic effects of E within MPOA neurons in its stimulation of maternal behavior.

First, there is good evidence that E binding to ER-alpha stimulates the transcription of the OT receptor (OTR) gene, with a resultant increase in OTR protein expression within the cell membrane of MPOA neurons, as well as within neurons in other brain regions (Champagne, Diorio, Sharma, & Meaney, 2001; Champagne, Weaver, Diorio, Sharma, & Meaney, 2003; Kremarik et al., 1995; Young, Wang, Donaldson, & Rissman, 1998). Importantly, OTR messenger ribonucleic acid (mRNA) increases within the MPOA near the time of parturition in rats (Meddle et al., 2007). Therefore, one aspect of E stimulation of maternal behavior is that it prepares MPOA neurons to be responsive to OT input from the PVN near the time of parturition. By increasing OTR levels in MPOA neurons, this genomic effect of E presumably allows OT to stimulate those MPOA neural circuits necessary for the onset of maternal behavior.

It is unlikely that the genomic effect of E on OTR synthesis is the only mechanism through which E acts to stimulate maternal behavior. Many genes contain estrogen response elements in their promoter regions, and E action on the MPOA is likely to influence the synthesis of many proteins that affect how MPOA neurons operate. One prominent example is the effect of E on the expression of Fos proteins. As I have already reviewed, E, particularly on a background of progesterone withdrawal, stimulates increased Fos expression in MPOA (Sheehan & Numan, 2002), and evidence indicates that E induces Fos expression via a genomic mechanism of action (Gruber, Gruber, Gruber, Wieser, & Huber, 2004; Hyder, Chiappetta, & Stancel, 1998). Since Fos proteins themselves serve as transcription factors that activate late-responding genes, which produce additional proteins, it appears likely that E gives rise to multiple genomic effects, resulting in the production of a variety of proteins, in addition to OTRs, that alter the phenotype of MPOA neurons. These effects then allow MPOA neurons to become functionally active and responsive to pup stimuli. A simplified summary of this proposal is shown in Figure 5.4.

In an interesting study, Lonstein, Greco, De Vries, Stern, and Blaustein (2000) found in postpartum rats during the maintenance phase of maternal behavior that about 40% of MPOA neurons that express cFos also contain ER-alpha. The authors raise the possibility that the display of established maternal behavior during the postpartum period in rats may be influenced by ER-alpha in the MPOA and other brain regions. This proposal seems unlikely because ovariectomized postpartum rats continue to display maternal behavior (see Chapter 3 of

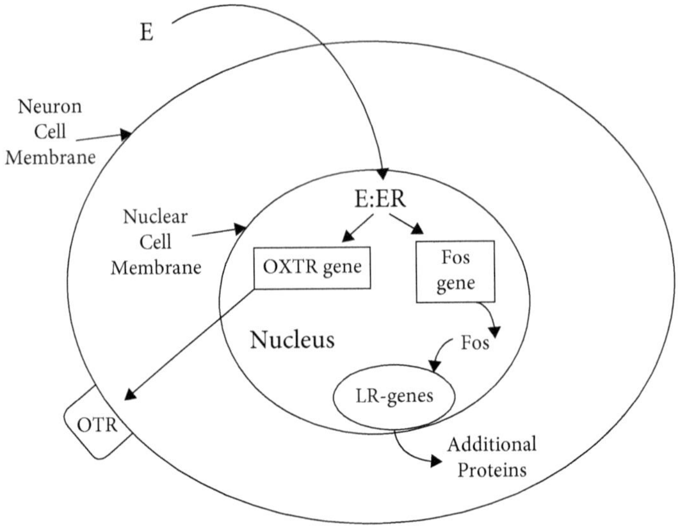

Figure 5.4. A simplified view of the effects of estradiol on medial preoptic area (MPOA) neurons through a genomic mechanism of action. Estradiol (E) passes through the cell and nuclear membranes and binds to intracellular estrogen receptors (ER) in the nucleus, forming the E–ER complex. The E–ER complex then affects the transcription of genes that contain estrogen response elements. The figure shows two of the genomic effects that estradiol exerts. Activation and transcription of the oxytocin receptor gene (OXTR gene) results in the synthesis of oxytocin receptors (OTR) that are incorporated into the cell membrane of the neuron. This effect would increase the responsiveness of MPOA neurons to oxytocin. Fos gene (an immediate early gene) transcription can also be activated by the E–ER complex, resulting in the increased synthesis of Fos proteins within the nucleus of the cell. Fos proteins, themselves, act as transcription factors to activate the transcription of a variety of late-responding (LR) genes. These LR genes then produce additional proteins that are capable of altering the structure and function of MPOA neurons. See text for details.

this volume). However, I do not want to absolutely exclude their proposal since even without the ovaries, E can be synthesized in the brain (neuroestrogens; Cornil et al., 2015). In addition, ligands other than E may be able to bind to and activate ERs (Numan & Stolzenberg, 2009; Stolzenberg & Numan, 2011). However, I would like to tentatively suggest that while cFos and ER-alpha co-localization may be labeling a subpopulation of MPOA neurons that are involved in maternal behavior, ER-alpha may only be utilized during the onset of maternal behavior and may not be important for its maintenance. More specifically, a tentative proposal is that the genomic effects of Fos, stimulated by the E–ER complex, along with the effects of the E–ER complex on the expression of OTRs, are important for the onset of maternal behavior, while pup-induced Fos expression within these same neurons, independent of ER-alpha activity, is involved in the maintenance of maternal motivation. However, the idea that ER-alpha may be activated by neuroestrogens, or by neurochemicals other than estrogens, remains an attractive possibility (see Chapter 11 of this volume). To effectively test the hypothesis of Lonstein et al. (2000) that the maintenance of maternal behavior in rats in influenced by ER-alpha in MPOA neurons, one would have to examine the effects on maternal behavior of specifically and selectively suppressing ER-alpha function in postpartum rats. For example, would ER-alpha antagonist injections into the MPOA alter the expression of established maternal behavior?

Other hormones in addition to E are involved in stimulating the onset of maternal behavior at parturition in rats. Although the neural site where progesterone acts to influence maternal behavior has not yet been determined (Numan, 1978; Numan et al., 1999), there is good evidence that the MPOA is a site where lactogens act to stimulate maternal behavior. Bridges, Numan, Ronsheim, Mann, and Lupini (1990) treated steroid-primed ovariectomized virgin female rats with bromocriptine to block E-induced endogenous prolactin release and injected either prolactin or saline into the MPOA. Prolactin injections into MPOA facilitated a short-latency onset of maternal behavior, with such females showing full maternal behavior after 1 to 2 days of pup stimulation, while the saline injected rats responded maternally after 6 days of pup stimulation. A similar facilitation of the onset of maternal behavior was observed when rat placental lactogens were injected into MPOA (Bridges et al., 1996). Finally, Bridges et al. (2001) reported that the injection of a prolactin receptor (PrlR) antagonist (S179D-PRL) into the MPOA disrupted the onset of maternal behavior in ovariectomized virgin rats treated with a steroid hormone regimen that induced short-latency maternal behavior in control females. S179D-PRL would prevent the ability of prolactin to bind to its receptor. The only critique I have of these studies is that prolactin, placental lactogens, and S179D-PRL were only injected into MPOA and not into nearby control sites. Therefore, the site-specificity of the results cannot be absolutely assured.

PrlR is a cell membrane bound receptor, and two major forms exist, a long form (PrlR-L) and a short form (PrlR-S). These two isoforms, produced from the same PRLR gene, do not differ in their extracellular prolactin binding domains, but they do differ in the length and composition of the amino acids in their intracellular domains (Bole-Feysot, Goffin, Edery, Binart, & Kelly, 1998). Importantly, both receptor types are expressed in MPOA (Bakowska & Morrell, 1997; 2003; Pi & Grattan, 1998a, 1998b). Prolactin seems to exert its cellular effects via two different mechanisms, a long-acting genomic mechanism and a short-acting cytoplasmic mechanism, with the former being regulated by PrlR-L and the latter by PrlR-S. The most studied cell signaling pathway through which prolactin produces its effects is the Janus kinase/signal transducer and activator of transcription (JAK/STAT) pathway (Bole-Feysot et al., 1998; Brown, Kokay, Herbison, & Grattan, 2010; Devi & Halperin, 2014). When prolactin binds to PrlR-L, but not to PrlR-S, the STAT5 protein is ultimately phosphorylated (pSTAT5). pSTAT5, which is a cytoplasmic protein, then becomes activated and translocates to the cell's nucleus to alter the transcription of target genes. In contrast, there is evidence that when prolactin binds to PrlR-S, it can stimulate intracellular signaling cascades that involve the activation of protein kinase C (PKC). Within neurons, it has been shown that such activation of PKC can result in the phosphorylation of ion channels, which then results in excitatory postsynaptic currents (depolarizations; Belugin et al., 2013; also see Brown, Piet, Herbison, & Grattan, 2012).

With this background, it is interesting to speculate that prolactin action on MPOA neurons involved in regulating the onset of maternal behavior in rats occurs by a dual track mechanism. During late pregnancy, prolactin action of PrlR-L may exert transcriptional effects on target genes, which remain to be identified, resulting in the production of specific proteins that alter the structure and function of these maternally relevant neurons. Therefore, in addition to the genomic effects of the E–ER complex and Fos proteins, prolactin and lactogens may also exert genomic effects that alter MPOA function. Second, as a result of the acute rise in prolactin near the time of birth, the neural activity and output of maternally relevant MPOA neurons may be potentiated via prolactin action on PrlR-S.

Given that E, prolactin, and Fos proteins appear to exert genomic effects that influence the structure and function of MPOA neurons, what kinds of proteins may be produced by these transcriptional effects? In addition to effects on the expression of OTRs and Fos proteins, E, prolactin, and Fos proteins could act on the genetic machinery of MPOA neurons to (a) affect the synthesis of neurotransmitter receptors (other than the OTR) so that MPOA neurons become more responsive to certain inputs, particularly with respect to pup stimuli; (b) increase the synthesis of neurotransmitters or neuromodulators in MPOA neurons so that the output of MPOA neurons can appropriately influence the neurons to which the MPOA projects; and (c) influence the length of dendrites and dendritic spines emanating from MPOA neurons so that they become more receptive to particular inputs (see Keyser-Marcus et al., 2001). Future research will be required to determine the specific cause–effect relationships between the actions of E, prolactin/lactogens, and Fos proteins and the induced changes in MPOA structure and function that are necessary for the onset and maintenance of maternal behavior.

Due to the fact that E increases the expression of OTRs within MPOA, is there evidence that OT actually acts on OTRs in MPOA to stimulate the onset of maternal behavior in parturient primiparous rats? This question has been answered in the affirmative by Pedersen, Caldwell, Walker, Ayers, and Mason (1994). An OTR antagonist (OTA) was injected bilaterally into the mother's MPOA after the birth of the first pup. Each pup was subsequently removed from the mothers after they were born. Approximately 1 hour after the birth of all pups, each female was presented with eight pups, and their maternal behavior was examined over a 30-minute period. OTA, in comparison to control saline injections, severely disrupted to onset of retrieving and nursing behavior. None of the nine rats that received OTA in MPOA retrieved pups and only three of nine were observed to crouch over pups. In contrast, all eight of the females that received saline injections into MPOA retrieved and nursed their pups. The authors also noted that misplaced injections OTA outside the MPOA did not disrupt maternal behavior.

These results should not be taken to mean that OT action on the MPOA is sufficient to stimulate the onset of maternal behavior in parturient rats. Indeed, I will show that OT acts at multiple sites to stimulate maternal behavior at parturition. These results do indicate that OT action on MPOA is necessary for the onset of maternal behavior. As far as I am aware, no reliable study exists that shows that OT action at a single neural site can stimulate maternal behavior in rats, except for the study by Yu, Kaba, Okutani, Takahashi, and Higuchi (1996), whose findings suggested that OT injection into the olfactory bulbs stimulates maternal behavior. But recall that I offered a reinterpretation of their results in Chapter 4, an interpretation in line with the proposal that OT acts at multiple sites to stimulate the onset of maternal behavior.

The MPOA and Maternal Behavior in Sheep, Rabbits, and Mice

Perrin, Meurisse, and Levy (2007) have provided evidence that MPOA neural activity is necessary for the onset and maintenance of both the appetitive and consummatory aspects of maternal behavior in sheep. They injected lidocaine into the MPOA of parturient and postpartum ewes. Lidocaine blocks voltage-gated NA+ channels and would therefore block action potential production in MPOA neurons. In comparison to control ewes, the lidocaine-injected mothers completely lost interest in their lambs. These ewes did not nurse their lambs and during the separation–reunion test (see the introduction to Chapter 4 of this volume), they did not approach and spend time with their lambs. Further, Da

Costa, Broad, and Kendrick (1997) have reported that cFos expression increases in the MPOA of sheep during maternal behavior.

Recall that Kendrick et al. (1987) reported that intracerebroventricular administration of OT to E-primed sheep stimulated maternal behavior. In a subsequent study, Kendrick, Keverne, Hinton, and Goode (1992) examined whether direct bilateral infusions of OT into the MPOA of E-treated ovariectomized ewes could stimulate maternal behavior. In comparison to control infusions of Ringer solution into MPOA, OT infusions were found to decrease rejection responses toward the lamb, but such infusions did not increase acceptance behavior. The OT-treated ewes displayed decreases in withdrawal from the lamb, aggressive head butts, and protest vocalizations (high pitch bleats), but did not allow the lamb to nurse. The tests for maternal responsiveness lasted only 15 minutes, which makes one wonder if full maternal behavior would have been displayed if the OT-treated ewes were allowed a longer exposure to lambs. Kendrick et al. (1992), however, concluded that OT action on MPOA only stimulates partial maternal responsiveness in sheep (decreases in avoidance), but that OT probably acts at multiple neural sites to stimulate full maternal behavior in steroid-primed ewes. This conclusion has received support from a study by Da Costa, Guevara-Guzman, Ohkura, Goode, and Kendrick (1996). They found that the chronic infusion of OT bilaterally into the PVN of sheep that were primed with both progesterone and increasing doses of E was able to induce full maternal behavior during a 15-minute test in six of eight ewes, while none of the eight control females showed maternal behavior. Importantly, OT infusions into PVN also induced the release of OT into blood plasma. They suggested that OT action on the PVN exerted a positive feedback effect on PVN OT neurons, resulting in the coordinated release of OT into multiple brain sites as well as into the periphery. These results fit with the view that OT acts at multiple neural sites to stimulate maternal behavior in sheep. To solidify this interpretation, it would have been interesting to examine the effects of chronic infusions of OT into the MPOA of sheep that were primed with both progesterone and E.

Finally, the effects on maternal behavior of OTR antagonist injection into the MPOA of parturient ewes, or steroid-primed ewes that receive vaginocervical stimulation, have not been examined, perhaps because an effective antagonist for sheep has not been developed (Kendrick, 2000). Based on the findings of Kendrick et al. (1992), if an effective antagonist were to be applied to the MPOA, it is possible that an enhancement of rejection behavior due to an interference with OT action would preclude lamb acceptance.

To my knowledge, the effects of inactivation of MPOA neurons on pup-directed maternal behavior in rabbits has not been examined. However, cFos expression is increased in the MPOA of postpartum maternal rabbits (Aguirre, Meza, & Caba, 2017; Gonzalez-Mariscal, Jimenez, Chirino, & Beyer, 2009).

Moreover, direct application of E to the MPOA has been shown to induce maternal nest building behavior in ovariectomized rabbits that have been systemically treated with progesterone followed by withdrawal of this hormone (Gonzalez-Mariscal, Chirino, Rosenblatt, & Beyer, 2005). Further, electrical MPOA lesions disrupt nest building in pregnant rabbits (Basurto, Hoffman, Lemus, & Gonzalez-Mariscal, 2018).

Finally, there is excellent evidence that the MPOA is essential for maternal behavior in laboratory mice. First, cFos expression increases in the MPOA of virgin and postpartum mice during the display of maternal behavior (Okabe et al., 2013; Renier et al., 2016; Tsuneoka et al., 2013: Wu, Autry, Bergan, Watabe-Uchida, & Dulac, 2014). Further, Wei et al. (2018) have shown that intracellular Ca2+ levels increase dramatically in MPOA neurons of female mice during the retrieval of pups. An increase in intracellular Ca2+ ions, which can be derived from both the extracellular compartment and intracellular storage sites, is used as a marker of neuronal activation (Grienberger & Konnerth, 2012). Finally, electrical or excitotoxic amino acid lesions of the MPOA eliminate maternal behavior in virgin and postpartum female mice (Akther, Fakhrul, & Higashida, 2014; Tsuneoka et al., 2013). The most complete study was performed by Tsuneoka et al., and their injection of *N*-methyl-D-aspartic acid (NMDA; an excitotoxic amino acid) into MPOA was important because this procedure destroys MPOA neurons while sparing axons of passage. When maternal virgins and postpartum females received these lesions, their behavior switched from maternal behavior to infanticide. This effect is different from that which is observed in rats, where NMDA lesions of the MPOA disrupt retrieving and nursing behavior, but do not induce infanticide (Numan et al., 1988). Tsuneoka et al. attempted to localize the part of the MPOA that is necessary for maternal behavior in mice by producing lesions at various MPOA locations, and they identified the central part of the MPOA (cMPOA) as being essential for maternal behavior in mice. They contrasted these results with those in rats that have emphasized the importance of the dorsolateral MPOA for maternal behavior (see Figure 5.1). However, knife cuts severing the lateral connections of the MPOA also abolish cFos expression in parts of the MPOA that include cMPOA (see Figure 5.3). Therefore, it is possible that the critical cMPOA neurons involved in maternal behavior send their axons out of the MPOA to their target regions through a dorsolateral trajectory. In further support of this idea, the NMDA lesions produced by Numan et al. (1988) that disrupted maternal behavior in rats ablated MPOA neurons that included those in cMPOA.

In a complex study, Wei et al. (2018) showed, through the use of advanced transgenic and optogenetic methods, that the selective inhibition of ER-alpha–containing neurons in MPOA disrupted retrieval behavior in virgin and postpartum female mice. Similar results have been reported by Fang, Yamaguchi,

Song, Tritsch, and Lin (2018). These results indicate that ER-alpha containing neurons in the MPOA are important for maternal behavior in mice. Recall that in home-cage tests, the maternal behavior of laboratory mice can occur promptly without hormonal stimulation, but that the hormonal events of late pregnancy do function to enhance maternal motivation as measured with operant procedures (Hauser & Gandelman, 1985) or with T-maze retrieval tests (Numan & Insel, 2003; Stolzenberg & Rissman, 2011). It is interesting to speculate that a basal level of maternal responsiveness governed by the output of MPOA neurons occurs in virgin females, but that E action on ER-alpha within these MPOA neurons functions to boost maternal motivation by further potentiating the output of these neurons to particular target sites. In support, Fang et al. (2018) reported that selective optogenetic stimulation of ER-alpha-containing MPOA neurons enhances retrieval behavior in virgin and postpartum mice, as measured by retrieval latencies. One conclusion from these results is that even in the absence of E, direct stimulation of these neurons can boost maternal motivation and enhance retrieval behavior.

In line with the view that prolactin is not essential for home-cage maternal behavior in laboratory mice (see Chapter 3 of this volume), Buonfiglio et al. (2015) have reported that a transgenic mouse line with a neuron-specific deletion of the STAT5 gene (regulated by PrlR-L), which would eliminate STAT5 synthesis, showed normal maternal behavior at parturition and during the postpartum period. However, these findings do not rule out the possible involvement of PrlR-S in mouse maternal behavior. To investigate the role of prolactin action on the MPOA in the maternal behavior of parturient laboratory mice, Brown et al. (2017) injected an adeno-associated virus gene bilaterally into the MPOA of a transgenic mouse line. This adeno-associated virus contained a gene that would specifically delete the PrlR gene from MPOA neurons in these transgenic mice, thus preventing the synthesis of both PrlR-L and PrlR-S. These females were subsequently mated and their maternal behavior was observed postpartum while these females were in their home cages. These animals gave birth to normal sized litters and initiated normal pup-directed behaviors. They cleaned their pups and ate the placentas and retrieved their pups to the nest. However, over the next 24 hours, the mothers appeared to abandon their pups, which subsequently died. The authors suggested that the deletion of PrlRs from MPOA neurons, which prevented prolactin and placental lactogens to act there, abolished effective nursing behavior. In another aspect of this study, however, they found that some PrlR-containing MPOA neurons projected to the PVN. Since normal maternal behavior was initiated at parturition, these finding, taken together, lead me to an alternative interpretation. It is possible that the disruption of connections between PrlR-containing MPOA neurons and PVN OT neurons disrupted the milk-ejection reflex, rather than maternal behavior, and perhaps the pups died because they could not obtain milk. Future research will need to examine these possibilities more carefully (see Augustine et al., 2017; Augustine, Seymour, Campbell, Grattan, & Brown, 2018). It would be interesting to determine whether prolactin/lactogen action on PrlR-S within MPOA is involved in stimulating the milk-ejection reflex.

The potential involvement of Prl-responsive MPOA neuron inputs to PVN OT neurons has broader significance for those mammalian mothers in which the onset of maternal responsiveness is under hormonal control, such as rats. In these cases, such projections may activate OT release into the brain, as well as into the periphery, to influence both maternal behavior and the milk-ejection reflex. For laboratory mice, OT does not appear to be essential for maternal behavior in home cage tests. In fact, OT knockout mice give birth and show normal maternal behavior, but the milk-ejection reflex is blocked and their pups eventually die. This phenotype is very similar to that which Brown et al. (2017) observed after deletion of the PrlR gene from MPOA neurons. (However, under more challenging environmental conditions, outside the home cage, MPOA-induced OT release into the brain may also influence mouse maternal behavior.)

The Larger Neural Circuitry Within Which the MPOA Operates to Influence Maternal Behavior

Neurotransmitters/Neuromodulators in MPOA/vBST

To understand the larger neural circuitry within which the MPOA and vBST operate to influence maternal motivation, it is important to first identify some of the neurochemicals that are contained within MPOA/vBST neurons, the outputs of which may influence maternal behavior. One approach that has been taken to identify such maternally relevant MPOA neurochemicals is to determine those neurochemicals that co-localize with Fos expression during the occurrence of maternal behavior. In laboratory rats and mice, it is clear that the MPOA and vBST contain large numbers of glutamatergic and GABAergic neurons (Geisler, Derst, Veh, & Zahm, 2007; Lonstein & De Vries, 2000; McHenry, Rubinow, & Stuber, 2015; Tsuneoka et al., 2013; Wu et al., 2014). Glutamate and GABA are major amino acid neurotransmitters, with the former being excitatory and the latter being inhibitory. Lonstein and De Vries (2000) were the first to report that about 50% of neurons in MPOA/vBST that express Fos during maternal behavior in postpartum rats contain the inhibitory neurotransmitter GABA. Some of these neurons may be interneurons, while others may be MPOA/vBST projection neurons that inhibit their target regions.

In a careful anatomical study in mice, Tsuneoka et al. (2013) explored the presence of glutamate and GABA within neurons in cMPOA, since NMDA lesions of the central part of MPOA abolish maternal behavior in maternal virgin mice and postpartum mice. They found that about 75% of cMPOA neurons that express Fos during maternal behavior contain GABA. Interestingly, although cMPOA neurons contained a very high number of glutamate neurons, only about 6% of Fos-expressing neurons in cMPOA contained glutamate. This co-localization, however, was significantly higher than that observed in control female mice that were not engaging in maternal behavior. Since not all active neurons express Fos (Morgan & Curran, 1991; Wan, Liang, Moret, Wiensendanger, & Rouillier, 1992), and since the cMPOA is so important for the expression of maternal behavior, it appears likely that glutamatergic projections from cMPOA may also be involved in maternal behavior control.

If GABA and glutamate output from the MPOA/vBST influence maternal behavior, how might these neurotransmitters exert their effects? Figure 5.5 presents a basic model that may be applicable to most female mammals that depend on the physiological events of late pregnancy and parturition, such as rats, rabbits, and sheep, for the prompt occurrence of maternal responsiveness at parturition. Virgin females in most mammalian species with a uniparental maternal care system avoid and reject young, while parturient females in such species show immediate maternal responsiveness. It can be proposed the MPOA output neurons may serve a dual role. Some MPOA projections to target regions may suppress avoidance responses to infant stimuli while other projections may enhance the appetitive and consummatory aspects of maternal behavior. As shown in the Figure 5.5, these MPOA effects may be complex, requiring an understanding of the microcircuitry within the target regions to which MPOA neurons project. The idea behind the represented circuits is that the physiological events of late pregnancy and parturition alter the phenotype of MPOA neurons so that they become responsive to infant stimuli. In response to these stimuli, MPOA output neurons suppress avoidance behavior and promote maternal responsiveness. With respect to suppressing avoidance of infants, the MPOA effect may be direct, and MPOA GABA neurons may synapse on and inhibit the output neurons of what I have labeled as the defensive circuit. Alternatively, MPOA glutamate neurons may activate local inhibitory interneurons within the defensive circuit that then, in turn, depress the output of the defensive system. Similarly, MPOA glutamatergic projections may promote maternal motivation by directly activating the output neurons that regulate the appetitive and consummatory aspects of maternal behavior. Alternatively, MPOA GABAergic projections to the maternal motivational circuitry may inhibit inhibitory interneurons within these circuits, in this way disinhibiting the output neurons that allow for the expression of maternal behavior.

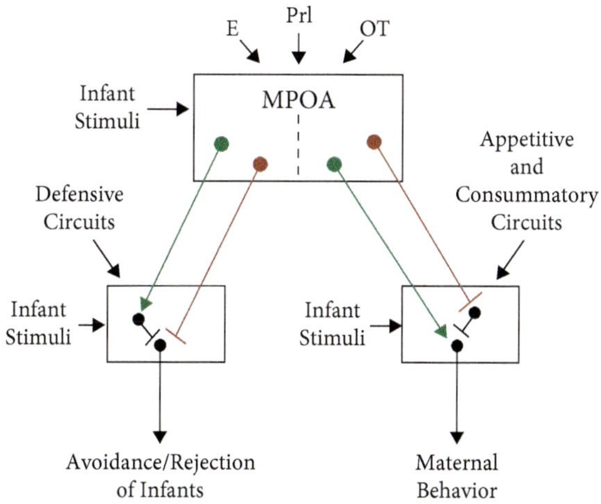

Figure 5.5. A proposed model through which the neural output of medial preoptic area (MPOA) neurons can stimulate the onset of maternal behavior. The functional activity of MPOA neurons is modified by the physiological events of late pregnancy and parturition, which includes the actions of estradiol (E), prolactin (Prl), and oxytocin (OT) on the MPOA. These effects allow MPOA neurons to become responsive to infant stimuli. It is proposed that the activation of MPOA output by infant stimuli has two major effects. MPOA neurons inhibit the responsiveness of defensive neural circuits to infant stimuli, so that infants are not rejected or avoided. MPOA output also promotes the responsiveness of appetitive and consummatory motivation neural circuits to infant stimuli, so that infants are approached and then subsequently cared for. Two major neurotransmitters utilized by MPOA neurons are GABA (shown in red) and glutamate (GLUT; shown in green), with the former being an inhibitory neurotransmitter, while the latter is excitatory. The MPOA can inhibit the output of the defensive neural circuit neurons through direct GABAergic inhibition of the output neurons in this general region. Alternatively, MPOA-GLUT neurons may indirectly suppress the output of defensive circuit neurons by activating inhibitory interneurons that then suppress the activity of the output neurons. Similarly, MPOA-GLUT neurons may directly stimulate the output of neurons in appetitive/consummatory motivation neural circuits. Alternatively, MPOA-GABA neurons may suppress the activity of inhibitory interneurons within these circuits, in this way indirectly increasing the responsive of the output neurons within appetitive/consummatory neural circuits to infant stimuli. Neurons ending in an arrow signify excitation, while those ending in a bar signify inhibition.
Source: Modified from Figure 5 in Numan and Stolzenberg (2009) with permission from Elsevier.

In addition to amino acid neurotransmitters, the MPOA region also contains several neuropeptides that may be involved in maternal behavior control. The MPOA contains many neurotensin-containing neurons in rats and mice (Geisler & Zahm, 2006; McHenry et al., 2017; Tsuneoka et al., 2013), and Tsuneoka et al. (2013) have reported that 50% of Fos activated neurons in cMPOA of maternal mice contain neurotensin. I am only aware of one study that has examined the role of neurotensin in pup-directed maternal behavior. In postpartum mice, Gammie, D'Anna, Gerstein, and Stevenson (2009) injected a neurotensin receptor (NTR) antagonist into the lateral ventricle, and this treatment did not affect retrieval behavior. There are several types of NTRs, the most prominent ones being type 1 (NTR1) and type 2 (NTR2). The antagonist that Gammie et al. used only blocked NTR1. If MPOA neurotensin neurons are involved in maternal behavior, it is possible that they could be exerting their effects on NTR2. Alternatively, NT action on both NTR1 and NTR2 may have to be blocked to disrupt maternal behavior.

The very anterior part of the PVN lies just dorsal to the posterior part of the MPOA (see Figure 4.1), and the most anterior part of the PVN has been referred to as the anterior commissural nucleus (ACN; Yoshihara et al., 2018). Although controversial, Yoshihara et al. have suggested that the ACN may actually be a subnucleus within the mouse MPOA, which is the reason I am discussing it in this section. (The controversy revolves around the issue of whether the ACN is part of the MPOA or part of the PVN.) The ACN contains many OT neurons (Otero-Garcia, Agustin-Pavon, Lanuza, & Martinez-Garcia, 2016; Tsuneoka et al., 2013), and many neurons in ACN also express Fos in maternally behaving virgin mice (Tsuneoka et al., 2013). However, at least in mice, Fos is not colocalized with OT in ACN during maternal behavior in virgins. Nevertheless, could these ACN OT neurons be important for maternal behavior? Recall that Insel and Harbaugh (1989) found that electrical lesions that involved the anterior PVN disrupted the onset of maternal behavior in rats. Tsuneoka et al. lesioned the ACN with NMDA and these lesions destroyed Fos-expressing ACN neurons but did not disrupt maternal behavior. In contrast, OT neurons in the ACN were resistant to the excitotoxic effects of NMDA. Two important points resulting from these findings are worth considering. First, just because neurons express Fos during maternal behavior does not mean they are involved in maternal behavior, since destruction of ACN neurons that express Fos during maternal behavior had no effect on the behavior. Second, the role of ACN OT neurons in the regulation of maternal behavior remains underdetermined. Although selective destruction of these neurons would probably not interfere with the maternal behavior of laboratory mice in home cage tests (but might have an effect under more challenging test conditions), such selective lesions might interfere with the onset of maternal behavior in rats, rabbits, and sheep. Perhaps the release of OT at parturition from

ACN could influence the onset of maternal behavior by acting on OTRs located on other neurons within MPOA.

A neuropeptide located in MPOA neurons that has received a lot of recent attention with respect to its role in regulating maternal behavior is galanin. Galanin neurons are located in cMPOA of both rats and mice (Cservenak, Kis et al., 2017; Tsuneoka et al., 2013; Wu et al., 2014). In mice, approximately 50% of MPOA neurons that express Fos during maternal behavior also contain galanin, and most galanin neurons also contain GABA (Tsuneoka et al., 2013; Wu et al., 2014). Since galanin can exert inhibitory postsynaptic effects via an action on certain galanin receptors (Constantin & Wray, 2016), it is possible that galanin reinforces the inhibitory postsynaptic effects of GABA. Most important, selective ablation of galanin neurons in the MPOA of laboratory mice disrupts maternal behavior in both virgin females and in postpartum mothers (Wu et al., 2014). The virgin females with such selective elimination of MPOA galanin neurons tended to display infanticide, while postpartum females with selective elimination of MPOA galanin neurons simply ignored their young. Control females in both groups showed maternal responses toward pups. Uncovering the neural inputs to MPOA galanin neurons, as well as the regions to which these neurons project, will shed light on the larger neural circuitry underlying maternal responsiveness (Kohl et al., 2018).

Amylin is a peptide that may also be involved in maternal behavior. Typically, amylin is released by the pancreas in concert with insulin and it has been shown to exert a suppressive effect on food intake (a satiety factor). However, amylin is also produced in the brain, and it increases within MPOA neurons of maternally behaving rats (Szabo, Cservenak, & Dobolyi, 2012). Amylin is present at very low levels in the MPOA of estrous cycling and late pregnant rats, but amylin mRNA and protein rise sharply by day 1 postpartum and continue to rise through day 9 postpartum, with these levels remaining high for the remainder of the postpartum period. Interestingly, a high percentage of amylin-containing MPOA/vBST neurons also contain cFos in maternally behaving postpartum rats, although many cFos neurons do not contain amylin. Importantly, amylin expression increases in the MPOA/vBST of sensitized virgin female rats who have shown maternal behavior for 2 days, while it does not increase in the MPOA/vBST of virgins that are exposed to pups but who have not shown maternal behavior (Szabo et al., 2012). These results show a strong correlation between increases in amylin expression in MPOA/vBST and the occurrence of maternal behavior, but it remains to be determined whether amylin-expressing MPOA neurons are actually involved in maternal motivation. Many physiological adaptations occur in maternal rats and other mammals, such as suckling-induced hyperphagia and the suppression of ovulation (Numan & Woodside, 2010). It is unlikely that amylin would be involved in the hyperphagia associated with suckling and lactation, since suckled rats eat more, but peripheral amylin suppresses food intake (Dobolyi, 2011). It is possible, however, that amylin within a certain population of MPOA neurons could be involved in the inhibition of ovulation since some brain regions to which certain MPOA neurons project are known to influence ovulation (Simerly & Swanson, 1988). Future research, therefore, needs to uncover whether MPOA amylin directly influences maternal motivation or, instead, affects some other function that is associated with the maternal condition.

Finally, there is another neuropeptide whose expression selectively increases in the MPOA of late postpartum female rats, toward the end of lactation (around days 15–20 of lactation): melanin-concentrating hormone (MCH; Costa et al., 2019; Rondini, Donato, Rodrigues, Bittencourt, & Elias, 2010). MCH is not present in MPOA of estrous cycling females, early postpartum females, or in postlactating female rats. In the MPOA during late lactation, MCH is present at high levels and colocalizes with GABA; like GABA, MCH exerts inhibitory effects on neurons (Sears et al., 2010). MCH-containing MPOA neurons are not located in the cMPOA or in the dorsolateral MPOA. Instead, they are located ventromedially in MPOA, close to the third ventricle. Importantly, an early lesion study showed the electrical lesions to this general region do not disrupt maternal behavior in early-postpartum rats (Noonan & Kristal, 1979).

In estrous cycling female rats and in male rats, MCH neurons are located primarily in the lateral hypothalamus and such neurons are involved in the control of food intake; the output of these neurons to other sites stimulates food intake (Sears et al., 2010). What might be the function of the selective increase in the expression of MCH in MPOA neurons toward the end of lactation? Recall from Chapter 3 that as the postpartum period progresses, maternal behavior declines. Since MCH-expressing neurons appear to respond to an organism's energy demands, it is conceivable that the increased energy drain that occurs as a result of milk production and lactation eventually stimulates the synthesis of MCH in the ventromedial MPOA, and perhaps an inhibitory action of MCH neurons on cMPOA neurons slowly suppresses maternal behavior and eventually results in the weaning of older pups (Rondini et al., 2010). If this proposal is accurate, MCH would be acting within the general region of the MPOA to suppress maternal responsiveness, as shown in Figure 5.6. There is some preliminary evidence to support this hypothesis. Early postpartum females quickly retrieve displaced pups, and such females also exhibit low levels of MCH in the ventromedial MPOA. However, after local injections of MCH into the MPOA, retrieval behavior is depressed in such females (Benedetto, Pereira, Ferreira, & Torterolo, 2014). It would be interesting to determine whether the inhibition of MPOA MCH neurons in late postpartum females would be able to stimulate pup retrieval in these females (cf. Pereira & Morrell, 2009).

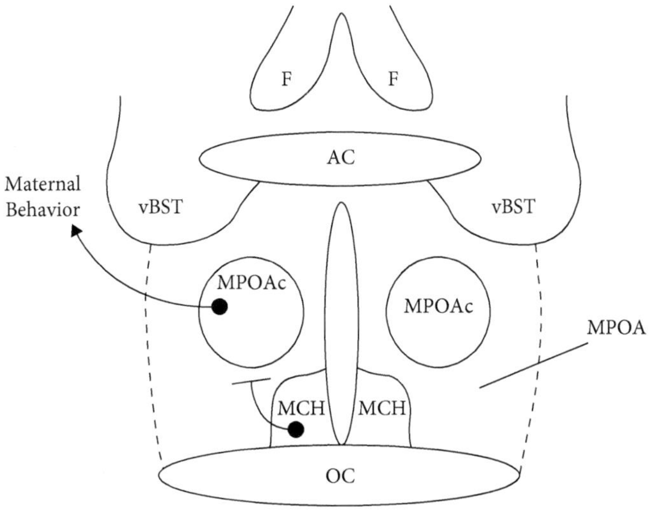

Figure 5.6. A potential neural circuit through which melanin-concentrating hormone (MCH)-containing neurons, located in the ventromedial part of the medial preoptic area (MPOA), may depress maternal behavior by inhibiting neurons located in the central part of the MPOA (MPOAc) that project to regions outside the MPOA to exert a positive influence on the occurrence of maternal behavior. Axons ending in a bar signify inhibition and those ending in an arrow signify stimulation/excitation. AC = anterior commissure; F= fornix; OC = optic chiasm; vBST = ventral bed nucleus of the stria terminalis.

It is also possible that MCH production during late lactation within the ventromedial MPOA contributes to physiological processes related to weaning. Since some of these MCH-expressing neurons project to the posterior pituitary, perhaps they depress suckling-induced OT release into the blood (the milk-ejection reflex; Costa et al., 2019). The potential behavioral and physiological effects of MCH are not mutually exclusive and may co-act in regulating the weaning process.

In a recent study, Pose et al. (2019) have reported that cFos is activated in the MPOA when postpartum female rats are exposed to either young pups or older pups. A more careful anatomical analysis, which was not performed in the Pose et al. study, might provide information on whether older, but not younger, pups activate cFos in MCH-containing MPOA neurons. Also of note, Pose et al. found that young pups activated cFos expression in vBST to a greater extent than did older pups.

In conclusion, MCH action within the MPOA may depress maternal behavior. The previously described evidence, however, does not exclude the possibility that

MCH neurons outside the MPOA, such as those in the lateral hypothalamus, may play a positive role in maternal behavior. There is some research, using mice with a null mutation of MCH and its receptor, that supports this possibility (Alhassen et al., 2019).

Neural Inputs to MPOA Relevant to Maternal Behavior

What are the neural inputs to the MPOA region that regulate its output to influence maternal behavior? I will tackle this question by discussing the role of OT, olfactory, somatic sensory, and processed multisensory inputs to the MPOA. I will end this section by analyzing the possible involvement of dopamine (DA) input to the MPOA in the regulation of maternal behavior.

Evidence has already been presented that the MPOA is one site where OT acts to stimulate the onset of maternal behavior. OT can influence MPOA neurons, where it acts on OTRs that are located there, as a result of release from PVN OT neurons (Knobloch et al., 2012; Otero-Garcia et al., 2016), particularly those located in the anterior PVN and ACN. The stimuli that could release OT onto MPOA neurons include parturition, suckling, and perhaps even ventral somatic sensory input from nuzzling pups in the absence of actual suckling (Cservenak, Keller et al., 2017). Finally, since some parts of the MPOA project to PVN OT neurons, it is very possible that, in response to stimulation from infants, the MPOA activates OT projections to several regions (Kohl et al., 2018; Numan & Numan, 1996; Simerly & Swanson, 1988). Therefore, MPOA projections to PVN-OT neurons might result in the positive feedback of OT onto MPOA neurons as well as the activation of PVN OT projections to regions outside the MPOA.

While olfaction is not essential for maternal motivation in rats, rabbits, and sheep, it is essential for maternal behavior in mice. However, in rats and rabbits, the valence of pup-related olfactory inputs changes from negative to positive near the time of parturition. (For additional information on rabbits, see Chirino & Gonzalez-Mariscal, 2015.) While olfactory inputs from pups inhibit maternal responsiveness in virgin rats and rabbits, the maternal behavior of primiparous rats and rabbits is no longer inhibited by such inputs, and postpartum females are actually attracted to pup odors. This valence switch serves two functions. First, infant olfactory inputs no longer inhibit maternal behavior. Second, attraction to pup odors, while not being essential for maternal behavior, probably helps the postpartum female to quickly locate her pups.

Do olfactory inputs act on the MPOA to influence maternal behavior? A possible route is shown in Figure 5.7. The accessory olfactory bulb (AOB), which receives vomeronasal input, projects directly to the medial amygdala (MeA), and the main olfactory bulb (MOB), which receives input from the main olfactory

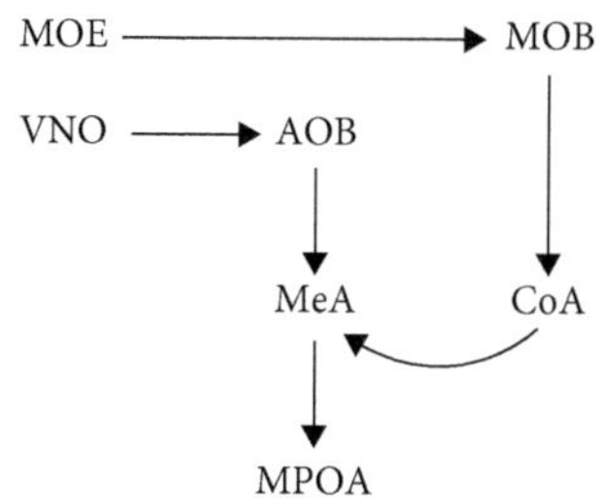

Figure 5.7. Neural pathways through which chemosensory inputs from the main olfactory epithelium (MOE) and the vomeronasal organ (VNO) can reach the medial preoptic area (MPOA). AOB = accessory olfactory bulb; CoA = cortical amygdala; MeA = medial amygdala; MOB = main olfactory bulb.

epithelium, projects to the cortical amygdala (CoA). (Refer to Figure 4.3 in Chapter 4 of this volume to see the locations of MeA and CoA in the rat brain.) Since CoA neurons project to MeA, MeA neurons are positioned to integrate sensory information from both olfactory systems (Keshavarzi, Power, Albers, Sullivan, & Sah, 2015; Scalia & Winans, 1975; Swanson & Petrovich, 1998). Further, there is evidence that both the MOB and the AOB project directly to the anterior part of MeA (Pro-Sistiaga et al., 2007). Importantly, one of the prominent projections of MeA is to the MPOA (Canteras, Simerly, & Swanson, 1995). Some projection neurons in the MeA contain GABA and others contain glutamate, and both of these populations project to the MPOA (Bian, 2013; Bian, Yanagawa, Chen, & Luo, 2008; Keshavarzi, Sullivan, Ianno, & Sah, 2014). An interesting proposal is that olfaction inhibits maternal behavior in virgin rats and rabbits because olfactory stimuli from infants activate GABA neurons in the MeA, which, in turn, project to and inhibit the output of the MPOA. As we will see in the next section, this hypothesis seems unlikely in that MeA neurons have been shown to influence other hypothalamic regions, outside the MPOA, to inhibit maternal behavior in virgin rats. However, it is certainly possible that MeA neurons are affected by the hormonal changes associated with pregnancy and parturition so that olfactory stimuli from pups activate MeA glutamate neurons that project to MPOA, which excites MPOA output so that the odors of infants become attractive (Numan, 2015; cf. Hong, Kim, & Anderson, 2014). It should be noted that olfactory stimuli stimulate cFos expression in MeA of postpartum maternal rats and that olfactory input also contributes to some of the cFos expression in the MPOA of maternal rats (Numan & Numan, 1995).

Although attraction to infant odors may facilitate maternal responses in rats, I want to reiterate that these inputs are not essential for maternal behavior. Not only does olfactory bulbectomy not interfere with the onset and maintenance of maternal behavior in rats, but direct depression of MeA neural activity also does not depress ongoing maternal behavior in postpartum rats (Numan et al., 2010).

Since olfaction mediated by the MOB is essential for maternal behavior in female mice, I would predict that depression of neural activity in CoA or other amygdala nuclei that receive inputs from MOB, such as MeA, in postpartum females would suppress maternal responsiveness. To my knowledge, only one study has provided some data relevant to this important issue. Chen et al. (2019) reported that optogenetic inhibition of a GABAergic cell population in the posterodorsal part of MeA reduced pup licking/grooming in both virgin and lactating laboratory mice, without affecting pup retrieval and nursing behavior. Clearly, this neuronal population is not necessary for the major appetitive and consummatory aspects of maternal motivation, but does influence the degree of pup grooming in virgin and lactating mice. Perhaps inhibition of this specific neuron population in the posterodorsal part of MeA disrupted the ability of maternal females to detect certain pup pheromones that stimulate maternal grooming of pups. More research along these lines, which investigates the potential roles of CoA and other parts of MeA in the maternal behavior of laboratory mice, should provide valuable information.

Stern (1996) has been a strong proponent of the idea that somatic sensory (tactile) inputs from pups are one of the sensory factors that influence maternal behavior. Suckling stimuli, of course, are one of the major forms of somatic sensory inputs that occur in postpartum females. A group of neurons in the posterior intralaminar complex (PIL) of the thalamus, located ventromedial to the thalamic medial geniculate nucleus, is activated by suckling in lactating rats (Cservenak, Keller et al., 2017). Not surprisingly, PIL receives strong somatic sensory inputs from the thoracic and lumbar regions of the spinal cord, and they can probably be activated by nonsuckling ventral tactile inputs in addition to suckling stimulation (Cservenak, Keller et al., 2017). A peptide referred to as tuberoinfundibar peptide 39 (TIP39), which appears to be an excitatory neurotransmitter, is expressed at very high levels in some PIL neurons of postpartum lactating rats, while it is expressed at very low levels in nonlactating females (Cservenak et al., 2013, Cservenak, Keller et al., 2017). In postpartum female rats, PIL-TIP39 neurons project to the MPOA/vBST and some of their axon terminals surround galanin neurons in MPOA (Cservenak et al., 2013, Cservenak, Kis et al., 2017b). Therefore, a neural pathway exists that would allow suckling stimuli and other ventral somatic sensory inputs from nuzzling pups to activate MPOA/vBST neurons (for a recent review, see Dobolyi, Cservenak, and Young, 2018). Are these inputs essential for maternal behavior? First, thelectomy (nipple removal) in postpartum mothers, which would prevent the pups from suckling, does not prevent the normal onset and maintenance of maternal behavior in rats (Numan & Numan, 1995; also see Lonstein, Simmons, Swann, & Stern, 1998; Walsh, Fleming, Lee, & Magnusson, 1996). Therefore, suckling stimulation, per se, is not essential for maternal responsiveness, although the thelectomy procedure does not rule out the importance of other ventral tactile stimuli associated with pups while the mother is hovering over them. In rabbits, suckling stimulation is also not essential for maternal motivation; however, it plays a major role in the regulation of a specific aspect of this behavior (i.e., timing). Normally, does crouch over the litter to nurse them inside the nest box for 3 to 5 minutes but thelectomized (i.e., nipple-removed) animals stay inside the nest box for a longer time than do nursing mothers (González-Mariscal, Melo, Parlow, Beyer, & Rosenblatt, 2000).

In a very interesting experiment, Cservenak et al. (2013) injected a long-acting antagonist to TIP39 into the MPOA of rats. Antagonism of TIP39 did not disrupt the onset and maintenance of maternal responsiveness, and this finding has also been confirmed in mice (Coutellier, Logemann, Rusnak, & Usdin, 2011). Importantly, however, antagonism of TIP39 in the MPOA of postpartum rats did prevent the display a CPP for a distinct cage compartment that had previously contained pups (Cservenak et al., 2013; the TIP39 antagonist was active during both the training and test phases of the CPP procedure). Control females preferred the pup-associated compartment during the test phase, while antagonism of TIP39 in MPOA prevented this preference, even though maternal behavior was normal during the training phase. These results suggest that TIP39 input to MPOA may be involved in some of the effects of maternal experience on subsequent maternal behavior. I will return to this issue when I discuss maternal memory later in this chapter.

Numan and Numan (1995) reported that neither thelectomy nor removal of the olfactory bulbs (both MOB and AOB) disrupted the onset and maintenance of maternal behavior in postpartum rats and also that neither procedure alone affected the expression of cFos in MPOA during maternal behavior. However, female rats with combined thelectomy and bulbectomy, although showing normal maternal behavior, did show a significant reduction in cFos expression in MPOA, although this cFos expression was still significantly higher than in postpartum females that were not exposed to pups and therefore were not displaying maternal behavior. Similar findings have been reported by Walsh et al. (1996). These findings should be interpreted in the context of the multisensory control of maternal behavior in rats. Many sensory stimuli are involved in the regulation of maternal behavior, and some of these stimuli exert their effects via affecting the functional activity of the MPOA. As long as there are some pup-related sensory inputs that can effectively activate the MPOA, the onset and maintenance of maternal behavior is normal. In this regard, TIP39 input to MPOA may be one of several sensory-related factors that influence ongoing maternal behavior in rats.

The medial prefrontal cortex (mPFC; see Figure 4.1) can receive processed sensory inputs from all sensory association neocortical regions (Numan, 2015;

Ongur & Price, 2000; Price, 2007). Therefore, mPFC projection neurons, which are excitatory and contain glutamate, can relay multimodal sensory inputs to their target regions. Importantly, mPFC neurons project to MPOA neurons (Balfour, Brown, Yu, & Coolen, 2006; Vertes, 2004; Wood et al., 2019), and inactivation of the mPFC disrupts maternal behavior in postpartum rats (Febo, Felix-Ortiz, & Johnson, 2010; Pereira & Morrell, 2011). One interpretation of these results is that when multisensory inputs to the MPOA are disrupted, maternal responsiveness is also suppressed, since many pup stimuli cannot drive the output of MPOA projection neurons. However, the mPFC projects to many brain regions in addition to MPOA (Numan, 2015), and therefore the fact that mPFC inactivation disrupts maternal behavior does not prove that mPFC inputs to MPOA are involved in maternal behavior. Interestingly, however, Vertes (2004) has shown that mPFC input enters the MPOA from a lateral direction. Perhaps lateral knife cuts of MPOA not only disrupt essential MPOA efferents, but also interfere with critical afferent input to MPOA.

Finally, in the rostral hypothalamus, there is a small nucleus called the anteroventral periventricular nucleus of the hypothalamus (AVPV). This nucleus is not part of the MPOA (Simerly, 1995), and it is located rostral to cMPOA, with its neurons being situated close to the third ventricle. Importantly, although separate from the MPOA, AVPV axons project to the MPOA (Simerly & Swanson, 1988; Scott, Prigge, Yizhar, & Kimchi, 2015). Three other characteristics of AVPV neurons are worthy of noting (a) many AVPV neurons contain ERs (Shughrue et al., 1997; Simerly, Zee, Pendelton, Lubahn, & Korach, 1997; Wang, Burger, Greenwald-Yarnell, Myers, & Moenter, 2018; Wintermantel et al., 2006); (b) E increases the excitability of AVPV neurons either through genomic or non-genomic mechanisms (Wang, DeFazio, & Moenter, 2016); and (c) a large population of AVPV neurons contain the enzyme tyrosine hydroxylase (TH). Since TH is involved in the synthesis of DA, it is usually considered that TH-containing AVPV neurons are dopaminergic. Importantly, AVPV TH neurons project to the MPOA (Scott et al., 2015).

In laboratory mice, Scott et al. (2015) have reported that 6-hydroxydopamine (6-HD) injections into the AVPV severely disrupted pup-stimulated maternal behavior in virgin female mice, but did not have this effect in lactating mice tested on day 4 postpartum. 6-HD is a neurotoxin that destroys TH-containing DA neurons. Virgin mice with 6-HD lesions of AVPV took longer to retrieve pups and did not retrieve all of the test pups to the nest area. One interpretation of these results is that AVPV DA neurons are required for the prompt onset of maternal behavior typically shown by virgin female laboratory mice. In virgin mice, perhaps pup stimuli activate AVPV DA input to MPOA, which then stimulates maternal behavior in these mice. Relevantly, MeA neurons project to AVPV in rodents and therefore olfactory inputs can reach AVPV (Canteras, Simerly, &

Swanson, 1995). Further, chemical stimulation of MeA increases extracellular DA levels in the MPOA of rodents (Dominguez & Hull, 2001).

It needs to be emphasized that lesions of cMPOA disrupt maternal behavior in both virgin and postpartum mice, and such mice show infanticide; importantly, NMDA lesions of the MPOA, which disrupt maternal behavior in mice and rats, do not damage AVPV neurons (Numan et al., 1988; Tsuneoka et al., 2013). Therefore, the effects of 6-HD lesions of AVPV on maternal behavior in mice are distinct from the effects of NMDA lesions of MPOA.

Some experiments conducted in rats may be relevant to the proposal that AVPV dopaminergic input to MPOA is involved in maternal behavior. There are two types of DA receptors, denoted as D1 or D2 receptors, and both of these DA receptors are located in MPOA (Bakowska & Morrell, 1995). With respect to the pregnancy termination model, the 15HO preparation represents a suboptimal hormonal stimulation model that only partially enhances a female rat's responsiveness to pup stimuli so that they show maternal behavior after 2 to 3 days of pup exposure, which is a sensitization latency shorter than that shown by virgin female rats (sensitization latencies average about 7 days), but is longer than the sensitization latencies shown by fully primed 15HO + E-treated rats (0 days). However, Stolzenberg et al. (2007) have reported that when a DA D1 receptor agonist is injected into the MPOA of 15HO female rats at the time of pup presentation (48 hours following the HO procedure), they behave like 15HO + E females and show full maternal behavior on their first day of pup exposure. It is as if DA stimulation of D1 receptors in MPOA was capable of substituting for the facilitatory effects typically exerted by E.

Before interpreting these results, one other finding needs to be considered: Results from a study by Numan, Numan, Pliakou et al. (2005) indicate that DA action on D1 receptors in the MPOA is not required for established maternal behavior in postpartum lactating rats during the maintenance phase of maternal behavior (cf. Miller & Lonstein, 2005). Therefore, DA action on D1 receptors in MPOA stimulates the onset of maternal behavior in rats that have been suboptimally primed with hormones, but is not essential for the continuance of the behavior once it has become established.

These combined results are not easy to interpret, but I can offer some speculations. In rats, E and other hormones, such as lactogens (see Brown, Herbison, & Grattan, 2015), may not only directly affect the MPOA, but may also act on the AVPV to stimulate DA input to MPOA. Therefore, DA input to MPOA from AVPV may be involved in the onset, but not the maintenance, of maternal behavior in rats. For laboratory mice tested in their home cages, the prompt maternal response of virgin female mice has a nonhormonal basis. Therefore, pup stimuli may directly activate AVPV DA input to MPOA to stimulate maternal behavior. However, for mice it appears that the postpartum female no longer requires AVPV TH neurons to show adequate maternal behavior during the maintenance phase of maternal behavior. The best conclusion is that for both rats and mice AVPV TH neurons, which are presumably dopaminergic, are important for aspects of the onset, but not the maintenance, of maternal behavior and that AVPV neurons are likely to exert this effect through their dopaminergic projections to the MPOA.

Other interpretations of these results are possible (Numan & Stolzenberg, 2009; Scott et al., 2015), particularly because AVPV DA neurons project to other brain regions in addition to MPOA. Therefore, much more research will be necessary to establish whether AVPV DA input to MPOA is actually involved in the onset of maternal behavior in rats and mice (and perhaps other mammalian species).

To conclude this section, Figure 5.8 shows a summary diagram of the inputs to the MPOA that may be involved in affecting MPOA output to its target regions in the regulation of maternal responsiveness. Of all the inputs to the MPOA that

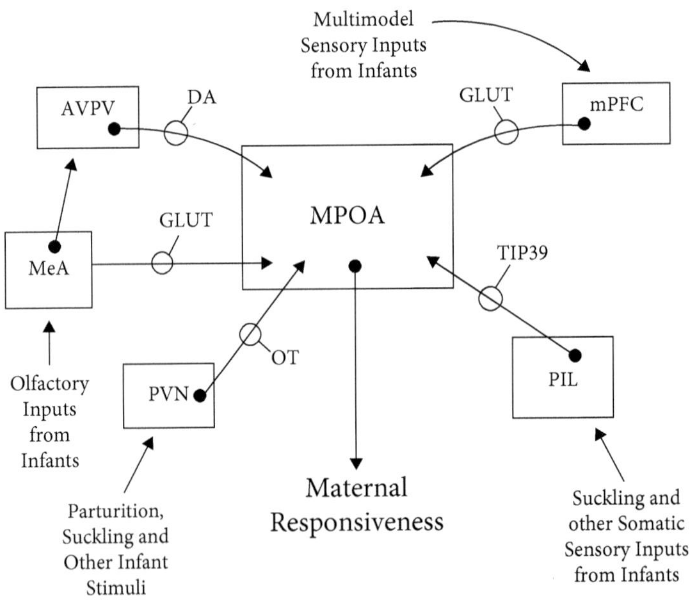

Figure 5.8. A summary diagram of the neural inputs to the medial preoptic area (MPOA) that may influence maternal behavior. AVPV = anteroventral periventricular nucleus of the hypothalamus; DA = dopamine; GLUT = glutamate; MeA = medial amygdala; mPFC = medial prefrontal cortex; PIL = posterior intralaminar complex of the thalamus; OT = oxytocin; PVN = paraventricular nucleus of the hypothalamus; TIP39 = tuberoinfundibular peptide 39. See text for details.

appear to regulate its output, OT input from the PVN seems to be particularly important in regulating the onset of maternal behavior in most mammalian species. From a comparative perspective, the potential involvement of mPFC inputs to MPOA appear to occupy a particularly important position for two reasons. First, such inputs are capable of providing multimodal sensory inputs from infants to the MPOA that are probably necessary for both the onset and maintenance of maternal behavior. Second, the mPFC provides a link that would allow cortical mechanisms to influence MPOA motivational circuits important for the appropriate regulation of maternal behavior. I will have more to say about this critical link in subsequent chapters.

The Defensive-Avoidance Circuit Active in Nonmaternal Virgin Female Mammals

In examining Figure 5.5, we can see that there are dual neural circuits that influence maternal behavior in species whose maternal behavior is triggered by the physiological events of late pregnancy and parturition, one that inhibits maternal responses in virgins and one that promotes maternal responsiveness in parturient females. Given that infant stimuli evoke avoidance and rejection responses in nulliparous females of these species, the question we are interested in at this point is the neural circuits that regulate such defensive responses. Most of the directly relevant research on this topic has been conducted on laboratory rats.

In virgin female rats and rabbits, olfactory input from pups has been shown to inhibit maternal responsiveness; olfactory input from pups has a negative valence that promotes avoidance responses. I have already described research that shows that olfactory inputs from both the MOB and AOB converge in the MeA, and I have suggested that in maternal females one population of MeA neurons may project to the MPOA so that females are attracted to pup odors (Figure 5.8). But in virgin females, another population of MeA neurons may project to defensive circuits that depress maternal behavior and promote avoidance responses. It appears that this population of MeA neurons projects to medial hypothalamic regions posterior to the MPOA, and these defensive/avoidance hypothalamic areas include the anterior hypothalamic nucleus (AHN) and the ventromedial hypothalamic nucleus (VMN). Refer to Figure 4.1 to see the general location of these nuclei.

Fleming, Vaccarino, and Luebke (1980) were the first to propose that in virgin female rats, olfactory inputs from pups converge on certain MeA neurons and activate them, giving rise to avoidance responses. In support of this idea, research has shown that either electrical or excitotoxic NMDA lesions of MeA facilitate the onset of maternal behavior in nulliparous rats (Fleming et al.,

1980; Numan, Numan, & English, 1993; Sheehan, Paul, Amaral, Numan, & Numan, 2001). Virgin females with MeA lesions do not avoid pups but instead tolerate their proximity, and they display sensitization latencies to the onset of maternal behavior of about 2 days, which is significantly shorter than the 6 to 7 day latencies shown by control females. It should be noted that females with MeA lesions are not anosmic in that they can locate a buried piece of candy (Numan et al., 1993). The MOB and AOB project to many brain areas, but olfactory projections to MeA in virgins may be involved in depressing maternal behavior by activating avoidance of pups, while not being essential for general olfactory sensitivity.

Canteras (2002) has described a medial hypothalamic defensive region that includes AHN and VMN. This region receives inputs from MeA, and the output of this defensive region promotes escape and avoidance responses to certain stimuli. Therefore, in virgin rats, it can be proposed that pup-induced, olfactory-driven inputs to MeA activate projections from MeA to AHN/VMN to elicit avoidance responses. In support of this view, research has shown that excitotoxic NMDA lesions of the AHN/VMN reduce sensitization latencies in nulliparous rats to about 2 days, mimicking the effects of MeA lesions (Bridges, Mann, & Coppeta, 1999; Sheehan et al., 2001). The VMN is a complex nucleus and Bridges et al. (1999) have indicated that NMDA lesions located in the ventrolateral part of VMN (VMNvl) were particularly effective in facilitating maternal behavior in virgin females. A very interesting finding relevant to the importance of projections from MeA to VMNvl in the inhibition of maternal behavior was reported by Sheehan et al. (2001). Nonmaternal nulliparous females received unilateral NMDA lesions of MeA and were exposed to pups. A subsequent cFos analysis of the brains of these nonmaternal females showed decreased cFos expression in AHN/VMN on the same side of the brain as the MeA lesion, and this depression in cFos expression was particularly dramatic in VMNvl. Since MeA projections to AHN/VMN are basically ipsilateral (Canteras et al., 1995), these results support the view that the MeA lesions blocked the conduction of some pup stimuli, most likely olfactory in nature, to AHN/VMN, resulting in decreased neural activation of this region, with a particular emphasis on VMNvl. Given that a MeA-to-AHN/VMN pathway appears to depress maternal behavior and promote pup avoidance in virgin rats, two mechanistic possibilities present themselves (Bian et al., 2008). Since some MeA neurons that project to the hypothalamus are glutamatergic, MeA neurons may directly excite AHN/VMN output to promote avoidance of pups. Alternatively, since other MeA neurons are GABAergic, such neurons may project to the AHN/VMN region to inhibit neurons that in turn inhibit the output of AHN/VMN neurons that mediate defensive responses. This latter mechanism would represent a process of disinhibition.

With respect to the role of VMNvl in promoting avoidance responses to conspecific stimuli, some very interesting work, although not dealing with maternal behavior, has been conducted in male mice. The VMNvl appears to contain two separate populations of neurons that are involved in either aggressive behavior or fear-related responses. When a mouse shows aggression toward a conspecific, one population of VMNvl neurons is activated, while when a mouse is attacked by another mouse and shows escape and avoidance responses a different population of VMNvl neurons is activated (Hashikawa, Hashikawa, Falkner, & Lin, 2017; Sakurai et al., 2016; Silva et al., 2013). Sakurai et al. have referred to the latter population of VMNvl neurons as social fear neurons (SFNs). Importantly, in a mouse that was showing aggression toward another mouse, selective stimulation of SFNs in VMNvl interrupted aggression and caused retreat and avoidance (Sakurai et al.).

It is interesting, therefore, to speculate that in virgin rats, MeA output may activate the social fear neuron population in VMNvl, causing avoidance of pups. Under certain conditions, however, MeA activation of VMNvl aggression neurons might be the mechanism underlying the small incidence of infanticide that occurs in virgin rats that are exposed to pups for the first time. Although the MeA projects to VMNvl (Gross & Canteras, 2012), Sakuai et al. (2016) have reported that MeA does not directly project to SFNs in VMNvl. This finding suggests the possibility the MeA GABA neurons might project to certain neurons in the VMNvl region that in turn inhibit VMNvl SFNs. Therefore, MeA GABAergic output to VMNvl may disinhibit, and therefore indirectly activate, VMNvl SFNs.

Given that a MOB/AOB-to-MeA-to-AHN/VMN neural circuit appears to cause a virgin female to avoid, and sometimes attack, pups, where might the AHN/VMN project to exert these effects? Most evidence points to excitatory glutamatergic projections from the hypothalamic defensive region to the dorsal part of the periaqueductal gray (PAG) in the midbrain in the promotion of a variety of defensive responses, such as escape and avoidance responses (Gross & Canteras, 2012; Silva et al., 2013). See Figure 4.1 for the general location of PAG. In fact, inhibition of the dorsal PAG suppresses defensive responses in mice that are confronted with an aggressive conspecific (Silva et al., 2013). The PAG has descending projections to the medullary reticular formation, which in turn influences spinal motor neurons that could regulate defensive responses.

There is some indirect evidence for the involvement of the dorsal PAG in the inhibition of maternal behavior that occurs in virgin rats. Sukikara, Mota-Ortiz, Baldo, Felicio, and Canteras (2010) exposed postpartum lactating maternal rats to either a neutral cloth or a cloth infused with cat odor (a potential predator olfactory signal). When retrieval tests were conducted, retrieval behavior was suppressed in the presence of the cat odor. NMDA lesions of the dorsal PAG blocked the suppressive effect of cat odor on maternal behavior. It would certainly be interesting to determine whether dorsal PAG lesions would facilitate maternal behavior in virgin females toward pups. The idea is that aversive odors, whether from a predator or from conspecific pups, may activate the dorsal PAG to promote defensive responses and depress maternal responses.

In thinking about how activation of a MOB/AOB-to-MeA-to-AHN/VMN-to-PAG circuit might prevent maternal behavior in virgin rats exposed to pups, the simplest explanation is that the activation of avoidance responses is incompatible with pup interaction and maternal behavior. However, it is additionally possible that some of the efferent connections of the defensive circuitry project back onto the MPOA and other neural regions involved in maternal motivation to directly inhibit maternal behavior. In this regard, it is important to note that the dorsal PAG projects to the MPOA (Rizvi, Ennis, & Shipley, 1992).

Figure 5.5 depicts the proposal that when the primiparous female's brain is primed by the physiological events of pregnancy and parturition, some of the outputs from the MPOA inhibit the processing of pup stimuli through the defensive circuit. There is not direct neurobehavioral evidence that supports this proposal, but there is strong anatomical evidence that shows that neural pathways exist that would allow such inhibition to occur. For rats, research has shown that MPOA/vBST efferents project to MeA, AHN, VMNvl, and PAG (Numan & Numan, 1996; Renier et al., 2016; Rizvi, Murphy, Ennis, Behbehani, & Shipley, 1996; Simerly & Swanson, 1988). Therefore, the properly primed MPOA/vBST is anatomically positioned to influence neural activity in all parts of the defensive circuit and could presumably act to suppress this circuit, in this way preventing avoidance and defensive responses to pup stimuli.

It should also be emphasized that OTRs are located in all nodes of the defensive circuit (see Table 4.1), as well as in the olfactory bulbs, and therefore OT action at these sites in parturient females could also depress avoidance responses.

Figure 5.9 shows an update of Figure 5.5, where the defensive system is indicated along with the proposed projections of the MPOA/vBST to the various nuclei that make up the defensive network. Future research, both anatomical and functional, needs to be performed to determine whether MPOA/vBST output actually inhibits the defensive circuit in its role as a regulator of maternal motivation. Further, the potential role of OT action on the defensive circuit in its facilitation of maternal behavior needs to be explored. With respect to sheep, since OT application to MPOA depresses the ewe's rejection responses toward a lamb (Kendrick et al., 1992), OT may be directly activating MPOA inhibition of the defensive circuitry.

Because the research on the defensive circuit has been performed on rodents, which are highly dependent upon olfactory input during social interactions, I have emphasized olfactory inputs to the defensive circuit in regulating its output. However, some other mammalian species are highly dependent upon

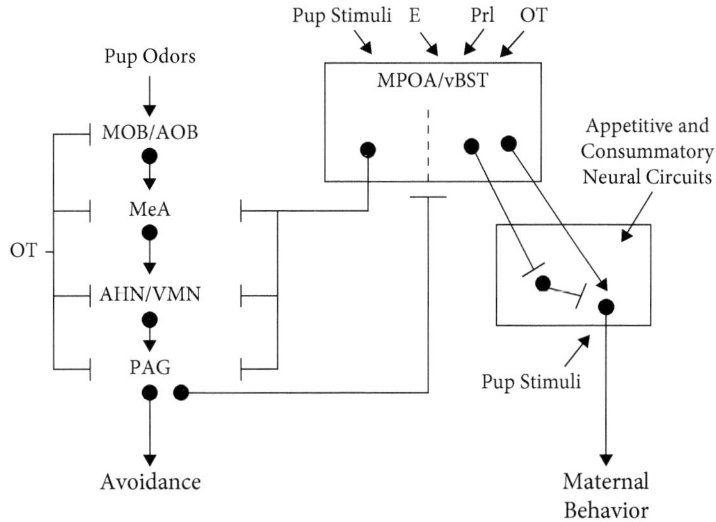

Figure 5.9. An elaboration of Figure 5.5, which outlines the defensive neural circuit that depresses maternal behavior in virgin female rats. Research has indicated that pup odors that activate the main and accessory olfactory bulbs (MOB and AOB, respectively) stimulate medial amygdala (MeA) neurons that project to defensive neural regions in the anterior hypothalamic nucleus (AHN) and the ventromedial hypothalamic nucleus (VMN). The neural projections of AHN/VMN to the periaqueductal gray (PAG) in the midbrain activate neurons that are proposed to have two effects. The downstream projections of the PAG, which ultimately influence motor neurons, trigger defensive responses such as avoidance and escape. In addition, projections of the PAG to the medial preoptic area and ventral bed nucleus of the stria terminalis (MPOA/vBST) may directly inhibit maternal motivation. This neural model also proposes that when MPOA/vBST is primed by the physiological events of pregnancy and parturition, so that maternal responsiveness is synchronized with the birth of the young, MPOA output inhibits the defensive circuit. The MPOA also activates appetitive and consummatory neural circuits, which then allows for the occurrence of maternal behavior. Oxytocin (OT) input, derived from the paraventricular hypothalamic nucleus, is also proposed to act at several nodes within the defensive circuit to depress defensive responses toward pups. Since OT exerts excitatory effects, it is proposed to inhibit defensive circuit neurons by activating inhibitory interneurons within the defensive circuitry. OT also acts on the MPOA to stimulate its output. Axon terminals ending in a bar signify inhibition, while those ending in an arrow represent excitation. E = estradiol; Prl = prolactin.

Source: Modified from Figure 6 in Sheehan, Cirrito, Numan, and Numan (2000) with permission from the American Psychological Association.

other stimuli, such as visual stimuli, in the regulation of their social behavior. How can we generalize my analysis of the research on the circuits that suppress maternal responsiveness in virgin females to most mammalian species with a uniparental maternal care system? The overarching take home message is that there are likely to be dual neural mechanisms that influence the occurrence of maternal behavior in most mammals, one that is active in virgin to suppress maternal responses, and one that is active in mothers to promote maternal behavior. It will be important to determine the detailed nature of the inhibitory network in species other than rodents.

For virgin female laboratory mice that show prompt maternal behavior when tested in their home cages, and in species where alloparental behavior naturally occurs in virgins, the defensive system must be relatively inactive even though such females have not been exposed to the physiological events associated with pregnancy and parturition. This probably results, in part, from the effects of experimental genetic selection or natural selection on the olfactory system that I described in the previous chapter. However, latent maternal inhibitory circuits may be present in such species and these circuits may become active under certain conditions (Hrdy, 2016; Mayer, Helton et al., 2019). For example, virgin female laboratory mice that retrieve pups in their home cages, but do not retrieve pups when tested in the novel context of a T-maze, demonstrate increases in cFos expression in the AHN in the T-maze compared to the home cage context. Further, the expression of cFos in AHN is higher in virgins than it is in postpartum females when tested in the T-maze context, and the latter females, but not the former, retrieve pups in the T-maze. cFos expression is also higher in the PAG of virgins tested in the T-maze when compared to the home cage context (Mayer, Helton et al., 2019). These findings are very important because they suggest that even when pup stimuli are not aversive, other environmental events may be able to tap into the defensive neural circuitry to depress maternal responsiveness.

The next step in my analysis, in reference to Figure 5.5, is to explore how MPOA/vBST output promotes maternal motivation.

MPOA Outputs That Promote Maternal Motivation

Introduction

Since the MPOA/vBST region is important for both the appetitive and consummatory aspects of maternal behavior in mammals, the question arises as to where the MPOA projects to influence these aspects of maternal behavior in mammals. Most research on this topic has been conducted on rodents and has focused on MPOA projection sites that influence the appetitive, reward-seeking aspects of maternal motivation; much less attention has been devoted to where the MPOA projects to influence the consummatory aspects of maternal behavior. The research clearly shows that MPOA interaction with the mesolimbic DA system controls the appetitive aspects of maternal behavior. Other evidence, although less conclusive, implicates MPOA interaction with PAG in the regulation of nursing behavior.

MPOA Interaction with the Mesolimbic DA System Regulates the Appetitive Aspects of Maternal Behavior: Introduction

There is excellent research that supports the proposal that MPOA projections to the ventral tegmental area (VTA; see Figure 4.1) in the midbrain contribute importantly to the regulation of the appetitive aspects of maternal motivation both at its onset at parturition and its subsequent maintenance during the postpartum period (Numan & Stolzenberg, 2009). The VTA contains DA, glutamate, and GABA neurons, and many of the glutamatergic and GABAergic neurons in the VTA are interneurons that can either stimulate or inhibit, respectively, the firing of VTA DA neurons (Dobi, Margolis, Wang, Harvey, & Morales, 2010). The projection of VTA DA neurons to the nucleus accumbens (NA) in the telencephalon (see Figure 4.1) forms a major component of the mesolimbic DA system (Numan & Woodside, 2010). The projection of VTA DA neurons to the NA has sometimes been referred to as the brain's reward system, since the release of DA from VTA terminals in the NA has been shown to be involved in many reward-seeking responses, such as food-seeking, the appetitive aspects of male and female sexual behavior, and appetitive maternal motivation (Numan, 2015; Numan & Woodside, 2010; Stolzenberg & Numan, 2011). Therefore, the mesolimbic DA system is a nonspecific motivational system that controls an organism's attraction to a variety of rewarding stimuli. I have been a strong proponent of the concept that the projection of maternally relevant MPOA neurons to VTA DA neurons that project to NA (MPOA-to-VTA-DA-to-NA) forms an intersection between a specific maternal appetitive motivational system and the nonspecific mesolimbic DA system (Numan, 2006; Numan & Woodside, 2010). MPOA activation of the mesolimbic DA system would therefore endow infant stimuli with rewarding properties that promote infant-seeking and infant-caretaking behaviors in the mother. Based on the responsiveness of the typical virgin female mammal to infants and the response of mothers to infants, one would conclude that the MPOA-to-VTA projection is functional in the latter and inactive in the former.

I have proposed a theoretical neural model of how MPOA interaction with the mesolimbic DA system controls appetitive maternal responses, such as pup retrieval (Numan, 2015; Numan & Stolzenberg, 2009; Numan & Young, 2016). In the discussion that follows, I will present the model in several parts, offering research that supports each part of the model. Figure 5.10 presents the initial

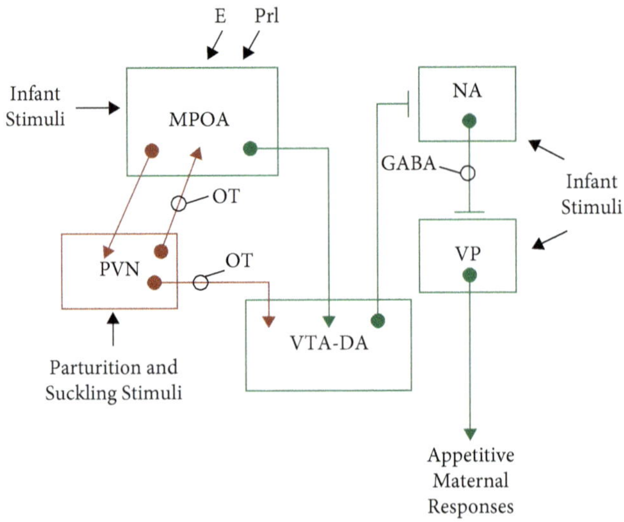

Figure 5.10. A neural model explaining how MPOA interactions with the mesolimbic dopamine (DA) system may regulate the appetitive aspects of maternal behavior. The parts of the figure shown in green are proposed to be essential for both the onset and maintenance of maternal behavior. The parts of the figure shown in red are essential for the onset of maternal behavior. When the medial preoptic area (MPOA) is properly primed by the physiological events of late pregnancy and parturition, it is capable of being activated by infant stimuli. Activation of MPOA inputs to ventral tegmental area (VTA) DA neurons stimulates the release of DA into the nucleus accumbens (NA). The model proposes that DA acts to inhibit NA GABAergic input to the ventral pallidum (VP), in this way increasing the responsiveness of the VP. Infant stimuli are shown as reaching both the NA and the VP. Inhibition of NA by DA allows infant stimuli to effectively activate the VP, whose projections promote the appetitive aspects of maternal behavior. With respect to the onset of maternal behavior at parturition, the basic neural circuits shown in green are supplemented by the circuits shown in red. In particular, MPOA activation of the paraventricular hypothalamic nucleus (PVN) stimulates oxytocin (OT) release into both the MPOA and VTA. The combined effects of the green and red circuits result in a surge of DA into NA, which is presumably essential for the initiation of appetitive maternal responses. Although not shown in the figure, OT action at the level of NA also promotes appetitive maternal motivation. Axons ending in an arrow signify excitation and those ending in a bar indicate inhibition. E = estradiol; Prl = prolactin.

Source: Modified from Figure 7 in Numan, Numan, Pliakou et al. (2005), with permission from the American Psychological Association.

components of the model. As background, I want to emphasize the anatomical and functional importance of the projection of the NA to the ventral pallidum (VP; see Figure 4.1). I will refer to this system as the NA–VP circuit. Medium spiny neurons (MSNs) in the NA send a GABAergic inhibitory projection to the VP. Therefore, within this circuit, increases in the neural activity of MSNs depress the activity of the VP, while inhibition of MSN output to VP should increase the responsiveness of VP neurons to excitatory inputs that it receives from other sources. The most important aspect of the model in Figure. 5.10, shown in green, depicts circuits that are essential for both the onset and maintenance of appetitive maternal motivation. After the MPOA has been properly primed with hormones (E and lactogens), it is capable of responding to infant stimuli (see Figure 5.8), which activate MPOA stimulatory projections to VTA DA neurons that, in turn, project to depress the output of NA MSN input to VP. In the model, note that infant stimuli not only act on the MPOA, but also provide excitatory inputs to both the NA and to the VP. Because MPOA activation of VTA-DA neurons is proposed to depress the output of NA, and therefore disinhibit the VP, the VP is capable of responding to infant stimuli when the MPOA is active. In the model, it is the output of VP that is necessary for appetitive maternal responses. In contrast, if the MPOA were not properly primed with hormones and remained unresponsive to infant stimuli, as in most nulliparous females, then NA activity would not be depressed. In this case, infant stimuli would excite NA, and the GABAergic inhibitory effects of NA on VP would reduce the responsiveness of VP neurons to these same infant stimuli with the result that appetitive caregiving responses toward infants would not occur.

The Mesolimbic DA System and Maternal Behavior

Strong evidence supports the involvement of the mesolimbic DA system in the control of both the onset and maintenance of maternal behavior in rodents. First, microdialysis and in vivo voltammetry studies show that DA is released into the NA during maternal behavior, and this release occurs both during pup retrieval and during nursing behavior (Afonso, Grella, Chaterjee, & Fleming, 2008; Afonso, King, Chatterjee, & Fleming, 2009; Afonso, Shams, Jin, & Fleming, 2013; Champagne et al., 2004; Hansen, Bergvall, & Nyiredi, 1993; Robinson, Zitzman, & Williams, 2011; Shnitko et al., 2017). Afonso et al. (2009) reported that extracellular DA concentrations in NA increased above baseline levels within the NA on day 1 postpartum in primiparous rats showing maternal behavior and in virgin female rats that were primed with stimulatory hormones and engaged in maternal behavior. In contrast, control nonhormone-primed virgin females did not show maternal behavior in response to pups, and DA levels in NA did not increase above basal concentrations. Importantly, exteroceptive stimuli received from pups that were placed in a Plexiglas container with holes also stimulated

DA release into NA of postpartum and hormone-treated virgins, but not in untreated virgin females (Afonso et al., 2013). This study shows that the actual performance of maternal behavior is not necessary to evoke DA increases in NA: If female rats are appropriately primed with hormones so that they would show maternal behavior, then pup stimuli alone can evoke DA increases in NA.

An interesting aspect of the studies by Afonso and colleagues was that basal levels of DA release in NA, prior to pup exposure, were higher in the nonmaternal than in the maternal females, but that upon pup exposure, DA increased above these baseline levels only in the maternal female rats. It has been proposed that high tonic levels of DA in NA may act to prevent a burst of DA release, in response to stimuli, from VTA axon terminals in NA (Mikhailova et al., 2016). Further, the research by Shnitko et al. (2017) indicates that the DA transporter is more effective in postpartum rats compared to virgins, which would allow for a faster reuptake of DA that has been released at synapses in NA, and this mechanism may contribute to the lower baseline levels of DA observed in NA of maternal rats. Perhaps these mechanisms underlie the findings of Afonso and colleagues. Such mechanisms would promote a greater extracellular DA signal to background ratio in maternal rats but not in virgins in response to pup stimuli.

What is the experimental evidence that VTA-DA neurons are involved in the appetitive aspects of maternal behavior? Numan, Stolzenberg, Dellevigne, Correnti, and Numan (2009) injected baclofen into the VTA of primiparous postpartum rats. Baclofen is a GABA-B receptor agonist that causes neural inhibition. Importantly, GABA-B receptors are located on VTA-DA neurons, and GABA action at these receptors inhibits these neurons (Edwards et al., 2017). Baclofen injections into the VTA suppressed retrieval behavior without interfering with nursing behavior, suggesting the involvement of VTA neurons in the appetitive, but not the consummatory aspects of maternal motivation. In support of the idea that baclofen produced its effects by blocking VTA-DA neurons, similar results on maternal behavior to those just described were observed by Hansen, Harthon, Wallin, Lofberg, and Svensson (1991) after injections of 6-hydroxydopamine, a neurotoxin that destroys DA neurons, into the VTA of postpartum lactating rats.

Byrnes et al. (2011) studied pup-stimulated maternal behavior in virgin female rats. Virgin females received injections of a chemical into VTA that chronically stimulated VTA-DA neurons, while control females received injections of an inactive compound. Chronic stimulation of VTA-DA neurons resulted in sensitization latencies of 4 days, which were significantly shorter than the 8-day sensitization latencies displayed by the control females.

In addition to pup retrieval, the VTA is also involved in additional appetitive aspects of maternal behavior in rats. Using the CPP paradigm, Seip and Morrell (2009) reported that depression of VTA activity blocked the expression of a preference for a cage compartment that had previously been associated with pups and maternal behavior.

These results indicate that depression of VTA-DA activity suppresses appetitive maternal responses in rats, while stimulation of these neurons can activate maternal motivation. Because VTA-DA neurons project to many regions in addition to NA (Breton et al., 2019; Swanson, 1982), subsequent research has asked whether selective DA action at the level of NA is important for maternal behavior in rats. Such a site of action is expected, since DA extracellular concentrations increase in NA during maternal behavior. There are two broad classes of DA receptors located in NA, the D1 DA receptor and the D2 DA receptor (Humphries & Prescott, 2010). Keer and Stern (1999) injected a nonselective DA receptor antagonist into the NA of postpartum rats and they found that the combined blockade of both D1 and D2 receptors disrupted retrieval behavior without disrupting nursing, emphasizing the involvement of DA input to NA in the appetitive aspects of maternal behavior. Numan, Numan, Pliakou et al. (2005) expanded on these results by examining the effects of selective D1 and D2 receptor antagonists on postpartum maternal behavior in rats. They found that the injection of a selective D1 receptor antagonist, but not a D2 antagonist, into the shell region of the NA (NAs) of postpartum lactating rats disrupted retrieval behavior, while leaving nursing behavior relatively unaffected. These combined results indicate that DA action on D1 receptors in NAs during the maintenance phase of maternal behavior is essential for the appetitive, but not the consummatory aspects of maternal behavior in rats.

Finally, and importantly, there is also evidence that the activation of D1 receptors in the NAs can stimulate the onset of maternal behavior in rats (Stolzenberg et al., 2007, 2010). Using the 15HO pregnancy termination model, where such females are not treated with E (see Chapter 3 of this volume), these researchers asked whether microinjection of a D1 agonist, but not a D2 agonist, into NAs could facilitate the onset of maternal behavior in these females that received suboptimal hormone stimulation. Females that received D1 agonist injections into NAs during pup exposure showed full maternal behavior on their first day of pup exposure, while females that received vehicle (water) or D2 agonist injections into NAs showed maternal behavior after 2 to 3 days of pup exposure, which is the standard latency that is observed in 15HO females. These results indicate that injections of a D1 receptor agonist into NAs can result in sensitization latencies similar to that normally shown by 15HO + E females: D1 activation in NAs was able to substitute for the absence of E treatment in 15HO females. We interpreted these results in the following way: Since 15HO females received partial hormone priming, pup stimuli may have been able to only weakly activate MPOA input to VTA, resulting in a mild release of DA in NAs, causing sensitization latencies of 2 to 3 days. However, by adding a D1 agonist to NAs, the amount of D1 activation is presumed to be equal to that which would occur in 15HO + E females. Once the female retrieves her pups, then the nuzzling pups also activate the consummatory aspect of maternal behavior (nursing behavior).

Figure 5.10 indicates that DA acts on NA to depress NA GABAergic projections to VP, in this way increasing the excitability and output of the VP to promote the appetitive aspects of maternal behavior. We can now add further specifications to this proposal by stating that the evidence shows that DA acts specifically on D1 receptors to depress NAs inhibition of VP. Although DA binds to and activates D1 receptors in NAs, this action functions to depress NA inhibitory control over VP, and it is VP output that is conceived as being essential for the appetitive maternal responses. There is good evidence to support this proposal. Bilateral suppression of NAs neural activity does not depress the normal onset and maintenance of maternal behavior in primiparous rats (Li & Fleming, 2003; Numan, Numan, Schwarz et al., 2005; Pereira & Morrell, 2011), while bilateral neuron-specific suppression of VP neural activity does disrupt retrieval behavior in postpartum rats (Numan et al., 1988; Numan, Numan, Schwarz et al., 2005). These results are consistent with the view that it is the output of the VP that is essential for appetitive maternal motivation in rats and that DA acts to suppress NAs activity, in this way releasing VP projection neurons from NAs MSN inhibition. The few studies performed on other species conform with the data for rats. NA lesions, performed during the postpartum period, do not disrupt the maternal behavior of California mice (*Peromyscus californicus*; Lee & Brown, 2007), and VP lesions performed during pregnancy disrupt the onset and maintenance of maternal behavior in laboratory mice (Akther et al., 2014).

In addition to research on maternal behavior, research on other types of appetitive reward-seeking responses has also begun to emphasize the role of VP output in the regulation of motivated behaviors (Chang, Smedley, Stansfield, Stott, & Smith, 2017; Faget et al., 2018; Fujimoto et al., 2019; Richard, Ambroggi, Janak, & Fields, 2016; Root, Melendez, Zaborsky, & Napier, 2015). The targets, and underlying mechanisms, through which VP projection neurons act to promote the appetitive aspects of maternal behavior, and other motivated behaviors, remains to be determined, and this should be an important focus of future research (see Faget et al., 2018, for some interesting recent research on these issues).

MPOA Interaction with the Mesolimbic DA System Regulates Maternal Motivation

The model depicted in Figure 5.10 indicates that the MPOA projects to and activates VTA-DA neurons that project to NA, and that this anatomical and functional relationship forms an intersection between a specific maternal motivation system and the nonspecific (general) mesolimbic DA motivational system. There is excellent anatomical support that the MPOA does indeed project to the VTA and is capable of activating, either directly or indirectly, VTA-DA neurons that project to NA (McHenry et al., 2017; McHenry, Rubinow, & Stuber, 2015; Numan & Numan, 1996; Simerly & Swanson, 1988). Further, ER-alpha containing MPOA neurons project to the VTA (Fahrbach, Morrell, & Pfaff, 1986; Tobiansky et al., 2016), as do MPOA neurons that express cFos during maternal behavior (Numan & Numan, 1997).

My laboratory has used the asymmetrical neural inactivation design to provide neurobehavioral evidence that the MPOA interacts with the mesolimbic DA system to regulate the appetitive aspects of maternal behavior in rats. This design is based on the facts that (a) the circuits shown in Figure 5.10 are bilateral (they occur on both sides of the brain), and their projections are also primarily ipsilateral in their organization (the right MPOA projects to the VTA on the right side of the brain; the right VTA projects to the right NA; the right NA projects to the right VP, and similarly, structures on the left side of the brain project to their targets on the left side of the brain); (b) depressing subcortical neural circuits on both sides of the brain has larger effects on motivated behaviors than does damaging the same neural circuits on only one side of the brain, which would leave the other side intact and able to regulate behavioral responses.

The results of the experiments from my laboratory are schematically depicted in Figure 5.11. Numan and Smith (1984) found that when postpartum lactating rats received unilateral lesions of the MPOA and VTA, severe disturbances in retrieval behavior, but not nursing behavior, occurred only when these lesions were located contralateral to one another (on opposite sides of the brain), but not when they were ipsilateral to one another. In a subsequent study, Numan, Numan, Schwarz et al. (2005) found that when unilateral neuron-specific inactivation of the MPOA was paired with unilateral neuron specific inactivation of the VP on the opposite side of the brain, severe retrieval deficits occurred in postpartum rats. However, if the unilateral MPOA and VP inactivations were on the same side of the brain, maternal behavior was relatively normal. These two studies show that when the critical circuits shown in Figure 5.11 are damaged on both sides of the brain, appetitive maternal responses are severely disrupted, but if these circuits are intact on one side of the brain, maternal behavior is relatively normal (also see Stack et al., 2002).

The research described in this section and the previous section provides anatomical and behavioral support for the proposal that the MPOA interacts with the mesolimbic DA system to control the appetitive aspects of maternal motivation and that this neural system is essential for both the onset and maintenance of such maternal responsiveness in rats and the few other species that have been investigated. An important question is the mechanism through which MPOA efferents to VTA activate VTA-DA neurons. This question is importantly related

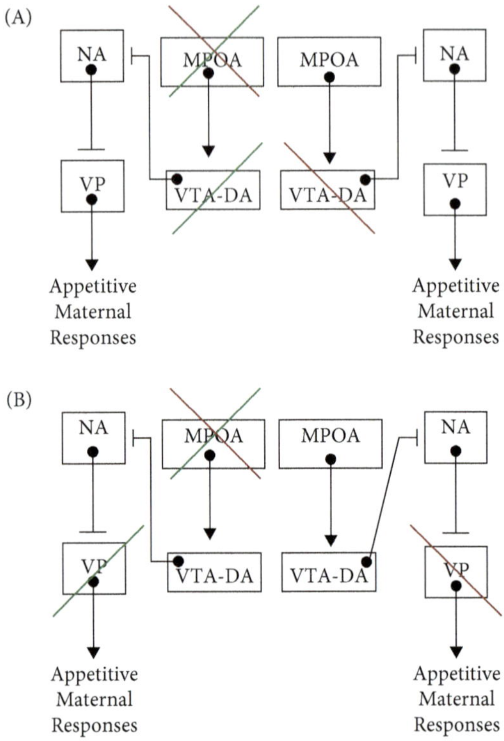

Figure 5.11. Evidence utilizing the asymmetrical neural inactivation design has provided evidence that the medial preoptic area (MPOA) interacts with the mesolimbic dopamine (DA) system to regulate the appetitive aspects of maternal behavior in postpartum laboratory rats. Plates A and B display the basic green circuit shown in Figure 5.10 as it is represented bilaterally in the rat brain. Green lines through neural regions indicate that these regions are inactivated ipsilateral to one another (on the same side of the brain). Red lines through neural regions indicate that these regions are inactivated contralateral to one another (on opposite sides of the brain). The neural disruptions shown in red interfere with the appetitive aspects of maternal behavior because MPOA interactions with the mesolimbic DA system are interrupted on both sides of the brain. The neural disruptions shown in green do not severely depress maternal behavior because MPOA interactions with the mesolimbic DA system are only interrupted on one side of the brain. Plate A shows that when unilateral inactivation of the MPOA is paired with a contralateral inactivation of the ventral tegmental area (VTA), appetitive maternal responses are disrupted. Ipsilateral inactivation of these regions is ineffective. Plate B shows that when unilateral inactivation of the MPOA is paired with a contralateral inactivation of the ventral pallidum (VP), appetitive maternal responses are similarly disrupted. Ipsilateral inactivation of these regions is ineffective. Axons ending in an arrow exert excitatory effects and those ending in a bar exert inhibitory effects.

Source: Modified from Figure 9 in Numan and Stolzenberg (2009) with permission from Elsevier.

to the neurotransmitters/neuromodulators contained in the projections of MPOA neurons to VTA.

As previously described in this chapter, a large proportion of MPOA neurons contain either glutamate or GABA, and some of these neurons also contain ER-alpha and express cFos during maternal behavior (Lonstein & DeVries, 2000; Tobiansky et al., 2013; Tsuneoka et al., 2013, 2017; Wu et al., 2014). Importantly, MPOA glutamatergic and GABAergic neurons project to VTA (Geisler, Derst, Veh, & Zahm, 2007; Tobiansky et al., 2013), and some MPOA projections to VTA terminate on VTA DA neurons that, in turn, project to NA (Tobiansky et al., 2016). Given these anatomical findings, one can speculate that MPOA neurons might activate VTA-DA input to NAs through two synergistic mechanisms. MPOA glutamatergic neurons might directly activate VTA-DA neurons, while MPOA GABAergic neurons might inhibit VTA-GABA interneurons that synapse on VTA-DA neurons, in this way releasing VTA-DA neurons from local inhibition (cf. Caba, Melo, Fleming, & Meza, 2019).

Recent attention has focused on the potential importance of MPOA neurons that contain the neuropeptide neurotensin in the activation of VTA-DA neurons. In rats, Geisler and Zahm (2006) have reported that neurotensin-containing MPOA neurons project to the VTA, and similar results have been reported for mice (McHenry et al., 2017). There are two types of NTRs, NTR1 and NTR2, and in laboratory mice it has been found that two-thirds of VTA-DA neurons contain NTR1 and that these VTA neurons project to NA (Woodworth, Perez-Bonilla, Beekly, Lewis, & Leinninger, 2018). Importantly, neurotensin has excitatory effects on VTA-DA neurons (Kempadoo et al., 2013), and selective stimulation of MPOA neurons that contain neurotensin and project to VTA promotes DA release in NA of mice. Interestingly, McHenry et al. (2017) found that 95% of MPOA neurons that contained neurotensin also contained GABA. Within some neurons in the brain, small molecule neurotransmitters, such as GABA, colocalize with larger neuropeptides, such as neurotensin (Nusbaum, Blitz, & Marder, 2017). Further, when neuropeptides and small molecule neurotransmitters are co-released, the small molecule neurotransmitter acts at the synapse where it is released, but the neuropeptide may diffuse further away from the site of release to affect nearby neurons (Nusbaum et al., 2017). With this understanding, it is interesting to speculate that when MPOA neurotensin/GABA neurons release these neurochemicals into VTA, GABA acts to depress the activity of VTA-GABA interneurons, while neurotensin acts directly on NTR1 located on VTA-DA neurons to directly stimulate DA release into NAs. A summary of the possible mechanisms through which MPOA output might activate VTA-DA input to NAs is shown in Figure 5.12.

The possibilities portrayed in Figure 5.12 are based on anatomical and physiological data, and the role of these neurochemically-specified MPOA

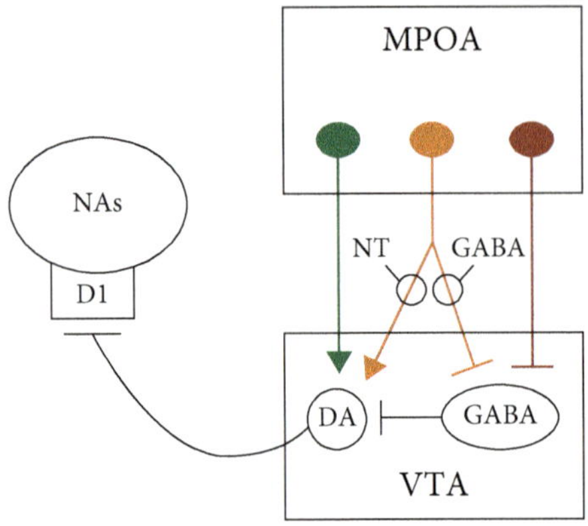

Figure 5.12. Potential neurochemical neural circuits through which medial preoptic area (MPOA) neurons may activate ventral tegmental area (VTA) dopamine (DA) neurons to stimulate DA release into the shell region of the nucleus accumbens (NAs). MPOA neurons that contain glutamate (GLUT) are shown in green, those that contain GABA are shown in red, and those that contain both GABA and neurotensin (NT) are shown in orange. The simplest mechanism is for MPOA-GLUT neurons to directly stimulate VTA-DA neurons. Alternatively, MPOA-GABA neurons may inhibit GABAergic interneurons in the VTA which then releases VTA-DA neurons from inhibition. MPOA neurons that contain both GABA and NT may exert a dual effect: GABA may inhibit the inhibitory interneurons within the VTA, while NT may directly stimulate VTA-DA neurons. DA action on D1 receptors in NAs is proposed to suppress the output of NAs. Axons ending in a bar are inhibitory, while those ending in an arrow are excitatory. See text for details and supporting evidence.

projections to VTA in maternal behavior regulation have not been fully explored. With respect to neurotensin, research on laboratory mice, although not based on careful behavioral observations, suggests that a null mutation of the neurotensin gene is not associated with a disruption of maternal behavior (Dobner, Fadel, Beitemeyer, Carraway, & Deutch, 2001). Also, as previously described in this chapter, Gammie et al. (2009) have reported that the intraventricular injection of a NTR1 antagonist does not disrupt pup retrieval in postpartum laboratory mice.

Recent research indicates that ER-alpha–containing MPOA neurons that project to VTA promote retrieval behavior in mice (Fang et al., 2018). These researchers provide evidence that such MPOA neurons activate retrieval behavior by inhibiting local VTA GABA interneurons, in this way releasing VTA-DA neurons from inhibition (see Figure 5.12). Therefore, these ER-alpha–containing MPOA neurons are presumably GABAergic.

Finally, there is also some preliminary evidence that MPOA galanin neurons, which co-localize with GABA, may project to VTA to influence the motivational aspects of maternal behavior in mice (Kohl et al., 2018). Perhaps galanin amplifies the inhibitory effects of MPOA GABA action within VTA. It appears likely that multiple mechanisms allow MPOA output to VTA to excite, either directly or indirectly, VTA-DA neurons that project to NAs, but the role of MPOA GABAergic neurons, along with their co-localized neuropeptides, may be of primary importance.

Oxytocin, the Mesolimbic DA System, and the Onset of Maternal Behavior

In my review of the role of OT in maternal behavior, I stressed that OT's primary role was to regulate the onset of maternal behavior and that OT neural systems were not essential for the maintenance of maternal behavior. However, I also noted some qualifications to this general statement (see the conclusion to the discussion of oxytocin and maternal behavior in Chapter 4 for a summary). During the postpartum period, OT does influence the quantity and quality of certain aspects of maternal behavior, and under challenging environmental conditions, OT neural systems appear to enhance maternal responsiveness via a dual effect of decreasing the mothers stress reactivity and by also enhancing maternal motivation. Figure 5.9 suggests some sites where OT might act to decrease fear-related responses, and I will develop this idea further in the next chapter. In this section, I want to present the research that indicates that OT action within the mesolimbic DA system is a route through which OT enhances the appetitive aspects of maternal motivation.

Figure 5.10 shows the oxytocinergic circuits, outlined in red, that appear to influence appetitive maternal motivation. These circuits are outlined in red to emphasize their essential role in the onset, but not the maintenance, of maternal behavior. The figure shows that parturition, suckling stimuli, and MPOA neurons can activate PVN-OT projections to MPOA and to VTA-DA neurons. I have already reviewed the research that OT acts on OTRs in MPOA to stimulate the onset of maternal behavior in rats. Importantly, Pedersen et al. (1994) have shown that when an OTR antagonist is injected into the VTA of parturient rats, the onset of maternal behavior is also severely disrupted.

Recent evidence has shown that OT action on the VTA can activate the mesolimbic DA system, stimulating DA release into NA. In rats and mice, PVN-OT neurons project to the VTA (Beier et al., 2015; Hung et al., 2017; Shahrokh, Zhang, Diorio, Grattan, & Meaney, 2010). Within the VTA, OTRs are located on DA neurons and on glutamate neurons (Peris et al., 2017; Xiao, Priest, Nasenbeny, Lu, & Kozorovitskiy, 2017), and application of OT to the VTA increases the excitability of VTA-DA neurons (Tang et al., 2014; Xiao et al., 2017). Assuming that some of the glutamate neurons in VTA that contain OTRs are interneurons that excite VTA-DA neurons, OT application to VTA may stimulate VTA-DA neurons either directly or via stimulating glutamatergic interneurons. Finally, microinjection of OT into the VTA of postpartum rats increases DA release into NA (Shahrokh et al., 2010).

Given this information, one can propose that at parturition, OT input to VTA-DA neurons acts in concert with MPOA stimulation of VTA DA neurons to cause a high level of DA release into NAs to stimulate the onset of maternal behavior (see Figure 5.10). Once maternal behavior becomes established during the postpartum period, such OT influences on VTA-DA neurons become less essential for the continuance of maternal motivation; in contrast, MPOA influences on VTA-DA neurons remain essential. However, under challenging environmental circumstances, activation of OT input to VTA (and to MPOA) may be necessary to enhance maternal motivation so that the mother will overcome environmental stressors to care for her young. Therefore, for all phases of maternal behavior, under stressful conditions, both MPOA and OT input to VTA-DA neurons that project to NAs may be important for proper and adaptive maternal responsiveness. OT input to MPOA is also presumably involved in this enhancement of maternal motivation.

Finally, at least for prairie voles, there is evidence that OT acts on OTRs in the NA to stimulate the onset of maternal responsiveness (Keebaugh & Young, 2011; Olazabal & Young, 2006a). Although not shown in Figure 5.10, as described in Chapter 4 (see Table 4.1), PVN OT neurons project to NA, where OTRs are located. Therefore, it appears likely that OT acts at each link in the MPOA-to-VTA-DA-to-NA circuit to influence maternal motivation in mammals, and MPOA activation of PVN OT neurons may participate in the release of OT at each of these sites. I will have more to say about the importance of OT action on NA later in this chapter.

Basolateral and Basomedial Amygdala Projections to the Ventral Pallidum and Appetitive Maternal Motivation

In Figure 5.10, I show that infant stimuli project to and stimulate both NA and VP and that MPOA-induced release of DA into NA acts (via D1 receptors) to suppress the inhibitory output of NA to VP, which allows the VP to respond to infant stimuli to regulate appetitive maternal motivation. The questions asked here are (a) What are the sources of input to NA-VP that allow these neural areas to receive processed infant stimuli? and (b) How does DA act on D1 receptors to suppress NAs responsiveness to infant stimuli? I have proposed a neural model, shown in Figure 5.13, that attempts to answer these questions (Numan, 2015; Numan & Young, 2016). The model shows that basolateral amygdala (BLA)/basomedial amygdala (BMA) neurons (see Figure 4.3 for the location of BLA and BMA within the amygdala) receive multimodal sensory inputs from infants and that these

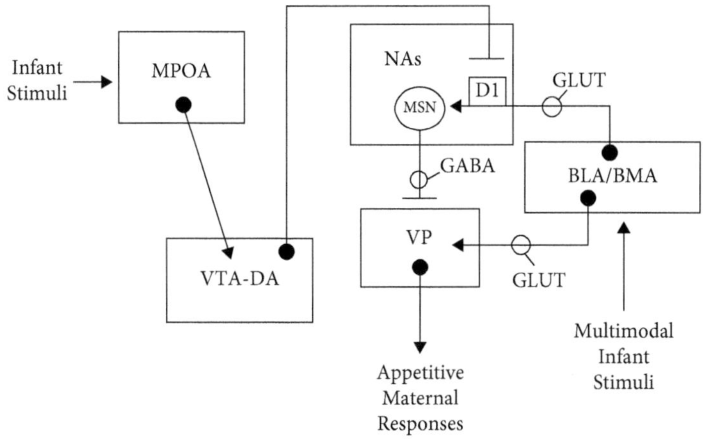

Figure 5.13. A more detailed neural model that explains how the nucleus accumbens shell (NAs)-ventral pallidum (VP) circuit is positioned to receive multimodal sensory inputs from pups and how dopamine (DA) acts on D1-DA receptors (D1) to suppress the excitability of the nucleus accumbens to infant-related stimuli. The basolateral and basomedial amygdala (BLA/BMA) provide multimodal sensory inputs to both the NAs and VP, and these neurons utilize glutamate (GLUT), an excitatory neurotransmitter. The model proposes that when the medial preoptic area (MPOA) activates DA release into the shell region of the nucleus accumbens, DA acts on D1 receptors that are localized on the axon terminals of BLA/BMA neurons that, in turn, synapse on GABAergic medium spiny neurons (MSN) in NAs that project to and inhibit VP. DA action on D1 receptors is proposed to exert presynaptic inhibition of BLA/BMA axon terminals, which reduces their excitatory input to the NAs. This effect reduces the ability of the NAs to inhibit the VP. The VP therefore becomes more responsive to sensory-related inputs from pups. VTA = ventral tegmental area. Axons ending in an arrow are excitatory and those ending in a bar exert inhibitory effects. See text for details and supporting evidence.

Source: Modified from Figure 5.8 in Numan (2015) with permission from Elsevier.

neurons, in turn, project to and excite the NAs and the VP. In support of this aspect of the model, BLA/BMA neurons receive olfactory, gustatory, auditory, visual, and somatic sensory inputs from the cortex, and BLA/BMA projection neurons use glutamate as their neurotransmitter (McDonald, 1998; Swanson & Petrovich, 1998). Further, BLA/BMA neurons on each side of the brain project to both NA and VP on each side of the brain, and these projections are primarily ipsilateral (Petrovich, Canteras, & Swanson, 2001; Petrovich, Risold, & Swanson, 1996). The model in Figure 5.13 also proposes that DA acts to decrease the responsiveness of NAs to infant stimuli by acting on presynaptic D1 receptors. Importantly, in the NAs of rats, D1 receptors are not only present on MSNs, but are also located on the axon terminals of glutamatergic inputs to MSNs (Dumartin, Doudnikoff,

Gonon, & Bloch, 2007). DA action on these presynaptic receptors results in presynaptic inhibition of glutamate release, which would selectively depress BLA/BMA excitation of NAs MSN inhibitory input to VP, while leaving BLA/BMA excitatory input to VP intact (Charara & Grace, 2003). Through this proposed mechanism, the VP would be disinhibited, allowing it to effectively respond to infant stimuli, which would allow for appetitive maternal motivation.

What is the behavioral evidence to support this model? First, and more generally, although it is well known that certain amygdala neurons, including those within BLA/BMA, are involved in fear-related responses, it is also well known that other BLA/BMA neurons are involved in appetitive motivational processes (Ferri et al., 2016; Gore et al., 2015; Janak & Tye, 2015; Numan, 2015): Separate populations of BLA/BMA neurons are involved in either reward-seeking behaviors or aversive/fear-related responses. It is the role of the BLA/BMA, and its projections to the NA-VP circuit, in appetitive motivation that fits with the model shown in Figure 5.13. With respect to maternal behavior, research from my laboratory has shown that bilateral inactivation of BLA/BMA neurons in postpartum rats disrupts retrieval behavior, while leaving the consummatory aspect of maternal behavior, nursing behavior, relatively unaffected (Numan et al., 2010). Relevantly, Lee et al. (2000) have reported that postpartum rats with lesions of BLA/BMA show deficits in learning an operant bar press response when pups are used as a rewarding stimulus; this same effect occurs in females with MPOA lesions. Finally, using an asymmetrical neuron-specific inactivation design, I have shown that unilateral inactivation of BLA/BMA disrupts retrieval behavior in postpartum rats if such an inactivation is paired with unilateral inactivation of the VP on the opposite side of the brain (Numan, 2015). By selectively interfering with the BLA/BMA-to-VP circuit on both sides of the brain, appetitive maternal motivation is suppressed.

I have already reviewed the evidence that D1 receptor agonist injections into NAs stimulate, while D1 receptor antagonist injections into NAs suppress, appetitive maternal responses. Whether DA is actually acting on the D1 presynaptic receptors shown in Figure 5.13, rather than on D1 receptors located on MSNs in NAs, to affect maternal motivation, remains to be determined.

It should be noted that OTRs are located in BLA/BMA (see Table 4.1). Whether OT acts in the amygdala of parturient females to enhance neural activity in BLA/BMA neurons that project to VP, in this way promoting the onset of maternal behavior, is an interesting question for future research (see Numan, 2012a, 2015; also see Chapter 11 of this volume).

The Involvement of the Lateral Habenula and the Laterodorsal Tegmental Nucleus in Maternal Motivation

The habenular nuclei, located in the dorsomedial part of the thalamus (sometimes referred to as the epithalamus; see Figure 4.3), are composed of the medial and lateral habenula (LHb). A body of research indicates that selective damage to LHb neurons disrupts the onset of maternal behavior in rats but does not appear to affect the maintenance of established maternal behavior, although more research needs to be done to validate the latter statement (Corodimas, Rosenblatt, Canfield, & Morrrell, 1993; Corodimas, Rosenblatt, & Morrell, 1992; Matthews-Felton, Corodimas, Rosenblatt, & Morrell, 1995). Importantly, MPOA projection neurons can ultimately reach the LHb via MPOA projections to the lateral preoptic area (LPOA; see Figure 4.2): Anatomical evidence exists for an MPOA-to-LPOA-to-LHb pathway (Barker et al., 2017; Matthews Felton, Linton, Rosenblatt, & Morrell, 1999b; Numan & Numan, 1996). Many LPOA neurons that project to the LHb contain glutamate and are therefore excitatory, although some are also GABAergic and inhibitory (Barker et al., 2017). Hence, it is possible that some MPOA projection neurons activate LPOA glutamate neurons that, in turn, excite LHb neurons to contribute to the activation of the onset of maternal behavior in rats.

The proposed positive role of the LHb in the onset of maternal behavior is at odds with a large body of research that indicates that the output of the LHb is involved in promoting avoidance responses to aversive stimuli. A population of LHb neurons that contain glutamate projects to and excites GABAergic neurons in the rostromedial tegmental nucleus (RMTg), a nucleus that is located immediately caudal to the VTA (see Figure 4.1), and is sometimes referred to as the tail region of the VTA (Brinschwitz et al., 2010; Fakhoury, 2018; Jhou, Geisler, Marinelli, Degarmo, & Zahm, 2009; Kaufling, Veinante, Pawlowski, Freund-Mercier, & Barrot, 2009; Stamatakis & Stuber, 2012). Further, RMTg-GABA neurons project to and inhibit VTA-DA neurons to promote avoidance behavior (Brown et al., 2017; Stamatakis & Stuber, 2012). Given this analysis, since LHb activation of RMTg neurons inhibits VTA-DA neurons, one might assume that LHb output should inhibit, rather than facilitate, maternal behavior.

However, the LHb is composed of several populations of neurons with different projections. One can divide the LHb into a medial part (LHbm) and a lateral part (LHbl), and Lonstein et al. (1998) have observed that cFos is expressed exclusively in the LHbm of maternally behaving postpartum rats. Further, many LPOA glutamatergic neurons project to LHbm, while some LPOA GABAergic neurons project to LHbl (Barker et al., 2017). Goncalves, Sego, and Metzger (2012) have reported that in rats, LHbm neurons project to the VTA, while LHbl neurons project to the RMTg, and there is evidence that certain LHb glutamatergic neurons (presumably in LHbm) can stimulate VTA-DA neurons that project to NA (Brown & Shepard, 2016; Omelchenko, Bell, & Sesack, 2009). Finally, there is some behavioral evidence that suggests that when LHb neural activity shifts from LHbl dominance to LHbm dominance, behavior switches from avoidance responses to approach responses (Gao, Groenewegen, Vanderschuren, & Voorn, 2018).

Given the previously described research, I would like to propose a hypothetical circuitry through which the LHb may be involved in stimulating the onset of maternal behavior in rats. Figure 5.14 shows that when "maternally relevant" MPOA output neurons stimulate the LPOA, these MPOA neurons activate

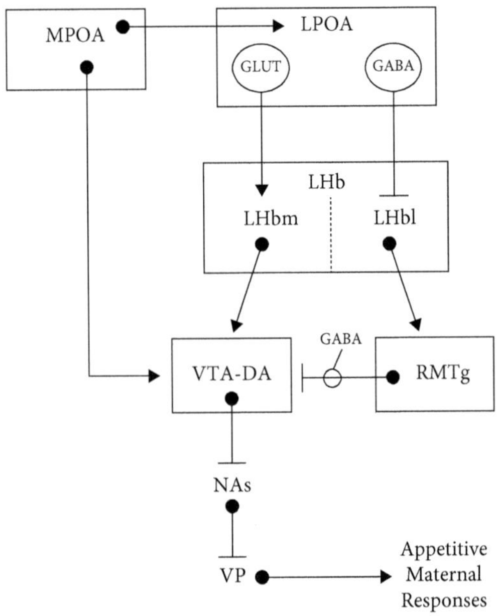

Figure 5.14. A neural model that proposes how a medial preoptic area (MPOA)-to-lateral preoptic area (LPOA)-to-lateral habenula (LHb) neural circuit may be involved in stimulating the onset of maternal behavior. MPOA neurons are depicted as stimulating both LPOA glutamatergic (GLUT) and LPOA GABAergic neurons. Stimulation of LPOA GLUT neurons activates projections which, in turn, excite neurons in the medial part of the LHb (LHbm). These LHbm neurons, in turn, excite ventral tegmental area (VTA) dopamine (DA) neurons that project to and inhibit the nucleus accumbens, which releases the ventral pallidum from inhibition. In contrast, MPOA excitation of LPOA GABAergic neurons activates projections, which inhibit neurons in the lateral part of the LHb (LHbl). Since the output of LHbl neurons is shown as activating the rostromedial tegmental nucleus (RMTg), which inhibits VTA-DA neurons, this pathway releases VTA-DA neurons from RMTg inhibition. The combined effects of exciting LHbm and inhibiting LHbl results in a strong activation of VTA-DA neurons and a surge of DA into NAs. Axons ending in an arrow exert excitatory effects and those ending in a bar exert inhibitory effects.

both LPOA glutamatergic neurons that excite the LHbm and LPOA GABAergic neurons that inhibit the LHbl. This proposed circuitry would result in a net excitation of VTA-DA neurons that project to NAs.

Interestingly, although most LHb neurons are glutamatergic, a recent study by Zhang et al. (2018) has identified a group of GABAergic neurons in LHbm that project directly to the RMTg. Therefore, it is also possible that MPOA-to-LPOA activation of certain LHbm neurons directly inhibits RMTg, which would suppress RMTg inhibition of VTA-DA neurons.

It is instructive to relate Figures 5.10 and 5.14 to the involvement of PVN-OT-to-VTA-DA projections and LHbm-to-VTA-DA projections in maternal behavior. Since OT and the LHb have been proposed to be mainly involved in the onset, rather than maintenance, of maternal behavior under standard laboratory conditions, it is possible that the synergistic effects of MPOA input to VTA-DA neurons combined with OT and LHbm stimulation of VTA-DA neurons, provides a critical threshold surge of DA release into NAs that is necessary for the onset of maternal behavior. Once maternal behavior becomes established, the singular activation of VTA-DA neurons by the MPOA may be all that is necessary to maintain maternal motivation for the remainder of the postpartum period, at least under standard nonstressful laboratory conditions.

Finally, and briefly, the laterodorsal tegmental nucleus (LDTg; see Figure 4.1), located in the mesopontine tegmentum, caudal and dorsal to the VTA, sends an excitatory input to VTA-DA neurons that project to NA, and activation of this pathway contributes to burst firing of action potentials in VTA-DA neurons (Forster & Blaha, 2000; Lodge & Grace, 2006; Omelchenko & Sesack, 2005; Steidl et al., 2017). Relevantly, Numan and Numan (1991) have reported that knife cuts posterior to the VTA disrupt maternal behavior in postpartum rats. These knife cuts severed many neural pathways, and the neural projection from LDTg to VTA would be among the pathways that would have been disrupted. Interestingly, anatomical studies also indicate that the MPOA projects to mesopontine regions near the LDTg (Numan & Numan, 1996; Simerly & Swanson, 1988). Given this analysis, future research should specifically focus on the role of the LDTg in maternal motivation.

In summary, there are a variety of routes and mechanisms through which MPOA projection neurons could regulate appetitive maternal motivation by directly or indirectly activating VTA-DA neurons that project to NAs. Although the MPOA-to-VTA circuit is likely to be the core circuit necessary for appetitive maternal motivation, it is likely that this core circuit is supported by additional circuits, which would include MPOA projections that influence PVN-OT neurons, LHbm neurons, and perhaps the LDTg.

The MPOA and the Consummatory Aspects of Maternal Motivation

Inactivation of MPOA neurons not only depresses the appetitive components of maternal behavior in rodents but also decreases nursing behavior. However, the effects of disrupting MPOA function on nursing behavior in rats are variable: Some studies have reported a total elimination of nursing behavior, while other studies have reported that although interference with MPOA activity depresses nursing behavior, this consummatory aspect of maternal behavior is not totally eliminated (Numan & Insel, 2003; Bosch & Neumann, 2008).

The suppression of nursing behavior in postpartum females with MPOA inactivation is difficult to interpret. If a female is not interested in infants because of a lack of appetitive motivation, then she may not approach and interact with them for a sufficient amount of time to allow suckling stimulation, and other ventral somatic sensory inputs from infants, to reflexively elicit nursing behavior. In this case, the suppression of nursing behavior in females with MPOA inactivation may be an indirect effect of the primary effect of a lack of attraction to and interest in infants. However, it is also possible that, in addition to an appetitive population of MPOA neurons that affects the mesolimbic DA system, there is an additional population of MPOA neurons that influences the consummatory aspects of maternal behavior. There is some anatomical evidence that supports this latter possibility.

Once a postpartum rodent has retrieved her pups to a nest, she will begin to nurse them. Initially, the female hovers over her pups while grooming them, but she subsequently becomes quiescent and immobile and engages in the crouch nursing posture (Stern & Johnson, 1990). The crouch nursing posture seems particularly important for allowing young pups, during the early postpartum period, to attach to the mother's nipples and suckle, in this way obtaining milk (Lonstein & Stern, 1997a).

Previous in this chapter in the subsection on the defensive-avoidance circuit active in nonmaternal virgin female mammals, I described the role of the dorsal PAG in the midbrain in the promotion of a variety of defensive responses, such as those involved in escape and avoidance. However, the PAG influences a variety of reflexive-like behaviors through its projections to the lower brainstem and spinal cord, each of which is likely mediated by distinct PAG circuits (Numan, 2015). In this regard, Lonstein and Stern (1997a, 1997b) have shown that the ventrolateral part of the PAG (vlPAG) is involved in the crouch nursing posture in rats. Lesions of the vlPAG, although not eliminating the nursing crouch, significantly decrease its duration, while leaving other aspects of maternal behavior in postpartum rats intact. Since the MPOA projects to the vlPAG, this could be a route over which MPOA projection neurons influence certain aspects of the consummatory components of maternal behavior (Stack et al., 2002). It is interesting to speculate that certain MPOA projections may stimulate the vlPAG to prolong the duration of the quiescent crouch nursing posture, while other MPOA neurons may inhibit the vlPAG to terminate nursing behavior.

It is obvious that most research has been done on MPOA interactions with the mesolimbic DA system in the regulation of appetitive maternal motivation, and much more research is needed to fully understand the neural regulation of consummatory maternal responses. From one point of view, this imbalance in research focus is understandable. A lack of appetitive maternal motivation would result in maternal neglect of offspring. However, if there is a deficit in consummatory maternal motivation, infants may not be fed properly. In the next section, I will also show that the consummatory aspects of maternal behavior have reinforcing effects that appear to be essential for the maintenance of maternal behavior during the postpartum period, as well as for the phenomenon that has been referred to as maternal memory (see Chapter 3 of this volume).

Neural Plasticity Within Maternal Brain Circuits, the Maintenance of Maternal Behavior, and Maternal Memory

In Chapter 3, I reviewed the evidence for rats and sheep that hormones and OT act to stimulate the onset of maternal behavior at parturition, but once maternal behavior becomes established during the early postpartum period, the subsequent maintenance of maternal responsiveness does not require continued hormonal and OT control (rabbits may be an exception, since PVN lesions disrupt the maintenance of maternal behavior in some rabbits; see Chapter 4 of this volume). These results indicate that when the hormone- and OT-primed female interacts with her young at parturition, some enduring modifications occur in the neural circuitry that regulates maternal motivation so that infant stimuli can subsequently activate maternal circuits in the absence of continued hormonal and OT mediation. In Chapter 3, I also proposed that the processes underlying the maintenance of maternal behavior may be similar to those that occur in the phenomenon that has been referred to as maternal memory in rats: Once a critical amount of mother–infant experience occurs in primiparous rats at parturition, subsequent episodes of maternal behavior, even after prolonged periods of mother–infant separation, become relatively emancipated from endocrine control.

These results show that the neural circuits that regulate maternal motivation in many mammals are not fixed and inflexible but can, instead, be modified by maternal experience with the result that an enduring mother–infant bond becomes established. In considering the potential neural sites that may be modified by maternal experience, it is worthwhile to consider the dual neural circuits that influence maternal responsiveness in most mammals: The defensive-avoidance circuit that suppresses maternal behavior in virgins and the excitatory circuit that stimulates maternal motivation (see Figure 5.5). It is possible that maternal experience with infants during the immediate postpartum period acts to downregulate the defensive system while also strengthening certain synapses within the excitatory system and that these neural modifications allow maternal behavior to occur during the maintenance phase of maternal behavior without the continued need for hormonal and OT mediation.

Although maternal memory occurs in rats, it does not occur in all mammalian species (Numan & Insel, 2003). Sheep provide a good example. Although the maintenance of maternal behavior in sheep does not appear to require hormones and OT (see Chapter 3 of this volume), a maternal memory process similar to that observed in rats does not occur. Estrous-cycling multiparous ewes, like nulliparous ewes, will not care for lambs, while estrous cycling multiparous rats show heightened maternal responsiveness in comparison to their nulliparous counterparts (see Chapter 3 of this volume). As described in Chapter 3, this difference between rats and sheep with respect to maternal memory formation makes evolutionary sense, given the social environments within which these two species live. For those species, such as sheep, where the maintenance of maternal behavior does not require hormones and OT, but a maternal memory process does not occur, it can be proposed that any neural modifications that occurred as a result of maternal experience at parturition, modifications that facilitate maternal responsiveness during the maintenance phase of maternal behavior, are reset to the nulliparous state once the young are weaned.

For the nonhormonal maintenance of maternal behavior and maternal memory to become established in rats, full interaction with pups for a critical amount of time at parturition is required (Jakubowski & Terkel, 1986; Orpen & Fleming, 1987). Exteroceptive stimulation from pups (sights, sounds, odors) is not sufficient; the mother must engage in the consummatory act of nursing behavior for an enduring mother–infant bond to form. Although suckling stimulation per se is not essential since maternal behavior is maintained normally in thelectomized postpartum rats (Numan & Numan, 1995), as is maternal memory (Bridges, 1975), proximal somatic sensory tactile inputs from pups appear to be necessary for maternal experience with pups to establish a strong mother–infant bond that persists in the absence of hormonal mediation (Morgan, Fleming, & Stern, 1992).

In an important series of experiments, Fleming's group has shown that DA and OT action at the level of the NAs is necessary for maternal memory formation in rats. In these studies, primiparous rats were allowed 1 hour of postpartum full maternal interaction with pups, and then the pups were removed. Immediately thereafter, DA receptor antagonists (D1, D2, or D1 + D2 antagonists), an OTA, or control solutions were injected into NAs. Ten days later, all females were re-exposed to pups and sensitization latencies for the reinitiation of maternal behavior were observed. The results showed that the combined blockade of both D1 and D2 DA receptors (D1Rs and D2Rs) disrupted maternal memory formation, while blockade of either receptor alone was ineffective (Parada, King, Li, & Fleming, 2008). OTA injections alone into NAs also blocked maternal memory formation (D'Cunha, King, Fleming, & Levy, 2011). These results indicate that the stimulation of D1Rs or D2Rs, and OTRs, in NAs are necessary for the establishment of a long-term mother–infant bond in rats, as measured by the establishment of maternal memory.

Given the knowledge of the neural circuitry underlying the onset of maternal behavior in rats, I have presented a neural model that attempts to explain the results from Fleming's research group (Numan, 2015; Numan & Young, 2016). The model is shown in Figure 5.15. The basic idea behind the model is that DA and OT input to the NAs, in part stimulated by the output of the MPOA during the early postpartum period, results in a massive inhibition of NAs MSN inhibitory input to the VP. Such an inhibition of NAs results in a super-excitation of VP by BLA/BMA input, which is proposed to strengthen BLA/BMA-to-VP synapses via an activity-dependent facilitation mechanism, similar to long-term potentiation. Subsequently, when hormones and OT are no longer acting on the MPOA, MPOA output to the mesolimbic DA system in response to pup stimulation, in conjunction with DA action on D1 receptors alone, is effective in maintaining maternal motivation due to the strengthened BLA/BMA-to-VP synapses, allowing pup stimuli to evoke appetitive maternal responses.

I have already described the evidence that supports the core neural circuitry shown in Figure 5.15. In addition, there is evidence that D2 DA receptors and OTRs can form heteromers on NAs MSNs, and that the combined effect of these heteromers, when activated by DA + OT, is to inhibit the MSNs on which they are located (Humphries & Prescott, 2010; Romero-Fernandez, Borroto-Escuela, Agnati, & Fuxe, 2013). Therefore, the inhibitory action of OT and DA on OTR-D2 receptors on MSNs that project to VP, coupled with the inhibitory action of DA on D1 presynaptic BLA/BMA terminals that would normally stimulate MSNs, all act to depress the inhibitory output of NAs to VP, rendering the VP super-receptive to excitation by BLA/BMA inputs.

The model shown in Figure 5.15 is very preliminary and requires much more research to validate it. For example, it is possible that OT exerts unique effects in NA that are independent of an action on the proposed OTR-D2 receptor heteromers. However, the model does make two important points. First, maternal experience-induced neural plasticity within the neural circuits that regulate maternal behavior can generally be conceived as being able to enhance

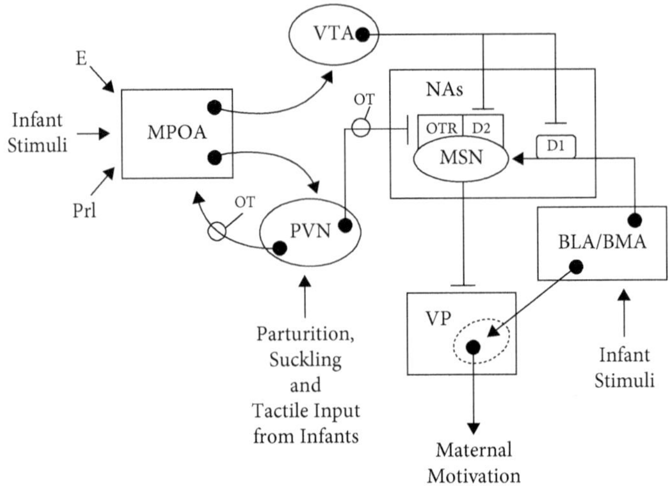

Figure 5.15. A model that describes the synaptic plasticity events that may occur within the mesolimbic dopamine (DA) system to give rise to the nonhormonal maintenance phase of maternal behavior in nonhuman mammals and to the process referred to as maternal memory in rats. At parturition, DA and oxytocin (OT) input to the shell region of the nucleus accumbens (NAs), in part stimulated by the projections of the hormone-primed medial preoptic area (MPOA) to both ventral tegmental area (VTA) DA neurons and paraventricular hypothalamic nucleus (PVN) OT neurons, results in a massive decrease in the excitability of NAs medium spiny neurons (MSN). This process dramatically decreases NAs MSN inhibitory input to the ventral pallidum (VP), which allows the VP to be superexcited by infant-induced basolateral and basomedial amygdala (BLA/BMA) neural input. Such superexcitation strengthens the synapses (shown as a dashed circle) between BLA/BMA and VP. The decrease in the excitability of NAs MSNs is proposed to be due to DA action on presynaptic D1 receptors, which depresses excitatory input to MSNs, combined with the direct suppression of MSNs by the action of OT and DA on heteromers formed by D2 DA receptors and oxytocin receptors (OTR). Subsequently, during the maintenance phase of maternal behavior, when hormones and OT are no longer essential for ongoing maternal behavior, infant-induced stimulation of MPOA output to the mesolimbic DA system, in conjunction with DA action on presynaptic D1 receptors alone, is effective in maintaining maternal motivation because of the strengthened BLA/BMA-to-VP synapse, which allows infant stimuli to evoke appetitive maternal responses. Axons ending in an arrow exert excitatory effects and those ending in a bar exert inhibitory effects. E = estradiol; Prl = prolactin.

Source: Modified from Figure 5.17 in Numan (2015), with permission from Elsevier.

maternal motivation. Second, the DA mechanisms involved in the synaptic strengthening that results in an enduring mother–infant bond are not identical to those that regulate ongoing maternal motivation. As I have already reviewed, DA action on D1, but not D2, DA receptors is essential for both the onset and maintenance of appetitive maternal behavior in rats, but the work of Fleming and her colleagues shows DA action on D1 and D2 receptors is involved in maternal memory formation. Therefore, at parturition, DA action on D1 receptors stimulates maternal behavior onset, but once consummatory maternal responses occur, the added action of DA on D2 receptors in NAs, combined with OT action at that site, contribute to synaptic plasticity within the maternal brain. Therefore, DA motivational mechanisms within NAs are partially distinct from DA reinforcement mechanisms that are involved in synaptic plasticity, with D2 receptor activation being involved in the latter, but not the former, process. In this regard, although DA release into NA is not essential for nursing behavior in postpartum rats (see the previous subsection on the mesolimbic DA system and maternal behavior), the performance of consummatory nursing behavior during the early postpartum period, which is associated with DA increases in the NA (Champagne et al., 2004; Hansen et al., 1993), may activate a reinforcement mechanism that is necessary for the subsequent enhancement of maternal motivation.

There is also evidence that maternal experience may modify the MPOA to enhance future maternal responsiveness. Maternal experience may alter the structure and function of MPOA neurons so that they are more responsive to infant-related neural inputs even in the absence of continued hormonal priming. Shams et al. (2012) compared the dendritic morphology of MPOA neurons in virgin female rats without maternal experience and parturient primiparous rats that had 24–36 hours of maternal experience. Maternal experience was associated with an increase in dendritic branching within those dendrites located close to the neuronal cell body. Such a modification would allow MPOA neurons to be more affected by excitatory (as well as inhibitory) inputs. However, since a parturient group without maternal experience was not included in this study, it is not clear whether the morphological changes observed in the MPOA were due to maternal experience per se or, instead, to the physiological events associated with pregnancy and parturition.

Some interesting work on the effects of maternal experience on MPOA functional organization has been conducted on laboratory mice. Virgin female laboratory mice show near immediate maternal behavior toward pups when tested in their home cages, but unlike parturient female mice, they are not likely to retrieve pups in a novel T-maze (Stolzenberg & Mayer, 2019; Stolzenberg & Rissman, 2011; also see Chapter 3 of this volume). However, if virgin mice are allowed 4 days (2 hours/day) of maternal experience caring for pups in their home cages, then they are just as likely as postpartum females to retrieve pups in a novel T-maze. These results indicate that maternal experience with pups can substitute for the physiological events associated with pregnancy and parturition to allow the virgin female laboratory mouse to care for pups under challenging environmental conditions. Since anxiety-related responses, as measured in the elevated plus maze, were not affected by maternal experience, Stolzenberg and Rissman (2011) interpreted this maternal experience effect in virgins as being the result of an enhancement in maternal motivation.

In a subsequent study, Stolzenberg, Stevens, and Rissman (2012) reported that OT expression is increased in the MPOA of virgin female mice that received 4 days of maternal experience in comparison to virgins that received only 2 days of maternal experience. The former group, but not the latter, retrieved pups in a novel T-maze. To measure OT expression, tissue punches were used to dissect out the MPOA and OT mRNA levels were measured. In examining the brain atlas coordinates that were used to make the MPOA tissue punches, it seems clear that the ACN, which is equivalent to the anterior part of the PVN, that lies dorsal to the posterior part of the MPOA proper was included in their punches. Therefore, I would conclude that OT expression was increased by maternal experience in the anterior PVN, rather than in the MPOA proper (unless one considers the ACN as part of the MPOA; see the previous discussion in this chapter of neurotransmitters/neuromodulators in MPOA/vBST). Such an increase in OT expression might have been able to cause an increase in maternal motivation as a result of oxytocinergic stimulation of OTRs on nearby MPOA neurons that project to the VTA and/or by direct OT projections from anterior PVN to VTA DA neurons (see Figure 5.10).

In rats, Amico, Thomas, and Hollingshead (1997) have shown that the hormonal events of pregnancy and parturition can increase OT mRNA in the PVN. The work of Stolzenberg et al. (2012) show that maternal experience in virgin mice can produce the same effect. Maternal experience can substitute for pregnancy hormones to induce OT increases in PVN, which presumably increases maternal motivation.

Direct experimental evidence that maternal experience-induced increases in OT release into MPOA may increase maternal motivation has been provided by Okabe et al. (2017). Virgin female mice were allowed either 20 minutes or 2 hours of maternal experience with pups in their home cages. Four days later, these females were re-exposed to pups in their home cages and latencies to retrieve three pups were recorded. The 2-hour maternal experience group showed significantly shorter retrieval latencies than did the 20-minute exposure group (2 minutes vs. 7 minutes, respectively) during the re-exposure session. Although this is not a dramatic effect, it does indicate that the long-exposure females were more eager to care for their pups. Also note that these tests were conducted in the home cage situation. Most important, when an OTR antagonist was injected into the MPOA during the initial exposure phase, but not during the re-exposure phase, the effect of the long maternal experience on subsequent maternal motivation was abolished. It is worth pointing out that the OTR antagonist did not affect maternal behavior during the initial pup exposure/experience condition, but did affect the translation of maternal experience into enhanced maternal motivation 4 days later. A speculative interpretation of these results can be proposed: An initial long period of maternal experience results in the release of OT from the anterior PVN into the MPOA. OT stimulation of MPOA output to VTA DA neurons may ultimately strengthen synapses with the mesolimbic DA system (BLA/BMA-to-VP) allowing for enhanced appetitive maternal motivation at a future time point (see Figure 5.15).

In summary, maternal experience effects at the level of the MPOA may enhance future maternal motivation by (a) making the MPOA more receptive to sensory input from pups and (b) activating OT input to MPOA, which, in turn, affects synaptic plasticity within the mesolimbic DA system.

Maternal experience may act to enhance future maternal behavior not only by potentiating the sensitivity of the MPOA to infant stimuli and its ability to activate the mesolimbic DA system, but also by potentiating the ability of MPOA output to depress the defensive-avoidance circuit that normally depresses maternal behavior in most inexperienced mammalian females. There is not much work that has investigated this likely possibility (see Numan & Insel, 2003). Research by Mayer, Helton et al. (2019) lends some support for this additional mechanism through which maternal experience might exert its effects on maternal responsiveness.

I will end this section by asking an interesting question: Are the maternal experience-induced mechanisms that regulate the nonhormonal maintenance of maternal behavior identical to those that regulate the formation of maternal memory? In the previous subsection on neural inputs to MPOA relevant to maternal behavior, I described the research of Cservenak et al. (2013). In postpartum rats, neurons in the PIL of the thalamus send a TIP39 projection to the MPOA, and PIL neurons are responsive to somatic sensory inputs from pups (see Figure 5.8). Cservenak et al. injected a long-acting TIP39 antagonist into the MPOA. This treatment did not disrupt the onset and maintenance of maternal behavior, but did prevent the formation and/or retention of a CPP for a distinct cage compartment that was associated with pups. I would like to propose an interesting follow-up study, which would use a short-acting TIP39 antagonist (see Cservenak et al., 2010), injected into the MPOA either during the training or retention trials of the CPP paradigm, to determine whether TIP39 input to the MPOA was necessary for the formation or retention of a CPP. If a TIP39 antagonist injection into MPOA blocked the formation, but not the retention, of a CPP, it would suggest that TIP39 action at the level of the MPOA, induced by ventral somatic sensory stimulation, may also be involved in maternal memory formation.

The formation of a CPP in lactating rats for a pup-associated context shares characteristics that are similar to the formation of maternal memory (Fleming, Korsmit, & Deller, 1994). Like maternal memory formation, the formation of a CPP requires full interaction with pups during the training sessions, and the activation of D1 and D2 DA receptors are also involved. An interesting question, therefore, is whether blockade of TIP39 action on the MPOA during an initial exposure to pups during the early postpartum period would block the formation of maternal memory so that when the affected females are re-exposed to pups after several weeks of separation from pups, an enhancement of maternal responsiveness would not be detected.

Since TIP39 antagonism in the MPOA does not disrupt the maintenance of maternal behavior, if it did disrupt maternal memory formation, such a finding would suggest that the mechanism underlying the maternal experience-induced nonhormonal maintenance of maternal behavior is at least partially distinct from the maternal experience-induced formation of maternal memory. Future research should be aimed at investigating this important issue. One possibility is that there are difference degrees of synaptic modifications within parental circuits that underpin either the maintenance of maternal behavior or the formation of maternal memory. A stronger enhancement of certain synapses within circuits that mediate maternal motivation and/or a stronger depression of certain synapse within circuits that inhibit maternal behavior may be required for maternal memory formation than for the nonhormonal maintenance of maternal behavior.

General Conclusions

This chapter described in detail the specific neural circuits and their neurochemical make-up, which have been shown to regulate the onset and maintenance of maternal behavior in nonhuman mammals. The research reviewed primarily explored those circuits involved in appetitive maternal motivation, with an emphasis on subcortical brain regions. Most of this research has been performed on rodents, but, as I will show in Chapter 8, these circuits are also involved in the parental motivation of humans. This result is not surprising since maternal behavior is a defining characteristic of all mammals, and therefore one would predict that evolutionarily conserved core subcortical circuits influence maternal motivation in all mammals.

For the typical female mammal that displays a uniparental maternal care system, nulliparous females avoid or reject infants, while primiparous parturient females care for infants. For these females, the physiological events associated with pregnancy and parturition act at crucial neural sites to depress avoidance responses to infants and to enhance maternal motivation, and this chapter described what we know about the neural circuits that depress maternal behavior in virgins and those that stimulate maternal behavior in mothers.

The MPOA was shown to be essential for both the appetitive and consummatory aspects of maternal behavior, and I described the larger neural circuitry within which the MPOA is embedded in its control of these two aspects of maternal motivation, with an emphasis on appetitive processes.

With respect to the stimulation of the appetitive aspects of maternal motivation, MPOA interaction with the mesolimbic DA system was shown to be crucial. In addition, synaptic plasticity within this critical circuit, which is triggered by consummatory maternal responses, appears to be involved in establishing the hormone/OT-independent maintenance phase of maternal behavior and in the establishment of maternal memory for those species in which a maternal memory process has been shown to occur.

Finally, the research I have examined has stressed the importance of the monoamine neurotransmitter, DA, in the regulation of maternal behavior. However, other monoamines are undoubtedly also involved, and a recent body of literature has begun to explore the importance of serotonin neural systems (Numan, 2015; Pawluski, Li, & Lonstein, 2019). One of the ways in which serotonin may affect maternal motivation is through its effects on the mesolimbic DA system (Pawluski et al.). Additionally, in Chapters 9 and 10, I will present an analysis of the involvement of serotonin neural systems in the developmental aspects of maternal behavior.

6

Anxiety Reduction and Maternal Aggression in Postpartum Nonhuman Mammals

Introduction

Up to this point, I have been concentrating on the hormonal and neural mechanisms that underpin infant-directed maternal responses in nonhuman mammalian females. However, the maternal condition is also associated with other important behavioral changes that are not infant-directed, but serve adaptive functions to ensure the survival of the mother's offspring. Several observational and experimental studies have indicated that postpartum mammals show a decrease in fearfulness and anxiety-related behaviors that co-occurs with an increase in aggression toward conspecific intruders that approach the mother's infant(s) (Agrati & Lonstein, 2016; Lonstein, 2007; Lonstein & Gammie, 2002; Numan, 1994; Numan & Insel, 2003; Numan & Woodside, 2010; Ostermeyer, 1983). The purpose of this chapter is to understand the nature of these changes and how they are related to one another. Most of the careful experimental research on this subject has been conducted on rodents. Initially, I will use the terms *fearfulness* and *anxiety* as if they represent the same process. However, later in this chapter, in the section on the neural circuitry of fear/anxiety and the mechanisms mediating its postpartum downregulation, I will present more accurate definitions of these terms.

A postpartum mother should be vigilant toward environmental factors that might harm her infants. She should have the emotional capability to cope with stressful or dangerous environmental challenges so that she can appropriately care for and protect her offspring. I will use postpartum maternal aggression as the prime example of how a mother meets a dangerous environmental challenge to safeguard her offspring. Research indicates that maternal aggression serves the purpose of protecting a mother's infant(s) from potentially infanticidal conspecifics (Hausfater & Hrdy, 1984; Lonstein & Gammie, 2002; Numan, 1994; Ostermeyer, 1983). If a mother does not effectively counter the approach of an intruder, her offspring are likely to be harmed and her reproductive success would decrease. Therefore, natural selection has undoubtedly resulted in the

The Parental Brain. Michael Numan, Oxford University Press (2020). © Oxford University Press.
DOI: 10.1093/oso/9780190848675.001.0001

evolution of an increase in aggressiveness toward strange conspecifics in postpartum mothers.

How might postpartum fearfulness and anxiety be related to postpartum maternal aggression? Figure 6.1 presents a hypothesis (Numan, 2010, 2017; Numan & Woodside, 2010). The figure proposes that there is an optimal level of anxiety-related behavior or fearfulness, which is associated with a proper level of attentive vigilance, which has a permissive effect on the occurrence of protective maternal aggression. If fearfulness is too low, resulting in a lack of concern for potential danger, a mother might not be vigilant enough to detect an approaching intruder, or she may be passive or indifferent to such an approach. But if a mother is too fearful, she may run away from an intruder instead of protecting her young. Therefore, a moderate amount of vigilance and fearfulness may be necessary for a mother to effectively protect her offspring.

There is some evidence to support the proposal shown in Figure 6.1. The elevated plus maze (EPM) is used to measure anxiety-related behavior in rodents, and the amount of time that an individual spends in the open (unprotected) arms of the maze is inversely related to the anxiety state of the organism. Ragan and Lonstein (2014) selectively bred rats to obtain two separate populations. One population of rats was selectively bred to display high anxiety-related behavior when they were virgins, and the other group was bred to show low anxiety-related behavior. These virgins were then mated, and their postpartum level of anxiety was measured in the EPM. Interestingly, the females from these two populations did not differ in their anxiety levels during the postpartum period; the results indicated that the postpartum condition was associated with a

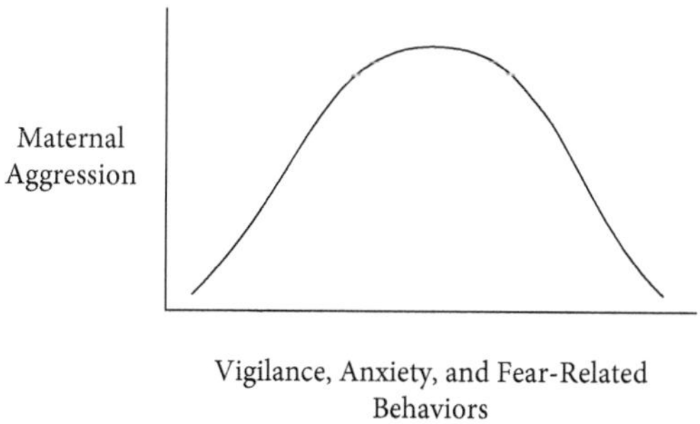

Figure 6.1. A proposed relationship between postpartum anxiety/fear and the ability to display effective maternal protective responses as measured by the occurrence of maternal aggression. The graph proposes that a moderate amount of vigilance, fearfulness, and anxiety allows for effective maternal aggression, while low levels or high levels of these emotional characteristics hinder the expression of maternal aggression.

decrease in anxiety in the high anxiety females and an increase in anxiety in the low anxiety females. These results can be interpreted as evidence that an optimal level of anxiety-related behavior is reached as a result of the physiological and behavioral (mother–pup interactions) changes that occur in parturient females. In a related study, Neumann, Kromer, and Bosch (2005) selectively bred rats so that virgin females showed extremely low levels of anxiety in the EPM. In their particular selectively bred strain, this low level of anxiety did not increase during the postpartum period, and these low anxiety rats also showed low levels of postpartum maternal aggression when compared to postpartum females that showed higher levels of anxiety.

In comparing these two studies, it appears that the postpartum condition was able to modify anxiety levels in the Ragan and Lonstein (2014) study, but not in the Neumann et al. (2005) study and that the extremely low levels of postpartum anxiety in the latter study were associated with very low levels of postpartum maternal aggression. Given that the postpartum anxiety levels were equivalent in the Ragan and Lonstein study, I would predict that postpartum maternal aggression between these two strains would have also been equivalent, but this possibility was not tested.

The results of Neumann et al. (2005) suggest that very low levels of fearfulness and presumed vigilance during the postpartum period are associated with low levels of maternal defense of offspring (the left end of the X-axis in Figure 6.1). The focus of this chapter, however, will be on the following proposition: When anxiety and fearfulness are extremely high so that the postpartum female is not equipped to cope with danger, then maternal aggression may be suppressed (the right end of the X-axis in Figure 6.1). In testing this proposition, it will be necessary to show that when postpartum anxiety is experimentally increased above optimal levels, maternal aggression will be suppressed. In undertaking this goal, I will outline the neural circuits that underpin maternal aggression and anxiety-related behavior, and I will then examine whether these circuits interact so that high activity in anxiety circuits acts to suppress maternal aggression circuits. Note that these two circuits are separate, and the hypothesis I am testing states that supernormal activity within anxiety-related circuits suppresses the activity of maternal aggression circuits. Therefore, an intervention directly within the neural circuits that regulate maternal aggression should leave postpartum anxiety levels intact. However, interventions that increase activity in anxiety circuits above a presumed optimal level should suppress activity in the maternal aggression circuits.

The time courses of the increase in maternal aggression and decrease in anxiety in postpartum rodents are virtually identical, both occurring during the first 2 weeks postpartum, and recent contact with pups is necessary for both events to occur (Gammie & Lonstein, 2006). The evidence indicates that the hormonal and other physiological events associated with pregnancy and parturition set up a postpartum maternal state that enables maternal contact with infants to maintain maternal anxiety reduction and heightened maternal aggression during the early postpartum period, a period during which infants are most vulnerable (Numan & Insel, 2003). These correlations are suggestive of a causal relationship between decreased fearfulness and increased aggression.

In this chapter, I will also present evidence that shows that too much anxiety and behavioral stress reactivity may also directly depress pup-directed maternal behavior. Therefore, high behavioral stress reactivity may not only interfere with maternal aggression, but when severe enough may also disrupt mother–infant interactions (see Klampfl & Bosch, 2019, for a recent review of these topics).

Behavioral Characteristics of the Postpartum Reduction in Fearfulness

Fleming and Luebke (1981) were the first to draw attention toward the occurrence of decreased fearfulness during the postpartum period. In a variety of tests that are used to measure anxiety-related behaviors, such as emergence from a start box into a novel open field, they showed that virgin female rats were more anxious or fearful than were primiparous rats that were tested during the early postpartum period: The postpartum rats emerged into the open arena with a shorter latency than did the nulliparous females. Similar results were reported by Bitran, Hilvers, and Kellogg (1991) in the EPM test, where postpartum females spent more time in the open arms than did virgins. It is important to emphasize that these tests for anxiety-related behavior in postpartum females were examined when pups were not present in the novel environment. This is an important consideration, because, as I outlined in Chapters 3 and 4, if pups are present in a novel environment, such as an open field or an EPM, then any increase in the exploration of a fear-provoking region in postpartum females in comparison to naïve virgins could be due to a decrease in fearfulness, an increase in maternal motivation, or both. Therefore, in my analysis of anxiety reduction during the postpartum period, I will emphasize those studies that test anxiety-related behavior in mothers without pups being present in the novel environment. An increase in the exploration of a novel and potentially threatening environment by postpartum females under such conditions, in comparison to virgins, shows that a general reduction in fearfulness/anxiety to a variety of novel situations occurs in postpartum females.

With respect to fearfulness, when a parturient female rat gives birth, two events occur: The aversive qualities of pup stimuli are eliminated, and there is also a reduction in general fearfulness and anxiety. Can these processes be separated? One approach to this question is to examine whether virgin females that are induced to show maternal behavior through the sensitization method also show a general reduction in anxiety-related behavior. This issue was addressed by Pereira, Uriarte, Agrati, Zulaga, and Ferreira (2005). Virgin female rats were either not exposed to pups or were sensitized through continuous pup exposure to show maternal behavior. A third group consisted of postpartum lactating females. These females were tested for anxiety-related behavior in the EPM (pups were not present in the maze). The sensitized maternal virgins and lactating females were tested after 7 days of maternal responsiveness. In comparison to the nonsensitized virgins, both the sensitized virgin and the postpartum females showed a reduction in anxiety and spent more time in the open arms. However, the lactating females showed a much greater reduction in anxiety than did the sensitized virgins: While the postpartum rats spent about 10% of their time in the EPM within the open arms, the sensitized virgins spent only about 3% of their time in the open arms. Also, 100% of the postpartum females entered the open arms while this was the case for only 57% of the sensitized virgins; none of the nonsensitized virgins entered the open arms. As I will soon show, recent proximal contact with pups appears to be important in causing a general reduction in fearfulness and anxiety in maternal females. While this effect can occur in sensitized virgins after prolonged contact with pups, the effect of mother–pup interaction in promoting a reduction in anxiety is much more potent in postpartum females. In the next section, I will show that a similar relationship occurs with respect to maternal aggression. Since sensitized females no longer find pup cues to be aversive, but they display much less anxiety reduction than do postpartum females, I conclude that these two processes are at least partly independent. There is something about the postpartum condition that not only eliminates the aversive qualities of pup stimuli, but also allows mother–infant contact to cause a dramatic decrease in the mothers' general fearfulness and anxiety, a decrease that is much larger than that observed in sensitized virgins.

Lonstein (2005), in an important study, analyzed in more detail the characteristics of postpartum anxiety reduction in rats. In the EPM, anxiety-related behavior was compared between naïve virgin (not exposed to pups and therefore not sensitized) and postpartum females. Pups were not present in the maze during the tests. The findings can be summarized as follows: (a) Postpartum females were less anxious than virgins on days 1 and 7 postpartum, but not later in the postpartum period (days 14 and 21 postpartum); (b) recent contact with pups was necessary for the decrease in anxiety in postpartum females. If the females were separated from their pups for 4 hours before being tested on the EPM, their anxiety level increased to that shown by virgins; (c) exteroceptive stimuli from pups were not able to maintain anxiety reduction in postpartum females; full proximal contact was necessary. When mothers were exposed to pups that were placed in a small wire cage for 4 hours before the mothers were tested in the EPM, which would provide the postpartum mother with visual, auditory, and olfactory stimulation from pups, but precluded somatic sensory tactile inputs from pups, the fearfulness of the mothers in the EPM reverted to the higher levels typical of virgins; (d) thelectomy (nipple removal) did not prevent anxiety reduction in postpartum females, indicating that suckling stimulation is not one of the proximal pup cues that causes the decrease in postpartum anxiety; and (e) ovariectomy or hypophysectomy did not prevent the typical reduction in anxiety in day 7 postpartum females.

Lonstein (2005) concluded that the physiological events of pregnancy set up a postpartum state that allows direct pup contact to maintain large decreases in anxiety-related behavior during the postpartum period. In a manner similar to the onset and subsequent maintenance of maternal behavior, it appears that the physiological factors associated with pregnancy and parturition subsequently allow direct pup contact to maintain postpartum anxiety reduction during the early postpartum period without the further need for continued hormonal stimulation. Lonstein emphasized the importance of ventral somatic sensory inputs from pups, but not necessarily suckling, as playing the primary role in maintaining postpartum anxiety reduction in rats.

Behavioral Characteristics of the Postpartum Increase in Aggression

In laboratory tests for aggression in rodents, an adult male or female intruder is placed in the home cage of a resident female, and the number of resident females fighting, the resident's latency to its first attack, and the number and duration of attacks by the resident are recorded. For a variety of rodent species, lactating resident females have been found to be much more aggressive than nonlactating resident females (Lonstein & Gammie, 2002; Numan & Insel, 2003).

In rats and mice, maternal aggression is high during the first 2 weeks of the postpartum period, and many of the characteristics of postpartum maternal aggression mirror the characteristics of postpartum anxiety reduction. First, aggression toward an intruder is much higher in postpartum females than it is in nonmaternal virgin females and sensitized virgins (Erskine, Barfield, & Goldman, 1980b; Ferreira, Pereira, Agrati, Uriarte, & Fernandez-Guasti, 2002; Martin-Sanchez et al., 2015). As an example, in rats, Ferreira et al. (2002) found that nonsensitized virgins did not show any aggressive responses toward an intruder in a 5-minute test, and sensitized virgins displayed only 3 attacks toward the intruder. In contrast, postpartum mothers showed 17 aggressive attacks toward the intruder. Second, postpartum hypophysectomy does not disrupt the high levels of aggression shown by postpartum mothers (Erskine, Barfield, & Goldman, 1980a). Finally, although short-term removal of young pups from the postpartum female's cage does not depress maternal aggression, long-term removal (5–24 hours) does (Erskine, Barfield, & Goldman, 1978; Ferreira & Hansen, 1986; Stern & Kolunie, 1993; Svare & Gandelman, 1973). Such a depression of aggression is reversed if pups are returned to the home cage several minutes prior to the aggression test.

In a manner similar to postpartum anxiety reduction, it appears that the physiological factors associated with pregnancy and parturition set up a postpartum maternal state that then allows pup stimuli to maintain high levels of aggression during the early postpartum period in the absence of the need for continued hormonal mediation (Numan & Insel, 2003). What is the nature of pup stimuli that maintain maternal aggression in postpartum females? For rats, the same stimuli that maintain postpartum anxiety reduction seem to also maintain high levels of postpartum maternal aggression. Ventral somatic sensory inputs from pups (but not necessarily suckling stimulation) are important (Mayer et al., 1987; Stern & Kolunie, 1993). For mice, in contrast to rats, ventral somatic sensory inputs from pups that include suckling stimulation during the early postpartum period are necessary for the maintenance of high levels of maternal aggression (Svare & Gandelman, 1976).

Opposing Roles of Oxytocin and Corticotropin-Releasing Factor in Anxiety-Related Behaviors

In Chapter 4, data were reviewed that showed that central oxytocin (OT) systems not only promote the onset of maternal behavior, but also exert anxiolytic effects. Therefore, the increased central release of OT into the brain as a result of mother–infant interactions during the postpartum period likely plays a role in the anxiety and fear reduction that occurs during this time, and this effect may help the mother cope with environmental challenges to care for and protect her offspring (Neumann, 2008; Neumann & Slattery, 2016). In support of this view, Neumann et al. (2000) showed that intracerebroventricular (ICV) administration of an OT receptor antagonist (OTA) to postpartum lactating rats significantly enhanced their anxiety-related behavior as measured in the EPM.

Corticotropin-releasing factor (CRF) is well known for its neuroendocrine effects in regulating the physiological stress response (Ulrich-Lai & Herman, 2009). Through several neural pathways, stressful, fearful, and anxiety-provoking stimuli activate CRF neurons located in the paraventricular nucleus (PVN) of the hypothalamus. The PVN also contains OT neurons that project to both the brain and the posterior pituitary (neural lobe), but CRF-containing neurons in the PVN represent a separate population from PVN-OT–containing neurons; they do not co-localize within the same neurons (Dabrowska et al., 2011). Some PVN-CRF neurons project to the median eminence where CRF is released into the pituitary portal veins to reach the anterior pituitary, where CRF stimulates the release of adrenocorticotropic hormone (ACTH) into the systemic blood supply. ACTH then stimulates the release of the steroid hormones, cortisol, or corticosterone from the adrenal cortex. One effect of circulating adrenal steroids is to stimulate gluconeogenesis by the liver, which raises blood glucose levels to provide an enhanced energy source to help an organism cope with a stressful environment. Since these adrenal steroids raise glucose levels, they are referred to as glucocorticoids.

In contrast to OT, CRF has anxiogenic behavioral effects (Adamec & McKay, 1993; Dunn & Berridge, 1990; Zhang et al., 2017). The fear and anxiety-inducing properties of CRF are due, in part, to its actions on neurons within the brain. PVN-CRF neurons not only project to the median eminence, put also project centrally to reach neural regions such as the central nucleus of the amygdala (CeA) and the bed nucleus of the stria terminalis (BST; Zhang et al., 2017). Furthermore, there are additional neural populations within the brain, outside the PVN, that produce CRF, which is used as a neurotransmitter or neuromodulator within central brain circuits. For example, the CeA contains a major population of CRF-containing neurons that project to the BST (Asok et al., 2018; Sakanaka, Shibasaki, & Lederis, 1986).

What is the evidence that the central release of CRF produces anxiogenic effects? Adamec and McKay (1993) found that ICV administration of CRF decreased the amount of time that rats explored the open arms of the EPM (also see Dunn & Berridge, 1990). It is worth noting, however, that systemic corticosterone administration to rodents also exerts fear and anxiety-inducing effects (Korte, 2001). Therefore, when one injects CRF into the cerebral ventricles, the peptide may ultimately activate anterior pituitary release of ACTH, and the resultant increase in glucocorticoids may be the factor that directly exerts an anxiogenic effect. This proposal is plausible because glucocorticoids can enter the brain to act on glucocorticoid receptors within certain neurons (Myers, McKlveen, & Herman, 2014). Importantly, Adamec and McKay showed that ICV CRF injections exerted an anxiogenic effect when hypohysectomized rats were tested in the EPM, supporting a direct influence of CRF action within the brain in promoting anxiety-related behavior. Given these findings, how do high levels of systemic corticosterone cause anxiety? There are probably multiple routes through which enhanced glucocorticoid action on its receptors in the brain produces heightened anxiety. One important mechanism is that within certain brain regions, such as CeA, corticosterone increases the expression of CRF mRNA (Schulkin, 2011; Shepard, Barron, & Myers, 2000). Therefore, high levels of glucocorticoids may stimulate the synthesis of CRF at certain brain sites and the central release of CRF from these sites may then induce a state of anxiety and fearfulness.

Given that the postpartum state is associated with decreased fearfulness/anxiety, one might expect that the central CRF system is relatively inactive during this period. If there is an inverse relationship between anxiety and maternal aggression, it can be proposed that increasing the functional activity of central CRF systems would increase anxiety and decrease maternal aggression. To begin an analysis of this proposal, I will review supportive research that employs ICV injections of CRF. In postpartum lactating rats, Klampfl, Neumann, and Bosch (2013) found that ICV administration of CRF into the lateral ventricle increased anxiety as measured in the EPM and decreased maternal aggression. Gammie, Negron, Newman, and Rhodes (2004) similarly reported that ICV administration of CRF decreased maternal aggression in lactating mice, but they did not examine potential effects on anxiety-related behavior. Gammie et al. proposed, however, that the central release of endogenous CRF is probably suppressed during early lactation and that this effect decreases fearfulness, which, in turn, potentiates maternal aggression.

The evidence, up to this point, indicates the opposing roles of OT and CRF on anxiety. I have just shown that an increase in anxiety due to the central administration of CRF appears to be associated with a decrease in maternal aggression. Is there any evidence that a decrease in the central actions of endogenous OT, which also results in an anxiogenic effect in postpartum rats (Neumann et al., 2000), would decrease maternal aggression? The PVN is the major source of OT neural projections within the brain and contains magnocellular (large cell bodies) and parvocellular (small cell bodies) neurons. Research indicates that magnocellular PVN-OT neurons project to forebrain targets, which include the amygdala (Knobloch et al., 2012). Within this context, it is instructive to examine conflicting findings with respect to the effects of PVN lesions of maternal aggression. Consiglio and Lucion (1996) performed electrical PVN lesions on day 5 postpartum in rats. Similar to the findings of Numan and Corodimas (1985), these lesions did not affect the pup-directed maintenance of maternal behavior, but they dramatically decreased maternal aggression when tested on days 7 and 9 postpartum. The electrical lesions of the PVN destroyed both magnocellular and parvocellular PVN neurons. Giovenardi, Padoin, Cadore, and Lucion (1998) injected ibotenic acid into the PVN of postpartum rats. Ibotenic acid, like *N*-methyl-D-aspartic acid (NMDA), produces neuron-specific damage without destroying fibers of passage. The findings indicated that such PVN lesions had no effect on postpartum maternal aggression. Significantly, the ibotenic acid lesions destroyed parvocellular PVN neurons, but left the magnocellular neurons intact. One interpretation of the findings from these two studies is that it may be necessary to destroy magnocellular PVN-OT neurons and their projections to the forebrain, and perhaps to other parts of the brain, to disrupt maternal aggression. It would be interesting to know if PVN electrical, but not ibotenic acid, lesions of the PVN would have also resulted in an anxiogenic effect. Finally, OT release from PVN cell bodies and dendrites exerts a positive feedback effect on OT release into the brain (see Chapter 4 of this volume). Bosch, Meddle, Beiderbeck, Douglas, and Neumann (2005) found that the administration of an OTA into the PVN, which would suppress OT release from the PVN to its forebrain targets, significantly depressed maternal aggression in postpartum rats. From these findings, it makes sense that PVN-OT projections to the forebrain promote maternal aggression and that this effect might be indirect, and mediated by the anxiolytic properties of the central actions of OT.

The evidence I have reviewed supports the hypotheses that an upregulation of CRF and a downregulation of OT increases anxiety, which may then decrease maternal aggression. These findings conform with the proposal that too much anxiety decreases maternal aggression. However, it is certainly possible that CRF and OT produce their effects by acting directly on independent anxiety/fear-related neural systems and maternal aggression neural systems. That is, CRF may directly activate fear/anxiety circuits and directly inhibit aggression circuits, while OT may directly inhibit fear/anxiety circuits and directly activate aggression circuits.

Therefore, in the following sections, I will critically evaluate research related to the proposal that high activity within anxiety/fear-related neural circuits may suppress activity with within the neural circuits that regulate maternal aggression. To do this, I will first describe what we know about the neural circuits that regulate maternal aggression, followed by a review of the circuits that underpin anxiety and fear-related behaviors. Finally, I will explore how these circuits might interact and whether this interaction includes a suppressive effect of high activity within anxiety circuits on maternal aggression circuits.

The Neural Circuitry of Maternal Aggression

In this section, I will describe the neural circuits that have been implicated in the regulation of maternal aggression in postpartum rodents. As this analysis proceeds, it will be helpful for the reader to refer to Figure 6.2, which presents a summary diagram of the relevant data.

With respect to the sensory factors that are necessary for a postpartum female rodent to detect an intruder and react aggressively, evidence indicates that intruder-related olfactory stimuli that activate the main olfactory and vomeronasal systems are involved (Numan, 1994). For example, Ferreira, Dahlof,

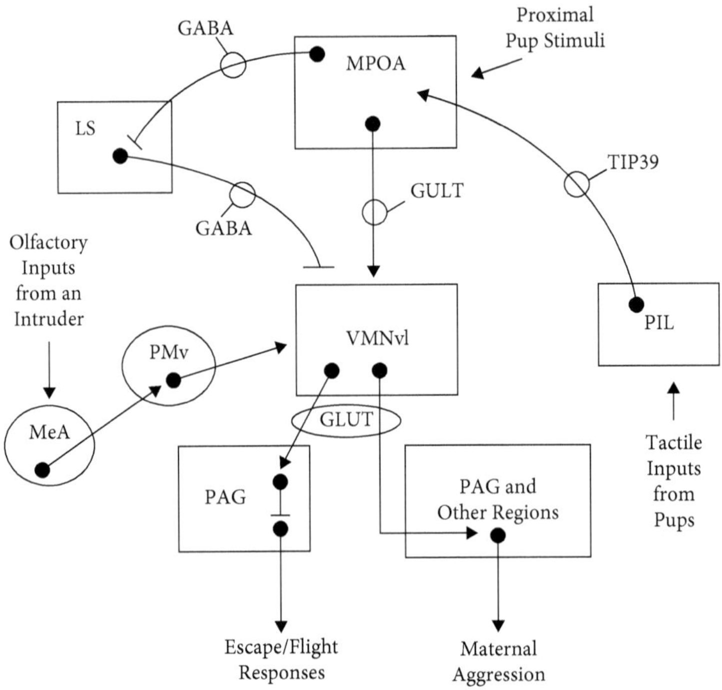

Figure 6.2. A summary diagram of the neural circuitry that promotes the occurrence of maternal aggression in postpartum rodents. Olfactory stimuli from an intruder at the mother's nest site activate the mother's medial nucleus of the amygdala (MeA). A MeA-to-ventral premammillary nucleus of the hypothalamus (PMv)-to-ventrolateral region of the ventromedial nucleus of the hypothalamus (VMNvl) neural circuit activates VMNvl neurons. VMNvl neurons project to the periaqueductal gray (PAG) in the midbrain and to other regions in the brainstem in order to depress fear-related escape/flight responses while also promoting the occurrence of maternal aggression. There is also evidence that a GABAergic projection of the lateral septum (LS) to VMNvl neurons depresses maternal aggression. Proximal contact with pups is necessary for the maintenance of maternal aggression. Some medial preoptic area (MPOA) neurons may enhance maternal aggression and it is possible that proximal pup stimuli activate these neurons. For example, the mother's reception of ventral somatic sensory inputs from pups may occur, in part, through projections from tuberoinfundibular peptide 39 (TIP39) containing neurons in the posterior intralaminar complex (PIL) of the thalamus to the MPOA. The output of the MPOA may enhance maternal aggression through two hypothetical routes. MPOA glutamate (GLUT) projections to VMNvl may directly stimulate maternal aggression while MPOA GABAergic projections to the LS may indirectly promote maternal aggression by suppressing the inhibitory effects of the LS. Direct MPOA projections to PAG (not shown) may also influence maternal aggression. Axons ending in an arrow exert excitatory effects while those ending in a bar are inhibitory.

and Hansen (1987) showed that postpartum destruction of the olfactory system within the nasal cavity virtually eliminated the occurrence of maternal aggression, without affecting pup-directed maternal care in lactating rats. Therefore, the detection of olfactory stimuli emitted by an intruder appears to be essential to elicit a maternal aggressive response. As shown in Figure 5.7, both primary and vomeronasal olfactory inputs converge on the medial amygdala (MeA). Unger et al. (2015), using a chemogenetic approach in postpartum mice, showed that the ablation of a specific subpopulation of neurons within MeA suppressed maternal aggression in postpartum mice, without affecting pup-directed maternal behavior (also see Hong, Kim, & Anderson, 2014). The disruption of these MeA neurons, which reduced maternal aggression, did not affect postpartum anxiety levels, as measured in the EPM.

Two aspects of these studies by Ferreira et al. (1987) and Unger et al. (2015) are worthy of consideration. First, the neural circuits that regulate maternal aggression are at least partially independent from the circuits that regulate pup-directed maternal behavior because olfactory ablation or ablation of a specific population of MeA neurons disrupted the former, but not the latter, behavior. Second, when one directly interferes with the circuits underlying maternal aggression, reduced postpartum fearfulness remains intact as long as sufficient proximal pup contact is maintained. More specifically, even if postpartum anxiety reduction is maintained, direct interference with maternal aggression neural circuits will suppress maternal aggression.

Which brain regions might be involved in receiving intruder-related olfactory inputs that then trigger maternal aggressive responses? Hansen (1989) was the first to show that neuron-specific damage to the ventrolateral region of the ventromedial nucleus of the hypothalamus (VMNvl; see Figure 6.4) disrupted maternal aggression in postpartum rats without affecting pup-directed maternal behavior. In my discussion of defensive-avoidance circuit active in nonmaternal virgin female mammals in Chapter 5, I noted that the VMN is a complex nucleus that contains separate populations of neurons involved in either aggression or fear-related (avoidance/escape) responses. Hansen's lesions presumably damaged both of these populations, but in postpartum rodents the fear-related neurons are presumably less active and the direct damage to the aggression-related neurons was the likely factor that disrupted maternal aggression.

More advanced chemogenetic and optogenetic techniques in mice have indeed shown that aggression-related neurons in VMNvl are essential for maternal aggression. Hashikawa et al. (2017) identified a specific population of neurons in VMNvl that display increased neural activity during maternal aggression in mice. Selective inhibition of these neurons in postpartum mice disrupted maternal aggression while not interfering with pup retrieval. Further, while virgin female mice show very little aggression toward an adult conspecific intruder, optogenetic stimulation of VMNvl aggression-related neurons in virgins increased aggression toward an intruder. These findings indicate that the output of a particular population of neurons in VMNvl is essential for aggressive responses toward an intruder in female mice and that the activity of these neurons is upregulated in the postpartum female.

Motta et al. (2013) have provided evidence concerning the particular neural route through which intruder-related olfactory stimuli may activate VMNvl to elicit aggressive responses in postpartum rats. They note that the ventral premammillary nucleus (PMv) of the hypothalamus, a nucleus that is located immediately caudal to VMN (see Figure 4.1), receives strong inputs from MeA (Canteras et al., 1995). They found that neuron-specific lesions of PMv virtually eliminated maternal aggression while not affecting pup-directed maternal care in postpartum rats. They suggest that olfactory inputs activate MeA and the following neural pathway may then trigger maternal aggression (refer to Figure 6.2): MeA-to-PMv-to-VMNvl. Note that PMv projects strongly to VMNvl (Canteras, Simerly, & Swanson, 1992).

While olfactory inputs ultimately activate VMNvl aggression neurons to promote maternal aggression, evidence exists that the lateral septum (LS; see Figure 4.1 for the general location of LS in the telencephalon) exerts an inhibitory control over VMNvl aggression-related neurons. Lee and Gammie (2009) have reported that experimentally induced increases in LS activity suppress maternal aggression in postpartum mice. Wong et al. (2016), although not examining maternal aggression, examined the role of an LS-to-VMNvl projection in the regulation of male aggression during a resident–intruder test in mice. Most of the LS output neurons are GABAergic. Using optogenetic techniques, they stimulated the LS GABAergic projection to VMNvl aggression-related neurons and found that such stimulation suppressed aggressive responses that the resident male mouse would normally show toward a conspecific intruder. Further, pharmacological inactivation of the LS increased the resident's aggressive responses. In comparing these two studies, it can be proposed that the GABAergic output of the LS to VMNvl normally suppresses aggressive responses and that the activity of this pathway is normally downregulated in postpartum rodents to allow for high levels of maternal aggression.

There is some evidence that the medial preoptic area (MPOA) is also involved in maternal aggression. Arrati, Carmona, Dominguez, Beyer, and Rosenblatt (2006) injected GABA receptor agonists into the MPOA of postpartum lactating rats to temporarily inactivate MPOA neurons. This treatment disrupted pup-directed maternal behavior, as one would expect, but also disrupted maternal aggression. The disruption of maternal aggression, however, was not likely due to an interference with proximal tactile inputs from pups because the females were only briefly separated from pups before the intra-MPOA injections and aggression tests, and control females injected with saline showed normal maternal aggression. Recent work by Klampfl et al. (2018) has provided additional evidence for a role of the MPOA in maternal aggression. It is interesting to speculate that one population of MPOA neurons may be involved in stimulating pup-directed maternal behavior, for example, through projections to the ventral tegmental area, while a separate population may be involved in activating maternal aggression. Importantly, MPOA neurons project to VMNvl and to LS (Numan & Numan, 1996; Simerly & Swanson, 1988). Perhaps MPOA GABAergic inhibitory projections to LS are involved in potentiating maternal aggression in postpartum rodents, while MPOA glutamatergic projections to VMNvl aggression neurons activate the output of these neurons. Future studies using optogenetic and chemogenetic techniques could be utilized to test whether separate and distinct MPOA neural projections are involved in either pup-directed maternal behavior or in maternal aggression.

Given that the output of VMNvl appears essential for maternal aggression, where do these neurons project to activate aggressive responses? The VMNvl projects strongly to the periaqueductal gray (PAG) in the midbrain (Canteras, Simerly, & Swanson, 1994), and research on aggressive behavior in cats has provided good evidence that glutamatergic projections from the medial hypothalamus, which includes VMNvl, to the PAG are involved in aggressive behavior (Numan, 2015). However, the research on rodents is not as clear. Although large lesions of the PAG decrease aggression in rats, this effect is temporary, suggesting that other VMNvl target regions are involved (Mos et al., 1983). The PAG is a large and functionally heterogeneous brain region: PAG projections to the medulla are involved in a variety of relatively reflexive motor responses, such as the crouch nursing posture, penile erections, behavioral immobility, and the sexual receptivity lordosis response (Numan, 2015). In Chapter 5, I reviewed the evidence that certain parts of the PAG, rather than being involved in aggression, are involved in a variety of defensive responses, such as avoidance and escape/flight responses. One possibility is that certain glutamatergic projections from VMNvl neurons activate inhibitory interneurons in PAG, which then suppress the activity of PAG output neurons that regulate escape/flight responses. Such a circuitry may exert a permissive effect on aggression, while other projections from aggression-related VMNvl neurons to the brainstem, which have yet to be determined, may directly activate aggressive responses (see Roberts & Nagel, 1996). However, since the PAG is a functionally heterogeneous region, I do not want to rule out the possibility that some VMNvl projections to PAG may directly activate aggressive responses in rodents.

Importantly, "maternally relevant" MPOA neurons project to PAG (Numan & Numan, 1996, 1997), so this may be another route through which MPOA output could influence maternal aggression (this potential pathway is not shown in Figure 6.2).

One final body of literature is relevant to the neural circuitry of maternal aggression. Hansen and Ferreira (1986) reported that electrical lesions of peripeduncular nucleus, which lies ventral to the medial geniculate nucleus of the thalamus, virtually abolished maternal aggression in postpartum rats without interfering with pup-directed maternal behavior (also see Factor, Mayer, & Rosenblatt, 1993). Note that the peripeduncular nucleus overlaps with the posterior intralaminar complex (PIL) of the thalamus (see the subsection in Chapter 5 on the neural inputs to MPOA relevant to maternal behavior). The PIL receives strong somatic sensory inputs from the spinal cord, and PIL neurons are likely to be activated by ventral tactile inputs from nuzzling pups. Indeed, Fos expression is increased in PIL when a postpartum rat is in proximal contact with her pups (Cservenak et al., 2013). In postpartum rats, the expression of tuberoinfundibular peptide 39 (TIP39) increases in PIL during lactation. TIP 39-containing PIL neurons, which are excitatory, project to MPOA and to PVN-OT neurons (Cservenak et al., 2013; Cservenak, Keller, et al., 2017; Cservenak, Kis, et al., 2017). Based on the assumption that lesions of the peripeduncular nucleus damaged PIL in postpartum rats, how might damage to this system disrupt maternal aggression without affecting pup-directed maternal behavior? One possibility is that such lesions disrupted PIL-TIP 39 input to maternal aggression neurons in the MPOA. Recall that Cservenak et al. (2013) injected a long acting TIP 39 antagonist into the MPOA, and this treatment did not affect pup-directed maternal behavior in postpartum rats. Maternal aggression was not measured in this study, and it certainly would be interesting to know if the antagonist treatment would have disrupted maternal aggression. Since ventral somatic sensory inputs to the mother from nuzzling pups appear essential in maintaining maternal aggression, perhaps these inputs reach the particular set of MPOA neurons involved in maternal aggression to stimulate their activity. Perhaps these MPOA neurons, in turn, inhibit the LS and/or activate VMNvl, which then promotes maternal aggression.

The potential involvement of PIL-TIP 39 neurons in maternal aggression may also include the stimulation of PVN-OT release. This might be one route through which ventral somatic sensory inputs from pups activate OT release. The effects of such OT release on maternal aggression may be indirect. Since OT exerts anxiolytic effects, the PIL-TIP 39 excitatory pathway to PVN-OT neurons may be primarily related to postpartum anxiety reduction, which then exerts a permissive effect on the occurrence of maternal aggression. Relevantly, there is evidence that ICV administration of TIP 39 exerts an anxiolytic effect when male rats are tested on the EPM (LaBuda, Dobolyi, & Usdin, 2004; also see Coutellier, Logemann, Kuo, Rusnak, & Usdin, 2011).

The Neural Circuitry of Fear/Anxiety and the Mechanisms Mediating Its Postpartum Downregulation

In analyzing the nature of fear and anxiety, researchers usually define fear-related behaviors as defensive responses (immobility, flight/escape, avoidance) that occur in the presence of a threatening stimulus, while anxiety-related behaviors are usually defined as defensive responses that occur during environmental conditions that predict the potential occurrence of a threatening event before it is present (Calhoon & Tye, 2015; Davis, Walker, Miles, & Grillon, 2010; Tovote, Fadok, & Luthi, 2015). Testing behavioral responses on the EPM, therefore, is used to measure an organism's anxiety state, since exploration of the unprotected open arm areas of the maze is considered as exposing the organism to potential danger. The postpartum condition is associated with decreased anxiety as tested in the EPM. With respect to conditions where an intruding conspecific approaches a mother's offspring in her home cage, and where maternal aggression can occur, this event probably presents the mother with both an imminent and potential threat. Such resident–intruder tests are typically used in the laboratory to examine aggression, and these tests usually consist of 5 to10-minute sessions. It can be assumed that an intruder would create a state of sustained anxiety and fear in a female resident, but I have proposed that the postpartum reduction in fearfulness and anxiety exerts a permissive effect that allows the mother to defend her offspring by displaying aggression toward the intruder, rather than fleeing from the intruder.

It is well known that the CeA (see Figure 4.3) plays an essential role in the mediation of anxiety-related behaviors in rodents and other species (Calhoun & Tye, 2015; Numan, 2015; Tovote, Fadok, & Luthi, 2015). The CeA receives a variety of sensory inputs from the basolateral amygdala, and CeA neurons project to other neural sites to mediate defensive responses. CeA can be divided into a lateral part (CeAl) and a medial part (CeAm). Since CRF activity in the brain has been shown to be anxiogenic, note that there is a dense population of CRF-containing neurons in CeAl (Fadok et al., 2017). In addition, these CeAl CRF neurons project to the dorsal part of the bed nucleus of the stria terminalis (dBST; Asok et al., 2018; Sakanaka, Shibasaki, & Lederis, 1986; see Figure 5.1). (Note the distinction between dBST and vBST, since the latter area has been shown the play a positive role in pup-directed maternal behavior.) Research indicates that this CeAl CRF projection to the dBST is involved in mediating a sustained fear and anxiety state (Asok et al., 2018; Davis et al., 2010; Jasnow, Davis, & Huhman, 2004; Pomrenze et al., 2019), and Sahuque et al. (2006) have found that CRF injections into dBST of rats exerted an anxiogenic effect as evaluated on the EPM. Since CRF has excitatory neurophysiological effects (Blank et al., 2003), it can be concluded that CRF activates a specific group of neurons in dBST to produce an anxiogenic effect.

To generalize these results to the occurrence of postpartum anxiety reduction in rodents, one could propose that this CRF pathway between the CeA and dBST is downregulated during the postpartum period such that CRF is released at lower levels into dBST of postpartum rats, compared to virgin females, under anxiety-provoking situations. Is there any evidence to support this point of view? Klampfl, Brunton, Bayerl, and Bosch (2014) injected CRF receptor (CRFR) agonists or antagonists into the BST of postpartum rats, with injection sites that included the dBST (cf. Klampfl & Bosch, 2019) and found that agonist injections increased anxiety in the EPM, while antagonist injections further reduced the anxiety of such females. These authors also report that CRFRs are located in BST, but that their expression is not decreased when lactating females are compared to virgins. Therefore, they suggest that the anxiolytic postpartum state is probably related to decreased release of CRF into BST by fear-provoking situations.

In addition to CRF projections from CeAl to dBST, PVN-CRF neurons also project to BST (Zhang et al., 2017). Melon, Hooper, Yang, Moss, and Maquire (2018) have reported that normal postpartum mice spend more time in the open arms of an EPM than do virgin mice. They also found that selective chemogenetic activation of PVN-CRF neurons in postpartum mice caused their anxiety levels to increase so that their behavior in the EPM was not different from that of virgins. These results indicate that experimentally induced supernormal activity in the PVN-CRF system exerted an anxiogenic effect in postpartum mice. This effect could be due to PVN-CRF projections to dBST, but it is also possible that PVN-CRF activation of ACTH release, with a concomitant increase in plasma corticosterone levels, increased the synthesis of CRF in CeAl (see the previous section in this chapter on the opposing roles of oxytocin and corticotropin-releasing factor in anxiety-related behaviors).

If CRF action at the level of dBST causes an increase in anxiety-related behavior, what might be the underlying operating mechanisms that mediate this effect? The BST is a complex neural region composed of many separate nuclear groups (Dong, Petrovich, & Swanson, 2001). With respect to the dBST, as shown in Figure 6.3A, it is divided into a dorsolateral region (dlBST) and a dorsomedial region (dmBST). In terms of anatomy, CeAl projects strongly to dlBST (Dong et al., 2001). In turn, the dlBST projects to both dmBST and the PAG (Dong, Petrovich, Watts, & Swanson, 2001). Finally, dmBST projects strongly to the lateral hypothalamus (LH; Dong & Swanson, 2006). Most neurons in BST are GABAergic and therefore exert postsynaptic inhibitory effects (Daniel & Rainnie, 2016). These general anatomical relationships are outlined in Figure 6.3B.

In an important study using mice, Kim et al. (2013) demonstrated that optogenetic stimulation of dlBST increased anxiety-related behavior in the EPM, while inhibition of this region resulted in decreased anxiety. In contrast, specific stimulation of dmBST projections to the LH exerted an anxiolytic effect in the EPM. Given these findings, how can we explain some of the anxiety-inducing properties of CRF? I would like to present a rudimentary neural circuitry that may be involved in such CRF actions. CRF input to dlBST probably excites these neurons. The projections of dlBST to PAG may activate PAG output circuits that promote defensive responses such as escape and avoidance. This effect probably occurs via a disinhibitory mechanism, as shown in Figure 6.3B. Further, dlBST inhibition of dmBST output suppresses the antianxiety effects exerted by this region (Kaneko et al., 2016; Nagano et al., 2015), and I am suggesting that this relationship represents CRF activation of dlBST, which, in turn, inhibits the output of dmBST.

How might the output of dmBST to the LH result in fear and anxiety reduction? This is an important question, since activity in this circuit, presumably due to decreased CRF action on dlBST, may be an important component of postpartum anxiety reduction. In an anatomical study in rats, Hahn and Swanson (2015) have explored the neural connections of the LH region that lies just lateral to the VMN of the hypothalamus (see Figure 4.3). They found that this LH region receives strong inputs from dmBST, and this LH region, in turn, provides a major input to all parts of the VMN. They also note that this LH region is composed of both glutamatergic and GABAergic neurons. Like the dBST, the VMN is a complex nucleus composed of many separate regions. Figure 6.4 presents a more detailed anatomical delineation of the VMN, which consists of ventrolateral (VMNvl), central (VMNc), and dorsomedial (VMNdm) components.

The VMNvl contains two populations of intermingled, but separate, neurons: Aggression-related neurons and social fear neurons (Hashikawa et al., 2017; Sakurai et al., 2016), and selective stimulation of VMNvl social fear neurons causes a resident male mouse to flee from, rather than attack, a conspecific intruder. Since VMNvl social fear neurons project to PAG (Sakurai et al., 2016), this could be the route through which these neurons activate escape and avoidance responses to social stimuli. It is interesting to speculate that dmBST input to LH results in LH inhibition of social fear neurons, and this could be one mechanism through which this connection causes a reduction in anxiety related behavior, particularly in response to social stimuli.

In contrast to VMNvl, VMNdm efferents promote defensive fear responses to stimuli other than those that arise from conspecific social stimuli (Kunwar et al., 2015; Wang, Chen, & Lin, 2015). For example, stimuli from predators appear to promote fear and anxiety-related responses as a result of activating VMNdm projections to the PAG, while conspecific fear-inducing stimuli promote fearfulness by activating VMNvl social fear neuron projections to PAG (Gross & Canteras, 2012). Therefore, the PAG appears to be a common outpost for the activation of defensive responses (immobility, escape/flight, avoidance), but different external stimuli that activate PAG defensive circuits do so through

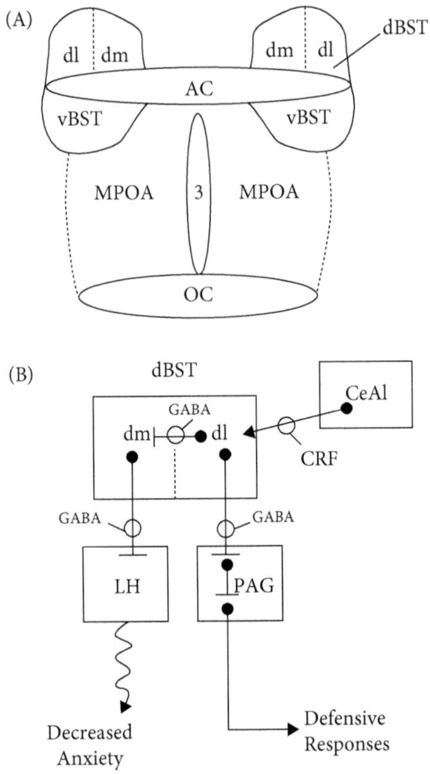

Figure 6.3. (A) A depiction of the nuclear components of the bed nucleus on the stria terminalis (BST) on a frontal section through the rodent brain at the level of the medial preoptic area (MPOA). The ventral BST (vBST) lies ventral, and the dorsal BST (dBST) lies dorsal, to the anterior commissure (AC). The dBST is divided into a dorsolateral (dlBST) and dorsomedial (dmBST) region. See Figure 5.1 for a complete frontal section through this brain region.(B) Neural interactions between the dBST and other brain regions. Corticotropin-releasing factor (CRF) containing neurons in the lateral part of the central nucleus of the amygdala (CeAl) project to and stimulate the dlBST. dlBST GABAergic neurons project to both the periaqueductal gray (PAG) and to the dmBST. The projections of the dlBST to the PAG promote anxiety- and fear-related responses by stimulating PAG output through a process of disinhibition. GABAergic projections from the dmBST to the lateral hypothalamus (LH) exert an anxiolytic/fear-reducing effect (the squiggly line leaves the mechanism of this effect as undefined; refer to Figure 6.5 for the potential underlying mechanism). Since dlBST GABAergic projections to dmBST inhibit this latter region, this is another route through which CRF activation of dlBST promotes anxiety and fearfulness. Axons ending in a bar are inhibitory and those ending in an arrow are excitatory. 3 = third ventricle; OC = optic chiasm.

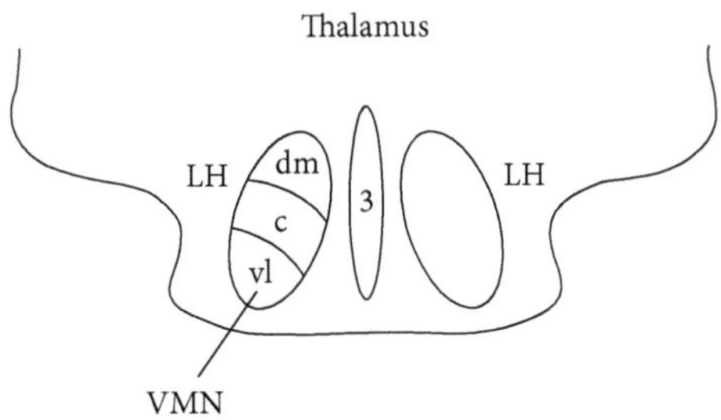

Figure 6.4. A detailed view of the anatomical organization of the ventromedial nucleus of the hypothalamus (VMN), which consists of neurons located in ventrolateral (vl), central (c), and dorsomedial (dm) regions. 3 = third ventricle; LH = lateral hypothalamus.

different neural routes. Finally, CeAm neurons, which can be activated by a variety of external stimuli, project directly to the PAG to cause behavioral immobility in response to fear-inducing stimuli. Since CeAm neurons are GABAergic, they stimulate the output of PAG defensive circuits through a process of disinhibition: They inhibit PAG inhibitory neurons that would normally suppress the output of PAG (Tovote et al., 2016).

The neural circuits underlying anxiety and fear response are much more complex than that which I have described, but the circuits that I have described provide a foundation for understanding aspects of the basic circuitry involved in generating fear and anxiety response to a variety of external stimuli that can pose a threat to the organism exposed to such dangerous stimuli. Figure 6.5 presents a summary of this basic circuitry. In viewing this figure, one can begin to understand how increased CRF activity in the brain can produce an anxiogenic effect. CRF activation of dlBST promotes anxiety in at least two ways: (a) by directly activating dlBST projections that stimulate the output of PAG defensive circuits through a process of disinhibition and (b) by directly activating dlBST neurons that depress the output of the dmBST and therefore depress the ability of dmBST to reduce anxiety through its projections to LH, as shown in Figure 6.5. It can be proposed that postpartum anxiety reduction is partly the result of decreased release of CRF into dlBST.

OT exerts anxiolytic effects, and mother–infant interactions that stimulate OT release into the brain have been shown to be involved in postpartum anxiety reduction. Where might OT act within the circuits shown in Figure 6.5? OT appears to exert an anxiolytic effect by suppressing the output of the CeA and the PAG. As shown in Table 4.1, OT axon terminals and OT receptors (OTRs) are located in the amygdala (including MeA and CeA) and in PAG. In nonlactating female rats, Bale, Davis, Auger, Dorsa, and McCarthy (2001) reported that microinjection of OT into CeA decreased anxiety-related behavior. Knobloch et al.

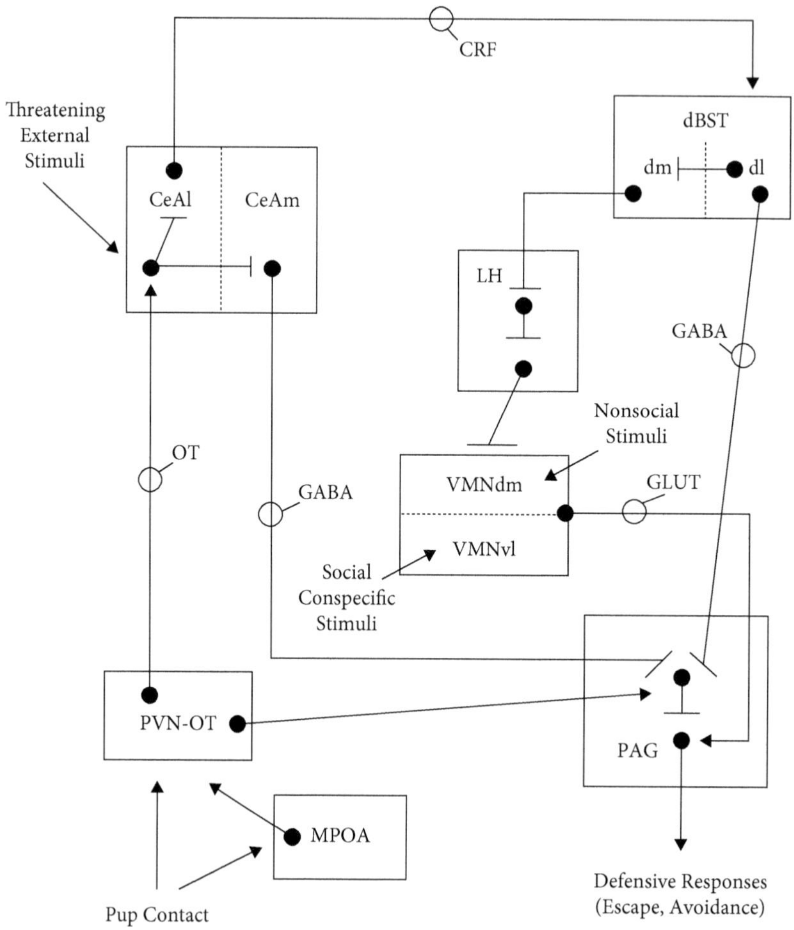

Figure 6.5. A detailed neural model of how corticotropin-releasing factor (CRF) promotes anxiety and fearfulness while oxytocin depresses anxiety and fearfulness to a variety of threatening/dangerous stimuli. Fear/anxiety-related neurons are located in the ventrolateral region and the dorsomedial region of the ventromedial nucleus of the hypothalamus (VMNvl and VMNdm, respectively). VMNvl "fear" neurons respond to threatening conspecific social stimuli, while VMNdm "fear" neurons respond to threatening nonsocial stimuli (stimuli that do not derive from conspecifics). The projections of the VMN excite periaqueductal gray (PAG) output neurons to stimulate escape and avoidance responses. The lateral hypothalamus (LH) is shown as exerting anxiolytic effects by inhibiting the VMN. CRF-containing neurons in the lateral part of the central nucleus of the amygdala (CeAl) are activated by a variety of threatening stimuli. These neurons project to and excite the dorsolateral bed nucleus of the stria terminalis (dlBST) to promote anxiety and fear-related responses. The output of the dlBST promotes these defensive responses by activating (through a process of disinhibition) the PAG,

(2012; also see Huber, Veinante, & Stoop, 2005) found that OT action at the level of CeA activates inhibitory neurons in CeAl, which, in turn, suppress the output of CeAm to PAG, in this way suppressing a fear response (behavioral immobility) to a stimulus that had been associated with shock in virgin female rats. In an important study in postpartum rats, Rickenbacher, Perry, Sullivan, and Moita (2017) demonstrated a similar effect. These researchers paired a peppermint odor with foot shock, so that the odor would become a conditioned aversive stimulus. On days 4 to 6 postpartum, when this odor was presented to females that were separated from their pups, the females displayed behavioral immobility, a conditioned fear response. However, if the pups remained with the postpartum female, they did not freeze to the conditioned fear stimulus, but instead these females were active and they pushed cage bedding in the direction of the odor source in an attempt to eliminate it. Significantly, when mothers were with their pups, but an OTR antagonist was injected into CeAl, the mothers showed behavioral immobility to the conditioned fear stimulus. These results suggest that contact with pups promotes OT release into CeAl, which results in an anxiolytic effect (suppression of a conditioned fear response) and that this effect allows the mother to show responses presumably meant to protect her offspring. This OT effect is likely the result of OT stimulation of CeAl neurons, which, in turn, inhibit the projection of CeAm to PAG. It would be very interesting to know whether OT also activates inhibitory interneurons in CeAl, which then inhibit the output of CRF neurons in CeAl that project to dlBST (see Figure 6.5).

These results indicate a complex role of CeAl in anxiety and fearfulness. GABAergic neurons in CeAl that project to CeAm depress anxiety, while CRF neurons in CeAl that project to dBST promote anxiety.

Figure 6.5. Continued.
and by inhibiting the output of the LH. The medial part of the central nucleus of the amygdala (CeAm) also increases anxiety and fearfulness by directly projecting to and activating the output of the PAG (through a process of disinhibition). Anxiety reduction in postpartum females, which results from proximal contact with pups, activates OT release at several nodes in the depicted fear circuits in order to depress fearfulness and anxiety. OT may activate inhibitory interneurons in CeA to suppress the output of both CeAl-CRF neurons to dlBST and the output of CeAm to PAG. OT is also proposed to activate inhibitory interneurons in the PAG, which suppresses PAG output. Proximal contact with pups may activate OT release into these sites by directly stimulating OT neurons in the paraventricular hypothalamic nucleus (PVN), and/or by activating a medial preoptic area (MPOA)-to-PVN-OT pathway. Axons ending in an arrow are excitatory, and those ending in a bar are inhibitory. See the text for the research that supports aspects of this neural model.

With respect to PAG, Lonstein, Simmons and Stern (1998) reported that electrical lesions of the PAG in lactating rats further reduced their anxiety levels as measured in the EPM. Control lactating rats were less anxious that virgins (as expected), but postpartum females with PAG lesions were less anxious than both virgins and intact lactating females (also see Miller, Piasecki, Peabody, & Lonstein, 2010). Further, Fiqueira, Peabody, and Lonstein (2008) found that microinjection of an OTR antagonist into PAG of postpartum rats increased their anxiety in the EPM, while OT infusion in PAG, after postpartum female rats were separated from their pups for 4 hours, decreased anxiety in the EPM compared to similarly pup-separated females that received control injections of saline into PAG. These results therefore emphasize at least two sites, CeA and PAG, where OT acts to exert general anxiolytic effects in postpartum rats. Endogenous OT release into the brain of lactating females presumably results from somatic sensory inputs from pups during mother–infant interactions. Such pup stimuli may activate OT release into the mother's brain through direct excitatory input to PVN-OT neurons and as a result of pup stimuli activating MPOA excitatory input to PVN-OT neurons (see Figure 6.5).

There is also some indirect evidence that OT action at the level of dlBST may also suppress anxiety in rats. First, dlBST neurons express high levels of OTRs (Dabrowska et al., 2011; see Table 4.1). Martinon and Dabrowska (2018) have found that OT axon terminals from PVN project to the dlBST and that these OT axon terminals contain presynaptic CRFRs, and it is likely that CRF action at this site exerts presynaptic inhibition. Blocking these CRF presynaptic receptors enhanced the release of OT into dlBST. An interesting aspect of the study by Martinon and Dabrowska is that there may be direct interactions between CRF neurons and OT neurons. CRF may enhance anxiety not only by directly stimulating anxiety circuits but also by inhibiting OT release. It would be interesting to know whether, in a manner similar to OT's action in CeA, whether OT action in dlBST activates inhibitory interneurons, which would then suppress the output of dlBST neurons to dmBST and to PAG.

Critical Evaluation of the Hypothesis That Decreased Activity Within Fear/Anxiety Neural Circuits Exerts a Permissive Effect on the Occurrence of Maternal Aggression by Releasing Aggression Circuits from Inhibition

Several pieces of evidence suggest that manipulations within the neural circuits that regulate fear and anxiety can influence maternal aggression. Increases or decreases in the activity of these circuits in postpartum rodents are correlated with either decreases or increases in maternal aggression, respectively. First, decreases in endogenous OT activity in CeA, via site specific injections of an OTR antagonist, increases fear-related responses in postpartum rats (Rickenbacher et al., 2017). Bosch et al. (2005) have reported that OT is released into CeA of postpartum rats during episodes of maternal aggression and administration of an OTR antagonist to CeA of postpartum rats was found to significantly decrease maternal aggression. One interpretation of these results is that OT action on CeA decreases anxiety and fearfulness to an optimal level and that such a decrease results in the release of maternal aggression circuits from inhibition by supernormal activity within anxiety/fear circuits.

Klampf et al. (2014) reported that administration of CRFR antagonists to the BST of postpartum rats, which included injection sites in dBST, decreased anxiety in the EPM and also increased maternal aggression. Therefore, even though postpartum rats normally show decreased fearfulness and increased aggression toward intruders, manipulations within dBST, which further decrease anxiety in the EPM, were associated with a further enhancement of maternal aggression.

The results of Klampf et al. (2014) suggest that increased activity across a CeAl-CRF pathway to dlBST increases anxiety and fearfulness, which, in turn, suppresses maternal aggression. Through which interactive pathways might this occur? In reference to Figure 6.5, perhaps CRF-induced activity within dlBST decreases the ability of the LH to suppress social fear neurons in VMNvl and other defense-related neurons in VMNdm. Increased activity in these fear-related neurons may, in turn, directly or indirectly suppress the activity of VMNvl aggression related neurons, leading to both increased fearfulness and decreased maternal aggression. Alternatively, perhaps a population of LH glutamatergic neurons directly activates VMNvl aggression neurons, and activity within this aggression circuit is inhibited by CRF action at the level of dlBST.

The LS inhibits aggression by inhibiting VMNvl aggression-related neurons (see Figure 6.2). In this context, D'Anna and Gammie (2009) reported that microinjection of CRF into LS suppressed maternal aggression in postpartum mice. Since Zhang et al. (2017) found that PVN-CRF neurons project to LS, perhaps fear/anxiety-inducing stimuli activate this projection, which allows such stimuli to inhibit maternal aggression. In support, there is evidence that CRF action within the LS produces anxiogenic effects (Anthony et al., 2014). Recent research also indicates that OT may inhibit the LS to reduce fear-related responses to a same sex conspecific in lactating mice (Menon et al., 2018). One possibility is that a group of "social fear" neurons exist in the LS, and their projections to VMNvl inhibit maternal aggression. Perhaps CRF input to the LS activates these neurons while OT input to the LS inhibits these neurons. Although I did not include the LS in Figure 6.5 (and Figure 6.6), it appears to be an additional important player in the interactive effects of OT and CRF in the regulation of postpartum anxiety and maternal aggression.

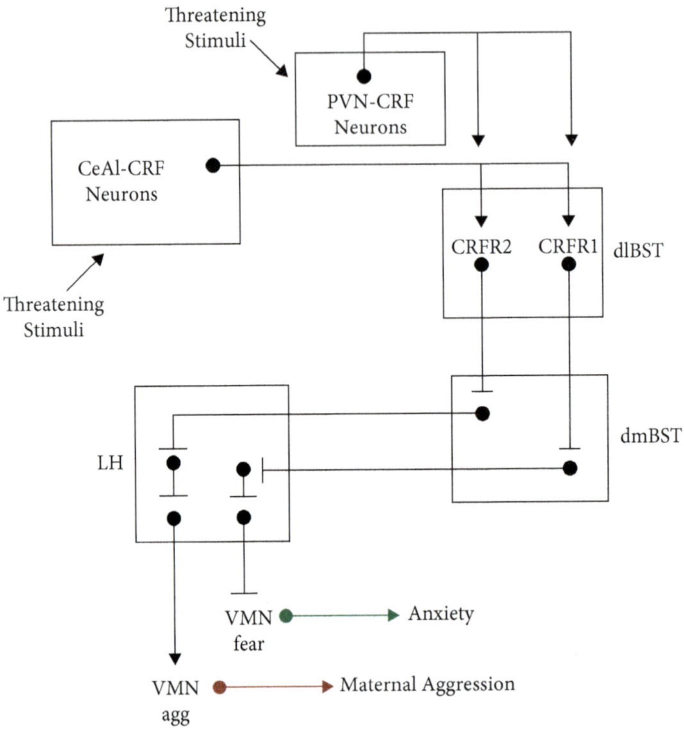

Figure 6.6. A hypothetical neural model which shows that when threatening/dangerous stimuli activate corticotropin-releasing factor (CRF) neurons in the lateral part of the central nucleus of the amygdala (CeAl) and in the paraventricular nucleus of the hypothalamus (PVN), these stimulated CRF neurons project to the dorsolateral bed nucleus of the stria terminalis (dlBST) to activate parallel neural pathways: CRF action on type 1 CRF receptors (CRFR1) activates pathways that enhance anxiety and fearfulness, while CRF action on type 2 CRF receptors (CRFR2) activates pathways that suppress maternal aggression. Very high levels of CRF release provoked by threatening stimuli would activate both pathways. Within the LH, increased activity within inhibitory interneurons, due to decreased inhibitory input from the dorsomedial part of the bed nucleus of the stria terminalis (dmBST), would act to suppress the output of neural systems in the lateral hypothalamus (LH), which either decrease anxiety or increase aggression. These effects would, therefore, increase anxiety and decrease maternal aggression. VMNagg = neurons in the ventromedial nucleus of the hypothalamus that promote aggression; VMNfear = neurons in the VMN that promote fearfulness and anxiety. Axons ending in an arrow are excitatory and those ending in a bar are inhibitory. Axons leaving VMN fear neurons are shown in green to indicate that they are facilitated by CRF, while those emanating from VMN aggression neurons are shown in red, indicating that their output is inhibited by CRF. See text for details and supporting evidence.

Therefore, these studies suggest that danger-provoking situations may inhibit aggression via two potential routes: (a) CeAl-CRF induced inhibition of the output of dmBST to an LH-to-VMN circuit, in this way blocking both the suppression VMN fear and anxiety-related neurons and the activation VMNvl aggression neurons, and (b) PVN CRF-induced activation of LS inhibitory circuits to VMN aggression neurons. One can conclude that the postpartum condition is typically associated with a downregulation of these CRF circuits and that OT neural pathways participate in this downregulation process.

I have already reviewed the evidence that the PAG lies downstream from the VMN, and its output mediates various defensive responses. VMN fear-related neurons activate, and other VMN neurons that promote aggression may inhibit, the output of PAG neurons that mediate fear-related responses. Lonstein, Simmons, and Stern (1998) reported that electrical lesions of the PAG not only resulted in an enhanced reduction in the anxiety of postpartum rats as tested in the EPM, but also further enhanced their maternal aggression, in comparison to control postpartum females. Since Figueira et al. (2008) found that OT acts on PAG to stimulate postpartum anxiety reduction, it would be interesting to determine whether an OTR antagonist injection into PAG would not only increase anxiety in postpartum rats, but would also act to decrease maternal aggression.

Although my analysis up to this point is complex, I have actually tried to simplify my discussion. Additional facts add further layers of complexity to the neural systems I have described. There are actually two types of CRF receptors where CRF acts to exert its effects: CRFR1 and CRFR2. Since CRF has a greater affinity for CRFR1 than for CRFR2 (Lukkes, Forster, Renner, & Summers, 2008), lower levels of CRF release should activate CRFR1, while higher levels of CRF release would be needed to fully activate the CRFR2. If the amount of CRF release at central synapses is positively correlated with anxiety levels, then one would predict that high levels of anxiety would be needed to activate CRFR2.

In relation to these two types of CRFRs within the LS, D'Anna and Gammie (2009) found that microinjections of CRF into the LS decreased maternal aggression in postpartum mice, and this disruption of maternal aggression by CRF was reversed by co-injection of a CRFR2 antagonist, but not a CRFR1 antagonist. What these findings suggests is that high levels of CRF release into LS, possibly derived from PVN-CRF neuron projections to LS, are needed to activate LS, which, in turn, inhibits maternal aggression (see Figure 6.2).

Further information on these two receptor types within the dBST has been provided by Klampfl et al. (2014). These researchers reported that administration of a CRFR2 agonist, but not a CRFR1 agonist, into a BST region that included the dBST, abolished maternal aggression in postpartum rats. In contrast, administration of a CRFR1, but not a CRFR2, agonist into this region increased anxiety as measured in the EPM. These results indicate that low levels of CRF release into dBST can produce an anxiogenic effect in postpartum rats, while higher levels of release are needed to suppress maternal aggression. How can we explain these effects? Figure 6.6 presents a hypothetical mediating mechanism. This figure, based on an elaboration of Figure 6.5 (with the added inclusion of VMN aggression-related neurons), hypothesizes that there are two populations of neurons in dlBST that serve to inhibit the output of dmBST neurons. One population contains CRFR1 and another population contains CRFR2. Activation of the dlBST CRFR1 population by lower levels of CRF would exert an anxiogenic effect, while activation of the dlBST CRFR2 population by higher levels of CRF would operate to suppress maternal aggression.

Figure 6.6, although speculative in nature, is instructive in the following general way. At certain neural regions that are downstream from CeAl-CRF neurons, aggression-related neural systems and anxiety-related neural systems may not directly interact. However, threatening stimuli at the level of CeAl-CRF neurons may activate parallel neural systems in dlBST, which act to increase anxiety and fearfulness and decrease maternal aggression. Threatening and stressful stimuli that activate PVN-CRF neurons that project to dlBST may similarly act to stimulate anxiety and decrease maternal aggression through the parallel neural pathways shown in Figure 6.6. Low levels of activation of CeAl-CRF and PVN-CRF output pathways may only increase anxiety, but higher levels of activity in these circuits by danger-provoking stimuli may both increase anxiety and decrease maternal aggression. The point to emphasize, however, is that danger-provoking stimuli are capable of increasing anxiety/fearfulness and decreasing aggression via a primary effect on CRF neural systems, which supports the proposal that too much CRF activity induced by danger-provoking stimuli can inhibit maternal aggression.

During the postpartum period, there appears to be a critical coordination of OT and CRF neural activity, which creates an optimal level of vigilance and protectiveness so that a mother can effectively care for her young under challenging environmental conditions, which would include dealing with the approach of a dangerous conspecific intruder (see Figure 6.1). Since OTRs are located in CeAl, dlBST, LS, and PAG, OT release into these sites, through the activation of PVN-OT neuron projections to these sites during mother–infant contact, may serve to fine tune and appropriately lower the potentially disruptive effects of high CRF activity on maternal aggression and other maternal behavioral coping strategies meant to help the mother deal with environmental challenges. In this way, one can understand the broader role of OT neural systems in the regulation of maternal behavior even during the postpartum maintenance phase of maternal behavior. Under standard, low-stress, laboratory conditions, OT's main role is to stimulate the onset of maternal motivation, and it is not essential for the continuance of maternal behavior, while under stressful conditions, OT plays an important role during all phases of maternal behavior by decreasing the mother's behavioral stress reactivity so that she can effectively care for her young.

The Effects of High Activity Within CRF Neural Systems on Infant-Directed Maternal Behavior

There may be a continuum of neural activity within CRF neural systems where a certain level of activity increases anxiety/fearfulness and a higher level decreases maternal aggression but does not suppress infant-directed maternal behavior. However, if CRF neural activity is further elevated, then infant-directed maternal behavior may also be depressed. Therefore, too much maternal stress reactivity may have broad deleterious effects, which include a suppression of mother–infant interactions. There is evidence that supports this proposition. Klampfl et al. (2014) reported that administration of a nonselective CRFR agonist into the lateral ventricle, which would stimulate both CRF1 and CRF2 receptors, disrupted nursing and retrieval behavior in postpartum rats. A similar disruptive effect on nursing behavior was observed after CRFR agonist injections into dBST. In mice, under normal conditions, PVN-CRF neuronal activity is low when maternal behavior occurs (Kim et al., 2019), and selective experimental stimulation of these neurons disrupts pup-directed maternal behavior (Melon et al., 2018). In marmoset monkeys, Saltzman and Abbott (2009) injected postpartum mothers with cortisol for 8 days. These were subcutaneous injections that entered the systemic blood supply and would affect all brain regions that contain glucocorticoid receptors. Such injections resulted in a type of maternal neglect such that the cortisol-treated mothers carried their infants for a significantly lower amount of time than did control mothers. While control mothers would retrieve their infants and then continuously carry them, the cortisol-treated mothers would repeatedly retrieve and then reject their infant (push them away), which resulted in a reduction in total time they carried, and therefore cared for their young. Perhaps this negative effect of cortisol on maternal behavior in marmosets was due to cortisol-induced increases in the synthesis and release of CRF within CeAl. Some support for this view comes from a subsequent report by Saltzman, Boettcher, Post, and Abbott (2011): ICV injections of CRF disrupted infant carrying in postpartum marmoset mothers. Future studies that vary the doses of CRF administered to mothers might be able to show that low doses increase anxiety, higher doses increase anxiety and decrease maternal aggression, and even higher doses also disrupt infant-directed maternal behavior.

How might very high levels of central CRF release depress infant-directed maternal behavior? In Chapter 5, it was proposed that PAG projections to MPOA may suppress pup-directed maternal behavior (see Figure 5.9). Perhaps this is a route whereby very high levels of CRF release into the brain suppress mother–infant interactions, as outlined in the circuits in Figure 6.5, which delineate how CRF can ultimately activate PAG. Alternatively, CRF may act directly on the MPOA to suppress mother–infant interactions, and there is recent evidence that supports this possibility (Klampfl et al., 2018). These proposals are interesting in the context of explaining the occurrence of abnormal maternal behavior that might occur under conditions where CRF or cortisol is *not* exogenously administered to mothers. Some mothers may be highly stress reactive, perhaps due to a reduction in the antianxiety effects of endogenous OT. Exposure to a stressful environment in such mothers might then cause abnormally high levels of CRF release, which then acts to suppress both maternal aggression and infant-directed maternal responses (see Klampfl et al., 2014, 2018).

Conclusions

The research presented in this chapter clearly shows that the regulation of stress reactivity during the postpartum period is an essential aspect of the occurrence of effective maternal care. Too much stress-induced fearfulness and anxiety, mediated, at least in part, by hyperactivity in central CRF systems, has deleterious effects on maternal competence, which decreases the ability of a mother to protect and care for her offspring. In this chapter, I have outlined the neural circuits underlying fear and anxiety-related behaviors, and the neural circuits underlying maternal aggression. I have also proposed neural models whereby hyperactive CRF neural systems can increase fearfulness/anxiety and decrease maternal aggression and infant-directed maternal behavior in postpartum rodents. Also emphasized was the essential role of central OT neural systems in downregulating the effects of CRF neural systems so that the postpartum female can cope with challenging environments to appropriately protect and care for her offspring. Clearly, maladaptive maternal behaviors are likely to occur during stressful environmental conditions if certain factors, such as early life experiences and/or genetic factors, disrupt the mechanisms that downregulate stress reactivity during the postpartum period. I will come back to this important issue when I discuss the development of maternal behavior in animals and humans.

The MPOA may not only be involved in pup-directed maternal motivation, but may also contribute to maternal aggression and to postpartum anxiety/fear reduction (Numan & Woodside, 2010). I have already described research that suggests the involvement of the MPOA in maternal aggression, and there is recent direct evidence to suggest that it is also involved in the general fear reduction that occurs during the postpartum period in rats. Klampfl et al. (2018) have reported that the administration of a CRFR agonist to MPOA increases anxiety, as measured in the EPM, in postpartum rats. Perhaps CRF activates inhibitory interneurons in MPOA, which, in turn, suppress MPOA excitatory projections to PVN-OT neurons, and inhibitory MPOA projections to VMN and PAG fear-related neurons (see Figures 5.9 and 5.10), in this way increasing general fearfulness. Therefore, the MPOA may be an additional crucial link in the neural circuits that allow proximal pup contact to decrease general stress reactivity in postpartum mothers, and this MPOA function may be suppressed by a hyperactive CRF neural system. These views suggest that the MPOA output may not only be involved in decreasing the aversive effects of novel infant stimuli, which is necessary for the immediate onset of maternal behavior in primiparous parturient females, but may also play a role in the general fear and anxiety reduction that occurs in postpartum females.

This chapter has focused on the role of OT and CRF neural systems in the regulation of maternal aggression and postpartum anxiety reduction. Not surprisingly, research has indicated that other hormonal and neurochemical systems influence these maternal adaptations. With respect to hormonal factors, proximal stimulation from pups in postpartum females, even in the absence of suckling stimulation, is likely to release prolactin into the mother's brain, and there is some evidence that such prolactin release may exert anxiolytic effects, perhaps by depressing the responsiveness of CRF neural systems (Donner, Bredewold, Maloumby, & Neumann, 2007). This hypothesis is contraindicated by the results that show that hypophysectomy does not eliminate anxiety reduction in postpartum rats (Lonstein, 2005). However, there is some evidence that prolactin is produced by neurons within the brain, and perhaps it is the actions of neuronal prolactin, rather than pituitary prolactin, that exert anxiolytic effects (Torner, 2008). The reader is referred to additional papers for further information on other neurochemical systems (Bayerl & Bosch, 2018; Klampfl & Bosch, 2019; Muroi & Ishii, 2019).

7

Alloparental Behavior and Paternal Behavior in Nonhuman Mammals

Introduction

Alloparental behavior and paternal behavior are both rare in mammals, but these behaviors do occur in about 3% to 5% of mammalian species. In these instances, pregnancy and parturition are not essential for the occurrence of parental behavior. In this chapter, I will describe the research that indicates that the subcortical neural circuits that regulate alloparental behavior and paternal behavior are basically the same as those that regulate maternal behavior in those mammals with a uniparental maternal care system. However, while the physiological events of pregnancy and parturition act on these circuits to stimulate the onset of maternal behavior in the vast majority of mammalian species, other factors activate these circuits for alloparental and paternal behaviors to occur.

Even when we examine rats, which display a uniparental maternal care system, research shows that the occurrence of maternal behavior is not rigidly tied to the physiological events of pregnancy and parturition. Nulliparous females can be sensitized to show maternal behavior, and the maintenance of maternal behavior in postpartum rats is free from hormonal control. These findings show that there are mechanisms that can allow infant stimuli to gain access to the parental neural circuitry in the absence of the hormonal events of pregnancy and parturition. Sensitized maternal behavior in rats, and in other species with a uniparental maternal care system, is unlikely to occur in nature. However, the fact that such sensitized maternal behavior can be experimentally induced suggests that the underlying mechanisms of the sensitization process may have been utilized by natural selection to create mechanisms that would allow other factors, outside the boundaries of late pregnancy and parturition, to induce prompt alloparental and paternal behaviors in certain species where the occurrence of these behaviors is adaptive.

The Parental Brain. Michael Numan, Oxford University Press (2020). © Oxford University Press.
DOI: 10.1093/oso/9780190848675.001.0001

The Laboratory Mouse as an Experimental Model of Allomaternal Behavior

Alloparental behavior is defined as the care of a conspecific infant by an individual that is not the genetic parent of the infant (Riedman, 1982). I want to elaborate on this definition by making a distinction between species that display a communal breeding strategy and those that display a cooperative breeding strategy. In communal breeders, many lactating females may share in the care of each other's infants (Lewis & Pusey, 1997). For example, communal nursing, where milk is shared between a mother's own pups and the young of another mother, occurs in wild house mice (*Mus musculus*: Weidt, Lindholm, & Konig, 2014). Clearly, this would be an example of alloparenting, but it occurs in females that have been exposed to the physiological events of pregnancy and parturition. Also recall that virgin wild (feral) house mice are infanticidal toward conspecific young. Therefore, maternal and allomaternal behavior in communal wild mice toward their own and alien young, respectively, is induced by the physiological events of pregnancy and parturition. Therefore, this aspect of allomaternal behavior is not the concern in this chapter.

The alloparental behavior that occurs in cooperatively breeding species, such as prairie voles, marmosets, and tamarins, is a concern of this chapter. In such species, it is common for a single male and female to do all of the breeding, while the nonbreeding male and female members that remain in the family group after they are weaned help the parents rear subsequent offspring (Diaz-Munoz, 2016; Solomon & French, 1997). In these cases, nonbreeding virgin females and males display caretaking behaviors toward their parents' younger offspring, indicating that such alloparental behavior can occur without the alloparents undergoing pregnancy and parturition.

Most strains of nulliparous female laboratory house mice, in contrast to their nulliparous feral counterparts, are spontaneously maternal when presented with pups from another mother under standard home cage testing conditions. As a result of inbreeding and selective breeding, the maternal responsiveness of these virgin females has been emancipated from strict control by the events of late pregnancy and parturition. For certain species, such as prairie voles, where alloparenting occurs in cooperatively breeding social groups, natural selection may have had effects similar to the experimental selection that has occurred in female laboratory mice, resulting in the evolution of alloparental behavior under natural conditions.

Therefore, the virgin female laboratory mouse may be an experimental model for understanding the mechanisms that regulate naturally occurring allomaternal behavior. There is excellent evidence that the neural circuitry that regulates allomaternal behavior in laboratory mice is similar to that which regulates maternal behavior in most postpartum female mammals. The evidence reviewed in Chapter 5 showed that medial preoptic area (MPOA) projections that activate the mesolimbic dopamine (DA) system regulate allomaternal behavior in laboratory mice. Further, in Chapter 4, I reviewed the evidence that experimental genetic selection has also removed olfactory inhibition over maternal behavior in virgin female laboratory mice. These two processes, produced by experimental genetic selection, appear to act together to create the allomaternal state in these mice: The defensive system has been downregulated while the maternal motivational system has been upregulated.

In Chapter 4, I also reviewed the research on the role of OT in the allomaternal behavior of laboratory mice. My analysis concluded that under standard home cage and nonstressful laboratory testing conditions, oxytocin (OT) neural systems were not essential for allomaternal behavior in female lab mice. However, under more challenging environmental conditions, OT neural systems were shown to play a positive role. In the following paragraphs of this section, I want to evaluate the proposal that OT neural systems may promote allomothering under stressful testing conditions in part by boosting maternal motivation in virgin female mice.

As reviewed in Chapter 5, paraventricular nucleus (PVN)-OT neurons project to and activate ventral tegmental area (VTA) DA neurons in laboratory mice. That could be one route through which OT boosts maternal motivation in virgin mice. In addition, PVN-OT input to MPOA may also boost maternal motivation in virgin lab mice, and there is evidence that the injection of an OT receptor (OTR) antagonist into MPOA decreases maternal motivation in mice under certain testing conditions (Okabe et al., 2017). Finally, in laboratory mice, OTRs are located in the nucleus accumbens (NAs) (Olazabal & Alsina-Llanas, 2016), and PVN-OT neurons project to NAs in mice (Otero-Garcia et al., 2016). Therefore, OT may act at multiple links in the MPOA-to-VTA-DA-to-NAs circuit to boost allomaternal motivation.

Based on this analysis, I would like to offer the following proposal. As a result of experimental genetic selection, the brain of the virgin female laboratory mouse has been modified so that pup stimuli have access to MPOA "maternal" neurons without the need for the physiological events associated with late pregnancy and parturition. Under nonstressful conditions, MPOA activation of the mesolimbic DA system may be all that is required for the expression of allomaternal behavior. Under more demanding environmental conditions, the recruitment of PVN-OT systems may further enhance the activity of MPOA interactions with the mesolimbic DA system to allow for effective allomaternal care. Perhaps in such allomaternal virgins, pup stimuli activate MPOA projections to both

VTA-DA neurons and PVN-OT neurons to enable full maternal responsiveness under a variety environmental conditions. All of these effects occur against a background of a downregulated defensive system, which eliminates aversive responses to olfactory-related infant stimuli.

Of course, an understanding of how allomaternal behavior is controlled in laboratory mice as a result of experimental genetic selection may, or may not, inform us about the way natural selection has modified the brain to allow for alloparental behavior in those species that show such behavior under natural conditions. Therefore, in the next section, I will examine some of the neural mechanisms that have been shown to underpin alloparental behavior in prairie voles, a cooperatively breeding microtine rodent species that shows such behavior in nature.

Alloparental Behavior in Prairie Voles

Under natural conditions, many prairie voles (*Microtus ochrogaster*) form cooperatively breeding social units. In these groups, an adult male and female form a pair bond after mating and engage in biparental care of their infants (maternal and paternal behavior). Importantly, most offspring in these groups do not disperse from their natal area once they are weaned, but instead remain in the group, and these virgin female and male offspring often help their parents rear subsequent litters (Getz, McGuire, Pizzato, Hofmann, & Frase, 1993; Kenkel, Perkeybile, & Carter, 2017; Lonstein & De Vries, 2001). Therefore, both allomaternal and allopaternal behaviors occur. Since these alloparents are caring for their younger siblings, such behavior probably increases their inclusive fitness, suggesting that kin selection may be involved in the evolution of alloparenting in prairie voles and other cooperatively breeding species. Although prairie vole alloparents are not lactating, they can be observed to approach and lick/groom their younger siblings, and they also huddle over them, which keeps the younger pups warm.

Several studies have examined the nature of alloparental behavior in prairie voles under laboratory conditions. In a typical study, prairie vole males and females are weaned at about 21 days of age, separated from their family group, and housed in groups of same-sex conspecifics. To test for alloparental behavior, a single prairie vole is placed in a novel cage and allowed to habituate to this cage for 15 to 45 minutes. After this habituation period, two unrelated prairie vole neonates are placed in the cage and the behavior of the experimental vole toward the pups is recorded over a 15-minute period. If the experimental male or female vole approaches, lick/grooms, and huddles over the pups, the vole is considered to show alloparental behavior. Voles are considered to be nonparental if they either ignore or attack the neonates. Research indicates that the age at which prairie voles are tested for alloparental behavior influences their responses to pups under these particular testing conditions, and that important sex differences exist. Most (80%–100%) virgin (sexually naïve) male prairie voles show alloparental behavior when tested as subadults (30 days of age) and as adults (60 days of age). In contrast, while most subadult virgin females display alloparental behavior, only about 50% of adult females do so (Lonstein & De Vries, 1999, 2001). Interestingly, experiential factors have been found to influence the degree of alloparenting displayed by adult virgin female prairie voles. If female prairie voles remain with their parents after weaning, even if younger siblings are not present, when they are tested for allomaternal behavior as adults, most females (80%) display allomaternal behavior toward unrelated neonates (Lonstein & De Vries, 2001). Perhaps age-related increases in stress reactivity influence allomaternal, but not allopaternal, behavior in prairie voles. Indeed, there is evidence that juvenile virgin female prairie voles display less anxiety-related behavior than do their adult counterparts (Olazabal & Young, 2005). Perhaps separation from the family group at 21 days of age until adulthood, which would not be typical under natural conditions, combined with being tested for parenting in a novel cage, has stress-inducing detrimental effects on alloparental behavior in virgin adult females, but not in their male counterparts. If some factor could boost maternal motivation (or decrease anxiety) in these adult virgin prairie voles that are weaned and separated from their family at 21 days of age, maybe more females would display allomaternal behavior. Just how remaining with their parents until testing enhances allomaternal behavior remains to be determined.

Significantly, Hayes and De Vries (2007) have also reported that about 50% of adult nulliparous female prairie voles show alloparental behavior. In contrast, and not surprisingly, 100% of parturient primiparous females are parental, even when tested with unrelated pups, which indicates that the physiological events of pregnancy and parturition boost parental motivation in female prairie voles. Additional results from the Hayes and De Vries study suggest that the central release of OT may have contributed to the enhanced maternal motivation in parturient prairie voles.

Olazabal and Young (2006a) have examined the importance of OT action on OTRs in NAs for allomaternal behavior in adult virgin female prairie voles. These nulliparous females were weaned at 21 days of age and raised in same-sex groups until they were tested for allomaternal behavior at about 60 days of age. Conforming with the data reviewed, about 50% of these virgins showed allomaternal behavior in a 15-minute test, while the remainder were not parental. Importantly, the allomaternal females had a higher density of OTR expression in NA than did the nonmaternal females. What these results suggest is that OT action in NA may be able to boost maternal motivation in a subset of virgin female prairies voles, allowing approximately 50% of these females to display allomaternal behavior. In support of this hypothesis, when an OTR antagonist was injected into NA of virgin adult female prairie voles, none of the females expressed allomaternal behavior. These results indicate that OT action on NA is necessary for alloparenting in adult female virgin prairie voles under these particular testing conditions (also see Keebaugh, Barrett, Laprairie, Jenkins, & Young, 2015). With respect to the previously described findings of Hayes and De Vries (2007), perhaps a surge in the central release of OT within the brain at parturition, co-acting with other physiological events, allows for the full expression of maternal responsiveness in all first-time prairie vole mothers. It is also possible that the expression of OTRs increases in the NA of parturient voles.

Ahern and Young (2009) examined the effects of additional early rearing experiences (being raised by both parents versus being raised by only a mother) on the display of allomaternal behavior in adult female prairie voles. The absence of the father did not increase the quantity of maternal behavior exhibited by mothers, with the result that pups who were only reared by mothers received less total parental care than did pups raised by both parents (Ahern, Hammock, & Young, 2011). When the allomaternal behavior of these offspring was examined when they were adults, the females that were raised by both parents showed much higher levels of alloparenting than did those raised only by their mother. However, the density of OTRs in NAs did not differ between these two groups, suggesting that other developmental factors contributed to the observed differences in allomaternal behavior. These findings might also be pertinent to the results of the study by Lonstein and De Vries (2001), who, as noted previously, showed that female prairie voles that remained with their parents after weaning displayed higher levels of allomaternal behavior in adulthood than females that were separated from their parents at weaning. Perhaps differences in OTR expression in the NA would not have been detected in these females as well. Therefore, the manner in which certain early-life experiences affect the neural systems that promote allomaternal behavior in prairie voles remains to be determined (cf. Perkeybile et al., 2019; also see Chapter 9 of this volume). One possibility is that certain early experiences affect the degree to which pup stimuli can access the MPOA and, therefore, the ability of the MPOA to activate PVN OT neurons.

An interesting question is whether OT is involved in the alloparental behavior of juvenile female prairie voles (see Schradin, Vuarin, & Rimbach, 2018, for an interesting analysis of age differences in the regulation of alloparental behavior in a variety of species). If such females exhibit less anxiety behavior during novel testing procedures, perhaps central OT systems are not needed to boost alloparental motivation (or to decrease anxiety). It would be interesting to determine whether the administration of an OTR antagonist to NA would interfere with the high levels of alloparenting shown by juvenile female prairie voles (cf. Keebaugh et al., 2015).

The role of OT neural systems in the alloparental behavior of male prairie voles is not as clear as that for females. Given the sex difference in the incidence of alloparental behavior in adult prairie voles, one might conclude that adult virgin male prairie voles would have more OTRs in NAs than adult virgin female prairie voles (when both sexes are weaned at 21 days of age and separated from their parents). However, in an early study (Insel & Shapiro, 1992), sex differences in the density of OTRs in the NA of adult virgin male and female prairie voles were not detected. In light of the findings of Olazabal and Young (2006), a more careful analysis of OTR expression in the NA of virgin male and female adult prairie voles should be undertaken. There is evidence that PVN OT release is increased during allopaternal behavior in adult virgin male prairie voles (Kenkel et al., 2012), and that the systemic administration of a nonpeptide OTR antagonist capable of crossing the blood brain barrier can suppress alloparenting in male voles (Kenkel et al., 2017; also see Bales, Kim, Lewis-Reese, & Carter, 2004). Interestingly, no one has examined whether the application of an OTR antagonist to NA could suppress allopaternal behavior in adult male prairie voles.

Results have recently been presented that are inconsistent with the view that OT action on OTRs in the brain is important for alloparental behavior in male prairie voles. In a preliminary study, Horie et al. (2019) developed a prairie vole mutant OTR gene (OXTR) knockout strain and found that adult mutant males displayed normal alloparental behavior. Receptor autoradiography indicated that the OTR protein was not detected in the brains of these mutant males, which included a complete absence of radioligand binding in NA. Unfortunately, the behavior of female prairie voles was not examined in this study.

How can we explain the Horie et al. (2019) findings, particularly in light of the findings of Kenkel et al. (2017)? Since the null mutation of the OXTR gene occurred throughout development, it is possible that some compensatory change allowed for the occurrence of alloparental behavior in the males of this mutant strain. V1a vasopressin receptors are present in the dorsal MPOA and ventral part of the bed nucleus of the stria terminalis (vBST) of prairie voles (Wang, Young, Liu, & Insel, 1997), and OT can bind to V1a vasopressin receptors. Therefore, perhaps endogenous OT action at this MPOA/vBST site compensated for a lack of action of OT on NA to enhance alloparental motivation (cf. the following section on alloparental behavior in marmosets and tamarins; also see Parker & Lee, 2001).

Surprisingly, no one has directly examined whether the MPOA is necessary for alloparental behavior in prairie voles. Only one study indirectly examined this issue. Kirkpatrick, Kim, and Insel (1994) found that cFos expression increased in MPOA during alloparental behavior in adult virgin male and female prairie voles (cf. Seelke et al., 2018). Significantly, the expression of Fos was greater in males than females, suggesting that pup stimuli may have greater access to

MPOA neural circuits in virgin males than in virgin females. It would be interesting to determine whether early rearing experiences can influence the degree to which pup stimuli can activate MPOA neurons in adult female prairie voles.

There is also evidence that DA neural systems are involved in the postpartum parental behavior of male and female prairie voles (Lonstein, 2002). With respect to alloparental behavior, Lei, Liu, Smith, Lonstein, and Wang (2017) found that extracellular DA levels increased in the NA of adult virgin male prairie voles during allopaternal behavior, and they also presented evidence that DA action on D1 receptors in NA promoted allopaternal care.

Although these findings are not fully conclusive, they certainly can be interpreted as indicating that when alloparental behavior occurs in prairie voles (particularly females), it involves pup-induced stimulation of MPOA input to the mesolimbic DA system coupled with MPOA activation of PVN-OT input to NA. Therefore, for those virgin females that express high levels of OTRs in NA, when they are tested under demanding conditions, it is likely that pup stimuli activate MPOA input to PVN-OT neurons that project to NA, in this way boosting maternal motivation to a level that facilitates alloparental behavior. This evidence supports the view that the neural systems that underlie naturally occurring alloparental behavior in prairie voles are similar to the neural circuits that underpin the hormone-stimulated maternal behavior that occurs in parturient female mammals that display a uniparental maternal care system. When alloparental behavior occurs, pup stimuli gain access to these circuits without the involvement of the physiological events of pregnancy and parturition, and I propose (as I did for laboratory mice) that this is because the MPOA has become an "open" system with respect to infant stimuli. By open system, I mean that the MPOA does not need to be primed by pregnancy hormones for pup stimuli to activate this region. To test this proposal, more research needs to explore the involvement of the MPOA in prairie vole alloparenting.

With respect to the regulation of alloparental that occurs under natural conditions in other rodent species, hardly any experimental research has been performed. Kalamatianos et al. (2010) compared forebrain OTR density in two species of African mole rats: the naked mole rat and the Cape mole rat. Naked mole rats, under natural conditions, live in colonies of up to 100 individuals. In these colonies, there is only one breeding female and one to three breeding males. The remaining members of the colony consist of nonbreeding subordinates that engage in various cooperative activities, which include alloparenting and burrow and nest construction. In contrast, Cape mole rats display a uniparental maternal care system, and after mating with a male, the female rears her offspring by herself. Kalamatianos et al. compared forebrain OTR binding densities in subordinate adult male and female naked mole rats with that detected in adult female Cape mole rats trapped during the nonbreeding season (they were not caring for pups). Importantly, they found high levels of OTR binding in NA of naked mole rats, but low to nondetectable binding levels in Cape mole rats. These correlational results are suggestive of the possibility that OT action on NA is important for alloparental behavior in subordinate adult naked mole rats (also see Mooney & Holmes, 2013). It would be interesting to determine whether OTR expression increases in the NA of Cape mole rat mothers during their breeding season.

Additional research on naked mole rats by Rosen, de Vries, Goldman, Goldman, and Forger (2008) explored the distribution of OT-immunoreactive axon terminals, derived from the PVN, in subordinate adult male and female naked mole rats and found OT projections to NA and to MPOA/vBST, suggestive of the possibility that OT input to these maternally relevant neural regions may contribute to the occurrence of alloparental behavior in male and female naked mole rat subordinates. However, in a preliminary study, Mooney, Douglas, and Holmes (2014) reported that the systemic administration of a nonpeptide OTR antagonist (only one dose level was administered), which is capable of crossing the blood–brain barrier, did not disrupt alloparental behavior in male and female adult subordinate naked mole rats, suggesting that the endogenous release of OT is not involved in the observed levels of alloparental behavior in this species. The OTR antagonist used in this study was the same as that which was used in the Kenkel et al. (2017) study, which was found to disrupt alloparental behavior in male prairie voles. Importantly, the dose administered in the Kenkel et al. study was much higher than that which was administered by Mooney et al. Perhaps a higher dose of the nonpeptide OTR antagonist would have had disruptive effects in the naked mole rat. This possibility gains in significance in light of the research that suggests that systemic administration of the nonpeptide OTR antagonist used in both of these studies (L-369,899) may not be as effective as generally believed in crossing the blood–brain barrier to reach critical neural sites (Smith, Freeman, Voll, Young, & Goodman, 2013). Clearly, more research is needed to determine whether OT is involved in alloparenting in naked mole rats. If OT is not important for alloparenting in naked mole rats, as it is in prairie voles, then such species differences would suggest that there are alternate routes that lead to the evolution of such behavior.

Alloparental Behavior in Marmosets and Tamarins

Marmosets and tamarins, observed under natural conditions and in captive groups, display a cooperative breeding social system. In this system, in addition to maternal behavior and paternal behavior, high levels of alloparental behavior occur: Nonbreeding juvenile, subadult, and adult individuals of both sexes that remain in the family group after they have reached independence help their parents care for more recently born offspring that are typically the younger siblings of the alloparents. These alloparents express a high level of interest and attraction to the parents' dependent offspring and aid their parents by carrying young infants and by provisioning them with food once they are weaned (Bales, Dietz, Baller, Miller, & Tardif, 2000; Digby, 1995; Yamamoto & Box, 1997; Yamamoto, Box, Albuquerque, & de Fatima Arruda, 1996). What is not clear from these observational studies is whether alloparents express spontaneous parental behavior toward young infants or whether they require a certain amount of exposure to young infants before they begin the show alloparenting. Perhaps a sensitization process occurs in these callitrichid alloparents, where exposure to young infants over a period of days while they are in their family group ultimately activates parental motivation in the alloparents. To determine whether a sensitization process is involved, or whether alloparental motivation is "spontaneous," experimental studies are necessary.

Experimental studies on alloparenting in New World primates have been conducted on the common marmoset (*Callithrix jacchus*). As reviewed in Chapter 3, although alloparenting occurs in common marmosets, the physiological events of late pregnancy further boost maternal motivation in parturient primiparous females. But what accounts for the baseline alloparental behavior that occurs in nonbreeding male and female marmosets? Interesting work on this issue has been performed by Pryce (1993). Female and male marmosets were examined when they were between 11 and 13 months of age, which is just before these individuals typically reach puberty. These marmosets were therefore classified as subadults. In one group of subjects, the marmosets were removed from their family group prior to the birth of their parents' new offspring. These individuals therefore did not receive any alloparental experience prior to being exposed for the first time to infant marmosets from their family group in a testing arena without other family group members being present. About half of these inexperienced subadult marmosets were observed to carry the infants. Another group of marmosets were allowed to receive experience in their family group after their parents' younger offspring were born. Pryce reported that virtually all marmosets that received such experience for less than 24 hours subsequently carried young infants from their family group when tested for alloparental behavior in the testing arena. Importantly, the alloparental experience obtained in the family group did not include infant carrying, because the parents prevented the subadults from doing so. Pryce suggests that tactile (touch-related) and olfactory inputs from infants for about 24 hours, while the subadults were in the presence of their parents, enhanced the alloparental motivation of the subadults.

An important consideration with respect to interpreting these results is that the occurrence of alloparental carrying behavior in marmosets is a two-way street. One aspect is the degree of alloparental motivation in the subadult monkeys that are exposed to infants. Another aspect is whether infants resist being carried by a potential alloparent. Zahed, Prudom, Snowdon, and Zeigler (2008) have reported that infant marmosets resist being carried by unfamiliar individuals. Therefore, it should be considered that the results of Pryce's (1993) data on inexperienced marmosets (only 50% carried infants) may have been influenced by the response of the infants to the potential alloparent. Pryce did not present any data relevant to this issue.

What can we conclude from these results? It appears that for about 50% of marmoset helpers, prior experience with infants is not required for the expression of alloparenting. However, brief exposure to young while the helpers are in the presence of their parents results in full alloparental motivation. Pryce (1993) concludes that evolutionary events have prepared cooperatively breeding marmosets to show relatively prompt alloparental behavior. Such preparation prevents the avoidance of young infants, while allowing only a short period of exposure to olfactory and tactile inputs from young infants, while in a family group, to promote full alloparental motivation. However, the response of infants to unfamiliar alloparents needs to be considered. It is possible that alloparental motivation in inexperienced marmosets may actually be higher than that reported by Pryce.

In a study of captive adult male nonbreeding common marmosets (21–48 months of age), with and without previous alloparental experience, Barbosa and da Silva Mota (2013) reported results that were similar to Pryce's (1993) findings. Males that had lived in family groups that contained younger siblings (experienced males) showed higher levels of alloparental behavior when exposed to an unrelated infant than did males that had lived in a family group that did not contain younger siblings. However, across four brief exposure tests to an unrelated infant (one per week, each of which lasted about 45 minutes), the alloparental behavior of the inexperienced males increased (also see Zahed et al., 2008).

Given these results, it seems clear that a baseline level of alloparental motivation exists in inexperienced nonbreeding marmoset helpers and that relatively brief exposure to infants further increases alloparental motivation. This conclusion is similar to that which I described for nulliparous laboratory mice in Chapter 3 (Stolzenberg & Rissman, 2011).

What physiological events may account for the high level of alloparental motivation that occurs either spontaneously, or after relatively brief exposures to young infants, in common marmosets? Roberts, Jenkins, Lawler, Wegner, and Newman (2001) have presented research suggesting the involvement of prolactin. Eight captive adult common marmosets (6 females and 2 males) were studied. None of these adults had prior experience with young. These individuals were selected for further study because, in pretests, they all retrieved and carried unrelated infants. It certainly would have been interesting to know what percentage of the total population that was pretested was represented by these individuals. These selected marmosets were then systemically injected with either bromocriptine (BC) or a control vehicle solution on different test days. BC inhibits prolactin release from the anterior pituitary and the results of this study showed that plasma prolactin levels were indeed undetectable in the BC-treated subjects, while plasma prolactin levels were detected in vehicle-treated marmosets. It was found that BC eliminated infant retrieval (carrying) in four of the eight marmosets, but the animals were given only 3 minutes to retrieve the infant before the test was terminated. If the animal did retrieve, its behavior was examined for an additional 10 minutes. For the 4 BC-treated monkeys that did retrieve infants, the duration of time that they carried the infant was significantly less than that observed in control females. These results suggest that prolactin may be involved in alloparental behavior in marmosets. However, BC is a D2 dopamine receptor agonist, and the animals that were injected with BC were hyperactive. Therefore, instead of disrupting alloparenting by inhibiting prolactin release, increased hyperactivity could have disrupted attentive processes, which disrupted retrieval behavior during short tests. In the rodent and rabbit studies that I reviewed in Chapter 3, the inhibitory effects of BC on the onset of maternal behavior in steroid-treated animals were reversed by the exogenous administration of prolactin. Until it can be determined that the disruptive effects of BC on alloparenting in marmosets can be reversed by exogenous prolactin, it is best to hold off on concluding that prolactin is involved in marmoset alloparenting. In fact, a correlational study that compared plasma prolactin levels with the occurrence of alloparenting in marmosets suggested that prolactin does not activate alloparental behavior (da Silva Mota, Franci, & de Sousa, 2006; also see Saltzman & Maestripieri, 2011).

It is interesting to compare alloparental behavior in inexperienced common marmosets with that which is observed in inexperienced adult nulliparous female prairie voles. In both of these groups, it appears that about 50% of the animals tested are spontaneously parental. For inexperienced virgin prairie voles, I reviewed the evidence that OT neural systems are involved in alloparenting. Is there any evidence that OT is involved in marmoset alloparenting? In a correlational study performed on captive family groups of marmosets, Finkenwirth et al. (2016) reported a positive correlation between alloparental care in male and female nonbreeding helpers and urinary OT levels. These investigators also examined proactive food sharing between alloparents and young infants in the family group, where the alloparent provides food to the younger sibling in the absence of such aid being solicited by the infant. They suggest that such proactive food provisioning by an alloparent toward a young sibling is a good measure of parental motivation. In this case, they also found a strong positive correlation between proactive food sharing by alloparents and urinary OT levels. To the extent that urinary OT levels also represent the central release of OT, these results suggest that OT neural systems may be involved in marmoset alloparental behavior.

As in other mammals, OT and vasopressin neurons are located in the PVN of marmosets (Wang, Moody, Newman, & Insel, 1997). In comparing the distribution of OTRs and V1a vasopressin receptors (V1aRs) in the brain of the common marmoset, Schorscher-Petcu, Dupri, and Tribollet (2009) found that V1aRs have a much wider distribution than do OTRs. In particular, OTRs and V1aRs are located in NA, but only V1aRs are located in MPOA. Recall the hypothesis of Freeman, Inoue et al. (2014) that I reviewed in Chapter 4. They suggested that in nonhuman primates, OT regulation of social behavior may involve its interaction with both OTRs and V1aRs. Since OT has a higher affinity for OTRs than for V1aRs, they suggest that low levels of endogenous OT would only affect OTRs, while higher levels of endogenous OT would affect both OTRs and V1aRs. Therefore, high levels of OT would presumably be able to affect both the NA and MPOA, while lower levels would only affect NA. The relatively high levels of urinary OT detected in family group marmoset helpers after the birth of new offspring (Finkenwirth et al., 2016), if representative of the release of OT within the brain, may have enhanced parental motivation by acting on these two critical sites.

Another aspect of the role of OT in parental motivation in marmosets should also be considered. Most mammals produce Leu8-OT, but common marmosets produce the variant Pro8-OT (French et al., 2016). Parreiras-e-Silva et al. (2017) have reported that Pro8-OT produces more long-lasting postsynaptic signaling effects at the OTR than does Leu8-OT, which suggests that the behavioral effects of OT may be more pronounced in marmosets. Perhaps the strong postsynaptic effects of Pro8-OT are involved in the relatively high levels of alloparental behavior in marmosets.

Due to the scarcity of the research on the neural control of alloparenting in marmosets, broad conclusions cannot be reached. Research indicates that alloparental motivation can occur after very brief exposure to infants and that OT neural systems, the MPOA, and NA may contribute to alloparental motivation in this species.

Conclusions on Alloparenting

In a comparison of laboratory virgin female mice (an experimental model of allomaternal behavior), and prairie vole and marmoset alloparents (natural models of allomaternal and allopaternal behavior), the overall evidence supports my proposal that the neural mechanisms that regulate maternal motivation in the typical female mammal that displays a uniparental maternal care system overlap with the neural mechanisms that control alloparental motivation. The quantity and quality of the evidence for this hypothesis is strongest for virgin laboratory mice and weakest for marmosets. The typical female mammal requires the physiological events of late pregnancy and parturition to modify the brain so that infant stimuli can gain access to maternal circuits. In alloparents, infant stimuli gain relatively prompt access to parental circuits without the need for pregnancy and parturition. Experiences within a family group also have important influences on the development of alloparental behavior.

In Chapter 5, I described dual subcortical neural circuits that influence maternal responsiveness in female mammals that exhibit a uniparental maternal care system—a defensive circuit that inhibits maternal behavior and promotes infant avoidance/rejection in nulliparous females and a parental circuit that regulates the appetitive and consummatory aspects of maternal behavior. When alloparental behavior occurs, either as a result of experimental genetic selection (virgin female laboratory mice) or natural selection (prairie voles, marmosets), it can be proposed that the defensive circuit has been downregulated and the parental circuit has been upregulated. However, latent defensive circuits may still exist in the brains of species that exhibit alloparenting, and these circuits may be activated under certain environmental conditions (see the section in Chapter 3 on hormones and maternal behavior in nonhuman primates and the subsection in Chapter 5 on the defensive-avoidance circuit active in nonmaternal virgin female mammals).

In addition to the current chapter, a recent review on alloparental behavior has been written, and the reader is referred to Glasper, Kenkel, Bick, and Rilling (2019) for additional information and insights.

Naturally Occurring Paternal Behavior

Introduction

In this section on paternal behavior, I want to focus on the mechanisms that regulate naturally occurring paternal behavior in those mammalian species that are not cooperative breeders, so that the occurrence of paternal behavior is not confounded by the high levels of alloparental motivation that exist in cooperatively breeding species. In addition, I will present experimental, rather than purely correlational evidence, to firmly establish causal relationships.

I will make the case that the subcortical neural circuits that regulate maternal and paternal motivation are basically the same, but that for paternal behavior to occur, factors outside the boundaries of pregnancy and parturition enable infant stimuli to gain access to these circuits. For both sexes in most mammals, it is highly probable that dual neural circuits regulate how an adult responds to infants, with one circuit regulating defensive and withdrawal responses and another that regulates parental responses. Since paternal behavior occurs naturally in only about 5% of mammalian species, it can be proposed that for the typical nonpaternal male mammal, the defensive circuit is dominant and paternal responsiveness is inhibited. For those mammalian species that exhibit paternal behavior under natural conditions, some factors other than pregnancy- and parturition-related events must activate the parental circuit and suppress the defensive circuit.

In line with these views, Wynne-Edwards and Reburn (2000) have proposed that it is unlikely the neural pathways that regulate paternal behavior in mammals are distinct from those that regulate maternal behavior. Since maternal behavior is common in mammals, while paternal behavior is rare, they argue that natural selection should have acted on existing maternal circuits so that such circuits would become operative in males under ecological conditions where the occurrence of paternal behavior would increase the survival of offspring.

An Evolutionary Perspective

Social monogamy occurs in about 10% of mammalian species (Lukas & Clutton-Brock, 2013). Although social monogamy is often associated with biparental care of offspring (both maternal and paternal behavior), such biparental care only occurs in about 50% of socially monogamous mammals, with the result that the prevalence of biparental, and therefore paternal, care occurs in about 5% of mammalian species. It is typically assumed that social monogamy evolved first for a male to obtain exclusive mating access to a single female under ecological and social conditions where males were unable to defend access to more than one female (Kleiman, 1977; Lukas & Clutton-Brock, 2013; Schacht & Bell, 2016). Paternal care is generally viewed as having evolved as a secondary adaptation, after the initial evolution of monogamy, under those conditions where the occurrence of paternal behavior alongside maternal behavior increased the reproductive success of both sexes (Stockley & Hobson, 2016).

Examining social monogamy and paternal behavior in apes supports the aforementioned views. The apes are divided into the lesser apes (gibbons and siamangs) and the great apes (gorillas, bonobos, chimpanzees, orangutans, and humans). Social monogamy and paternal behavior in great apes only occur in humans. In contrast, social monogamy without paternal behavior occurs in gibbons while both social monogamy and paternal behavior occur in siamangs (Fernandez-Duque, Valeggia, & Mendoza, 2009; Rafacz, Margulis, & Santymire,

2012). In a very interesting field study, Palombit (1996) found that the pair bonds between male and female partners in siamangs showed greater social cohesion than those observed in gibbons. Siamangs showed higher rates of affiliative interactions such as close proximity between the male and female pair, relaxed contact, embraces, and shared use of sleeping trees. Although paternal behavior does not occur in gibbons, siamang fathers begin carrying their infants once the infants are about 6 months of age (Rafacz et al., 2012). Two points are worth considering in this analysis of monogamy and paternal behavior in gibbons and siamangs. First, even when a male and female remain together in a common territory and may form an exclusive mating relationship, that should not be taken to mean that such a relationship represents a strong affiliative social bond. It is important to make a distinction between pair living (gibbons) and pair bonding (siamangs; Tecot, Singletary, & Eadie, 2016). Second, since male siamangs begin to show paternal behavior only after being exposed to their infant for about 6 months, sensitization processes may be involved in arousing paternal motivation in siamangs. The strong social bond between male and female siamangs, including close social proximity, may have afforded the male high levels of proximal contact with a young infant, which, in turn, may have activated the male's parental brain circuits through a sensitization process. It should also be considered, however, that the paternal motivation of male siamangs may be high at the birth of his young but that the female only permits the male to care for the infant after the infant reaches 6 months of age.

It is unlikely that sensitization processes are the route through which paternal behavior is activated in other mammalian species that show paternal behavior because in many of these species paternal behavior is shown near the time the infant is born. A good case in point is the paternal behavior shown by New World titi monkeys of the *Callicebus* genus. Titi monkeys live in monogamous social groups where biparental care of offspring occurs without alloparental behavior, at least when these monkeys are observed within their family groups (DeLuycker, 2014; Schradin, Reeder, Mendoza, & Anzenberger, 2003; Spence-Aizenberg, Di Fiore, & Fernandez-Duque, 2016). In these species, although the female nurses the young, in other respects it is the father that is the primary caregiver, and the father carries and transports its young infant 70% to 90% of the time until the infant reaches independence at about 5 months of age. Significantly, DeLuycker reported that a male titi monkey touched, sniffed, and licked its offspring within 3 minutes of its birth, and the male carried the newborn infant for most of the time beginning at 24 hours after its birth, only transferring the infant to the mother for nursing bouts. Paternal behavior in titi monkeys would seem to be an excellent candidate to explore the underlying physiological mechanisms that regulate prompt paternal behavior in a primate, but there is basically no research that has explored this important issue (cf. Schradin et al., 2003).

When examining only mammals, it is clear that maternal behavior is the typical parental care system. However, biparental care of offspring is the most common parental care system in birds. Also, although parental care is rare in fish and amphibians, when it occurs, depending on the species, one can observe either biparental care, uniparental maternal care, or uniparental paternal care (Brown, Morales, & Summers, 2010; Grone, Carpenter, Lee, Maraska, & Fernald, 2012; Roland & O'Connell, 2015). When parental care occurs in these species, it can include care of the eggs and hatchlings. In the paragraphs that follow, I want to present the limited research that has been conducted on the neural basis of parental behavior in fish, amphibians, and birds. This limited research suggests that the neural basis of parental behavior exhibits commonalities across vertebrates, which suggest that ancient and evolutionarily conserved neural systems appear to regulate parental behavior in mammalian and nonmammalian vertebrates.

Important research has been conducted on biparental care in the avian ring dove (*Streptopelia risoria*). In ring doves, both parents incubate their eggs prior to hatching. After the young have hatched, both parents also care for their squabs, and a dominant behavior is regurgitation feeding, where the parents regurgitate a milk-like substance (crop milk) into the mouths of begging young. Buntin, Berghmann, and Buntin (2006) reported increased Fos expression in the preoptic area (POA) of both the mother and the father when they were caring for their squabs. Slawski and Buntin (1995) found that neuron-specific lesions of the POA decreased parental feeding responses in birds of both sexes, and large lesions basically abolished parental behavior directed toward the squabs. With respect to incubation behavior, Komisaruk (1967) found that progesterone application to the POA induced incubation behavior in male and female ring doves.

Although most bird species display parental behavior, about 1% of bird species are brood parasites: They do not exhibit either maternal or paternal behavior. Instead, the females of these species leave their eggs in the nest of another species (Lynch, O'Connell, Louder, Balakrishnan, & Fischer, 2019). In an interesting study, Lynch et al. compared the gene expression patterns within the POA of the parasitic cowbird with the nonparasitic (parental) red-winged blackbird. Several differences in gene expression were found. In particular, the expression of the prolactin receptor gene was downregulated within the POA of cowbirds. Given the importance of prolactin action on the MPOA for mammalian maternal behavior, the downregulation of the prolactin receptor in the POA of cowbirds may contribute to the absence of parental behavior in this parasitic species. In support of this view, Buntin, Becker, and Ruzycki (1991) have reported that intracerebroventricular administration of prolactin facilitates parental behavior in biparental ring doves. See Smiley (2019) for a recent review on the role of prolactin in avian parental behavior.

The homolog of mammalian oxytocin in fish is the neuropeptide isotocin; mesotocin is the homolog of oxytocin in reptiles, amphibians, and birds (Gimpl & Fahrenholz, 2001). I am not aware of any definitive research on the role of mesotocin in the parental behavior of birds, although correlational supportive evidence has been presented by Chokchaloemwong et al. (2013). There is direct evidence that isotocin is involved in paternal behavior in fish. The teleost fish, *Amphiprion ocellaris*, is a predominantly paternal species, and males are the major caretakers of fertilized eggs. Males spend large amounts of time in the nest and fan eggs to aerate them. DeAngelis, Gogola, Dodd, and Rhodes (2017) found that the systemic administration of an isotocin receptor antagonist disrupted paternal caretaking activities in this species. The authors suggested that the antagonist was able to cross the blood–brain barrier, implying that central isotocin systems regulate paternal behavior in this species, although a peripheral site of action cannot be discounted. In support of a central site of action, DeAngelis, Dodd, Snyder, and Rhodes (2018) have reported that isotocin receptor expression is increased in the brains of parenting males (whole brain analysis) when compared to the levels expressed in nonparenting males that are not caring for fertilized eggs.

In the monogamous biparental cichlid fish, *Ametitlania nigrofasciata*, both parents care for their eggs and hatchlings by fanning eggs and by transporting hatched offspring in their mouths (O'Connell, Matthews, & Hofmann, 2012). If the female is removed so that the male is left alone with eggs and hatchlings, paternal behavior increases above the levels shown when both parents are present. These authors found increased Fos expression within isotocin neurons in the cichlid homolog of the mammalian PVN during paternal behavior. Importantly, the systemic administration of an isotocin receptor antagonist decreased paternal behavior.

In amphibians, there is anatomical evidence that hypothalamic mesotocin neurons, located in a region homologous to the mammalian PVN, project to the anterior POA, NAs, VTA, and to other extrahypothalamic sites (Gonzalez & Smeets, 1992, 1997). Perhaps isotocin neurons in fish and mesotocin neurons in amphibians project to these regions to promote parenting in a manner similar to that which I described for maternal mammals (but see Schulte & Summers, 2017).

In the studies that I just reviewed, I indicated that isotocin neurons in fish and mesotocin neurons in amphibians are located in a hypothalamic brain region that is homologous to the mammalian PVN. It is extremely important to note that this brain region consists of neurons that are part of the POA in these ancestral vertebrate species (also see Goodson, Evans, & Bass, 2003; Knobloch & Grinevich, 2014; Mennigen, Volkoff, Chang, & Trudeau, 2017). Therefore, in fish and amphibians, when parental behavior occurs, the role of the POA in parental behavior appears to have evolved in close anatomical and functional association with OT-like neuropeptides (isotocin; mesotocin). In birds and mammals, neurons containing OT-like neuropeptides are separated from the POA, moving slightly caudally to form the PVN (and SON), although remnants of POA OT neurons exist in the anterior PVN of mammals (also called the anterior commissural nucleus), in a region that lies dorsal to the caudal MPOA. It is also worth emphasizing that OT-like neuropeptides appear to be involved in the parental behavior of species (fish and amphibians) that do not lactate. Therefore, the involvement of OT-like neuropeptides in parental behavior predates its involvement in the milk-ejection reflex.

This brief and limited review indicates that both the preoptic region and oxytocin-like neuropeptides may be involved in parental behavior, including paternal behavior, in nonmammalian vertebrates. Therefore, expanding on the proposal of Wynne-Edwards and Reburn (2000), it is highly likely that ancient and evolutionarily conserved *parental* neural circuits are present in the brains of both sexes across vertebrate species. Depending on the adaptive consequences of engaging in parental activities, natural selection could act on these circuits so that parental behavior does not occur (parental circuits remain inactive), or, when parenting is adaptive, these circuits could be activated by particular external and/or internal events so that only maternal behavior, only paternal behavior, or biparental behavior occurs. Of course, these ancient circuits, which likely include the POA, OT-like neurons, brainstem DA neurons, the NA, and limbic forebrain regions, provide only an elemental and basic subcortical foundation for parental motivation. With the increasing social complexity that occurs in mammals, and especially in primates, additional neural elements, particularly cortical neural circuits, overlay this basic and ancient circuitry, and the interactions between cortical and subcortical circuits in the regulation of parental behavior will be emphasized when I discuss the human parental brain in the next chapter.

The most extensive research on naturally occurring paternal behavior in noncooperatively breeding mammals has been conducted on biparental rodents: the California mouse (*Peromyscus californicus*) and the dwarf hamster (*Phodopus campbelli*).

Paternal Behavior in the California Mouse

The seminal research on paternal behavior in the California mouse (a Peromyscus genus that is not closely related to the Mus genus) has been performed by Gubernick and his colleagues. In the wild, these mice form long-term monogamous pair bonds and exhibit biparental care of young. Although the male does not lactate, after the female's parturition he shows all the components of parental behavior that the mother displays. Paternal males retrieve displaced pups, engage in nest building, huddle over the pups in the nest to keep them warm, and lick/groom their young (Bester-Meredith, Burns, Conley, Mammarella, & Ng, 2017; Gubernick & Alberts, 1987; Gubernick, Schneider, & Jeannotte, 1994). Gubernick and Nelson (1989) tested the responsiveness of adult virgin male California mice toward an alien foster pup and found that only 19% of such males showed paternal responses, while the remaining virgin males either attacked or ignored the pup. In contrast, virtually all males (80%) that mated with a female partner and remained with her through parturition displayed paternal behavior toward an unfamiliar foster pup on day 1 postpartum (cf. de Jong, Korosi, Harris, Perea-Rodriguez, & Saltzman, 2012, who found that a higher percentage of virgin males are parental toward alien young than that reported by Gubernick & Nelson, 1989). These results suggest that some aspects of being paired with a female, copulating with her, and cohabitating with her through parturition converts a nonpaternal virgin male into a male that shows paternal responsiveness.

In an interesting study, Gubernick et al. (1994) examined the paternal responsiveness of male California mice after a variety of experiences. Virgin males that were prescreened for the absence of paternal responsiveness toward a foster pup were placed into the following independent groups: (a) The male copulated with a female, and then the female was immediately removed from the male's cage; (b) the male copulated with a female and remained with her for 24 hours, after which she was removed; (c) the male cohabitated with a female for 24 hours without mating with her, and therefore the female did not become pregnant; or (d) the male mated with a female and cohabitated with her through parturition. As previously noted, males in the last group were paternal. Most males (65%) living with a pregnant female (who had not yet given birth) ignored or attacked unfamiliar pups and did not show paternal behavior toward an unfamiliar pup until after the birth of their own offspring. However, about 35% of the males that copulated with their partner showed paternal behavior toward an alien pup after just 24 hours of postcopulatory cohabitation with the female. These results suggest that some factors associated with mating with a female and remaining with her through parturition activates paternal responsiveness in most males. However, copulatory stimulation and cohabitation with a female for as little as 24 hours is sufficient to activate paternal behavior in a minority of males.

What is it about mating and cohabitating with a female through parturition that promotes paternal responsiveness in most male California mice? One important factor is that chemosignals from the parturient female activate and maintain paternal responsiveness in the majority of male California mice during the first 3 days postpartum (Gubernick, 1990; Gubernick & Alberts, 1989; Gubernick et al., 1994). At parturition, if males are separated from their pups at birth, their paternal responsiveness to foster pups (tested on day 3 postpartum) declines. However, paternal behavior is maintained if such males are exposed to urinary pheromones from their female partner. In contrast, the maintenance of maternal behavior in postpartum females is not supported by the presence of stimuli from the male partner. Instead, maternal behavior is maintained only by the presence of pups; if the pups are removed at birth the female does not show maternal responses toward a foster pup on day 3 postpartum even if she remains with her male partner. Therefore, a clear sex difference exists. As in most female mammals, the presence of pups maintains postpartum maternal behavior. In males, however, chemosignals from the postpartum female are sufficient to maintain the male's paternal responsiveness, at least through day 3 postpartum. After day 3 postpartum, it appears that chemosignals from the female co-act with pup stimuli to maintain paternal responsiveness (Gubernick & Alberts, 1989).

What can we conclude from these studies? It appears that copulatory stimuli coupled with olfactory stimuli from the pregnant and parturient female partner transform a nonparental male into a paternal male in most California mice. It can be proposed that some alteration in the olfactory system (the main and/or vomeronasal system [VNS]) may be necessary for full paternal behavior to be displayed in this species. Similar to maternal behavior in rats, it appears that dual neural circuits regulate the occurrence of paternal behavior in California mice, one that inhibits paternal behavior and is dominant in virgin males and one that promotes paternal behavior and is active in males during the female partner's postpartum period. Perhaps exposure to female stimuli modifies the valence of pup-related odors in male California mice so that such odors no longer activate defensive neural circuits but instead activate appetitive paternal circuits. Curiously, I am not aware of any studies that have examined the effects olfactory bulbectomy on the paternal behavior of California mice. There is some evidence that a sensitization-like process may occur in virgin male California mice, in that repeated exposures to pups have been shown to enhance parental responses of virgin males (Horrell, Perea-Rodriquez, Harris & Saltzman, 2017). Perhaps, as in virgin female rats, repeated pup exposure familiarizes the male to the pup's odors, which decreases their aversive qualities, while also allowing pup stimuli to activate central parental neural circuits.

Since a substantial proportion of males switched from being nonparental to showing paternal behavior after mating with a female and cohabitating with her for 24 hours, besides the potential paternal stimulating effects of this experience per se, it is intriguing to speculate that the formation of a pair bond between the male and his partner may have in some way primed paternal neural circuits. As I will review in Chapter 11, there are many similarities in the neural circuits that regulate parental behavior and those the regulate pair bond formation (Numan & Young, 2016), so this idea is worthy of consideration.

Is there any evidence that hormones influence paternal behavior in California mice? A common hypothesis, based on correlational and some experimental evidence from several mammalian and bird species, is that testosterone levels should decline to prevent aggressiveness and allow for infant care behavior by fathers (Grebe, Sarafin, Strenth, & Zilioli, 2019; Lynn, 2016; Tecot & Baden, 2018; Wingfield, Hegner, Dufty, & Ball, 1990). However, Tecot and Baden (2018) have noted, in a cross-species analysis, that androgens can be positively, negatively, or unrelated to paternal behavior. With respect to experimental research on male California mice, Trainor and Marler (2001, 2002) found that testosterone promotes paternal behavior through its metabolic conversion to estradiol. Aromatase is the enzyme that converts testosterone to estradiol, and Trainor, Bird, Alday, Schlinger, and Marler (2003) also found that fathers had significantly more aromatase activity in the MPOA than mated males without pups. It is interesting to speculate that pheromonal and other stimuli from parturient mothers not only suppress the proposed olfactory inhibition of paternal behavior in California mouse fathers, but also activate aromatase expression in the MPOA, which then allows circulating levels of testosterone to be converted to estradiol within the MPOA, and estradiol action at this site may facilitate paternal motivation by allowing pup stimuli to activate certain MPOA neurons. In addition to directly enhancing paternal motivation, estradiol priming of MPOA neurons may promote the ability of MPOA efferents to suppress the defensive neural system. It certainly would be important to determine whether estradiol application directly to the MPOA of male California mice could promote paternal behavior under certain experimental conditions (such as during the latter part of the female partner's pregnancy), but, to the best of my knowledge, such an experiment has not been performed.

This focus on the MPOA in the paternal behavior of California mice supports the proposal that the neural basis of paternal behavior overlaps with the neural circuits that regulate maternal behavior. In fact, Fos expression increases in the MPOA of paternal California mice (de Jong, Chauke, Harris, & Saltzman, 2009), and Lee and Brown (2002) found that electrical lesions of the MPOA essentially eliminated maternal and paternal behavior in postpartum female and male California mice, respectively. In a subsequent study, Lee and Brown (2007) found that electrical lesions of either the MPOA or basolateral amygdala, but not the NA, disrupted paternal and maternal behavior in California mice. These findings match the results obtained in maternal rats and indicate that the wider neural circuitry that regulates maternal behavior matches that which regulates naturally occurring paternal behavior in a biparental rodent species. Although not evaluated by Lee and Brown (2007), I would predict that lesions of the ventral pallidum would disrupt maternal and paternal behavior in the California mouse.

Finally, in an important preliminary study, Horrell, Saltzman, and Hickmott (2019) compared the neurophysiological characteristics of single MPOA neurons in virgins and fathers (at day 7 postpartum), using an in vitro brain slice preparation. They provided convincing evidence that MPOA neurons in paternal California mice are under less synaptic inhibition than are the MPOA neurons of virgins. They suggested that this result may have been due to a decrease in the number of inhibitory synapses onto the MPOA neurons of fathers. Perhaps this was caused by decreased inhibitory input to the MPOA from the defensive neural circuit, and from the periaqueductal gray (PAG) in particular (compare to Figure 5.9).

Paternal Behavior in the Dwarf Hamster

The dwarf hamster (*P. campbelli*) displays a biparental behavioral phenotype similar to that of the California mouse. After mating, the male and female form a long-term monogamous pair bond and the male engages in intense paternal care during the postpartum period that matches that displayed by the female (Vella, Evans, Ng, & Wynne-Edwards, 2005). Vella et al. (2005) compared the response of adult virgin male dwarf hamsters and day 3 postpartum first-time fathers to an unfamiliar pup. While most of the first-time fathers displayed paternal behavior (retrieving a displaced pup to the nest), 70% of the virgin males either attacked or ignored the pup, while 30% exhibited paternal responses. These results indicate that there is a transition from nonparental to parental behavior when virgin males are compared to males that mated with a female and remained with her through day 3 postpartum.

Since estradiol has been shown to be important for paternal behavior in the California mouse, a series of studies was performed by Wynne-Edwards and her colleagues to explore the potential involvement of estradiol in the paternal behavior of the dwarf hamster. Reburn and Wynne-Edwards (1999) reported that blood levels of testosterone peaked in male hamsters on the day before their female partner's parturition. This correlational evidence conforms with the view that testosterone, and possibly its conversion to estradiol in the brain, may play a role in stimulating the onset of paternal behavior in this species. However, subsequent experimental evidence could not establish a causal relationship. First, castration of males during the female partner's pregnancy did not affect the normal postpartum paternal behavior of first-time fathers, although plasma levels of testosterone and estradiol were nearly eliminated by the gonadectomy (Hume & Wynne-Edwards, 2005). Second, the systemic injection of males with an aromatase inhibitor, which would block estradiol synthesis, beginning during the middle of female's pregnancy and continuing through day 6 postpartum, did not affect the male's paternal behavior (Hume & Wynne-Edwards, 2006). These authors therefore concluded that although the conversion of testosterone to estradiol may be important for paternal behavior in California mice, they could not find support for such a role in the dwarf hamster. Timonin, Cushing, and Wynne-Edwards (2008) suggest that although maternal behavior and paternal behavior may utilize common central nervous system parental circuits, the particular factors that activate these circuits in males to promote paternal behavior may vary across different biparental mammalian species.

The particular factors that promote paternal behavior in dwarf hamsters remain to be determined, and there is some recent evidence that contrasts with the aforementioned research, suggesting that a role for estradiol in the paternal responsiveness of this species should not be completely ruled out (Romero-Morales, Martinez-Torres et al., 2018).

It is also worth emphasizing that the percentage of adult naive virgin males in the California mouse and in the dwarf hamster that show paternal responses toward an unfamiliar pup varies across studies and can sometimes exceed 50% (de Jong et al., 2009; Romero-Morales, Martinez-Torres et al., 2018). Therefore, it should be considered that virgin males in certain biparental species may have been evolutionarily prepared to show relatively high levels of paternal responsiveness as virgins, with this basic level of responsiveness being further enhanced by mating and cohabitating with a female partner through the female's parturition. Whatever the mechanisms that activate parental neural circuits in naturally paternal males, what seems clear from the brain research on the California mouse is that the brain circuits that underpin paternal behavior overlap extensively with those that regulate maternal behavior in this biparental species, and these circuits match those that regulate maternal behavior in female mammals that exhibit a uniparental maternal care system.

Experimentally Induced Paternal Behavior in Mammalian Species That Naturally Exhibit a Uniparental Maternal Care System

Introduction

Paternal behavior does not occur in rats (*Rattus norvegicus*) and house mice (*Mus musculus*) in nature; these species typically exhibit a uniparental maternal care system and their mating system is not monogamous but instead is either polygynous or promiscuous. However, under laboratory conditions, paternal behavior can be experimentally induced in male laboratory rats and mice. Several studies have explored the neural basis of such paternal behavior. The reader might wonder why an examination of this issue is important. These studies are important because they show that when paternal behavior occurs in these laboratory species, it relies on the same neural mechanisms that regulate maternal behavior in rats and mice. What this research suggests, therefore, is that dormant parental circuits are present in the brains of male rats and mice, which supports the idea of a common parental circuitry in vertebrates. By understanding how these circuits can be experimentally activated, one might also gain further insight into the mechanisms that activate naturally occurring paternal behavior in such species as the California mouse and dwarf hamster, as well as in other naturally paternal mammals.

Paternal Behavior in Laboratory Rats

Adult virgin male laboratory rats can be induced to show parental behavior toward pups through the sensitization process that I described for nulliparous female rats. After several days of continuous exposure to pups, such males, although initially avoiding or attacking pups, will eventually display parental behavior, retrieving the pups, crouching over them, and licking/grooming the pups (Rosenblatt, 1967). When comparing the sensitization latencies of virgin males with those of virgin females, the latencies in male rats (about 11 days) tend to be longer than those observed in their female counterparts (about 7 days; Mayer, Freeman, & Rosenblatt, 1979). Males are more resistant to pup-stimulated parental behavior, but the behavior still eventually occurs.

Several investigators have attempted to stimulate parental behavior in adult virgin male rats by systemically treating gonadectomized males will hormone treatments that have been found effective in stimulating parental behavior in virgin female rats. Such treatments, which involve long-term estradiol exposure coupled with prolactin on a background of progesterone withdrawal, have been found to shorten sensitization latencies in virgin male rats in comparison to untreated control males (Lubin, Leon, Moltz, & Numan, 1972; Rosenblatt, Hazelwood, & Poole, 1996). Significantly, while such hormone treatments promote parental behavior in virgin females after 1 to 2 days of pup exposure (see Chapter 3 of this volume), such treatments promote parental behavior in virgin males after 3 to 4 days of pup exposure, conforming to the view that males are more resistant to the expression of parental behavior than are females. Supporting this view are findings from the aforementioned research that show that higher doses of hormones are needed to shorten sensitization latencies in males than in females.

Similar to female rats, the MPOA is also essential for the parental responsiveness that can be expressed in male rats: Lesions of the MPOA completely abolish the parental behavior that can be induced in male rats by hormone treatments, and this includes neuron-specific lesions of the MPOA with *N*-methyl-D-aspartate (Rosenblatt et al., 1996; Sturgis & Bridges, 1997). Furthermore, as in females, estradiol has been shown to stimulate parental behavior in steroid-primed male rats by acting on the MPOA (Rosenblatt & Ceus, 1998).

These finding suggest that in male rats, as in female rats, dual neural circuits regulate parental responsiveness: a defensive circuit and a parental circuit. Since the MPOA is important for parental behavior in females and males, it is likely that common neural circuits regulate parental behavior in both sexes, although the activation of such circuits appears to be under greater inhibitory control in males.

In Chapter 5, I described the defensive neural circuit in virgin female rats that opposes the expression of maternal behavior, and I indicated that this defensive circuit is, in part, activated by olfactory input from novel pup stimuli. Is there any evidence for a similar effect in male rats? Izquierdo, Collada, Segovia, Guillamon, and del Cerro (1992) have reported that electrical lesions of the bed nucleus of the accessory olfactory tract (BAOT) promote pup-stimulated parental behavior in male rats. Males with BAOT lesions exhibited sensitization latencies of 3 days while control males showed parental behavior only after 12 or more days of continuous pup exposure. The BAOT, which adjoins the medial amygdala (MeA), receives input from the accessory olfactory bulb (AOB), and it is interconnected with the MeA; it is part of the VNS (Izquierdo et al., 1992).

In an excellent review, Segovia and Guillamon (1993) described research that shows that there is a sexual dimorphism within the VNS of rats. For example, the volume and number of neurons in the AOB and BAOT are greater in males than in females. This difference, in part, may account for the fact that male rats are more resistant to pup-stimulated and hormone-stimulated parental responsiveness than are female rats. More specifically, the defensive neural circuit appears to be more dominant in male rats than in female rats.

Under natural conditions, paternal behavior does not occur in male rats or most other male mammals because such males are never exposed to the conditions, such as continuous exposure to young over long periods of time, or exposure to pregnancy hormones, which have been shown to activate parental responses in males under experimental laboratory conditions. But these experimental results suggest that latent parental circuits are probably present in the brains of all male mammals, which provide a substrate upon which natural selection could act. In mammalian species where a biparental care system is necessary for infant survival, natural selection would be able modify these latent parental circuits. Such a modification would create regulatory mechanisms where factors other than the physiological events associated with pregnancy termination, or continuous exposure to infants over long periods, would be able to activate parental circuits in fathers so that prompt paternal behavior occurs that coincides with their mate's parturition. Perhaps in naturally paternal males, mating with a female, forming a pair bond with the female, and remaining with a female through her pregnancy and parturition, in some way depresses the defensive circuit and activates the parental circuit. Perhaps copulation that is coupled with relatively long-term exposure to pheromones from the female mate modifies the VNS and the main olfactory system, so that novel infant odors no longer suppress parental behavior. These same copulatory and pheromonal stimuli may also activate parental motivation circuits. Since aromatase is present in the MPOA of rodents, including male rats (Tabatadze, Sato, & Wooley, 2014), a mechanism would also exist to allow circulating testosterone in males of some species to be converted to estradiol within MPOA to promote paternal behavior. These proposed processes seem to fit with the evidence I described with respect to some of the factors that are associated with naturally occurring paternal behavior in the California mouse.

Interesting work by Mennella and Moltz (1988b) adds support to the previously expressed views. They placed virgin male laboratory rats (Wistar strain) in a 4×2×3-foot arena that contained nest boxes and a late-pregnant female rat. After the female gave birth, most males attacked the young pups. However, if males first mated with a female and remained with her for 12 hours before being separated from her, when these males were subsequently placed in an arena with a novel late-pregnant female, 5 days prior to her parturition and 18 days after the male's copulatory experience, most males did not commit infanticide upon the birth of the female's young. These results indicate that copulation coupled with a 12-hour period of contact with the female mate inhibited infanticide in these male rats at a time that would have coincided with the birth of their own young (pregnancy in rats lasts 22–23 days). Mennella and Moltz (1988a) subsequently found that if the vomeronasal organ was removed from sexually naïve virgin male rats, then when they were placed in the arena with a pregnant female, 5 days before her parturition, they did not commit infanticide, as did control males, upon the birth of the female's young. Taken together, these findings can certainly be interpreted as showing that copulatory and postcopulatory stimuli suppress vomeronasal activation of the defensive circuit in male rats as indicated by the suppression of infanticidal behavior. The evolutionary significance of the fact that infanticide is suppressed near the time when the male's own young would have been born and cared for by the mother will be discussed in detail in the next section.

Paternal Behavior in Laboratory Mice

Virgin females in most strains of laboratory house mice show "spontaneous" maternal behavior when tested with foster young in their home cages, while feral virgin female house mice are infanticidal (they exhibit pup-directed aggression). In sharp contrast to this dichotomy, virgin male house mice of several inbred and outbred laboratory strains of house mice attack foster pups and therefore behave similarly to feral males (McCarthy, 1990; Palanza & Parmigiani, 1991; Soroker & Terkel, 1988; vom Saal & Howard, 1982). Therefore, selective breeding and inbreeding appear to have increased parental motivation in virgin female lab mice but not in their virgin male counterparts.

Certain experiences have been shown to modify the response of male laboratory mice to pups, switching them from infanticidal to parental. In an important series of experiments, vom Saal (1985) examined the response of male mice from the outbred CF-1 laboratory strain to unfamiliar foster pups. A majority of virgin males of this strain initially attacked test pups. However, if such males were allowed to mate with a female, from whom the males were separated after ejaculation occurred, when they were presented with foster pups 20 days later, the males did not attack the pups, but instead showed parental behavior (pup retrieval, pup grooming, and crouching over pups). Importantly, this switch from infanticide to parental behavior was exhibited only during a time period when the male's own young would have been born and subsequently nursed by the inseminated partner. That is, infanticide was inhibited and parental behavior occurred toward foster pups only between 3 and 7 weeks after the male's mating experience. Since pregnancy lasts about 20 days in mice and pups are weaned at about 28 days, this switch from pup-directed aggression to pup-directed paternal care coincides with the period when the male's own young would have been preweanling dependent offspring.

vom Saal's (1985) finding of a tight time-based inhibition of infanticide and promotion of parental behavior in male CF-1 laboratory mice at particular time points after mating appears to be a mechanism that prevents a male from killing his own offspring. Note that feral house mice exhibit a uniparental maternal care system and that natural paternal behavior does not occur. Wild mice typically live in groups called demes, which contain one dominant male and several breeding females. Since the dominant adult male does all of the mating, the young that are born are typically the male's own offspring. Therefore, in feral mice, this time-locked inhibition of infanticide after copulation with a female is likely a mechanism that prevents the male from killing his offspring rather than a mechanism that governs paternal behavior. As described in the previous section, a similar adaptive process appears to inhibit infanticide in male rats.

But why should sexually naïve virgin males attack pups? Hrdy (1979) has proposed that male infanticidal behavior has adaptive significance. If the dominant male in a deme is overthrown by a sexually naive intruding male or a subordinate male within the deme who subsequently takes over the deme, it may be adaptive for the intruder or the previously subordinate male to kill the pups that the breeding females are nursing, since these pups cannot be his offspring.

Since ovulation is inhibited throughout most of lactation, by killing the female's offspring, nursing will cease and sexual receptivity and ovulation will occur in the previously lactating females. Such male infanticide, therefore, will allow the usurping male to mate with the female sooner, rather than waiting until the young are weaned, in this way enhancing his reproductive success (vom Saal & Howard, 1982).

Although copulation per se inhibits infanticide in male CF-1 mice at certain time points after the mating experience, research has indicated that there are important strain differences in the types of experiences that suppress infanticide and promote parenting in male laboratory mice. For example, Palanza and Parmigiani (1991) reported that copulation per se was not sufficient to suppress male infanticide toward foster pups 20 day after mating in the outbred Swiss Webster strain of laboratory mice. Instead, infanticide was inhibited and parental behavior occurred only if a male mated with a female and cohabitated with her through day 18 of her pregnancy. Importantly, if a male did not copulate with a female, but did cohabitate with a pregnant female through day 18 of her pregnancy, infanticide was not inhibited. Palanza and Parmigiani proposed that copulation in this mouse strain serves as a priming mechanism that then allows cohabitation with a pregnant female to suppress pup-directed attack and promote pup-directed caretaking in male Swiss Webster mice.

Because of these strain differences, most studies that have examined the neural basis of paternal behavior in laboratory mice have utilized males that copulate with their female partner and then remain with her through her parturition, at which time paternal behavior typically occurs. Note, however, that in feral mice, such an experience (continuous cohabitation with a pregnant female through her parturition) would not occur within demes, which may be the reason why natural paternal behavior is not observed.

In an excellent series of experiments, Tachikawa, Yoshihara, and Kuroda (2013) examined some of the mechanisms that regulate paternal behavior in the C57BL/6 inbred strain of laboratory mice. They found that the majority (81%) of sexually naïve (virgin) male mice exhibited pup-directed aggression when tested with foster pups. In contrast, if males mated with a female and remained with her through her parturition, infanticide toward foster pups was completely inhibited, and 90% of these males were also parental. Significantly, removal of the vomeronasal organ prevented infanticide and promoted parental behavior in virgin male mice, while not affecting the paternal behavior shown by fathers (also see Trouillet et al., 2019; Wu et al., 2014). These results indicate that the vomeronasal organ is not necessary for parental behavior, but that the detection of pup pheromones by the vomeronasal organ in virgin males (see Isogai et al., 2018) activates infanticide and suppresses parental behavior. This finding coincides with the results of Mennella and Moltz (1988a), where removal of the vomeronasal organ suppressed infanticide in male rats. Through a mechanism similar to that which I described for virgin female rats, it appears that pup-related pheromones activate a defensive circuit in virgin male mice (and rats) that promotes either attack or avoidance of pups and inhibits parental behavior. It can be proposed that copulation and cohabitation with a female in some way suppresses the inhibitory effects of vomeronasal input on the paternal behavior of male mice. Interestingly, this vomeronasal inhibition of parental motivation does not occur in virgin female laboratory mice, but does occur in feral virgin female mice (see Chapter 4 of this volume). Therefore, selective breeding and inbreeding appear to have exerted sex-specific effects in laboratory mice, decreasing vomeronasal inhibition of parental behavior in females, but not in males. (As in female mice [see Chapter 4 of this volume], it is assumed that olfactory input from the main olfactory system, as opposed to the VNS, is essential for paternal behavior in mice; cf. Liu et al., 2013.)

To examine some of the neural circuits that are active in infanticidal and paternal male mice, Tachikawa et al. (2013) analyzed Fos expression within the brains of parental and nonparental males (also see Mayer, Crepeau et al., 2019). Males were exposed to pups in a wire mesh ball, which allowed the males to sniff and lick the pups, but prevented the males from biting pups. In response to this stimulus, naïve virgin males showed increased Fos expression in the anterior hypothalamic nucleus and the ventrolateral part of ventromedial hypothalamic nucleus. Such increased Fos expression in these parental inhibitory regions was presumably the result of vomeronasal organ activation of AOB stimulatory inputs to the MeA (Chen et al., 2019; Tachikawa et al.; Trouillet et al., 2019). In contrast, Tachikawa et al. reported that fathers exhibited increased Fos expression in the MPOA rather than in the anterior hypothalamic nucleus and the ventrolateral part of ventromedial hypothalamic nucleus. (Findings similar to those of Tachikawa et al. have been reported for paternal and nonpaternal male Mongolian gerbils with the additional finding that nonpaternal males also expressed increased Fos in PAG, which has been proposed to be an additional link in the defensive/rejection circuit, as reviewed in Chapter 5 and shown in Figure 5.9 (Romero-Morales, Cardenas et al., 2018).]

In reference to Figure 5.9, it appears that the defensive neural circuit is activated by pup stimuli in virgin males, while the output of MPOA parental circuits is activated in paternal male mice. One possibility is that copulation and cohabitation with a mate modifies the function of the male's MPOA, and the outputs of MPOA both enhance parental motivation and suppress the defensive circuit. It is also possible that copulation and cohabitation with a pregnant/parturient female in some way decreases the responsiveness of the vomeronasal organ/AOB to the pup pheromones that promote pup-directed aggression in virgin male mice (see Tachikawa et al., 2013), which then eliminates a potential inhibition of MPOA by the defensive circuit. These two possibilities are not mutually exclusive, and both processes probably occur. Interestingly, the pup pheromones that elicit pup-directed aggression in virgin male mice may actually be maternal pheromones that have been applied to the pups through maternal licking/grooming or through other maternal secretions (Isogai et al., 2018). Therefore, when a male copulates and cohabitates with a female throughout her pregnancy/parturition, perhaps the male's vomeronasal organ receptors habituate and become less responsive to pheromones that would elicit aggression in inexperienced virgin males. Note how this idea can be related to the previously described finding by Gubernick and his colleagues, where it was shown that maternal chemosignals can influence paternal behavior in California mice.

There is also more direct evidence that the MPOA is essential for paternal behavior in mice. Akther, Kakhrul, and Higashida (2014) allowed male laboratory mice to mate with a female and remain with her through her pregnancy and parturition, at which time the males typically displayed paternal behavior toward their offspring. However, sires that received electrical lesions of the MPOA after they mated with their female partner did not show paternal behavior upon the birth of their offspring. Similar results have been reported by Tsuneoka et al. (2015) after neuron-specific lesions of the MPOA with *N*-methyl-D-aspartate. Importantly, Wei et al. (2018) found that optogenetic activation of estrogen receptor-alpha–containing MPOA neurons in virgin male mice promoted pup retrieval behavior. The results of Wei et al. lead to two important conclusions. First, stimulation of MPOA output can activate paternal behavior in virgin male mice, suggesting that MPOA output suppresses defensive responses and promotes parenting. Second, since activation of MPOA neurons that contain estrogen receptors is involved in this facilitation of paternal behavior, the possibility exists that the conversion of testosterone to estradiol via aromatase activity within MPOA may play a role in paternal responsiveness in mice. Relevantly, Akther et al. (2015) have reported that aromatase levels are higher in the MPOA of paternal males than in virgins and that systemic treatment of fathers with an aromatase inhibitor suppressed paternal behavior in mice.

There is also evidence that MPOA galanin neurons are involved in paternal behavior in male mice, as they are in maternal behavior of female mice. Wu et al. (2014) found that selective ablation of MPOA galanin neurons in paternal mice decreased paternal behavior without inducing infanticide. These results suggest that once a male copulates with a female, and cohabitates with her through her pregnancy and parturition, these experiences suppress vomeronasal activation of the defensive circuit, so that pup-directed aggression will not occur, but that the output of MPOA galanin neurons is still necessary for the display of paternal motivation. In contrast to these findings, Kohl et al. (2018) have reported that selective optogenetic stimulation of MPOA galanin neurons that project to PAG (part of the defensive circuit), suppresses infanticide, without inducing parental behavior, in sexually inexperienced male mice. It can be suggested that there are two populations of MPOA galanin neurons. One population may be stimulated selectively by certain copulatory and postcopulatory experiences to contribute to the suppression of infanticide (vom Saal, 1985), while another population may be stimulated by mating and cohabitating with a female through parturition, with this population serving to enhance paternal motivation, perhaps through projections to the VTA (Kohl et al., 2018; also see Chapter 5 of this volume).

There is only limited evidence with respect to the broader neural circuitry that regulates the paternal motivation in mice. It has been found that electrical lesions of the ventral pallidum disrupt paternal behavior in laboratory mice (Akther et al., 2014) and that oxytocin may act on the NA to promote paternal behavior (Akther, et al., 2013; also see Wang, Wang, Wang, & Tai, 2018). Given this evidence, it appears that the larger neural circuitry that regulates paternal behavior matches the circuits that regulate maternal behavior. As described previously, similar evidence was obtained with respect to the larger neural circuitry that regulates naturally occurring paternal behavior in the California mouse. It is also clear that more research needs to examine the role of OT in the paternal behavior of rodents. In this regard, recent work on Mandarin voles (a naturally biparental species) has provided evidence that OT may act on the MPOA to enhance paternal behavior in this species (Yuan et al., 2019). Therefore, there is evidence from mice and Mandarin voles that OT may act on NA and MPOA to promote paternal behavior, which is similar to its effects on mammalian maternal behavior (cf. Sharma et al., 2019). I predict that future research will show that the VTA is an additional site of OT action in the regulation of paternal behavior, which would further confirm the similarities in the neural regulation of paternal and maternal behavior.

Conclusions on Paternal Behavior

My examination of the mechanisms that regulate paternal behavior in naturally paternal nonhuman mammalian species, along with the research on experimentally induced paternal behavior in nonhuman mammalian males that do not exhibit such behavior under natural conditions clearly indicates six important conclusions: (a) The subcortical neural circuits that regulate paternal behavior match those that regulate maternal behavior; (b) dual subcortical neural circuits are present in the brains of female and male mammals, one that inhibits parental behavior and promotes infant avoidance or infant-directed aggression and one that promotes parental motivation; (c) vomeronasal activation of infant-directed aggression appears to be crucially involved in the negative response of virgin

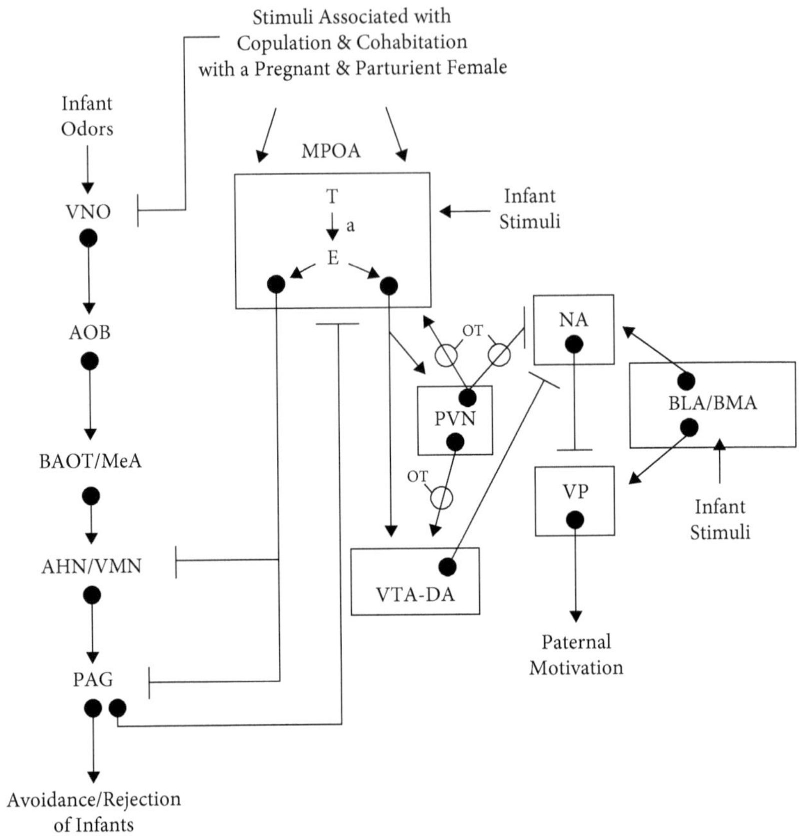

Figure 7.1. A summary diagram that proposes a neural mechanism through which paternal behavior can be induced in certain male nonhuman mammals. In the typical virgin male, infant odors activate the defensive circuit which gives rise to avoidance/rejection responses toward infants. This defensive circuitry is similar to that described for most virgin female mammals that use olfaction as a major sensory detection system (see Figure 5.9). For many rodent virgin males, the vomeronasal organ (VNO) is particularly important in detecting those infant pheromones that activate defensive responses. In biparental nonhuman mammals, when a male copulates with a female and cohabitates with her throughout her pregnancy and parturition, the male's response to infants is switched from aversion/rejection to acceptance and paternal care of young. The neural model proposes that copulatory stimuli and cohabitation with the female mating partner throughout her pregnancy and parturition, including exposure to the female's pheromones, modifies the responsiveness of the male's medial preoptic area (MPOA) to infant stimuli. For certain species, such as the male California mouse, the aromatase enzyme (a) may also increase within neurons of the male's MPOA near the time of the female's parturition. The conversion of testosterone (T) to estradiol (E) within MPOA may be an additional stimulus that increases the male's positive responses

male rats and mice to pups; (d) while the physiological events of pregnancy and parturition inhibit the defensive circuit and activate the parental circuit in females, other experiences, which involve copulating with a female partner and remaining with her through her parturition, promote paternal behavior in males; (e) for some species, the conversion of testosterone to estradiol within the brain may be involved in stimulating paternal responsiveness; and (f) the parental neural circuits that probably exist in the brains of all male mammals provide a substrate upon which natural selection can operate. When paternal behavior, along with maternal behavior, is necessary for infant survival, mechanisms will evolve that will allow particular external and internal factors to activate the parental circuitry, and inhibit the defensive circuit, in male mammals.

Based on the evidence I have reviewed with respect to paternal behavior, Figure 7.1 proposes some of the mechanisms and neural circuits that may underpin paternal behavior in those mammalian species that have received the most experimental investigation. This model is a modification of the maternal subcortical circuits depicted in Figures 5.9 and 5.10 and is derived primarily from data on rodents. In species where olfaction is not the main sensory modality that influences social behavior, modifications will undoubtedly occur in these other underlying mediating mechanisms. To reiterate the views of Timonin, Cushing, and Wynne-Edwards (2008), although maternal behavior and paternal behavior in most mammals may utilize common subcortical parental circuits, the particular factors that activate these circuits in males to promote paternal behavior may vary across different biparental mammalian species. However, the proposal that

---

Figure 7.1. Continued.

to infants. The stimulated output of the MPOA promotes paternal responsiveness by inhibiting the defensive neural circuitry and activating both the mesolimbic dopamine (DA) system and oxytocin (OT) neural systems originating from the paraventricular nucleus (PVN) of the hypothalamus. The operational system that promotes paternal behavior is similar to that which promotes maternal behavior in the typical female mammal. The main difference is that the physiological events of late pregnancy and parturition prime the MPOA in most female nonhuman mammals, while other stimuli (copulatory stimuli; female pheromones; the conversion of T to E) prime the MPOA in paternal males. Additional evidence also suggests that maternal pheromones, and perhaps copulatory stimuli, may directly depress the responsiveness of the male's VNO to infant odors. AHN = anterior hypothalamic nucleus; AOB = accessory olfactory bulb; BAOT = bed nucleus of the accessory olfactory tract; BLA/BMA = basolateral and basomedial amygdala; MeA = medial amygdala; NA = nucleus accumbens; PAG = periaqueductal gray; VMN = ventromedial nucleus of the hypothalamus; VP = ventral pallidum; VTA = ventral tegmental area. Axons ending in a bar are inhibitory and those ending in an arrow exert excitatory effects.

dual neural circuits exist in the brains of both sexes, regulating either avoidance/rejection of infants or parental responsiveness, is probably generally applicable. Further, in those species where sexually inexperienced males (and females) show high levels of parental motivation, such as in species where alloparenting is adaptive, evolutionary forces have probably acted to downregulate the defensive circuit and upregulate the parental circuit.

In addition to the current chapter, three recent reviews on the paternal brain have been written: Feldman, Braun, and Champagne (2019); Glasper, Kenkel, Bick, and Rilling (2019); Horrell, Hickmott, and Saltzman (2019). The reader is referred to these important papers for additional information and insights.

8

The Parental Brain in Humans

Introduction

Research on the brain regions involved in parental behavior in humans has primarily employed functional magnetic resonance imaging (fMRI) technology, and most of this research has explored the neural correlates of maternal behavior, although some research has also examined paternal behavior. In a typical study, participants, who are usually women, are placed in an fMRI scanner and exposed to various infant or control stimuli. The scanner detects brain regions that become active upon exposure to these stimuli by measuring an increase in the blood-oxygen–level dependent (BOLD) signal. This procedure relies on the following principle: As neural activity increases in a particular brain region, increases in blood flow to that region result in increases in oxygenated hemoglobin, which results in an increased BOLD signal that is detected by the scanner (Attwell & Iadecola, 2002). Therefore, the BOLD signal is an indirect measure of increases in neural activity. Further, fMRI measures of such increases in neural activity in response to particular stimuli are primarily correlational in nature and therefore cannot provide definitive proof with respect to cause–effect relationships. Therefore, wherever possible, and particularly with respect to subcortical brain regions, I will focus on the overlap between brain regions that become active when women (or men) view infant stimuli and those brain circuits that have been experimentally shown to regulate parental behavior, or other processes that could impact parental behavior, in nonhuman mammals. Such a convergent analysis is likely to uncover brain regions that regulate parental behavior and motivation and parent–infant attachment in humans.

Hrdy (2009) has provided strong evidence that allomaternal behavior was crucial for infant survival during early human evolution, and this may explain why human maternal motivation is relatively emancipated from the physiological events of late pregnancy and parturition. Such a process fits with the fact that women who choose to adopt infants can become perfectly normal parents (Grasso, Moser, Dozier, & Simons, 2009; Singer, Brodzinsky, Ramsay, Steir, & Waters, 1985). However, recall that for those nonhuman primates that are cooperative breeders and display allomaternal behavior, the endocrine events associated with late pregnancy and parturition still function to boost maternal

The Parental Brain. Michael Numan, Oxford University Press (2020). © Oxford University Press.
DOI: 10.1093/oso/9780190848675.001.0001

motivation (see Chapter 3 of this volume). Similar processes also appear to be operative in postpartum women (Fleming, Ruble, Krieger, & Wong, 1997; Glynn, Davis, Sandman, & Goldberg, 2016). It is likely, therefore, that when nulliparous women adopt infants, subsequent experience with infants boosts maternal motivation through a sensitization-like process, with this experiential effect ultimately substituting for the maternal motivation boosting effects of the physiological events associated with pregnancy and parturition. This process is analogous to what has been shown to occur in allomaternal laboratory mice, as reviewed in Chapters 3 and 5 (Stolzenberg & Mayer, 2019). An implication of these studies is that there may be a window of vulnerability for faulty parental behavior in nulliparous women exposed to challenging environmental conditions, and, similarly, for postpartum women with abnormal hormonal profiles associated with pregnancy and parturition (Numan & Insel, 2003). Under challenging and stressful environmental conditions, and prior to the maternal motivation–boosting effects of maternal experience with infants, faulty maternal behavior, such as infant neglect and infant abandonment, may occur. Recall that latent inhibitory defensive circuits that depress maternal behavior are likely to exist in the brains of species, including primates, that typically show high levels of alloparental behavior. Stressful and challenging environmental factors may activate such circuits, with the result that maternal motivation is depressed, even in postpartum females (Hrdy, 2016; Mayer, Helton et al., 2019). This understanding is important because it relates to the idea that faulty or less than adequate maternal behavior may still occur under certain circumstances in species such as humans where alloparental behavior occurs and where the baseline level of maternal motivation in nulliparae is typically higher than in those species that exhibit a uniparental maternal care system. In this chapter and in Chapter 10, which deals with the impact of developmental processes on the human parental brain, the fact that such defensive/rejection neural circuits that depress maternal behavior exist within the human brain will be shown to be significant.

In the analysis that follows, I will first examine the subcortical neural circuits that have been associated with human maternal behavior. Since most experimental research on maternal behavior in nonhuman mammals has involved subcortical mechanisms, this is where my comparison of such research with the human fMRI research will yield the most convincing evidence with respect to the proposal that such circuits are also causal to the regulation of human maternal behavior. I will then examine the cortical neural circuits, and their interactions with subcortical circuits, that have been correlated with maternal behavior in humans. The chapter will conclude with an examination of the paternal brain in humans.

Subcortical Neural Regions and Circuits Implicated in the Maternal Behavior of Women

Introduction

Figure 8.1 depicts a summary diagram from the research described in Chapter 5, showing the subcortical neural circuits that regulate maternal behavior in nonhuman mammals. Most of this research has been conducted on rodents, and the question we pose in this chapter is whether certain aspects of these subcortical circuits also underpin maternal motivation in women. Briefly, the diagram shows that when the medial preoptic area (MPOA) is properly primed so that it responds to infant stimuli, it activates ventral tegmental area (VTA)-dopamine (DA) projections to the nucleus accumbens (NA) and oxytocin (OT) release from the paraventricular nucleus (PVN). DA action at the level of NA suppresses NA inhibitory input to the ventral pallidum (VP), which allows the VP to become responsive to infant stimuli that are relayed to VP from positively valent basolateral amygdala (BLA)/basomedial amygdala (BMA) neurons. The output of the MPOA is also shown as inhibiting the defensive neural circuitry, which suppresses avoidance and rejection responses toward infant stimuli that are relayed to the defensive circuitry by negatively valent amygdala neurons. In this regard, recall that the amygdala contains separate populations of neurons that respond to either rewarding or aversive stimuli. In the typical nulliparous female, infant stimuli activate negatively valent amygdala neurons, and withdrawal from, or rejection of, infants is the outcome. In the parturient female, infant stimuli activate MPOA output and positively valent amygdala neurons, with the outcome that proactive voluntary appetitive maternal responses occur. With respect to the sites where OT acts to promote maternal behavior in rodents, the evidence showed that OT acts at each node in the MPOA-to-VTA to NA circuit, in this way fostering maternal motivation, and it also likely to suppress the output of elements within the defensive neural circuitry. Although there is not current research that shows that OT acts on BLA/BMA to promote maternal motivation in nonhuman mammals, since OT receptors (OTRs) are located in these amygdala regions in rodents, it is certainly possible that OT acts to promote BLA/BMA input to VP, as depicted in Figure 8.1 (Numan, 2012a; Numan & Young, 2016). The involvement of BLA/BMA, in addition to medial amygdala, is emphasized in Figure 8.1 because the BLA/BMA is essential for maternal motivation in nonhuman mammals and because, in both humans and rodents, a variety of infant stimuli, in addition to olfactory stimuli, influence maternal responsiveness, with the BLA/BMA region being a recipient of such multiple sensory inputs (see Chapter 5 of this volume).

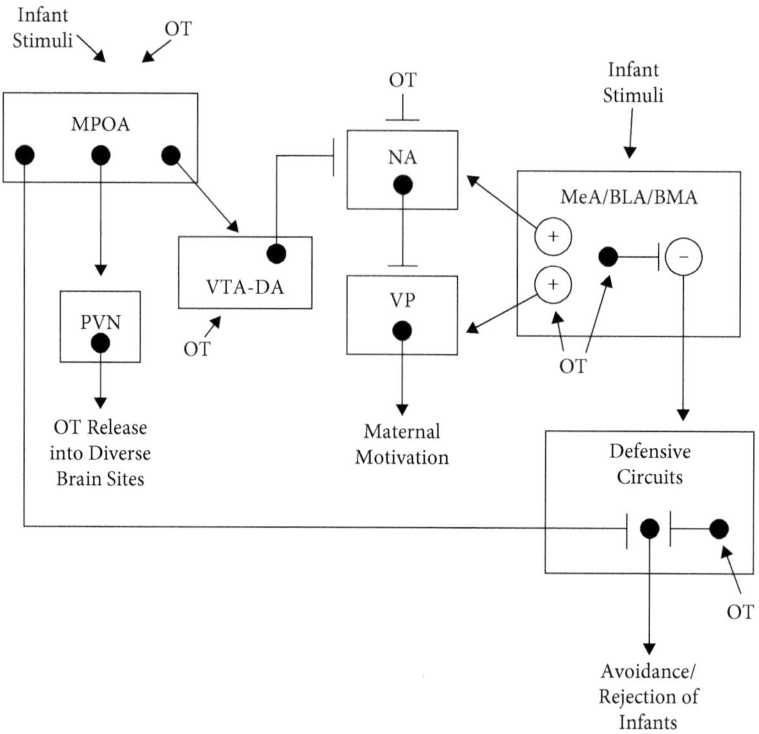

Figure 8.1. The subcortical neural circuitry, in abbreviated form, that has been shown to be involved in the regulation of parental behavior in nonhuman mammals, as described in detail in Chapter 5. The medial preoptic area (MPOA) responds to infant stimuli and promotes parental motivation by activating both dopamine (DA) neurons in the ventral tegmental area (VTA) and oxytocin (OT) neurons in the paraventricular nucleus of the hypothalamus (PVN). The output of the MPOA is also proposed to inhibit the defensive neural circuitry to depress avoidance and/or rejection of infants. Negatively valent (minus sign) neurons in the amygdala (medial amygdala [MeA]) and basolateral/basomedial amygdala (BLA/BMA) respond to infant stimuli and give rise to avoidance/rejection responses by projecting to the defensive neural circuitry. Positively valent (plus sign) neurons in the amygdala respond to infant stimuli and give rise to appetitive (reward-seeking) responses by projecting to the nucleus accumbens (NA) and ventral pallidum (VP). The projection of positively valent amygdala neurons to VP activates appetitive maternal motivation. DA and OT act to depress NA, which allows VP to effectively respond to these positively valent amygdala neurons. OT is proposed to act at several other sites within these subcortical circuits to enhance parental motivation and to depress the activation of defensive circuits. Neural effects ending in a bar are inhibitory and those ending in an arrow are excitatory.

Because of the important role of OT in maternal behavior, it is important to determine where subcortical OTRs are located in the human brain. Early studies employed in vitro autoradiography on postmortem human brain tissue using radioactive ligands to detect the presence of OT and arginine vasopressin (AVP) binding sites (Loup, Tribollet, Dubios-Dauphin, & Dreifuss, 1991; Loup, Tribollet, Dubois-Dauphin, Pizzolato, & Driefuss, 1989). In these studies, OT-binding sites were detected in the following regions that are pertinent to the neural circuits shown in Figure 8.1: MPOA, VP, and the substantia nigra pars compacta, including a region that adjoins the lateral part of the VTA. Although they did not specifically report it, the autoradiographs suggest that OT-binding sites may have also been present in the NA. AVP-binding sites were detected in BLA (high levels of OT release would be able to bind effectively to AVP receptors). Significantly, OT-binding sites were also detected in the anterior hypothalamic nucleus and the periaqueductal gray, which are parts of the defensive circuit in nonhuman mammals (see Chapter 5 of this volume).

In a more recent study that employed immunohistochemical procedures on human brain tissue (Boccia, Petrusz, Suzuki, Marson, & Pedersen, 2013), utilizing an antibody for OTRs, OTRs were detected in MPOA and BLA. This study also detected OTRs in the anterior hypothalamic nucleus/ventromedial hypothalamic nucleus (VMN) and central nucleus of the amygdala (parts of the defensive and fear circuitry; see Chapters 5 and 6 of this volume). Using postmortem brain tissue from women, Bethlehem, Lombardo et al. (2017) reported high expression of OTR messenger ribonucleic acid (mRNA) in hypothalamus, NA, and ventral midbrain, and moderate expression in the amygdala. Quintana et al. (2019) have reported high levels of expression of OTR mRNA in the striatum-pallidum region, MPOA, and PVN/supraoptic nucleus in postmortem human brain tissue; the olfactory bulbs also expressed OTR mRNA.

Although more research needs to be performed on the subcortical location of OTRs in the human brain, the evidence that does exist suggests that OT may exert effects at many of the key nodes depicted in Figure 8.1 to influence human maternal responsiveness. A major advance in our understanding of the role of OT in human parental behavior will occur once safe and selective radioactive nonpeptide ligands, capable of crossing the blood brain barrier and binding to OTRs, are developed to be used in conjunction with positron emission tomography (PET) scans (Insel, 2010). (PET scans detect emissions from radioactively labeled chemicals that are systemically administered to participants in an experiment. In some cases, these radioactively labeled chemicals are ligands that can bind to neurotransmitter or neuromodulator receptors in the brain.) Such developments will enable researchers to detect the location of OTRs in vivo in the human brain. Further, the detection of decreases in exogenously administered radioactive OT ligand binding when a mother is viewing infant stimuli within a PET scanner will inform us about the neural sites where endogenous OT is being released in response to infant stimuli.

In my review of associations between OT and human postpartum maternal behavior in this chapter, keep in mind the animal data that indicate OT neural systems are important for the onset of maternal behavior and also operate to promote competent postpartum maternal behavior under challenging environmental conditions. OT exerts these effects by both boosting maternal motivation and decreasing anxiety and stress reactivity.

Nulliparous Women

Based on the occurrence of allomaternal behavior during the course of human evolution, researchers have examined the brain regions activated when nulliparous women view different types of infant stimuli. Glocker et al. (2009) reported that when nulliparous women viewed either infant faces or faces of older children, the BOLD response in the NA was greater when infant photographs were viewed. These researchers proposed that infantile visual features may be inherently attractive to nulliparous women because of their ability to activate the mesolimbic DA reward system. Since such activation occurs in these women in the absence of pregnancy and parturition, these results support the idea of high levels of parental (alloparental) motivation in women.

What neural events might account for the increased BOLD response in the NA? Based on the neural model shown in Figure 8.1, it is likely that the enhanced NA BOLD response that occurred when viewing infant faces was associated with increased action potential frequency in VTA-DA axon terminals within the NA, as has been proposed by other researchers (Schott et al., 2008). However, I do want to emphasize a technical issue related to fMRI research. It is often the case that the spatial resolution of fMRI technology is not sensitive enough to differentiate nearby regions from each other. In some studies, the enhanced BOLD response that is designated as occurring in the NA may actually be occurring in the nearby VP. Related to this point, in some of the brain images depicted in the Glocker et al. (2009) study, the increased BOLD signal appears to be in the VP rather than in the NA (see Figure 2 in Glocker et al., 2009). Due to the importance of VP output for rodent maternal behavior, researchers should try to differentiate BOLD responses in NA from those that might be occurring in VP.

It needs to be re-emphasized that the mesolimbic DA system is part of a general motivational system and that its activation occurs in response to a variety of attractive and rewarding stimuli. Evolutionary factors may have allowed infant facial features to easily activate this general reward system, even in nulliparous women, but it is unlikely that the observed activation of the NA-VP circuit in the Glocker et al. (2009) study is identifying brain regions specifically related to maternal motivation. In relation to the nonhuman animal research, it would have been instructive to determine whether hypothalamic regions that included the MPOA were also activated in nulliparous women who viewed attractive infant faces. If the MPOA were activated under these conditions, such activation might represent the stimulation of specific maternal circuits that regulate allomaternal behavior in nulliparous women. MPOA input to VTA-DA neurons may have then caused the observed BOLD response within the NA-VP circuit.

Although infant facial images appear to be rewarding stimuli for nulliparous women, and these women may also express a general interest and attraction to infants, this is not the same as devoting one's life to the care of an infant. As indicated in the introduction to this chapter, although a baseline level of allomaternal motivation may be evident in nulliparous women, it is highly likely that the physiological events associated with the end of pregnancy coupled with postpartum maternal experience, or the maternal experience associated with mother–infant interactions in those nulliparous women who choose to adopt infants, operate to boost maternal motivation above the baseline levels that are observed in nulliparae who have little or no experience with infants. Given this understanding, is there any evidence that certain infant stimuli may actually activate negatively valent amygdala neurons in nulliparous women, the output of which engages withdrawal/rejection circuits?

As background, I want to present research by Kirsch et al. (2005) and Gamer, Zurowksi, and Buchel (2010) that was performed on men and did not involve infant stimuli. Kirsch et al. reported that when men viewed angry or fearful adult human facial expressions while in an fMRI scanner, the BOLD signal increased in the amygdala (specific amygdala regions were not differentiated). However, when such men first received intranasal administration of OT, which presumably entered the brain to affect central neural activity, then the amygdala activation associated with viewing these faces, indicative of a threatening situation, decreased. Gamer et al. (2005) found that intranasal treatment with OT decreased amygdala BOLD responses to fearful facial expressions, but increased the amygdala BOLD response to happy facial expressions. These findings may be interpreted in the following way: There are two populations of amygdala neurons, one that responds to aversive stimuli with a negative valence and another that responds to attractive stimuli with a positive valence. OT action within the brain may act to downregulate the amygdala aversion system while upregulating amygdala appetitive (reward-seeking) circuits (cf. Figure 8.1).

With respect to maternal responsiveness, infant cries usually signify infant distress. For the typical postpartum woman, such cries are likely to alert the mother who then engages in behavior to soothe her infant. However, for some women, infant cries may be annoying and may sometimes result in a hostile response toward the infant. Is it possible that infant cries are more likely to engage negatively valent amygdala neurons in nulliparous women, while such cries activate positively valent amygdala neurons in normal postpartum women? Riem et al. (2011) reported that infant cries activated the amygdala in nulliparous women and that intranasal administration of OT decreased this amygdala response (cf. Bos, Spencer, & Montoya, 2018). In an interesting study, Bakermans-Kranenberg, van Ijzendoorn, Riem, Top, and Alink (2012) trained nulliparous women to squeeze a handgrip gauge that measured the amount of force they were exerting. When these women listened to infant cries, they tended to exert excessive force on the handgrip. Importantly, intranasal application of OT decreased this excessive force response. One interpretation of these results is that nulliparous women find certain infant cries to be aversive, with such cries activating negatively valent amygdala neurons, and that intranasal application of OT acts to suppress the aversive qualities of infant cries by depressing the output of these negatively valent amygdala neurons.

In contrast to these findings in nulliparous women, Kim et al. (2011) reported that when mothers, at 1-month postpartum, listened to recordings of their own infant crying, the BOLD response increased in the amygdala to a greater extent than that which occurred when they listened to an unfamiliar infant cry, and this differential response was greater in women who were breastfeeding than in those who were not. To the extent that breastfeeding was associated with the enhanced central release of endogenous OT, these results suggest that OT may have enhanced the amygdala response to the cries of the mother's own infant. Further, when all mothers were considered, there was a positive correlation between measures of maternal sensitivity and the BOLD response to own-infant cries, relative to unfamiliar cries, in both the amygdala and VP. Behavioral measures of maternal sensitivity were taken outside the scanner at 3 to 4 months postpartum during observations of mother–infant interactions. Such measures included adaptive responses to the infant's communicative signals, positive affect, vocal clarity, supportive presence, consistency of style, and affectionate touch.

Although more work needs to be done on the issues discussed in this section, a preliminary interpretation is that infant cries may be annoying and evoke aversive responses in some nulliparous women by activating negatively valent amygdala neurons that project to avoidance circuits. In contrast, when postpartum women, who have been exposed to the hormonal events of late pregnancy and to endogenous central OT stimulation, form a selective attachment to their own infant, their infant's cries no longer activate amygdala aversion circuits, which are likely to be downregulated by OT, but instead activate positively valent amygdala neurons, that are likely to be upregulated by OT. The activation of such positively valent neurons may participate in neural circuits that promote aid-giving responses to cries that are indicative of their infant's distress, and an amygdala-to-VP circuit may be involved (see Figure 8.1). It is also likely that similar mechanisms are operative in adoptive mothers after a sufficient amount of maternal experience through interactions with their adopted infant.

Postpartum Women

Introduction

In the typical fMRI study that is designed to associate neural activations with maternal responsiveness in postpartum women, mothers are scanned while they are exposed to various infant stimuli, which can include recorded vocalizations of cries or laughter, photographs of infant faces, or videos of infants. Therefore, similar to the studies performed on nulliparous women, most studies have used infant auditory or visual stimuli. In this regard, the role of infant olfactory cues and the maternal brain activations that they may induce deserve more attention, since such stimuli influence maternal behavior in postpartum women (see Croy, Mohr, Weidner, Hummel, & Junge-Hoffmeister, 2019, for some recent research on this issue).

In many studies, a common procedure is to expose mothers to stimuli from their own infants and from unrelated infants (Rigo et al., 2019). Since humans form selective attachments to their own infants, it is assumed that those brain regions that show a greater BOLD response to own-infant stimuli relative to unfamiliar-infant stimuli are likely to contribute to the neural circuits that regulate maternal motivation and the strong mother–infant bond. As indicated by Rilling (2013), such a protocol may actually miss detecting brain regions that respond to generic infant stimuli, and such ignored regions may be importantly involved in affecting maternal responsiveness.

Another common procedure is to associate neural BOLD responses to own infant stimuli relative to unfamiliar infant stimuli with various measures of the quality of maternal responsiveness. To obtain measures of maternal responsiveness, mothers are observed, outside the scanner, interacting with their infants, and these observations are used to rate the mothers in terms of maternal sensitivity, also referred to as mother–infant synchrony, and maternal intrusiveness, which is considered to be measure of maternal insensitivity (Atzil, Hendler, & Feldman, 2011; Atzil et al., 2017; Musser, Laurent, & Ablow, 2012). Maternal sensitivity is usually defined as a mother's appropriate and consistent perception and reaction to her infant's communicative cues, such as cries, laughter, and facial expressions—cues that are indicative of the infant's emotional state. Maternal intrusiveness, sometimes considered as excessive maternal behavior, is usually defined as a mother's inappropriate interactions with her infant, such as playing with her infant when the infant wants to rest. Mothers that are intrusive typically exhibit a demanding maternal style that may interrupt their child's ongoing behaviors.

Subcortical Neural Activations

A recent study by Atzil et al. (2017) provides excellent support for the view that the subcortical circuits that regulate maternal behavior in nonhuman mammals (Figure 8.1) are also operative in postpartum women and that the strength of the connectivity between the neural regions within these circuits is positively correlated with measures of mother–infant synchrony. A unique aspect of this study was that it employed both PET scans and fMRI scans. In the first part of the study, while in a PET scanner, mothers watched video films of their infant or an unfamiliar infant playing alone. While watching these films, mothers were systemically injected with a radioactive ligand capable of crossing the blood brain barrier and binding to dopamine receptors. PET scan detection of decreases in radioactive ligand binding within a brain region was used to measure increased endogenous release of DA into that site, since endogenous DA competes with the radioactive ligand for receptor occupancy. Mothers who were designated as displaying high levels of mother–infant synchrony had greater release of endogenous DA into the NA when they watched their own infants in comparison to unrelated infants. In contrast, mothers that scored low on measures of mother–infant synchrony did not exhibit a differential DA response, with DA being released into NA under both viewing conditions. One interpretation of these results is that high-synchrony mothers formed a more selective bond with their infants than did the low-synchrony mothers (Atzil et al., 2017). Perhaps the strength of the selective bond that mothers formed with their infants influenced the quality and sensitivity of their maternal responsiveness. Since Glocker et al. (2009) provided evidence that DA is released into NA when nulliparous women view infant faces, a related interpretation of the Atzil et al. findings is that high-synchrony mothers, but not low-synchrony mothers, found images of their own infants to be more attractive than those of unfamiliar infants.

In the second part of the Atzil et al. (2017) study, the resting-state functional connectivity between the amygdala, MPOA, and NA-VP circuit was measured in high-synchrony and low-synchrony mothers. To do this analysis, positive correlations were computed between spontaneous BOLD responses in these regions while the mothers were resting in a scanner and not viewing any images. High-synchrony mothers showed greater intrinsic connectivity between these crucial regions (higher correlations between BOLD responses) than did low-synchrony mothers. These results suggest that the circuitry shown in Figure 8.1 demonstrates stronger intrinsic (baseline) connectivity in high-synchrony mothers.

An analysis of these results, in conjunction with the animal research reviewed in Chapter 5, suggests that postpartum maternal experience may have solidified the subcortical neural circuitry that regulates maternal motivation towards one's own infant to a greater degree in high-synchrony than in low-synchrony mothers. Perhaps activation of MPOA inputs to VTA-DA neurons that project to NA while mothers were interacting with their infants prior to the onset of the Atzil et al. (2017) study (the mothers were between 4 and 24 months postpartum) was more effective in strengthening the neural connections between the amygdala and the VP in high-synchrony mothers, which improved the quality of their maternal responsiveness to their own infants (see Figure 5.16). I presented a similar proposal when I discussed the work of Kim et al. (2011) in the previous subsection on nulliparous women. Please note that all the mothers in this study were raising their infants properly and that none of the mothers was abusive or neglectful. These results are simply showing that normal variations in the quality of maternal behavior displayed by postpartum women can be related to differences in the operation of the subcortical neural circuits that appear to underpin maternal responsiveness in both humans and animals. However, one can certainly infer from these findings that if the subcortical circuits that regulate maternal motivation were to completely break down, then extremely poor parenting, detrimental to infant survival, would be the result.

In addition to the research by Atzil et al. (2017), several other studies support the proposal that the subcortical neural circuits involved in human maternal behavior match those that regulate maternal behavior in other mammals (for a recent review, see Rigo et al., 2019). In a study by Strathearn, Fonagy, Amico, and Montague (2009), when primiparous postpartum women viewed faces of their own or unfamiliar infants, the BOLD responses in the hypothalamus and NA were greater when the mothers viewed their own infant faces, and the magnitude of these BOLD responses was positively correlated with plasma OT levels. These results are consistent with a hypothesis that when mothers view their own infants, in comparison to unrelated infants, there is a greater activation of MPOA inputs to PVN-OT neurons and to VTA-DA neurons that project to NA. Other studies have also reported that the hypothalamic BOLD response increases when postpartum mothers are exposed to stimuli from their own infants (Swain, 2011). What is not clear from these studies, in contrast to that of Atzil et al., is the extent to which the MPOA was specifically activated, since the fMRI procedures that were used did not distinguish between the contribution of individual hypothalamic nuclei to the total BOLD response.

In an anatomical study, Kim, Leckman, Mayes, Feldman et al. (2010) used a voxel-based morphometry procedure to analyze MRI scans of the brains of postpartum women, all of whom were breast feeding their infants. Voxel-based morphometry measures gray matter volume in brain regions, and it has been suggested that increases in gray matter volume may be indicative of increases in functional activity within the brain regions that are examined. Kim et al. obtained MRI scans from their participants at 2 to 4 weeks and 3 to 4 months postpartum and found that the gray matter volume in the hypothalamus, which included the MPOA, increased across these time points. Increases in gray matter volume also occurred in the VTA, amygdala, and VP. These results are consistent with the hypothesis that as maternal experience increases across the postpartum period, functional activity is strengthened across aspects of the neural circuitry shown in Figure 8.1.

With respect to the amygdala, and based on the rodent research that the BLA/BMA regulates appetitive maternal responses in rodents (Numan et al., 2010), Barrett et al. (2012) analyzed the amygdala BOLD response in mothers at 3 months postpartum while they viewed photographs of their own or unrelated infants. While viewing the photographs, the mothers also rated their emotional responses. The BOLD response in BLA/BMA was greater when mothers viewed their own infants, and mothers also reported feeling more positive when viewing their own infants (also see Rigo et al., 2019). Importantly, the BLA/BMA BOLD response was positively correlated with the intensity of the mothers' positive affect. Note how these results parallel those of Kim et al. (2011) with respect to the relationship between own-infant cries and the amygdala BOLD response in postpartum women. These results conform with the proposal that own-infant stimuli activate positively valent amygdala neurons in postpartum women.

Finally, in an interesting study, Atzil et al. (2011) recorded brain BOLD responses in mothers as they watched videos of their own or unfamiliar infants while in an fMRI scanner. Based on prior behavioral observations, the maternal style of these postpartum women was classified as being synchronous or intrusive. In an initial analysis, all mother showed greater BOLD responses in the amygdala and NA when they watched their own child in comparison to an unfamiliar child, although intrusive mothers demonstrated greater activations in the amygdala while watching their own child than did synchronous mothers. Importantly, however, the task-based functional connectivity between the amygdala and NA was greater in synchronous mothers when compared to intrusive mothers while watching their own infants. Task-based functional connectivity is a measure of positive BOLD responses between two neural regions during a participant's exposure to particular stimuli while in a scanner (note how this task-based functional connectivity measure differs from resting-state functional connectivity). For synchronous mothers that were viewing their infants, when the BOLD response increased in the amygdala, this was associated with increases in the NA BOLD response. In contrast, for intrusive mothers, the BOLD responses in amygdala and NA tended to be uncorrelated. Further, plasma OT levels were positively correlated with amygdala and NA activations in synchronous mothers, but not in intrusive mothers. My interpretation of these results is that there are two populations of amygdala neurons, one with a positive valence and one with a negative valence. For synchronous mothers, own-infant stimuli primarily activate positively valent amygdala neurons that project to the NA-VP circuit, and the connectivity within this circuit may be promoted by the central release of OT. For intrusive mothers, I would suggest that own-infant stimuli activate both amygdala populations, perhaps at different time points. Own-infant stimuli activation of negatively valent amygdala neurons that project to regions other than the NA-VP circuit, with projections that might affect aversion circuits that result in annoyance-related responses, might give rise to intrusive maternal responses and to other behaviors indicative of insensitive parenting. A potentially valuable implication of these findings is that even in healthy mothers there are normal variations in maternal behavior and differences in the extent to which appetitive maternal circuits and aversion/annoyance neural circuits are activated during mother–infant interactions may give rise to variations in maternal style.

Much more research needs to be done on the subcortical circuits that are associated with maternal behavior in women, and more attention needs to be focused on the MPOA. Novel fMRI procedures might be able to more clearly distinguish MPOA BOLD responses from other hypothalamic responses if the mother was required to engage in appetitive maternal responses. For example, if a mother were placed in a scanner while viewing a distant image of an infant or an inanimate object and instructed to press a button to bring the image closer to the her, I would predict that such behaviors would engage MPOA neural circuits only when infant stimuli were involved and that the MPOA BOLD response would be greater if the image was of the mother's own infant rather than an unfamiliar infant. Lonstein, Levy, and Fleming (2015) have made a similar suggestion. Further, more advanced MRI procedures could be utilized to clearly distinguish the preoptic hypothalamus from nearby neural regions (see Schindler et al., 2013).

An important study by Moll et al. (2012) is relevant to the involvement of the MPOA in positive social motivation and emotion in humans, which would include parental responsiveness. These researchers examined whether there are brain regions that are active during affiliative emotional states (feelings of warmth and tenderness toward kin) that are partially distinct from those that are active during other pleasant emotions. While in an fMRI scanner, women and men read short scenarios that were divided into the following types: (a) positive valence scenarios that were associated with kin and evoked feelings of warmth and tenderness (positive-affiliative: "You read a book to your child who fell asleep in your lap"); (b) positive valence scenarios that were not associated with kin and did not evoke feelings of warmth and tenderness (positive-nonaffiliative: "Your boss was impressed by your performance at work"); (c) neutral scenarios that did not give rise to strong pleasant feeling states and did not evoke feelings of warmth and tenderness (neutral: "You went to lunch with co-workers"). In comparison to neutral scenarios, all pleasant scenarios (affiliative and nonaffiliative) were associated with increased BOLD responses in NA. Significantly, positive-affiliative scenarios selectively activated the septal region and the preoptic area, while positive-nonaffiliative scenarios did not. Matching the proposal presented by Numan (2006) with respect to the interaction between the MPOA and the mesolimbic DA system in the control of the appetitive aspects of maternal behavior in rats (see Chapter 5 of this volume), Moll et al. suggest that the activation of the septal-preoptic region along with concurrent activation of the NA-VP circuit by positive-affiliative scenarios represents an interaction between a specific affiliative social motivational system and a more general reward-related motivational system. I will revisit this study by Moll et al. when I discuss the relevance of parental brain circuitry to the evolution of human social behavior in Chapter 11.

Taken together, the evidence presented clearly supports the view that the subcortical circuits that are involved in human maternal behavior match those that contribute to maternal behavior regulation in other animals. The evidence also supports the idea that dual subcortical neural circuits regulating maternal behavior exist in the human brain as they do in the brains of other mammals, one that promotes and one that interferes with maternal behavior. Although the positive circuit may be dominant in the human brain due to the importance of allomaternal behavior, the fact that inhibitory circuits also appear to exist suggests one route through which abnormal and sometimes pathological mother–infant interactions may occur: Experiential and/or genetic factors may upregulate aversion/rejection circuits with the result that faulty maternal behavior develops. I will explore this issue in more detail in Chapter 10, which reviews developmental processes that affect human maternal behavior.

Cortical Neural Regions and Circuits Relevant to the Maternal Behavior in Women

Introduction

Behavioral neuroscientists who study maternal behavior in nonhuman mammals have focused much of their research on subcortical mechanisms that regulate the behavior, and, as I have just shown, these subcortical circuits are also involved in human maternal responsiveness. In contrast, affective and cognitive neuroscientists who study the neural basis of maternal behavior in humans have tended to emphasize the involvement of cortical brain activity, and this focus is related to a primary interest in the thought processes and conscious feeling states that are associated with human maternal behavior, such as maternal love and maternal emotional and cognitive empathy. Since the purpose of conscious emotional and cognitive processes is to influence the way a mother reacts to her infant behaviorally, a major goal of this section is to link cortically mediated cognitive thought processes and emotional feeling states to subcortical behavioral circuits to understand how thoughts and emotions can be translated into appropriate maternal behavior.

Anatomical Overview of Human Cortical Regions Relevant to Maternal Behavior

Figure 8.2 shows a lateral and midsagittal view of the human brain and emphasizes the neural regions that have been implicated in the maternal behavior of women. In the lateral view, I have indicated the location of the major cortical lobes: frontal, parietal, occipital, and temporal. In the frontal lobe, I have identified the approximate locations of the dorsolateral and ventrolateral prefrontal cortex (dlPFC and vlPFC, respectively): Those parts of the lateral frontal cortex that lie anterior to the primary motor cortex. In the very ventral part of the vlPFC lies the inferior frontal gyrus (IFG). Located ventral and medial to the IFG, shown in pink, is the posterior orbitofrontal cortex (pOFC) on the ventral surface of the frontal lobe and the anterior insular (AI) cortex. Although I have shown the AI cortex on the lateral surface of the brain, in primates it is actually buried within the depths of the lateral fissure and can actually be observed only if one separates the overlying parts of the frontal and temporal lobes. That is, parts of the AI are located medial to the posterior part of the IFG. In the temporal lobe, I have indicated the superior temporal gyrus (STG) and the underlying superior temporal sulcus (STS). Located in the posterior part of the STG and the adjoining ventral posterior parietal lobe is an area referred to as the temporoparietal junction (TPJ).

In the midsagittal view, I am emphasizing medial regions of the prefrontal cortex (mPFC), and I am identifying particular areas by the Brodmann area numbers that they are typically associated with (Fuster, 2008). Within the anterior cingulate gyrus, which surrounds the anterior part of the corpus callosum, are located areas 24 (dorsal anterior cingulate cortex) and area 25 (ventral or subgenual anterior cingulate cortex [sgACC]). Area 32 within the mPFC lies rostral to areas 24 and 25 and is usually referred to as the pregenual ACC (pgACC). Areas 25 and 32 comprise parts of what is referred to as the ventromedial prefrontal cortex (vmPFC). Located anterior to areas 24, 25, and 32, on the medial surface of the prefrontal cortex are parts of areas 9 and 10, referred to as the dorsomedial prefrontal cortex (dmPFC; the lateral portions of these regions comprise parts of the lateral prefrontal cortex).

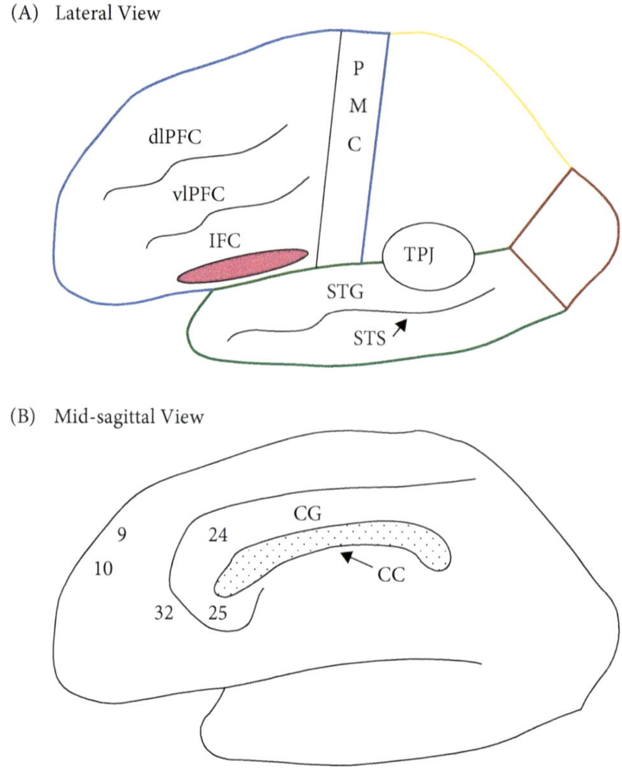

Figure 8.2. A lateral (A) and midsagittal (B) view of the human cerebral cortex that emphasizes those neural regions that have been implicated in regulating the maternal responsiveness of women. (A) The location of the major lobes located on the lateral surface of the cerebral cortex are outlined in different colors (occipital lobe = red; parietal lobe = yellow; temporal lobe = green; frontal lobe = blue). The prefrontal cortex is that part of the frontal lobe that lies rostral to the primary motor cortex (PMC). The lateral aspects of prefrontal cortex are divided into a dorsolateral (dlPFC) and a ventrolateral part (vlPFC). At the base of the vlPFC lies the inferior frontal gyrus (IFG). On the ventral surface of the PFC, ventromedial to the IFG, lies the posterior orbitofrontal cortex (pOFC), and medial to the posterior IFG, buried under this region, lies the anterior insular cortex (AI). The general location of the pOFC and AI is shown in pink. Within the temporal lobe, note the locations of the superior temporal gyrus (STG) and sulcus (STS). Finally, the temporoparietal junction (TPJ), outlined within a circle, is located at the junction of the posterior temporal lobe and the ventral parietal lobe. (B) The midsagittal view depicts the medial regions of the prefrontal cortex (mPFC), and important regions are identified by their Brodmann area numbers (see Fuster, 2008). Within the anterior part of the cingulate gyrus (CG), which surrounds the anterior part of the corpus callosum (CC), are located areas 24 (dorsal anterior cingulate cortex) and 25 (ventral or subgenual anterior cingulate cortex). Area 32 within the mPFC lies rostral to areas 24 and 25 and is usually referred to as the pregenual anterior cingulate cortex. Areas 25 and 32 comprise parts of the ventromedial PFC. Located anterior to areas 24, 25, and 32, within the mPFC, are parts of areas 9 and 10, which are referred to as the dorsomedial PFC.

Maternally Relevant Functions Associated With Select Cortical Regions

Potential Correlates of Maternal Love and Maternal Emotional Empathy

Emotional empathy refers to the ability of one person to vicariously experience the emotional state of another person, and such empathizing can involve both negative and positive emotional experiences (Bernhardt & Singer, 2012; Jabbi, Swart, & Keysers, 2007). In this section, I want to distinguish natural emotional experiences from emotional empathy. The former refers to our endogenous emotions triggered by rewarding or aversive events, while the latter refers to our ability to share in the natural emotional experiences of others. As indicated by Singer and Lamm (2009), emotional empathy relies on self-awareness and the distinction between oneself and another person. There must be mechanisms within the brain that distinguish between the sources of our emotional experiences. With respect to maternal states, when a mother looks at her sleeping baby or thinks about her infant, she may have the natural emotional experience of feeling love for her child. In contrast, if a mother sees her child in distress, she may emotionally empathize with the child and vicariously feel that distress herself. Such empathizing may result in caregiving responses to relieve the infant's distress. It should be obvious that if a mother does not love her child or does not empathize with her child, such deficits could lead to maternal neglect (cf. Lockwood, 2016).

fMRI research has indicated that the IFG-pOFC-AI regions display an enhanced BOLD response during a variety of natural and empathic feeling states and that the recorded BOLD response is positively correlated with the intensity of the emotional experience (Craig, 2009; Gu, Hof, Friston, & Fan, 2013; Jabbi & Keysers, 2008; Jabbi et al., 2007; Numan, 2015). Further, lesions of AI blunt emotional experiences in humans (Terasawa, Kurosaki, Ibata, Moriguchi, & Umeda, 2015). During natural and empathic feeling states the AI region is typically co-active with parts of the mPFC (areas 24, 25, 32; see Numan, 2015). Craig (2009) has proposed that the AI region is a sensory-like region that mediates natural emotional experiences and emotional empathy, while the ACC serves as a motor-related region that promotes motivated behaviors in response to these feeling states.

Neuroanatomical studies on nonhuman primate and human brains have defined the neural connections of the AI region (IFG-pOFC-AI), and some of these connections in relation to maternal feeling states and behavior are summarized in Figure 8.3 (Barbas, Zikopoulus, & Timbie, 2011; Hoistad & Barbas, 2008; McDonald, 1998; Nieuwenhuys, 2012; Ongur, An, & Price, 1998; Ongur & Price, 2000). In this diagram, I have related the cortical anatomy to subcortical circuits

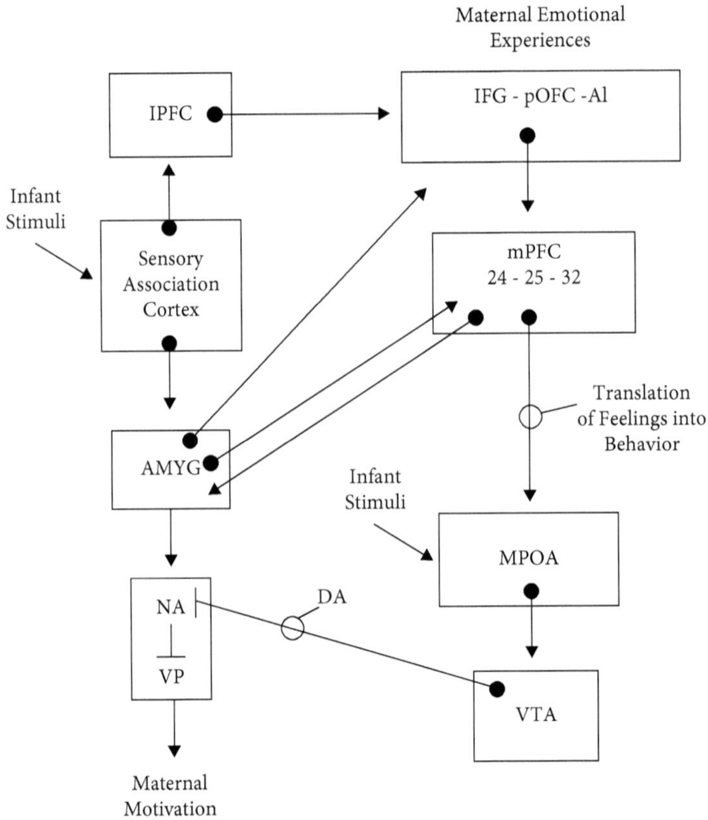

Figure 8.3. A neural circuitry model which links cortical with subcortical circuits to explain how maternal feeling states (maternal empathy and maternal love for her infant) might be translated into maternal behavior. Infant stimuli reach the anterior insular region, which includes the inferior frontal gyrus (IFG), posterior orbitofrontal cortex (pOFC), and anterior insula (AI) via projections from the amygdala (Amyg) and lateral PFC (lPFC). Neural activity within the AI region is proposed to give rise to emotional empathy and feelings of maternal love. The projection of the AI region to areas 24 25, and 32 of the medial prefrontal cortex (mPFC) is an important route, which allows these feeling states to be translated into maternal behavioral responses because these areas of the mPFC (particularly area 32) are proposed to activate the medial preoptic area (MPOA), which, in turn, activates the mesolimbic dopamine (DA) system. Note the important involvement of the amygdala, since it projects to several regions which influence maternal feeling states (cortical regions including the AI region and mPFC) and maternal behavior (subcortical regions: nucleus accumbens [NA] and ventral pallidum [VP]). Also observe the feedback of the mPFC to the amygdala, which would allow the mPFC to regulate amygdala reactivity to infant stimuli. Although not shown in this figure, infant- and MPOA-induced activation of paraventricular nucleus oxytocin release would be able to influence activity at several nodes in the described neural network. VTA = ventral tegmental area. Axons ending in an arrow are excitatory and those ending in a bar are inhibitory.

involved in maternal behavior in an attempt to show how maternal feeling states can be translated into maternal behavior (see Figure 8.1; also see Numan, 2015, 2017). A variety of positively or negatively valent sensory inputs can reach the AI region via projections from the amygdala or from the lateral PFC. Barbas et al. (2011) have proposed that amygdala projections to the orbitofrontal and insular regions may be a route over which various positive or negative emotions are subjectively experienced. The AI, in turn, projects to those mPFC regions in and around the anterior cingulate cortex. Importantly, and often overlooked, parts of the mPFC have extensive projections to the hypothalamus. In particular, area 32 of the mPFC demonstrates a dense projection to the MPOA in macaque monkeys; area 25 also projects to MPOA, but has a particularly dense terminal field within the VMN (Ongur, An, & Price, 1998). Significantly, Wallis, Cardinal, Alexander, Roberts, and Clarke (2017) have reviewed research that indicates that increased activity in area 32 is associated with positive affect in humans. Therefore, the connections from AI-to-mPFC-to-MPOA could be a link that connects certain cortically mediated feeling states with subcortical regions that regulate maternal behavior. More specifically, Figure 8.3 depicts a potential circuit through which feelings of maternal love, and a mother's ability to empathize with her infant's emotional state, could lead to appropriate goal-directed maternal responsiveness, such as aiding, protecting, and playing with her infant, via connections to MPOA. It is interesting to speculate that area 32 projections to MPOA might influence the positive aspects of maternal motivation (maternal sensitivity and mother–infant synchrony), while area 25 connections to the VMN might influence maternal intrusiveness or other types of negative maternal responses to certain infant stimuli that arouse negative affect in the mother.

It is worth emphasizing the potentially critical role of the basal amygdala nuclei. These nuclei (BLA/BMA) project not only subcortically to the NA-VP circuit, but also to cortical circuits that regulate emotional experiences. Therefore, the output of the amygdala is positioned to influence both maternal motivation and maternal emotion.

In Figure 8.3, I also show that mPFC regions in and around the ACC project back to the amygdala. Therefore, cortical mechanisms have the potential to influence the level of activity and output of the amygdala. I will return to this important neural connection when I describe the concept of emotion regulation. As I will show, this connection may be an important top-down regulatory mechanism that dampens negative emotions and anxiety mediated by the amygdala, which allows a mother to appropriately care for her infant under challenging and demanding situations.

In my previous discussion of subcortical neural activations, where I presented the subcortical circuitry associated with maternal responsiveness in postpartum women, I described the research of Atzil et al. (2017), which showed that high synchrony mothers showed greater intrinsic resting state functional connectivity between the amygdala, MPOA, and NA-VP circuit than did low synchrony mothers. Importantly, these researchers also examined cortical regions and found that the mPFC was strongly connected to these subcortical circuits in high synchrony mothers. This finding supports the proposed neural model presented in Figure 8.3.

Cognitive Empathy and Mentalizing

Appropriate social interactions not only involve our ability to share the emotions of another person (emotional empathy), but also to understand the causes of their emotions so that we can explain and respond effectively to their emotional state (Kanske, Bockler, Trautwein, Parianen Lesemann, & Singer, 2016; Schuwerk, Schurz, Muller, Rupprecht, & Sommer, 2017). *Mentalizing* is a general term that denotes our ability to understand the thoughts, intentions, and emotions of others. Cognitive empathy is that aspect of mentalizing used to refer to our ability to understand the emotional state of another person. Successful cognitive empathy would seem to be particularly important during mother–infant interactions with a preverbal infant. Ashar, Andrews-Hanna, Dimidjian, and Wager (2017) have made a distinction between emotional empathy, cognitive empathy, and empathic care (also see Singer & Lamm, 2009). Empathic care refers to one's desire to act prosocially toward another individual after we feel and understand their emotional state. For example, if a mother observes emotional distress in her infant, she may also feel that distress (emotional empathy) and understand the causes of that distress (cognitive empathy). The interaction between these two sources of empathy may then lead to empathic care, with the result that the mother behaves in an appropriate way to calm her infant and relieve its distress. For example, if a mother understands that her infant is crying because it is hungry, she would then feed her baby.

fMRI studies have indicated that the following neural regions are particularly important components of the mentalizing neural network: TPJ, STG/STS, and areas 9 and 10 in the dmPFC (Kanske et al., 2016; Molenberghs, Johnson, Henry, & Mattingley, 2016; Schurz & Perner, 2015; Walter et al., 2004). Although cognitive empathy (dmPFC, TPJ, and STG) and emotional empathy (IFG-pOFC-AI and its connections to areas 24, 25, and 32 of the mPFC) may represent separate neural networks, one might expect that they should interact to allow for the occurrence of appropriate empathic care, which would then lead to helping or prosocial behavior. Neuroanatomical tracing studies in rhesus monkeys (Barbas, Ghashghaei, Dombrowski, & Rempel-Clower, 1999; McDonald, 1998; Petrides & Pandya, 2007) have indicated that dmPFC areas 9 and 10 not only project to the STG but also project to the AI and to areas 24, 25, and 32 of the medial prefrontal cortex, and that the STG/TPJ also projects to areas 24, 25, and 32 in the mPFC. Further, the STS projects to the lateral and basal amygdala. These anatomical findings suggest that cognitive empathy neural systems can influence the output of emotional empathy systems. One can propose that once an appropriate understanding of another individual's emotional state (cognitive empathy) is appreciated, the emotional empathy system's connections to prosocial subcortical networks are more effectively activated, due to input from the cognitive empathy system, which then leads to an optimal behavioral expression of empathic care. This proposal, in relation to maternal responsiveness, is presented in Figure 8.4. The figure shows that in response to infant distress signals, such as crying, emotional empathy systems in the AI region are engaged. An understanding of the reasons why the infant is in distress, mediated by cognitive empathy systems (dmPFC, STG, TPJ), enhances the functional connectivity between the AI region and mPFC areas 24, 25, and 32. Strong activation of mPFC output, in turn, engages subcortical systems that regulate appropriate maternal responsiveness to the infant's distress. A study by Ashar et al. (2017) provides partial support, in a nonmaternal context, for the model shown in Figure 8.4. While in an fMRI scanner, human participants read biographies describing stories of suffering individuals. During these scanning sessions, each participant rated their empathic care, that is, their desire to help the suffering individual. The degree of empathic care was positively related to BOLD responses in the vmPFC and in subcortical regions such as the MPOA and NA.

Kanske et al. (2016) have noted that individuals can exhibit good emotional empathy, but poor cognitive empathy, or vice versa. Such individual differences may be related to the functional organization within each system. With respect to maternal behavior, one would predict normal activity within each empathy system, along with an appropriate degree of functional connectivity between the two systems, would be essential for appropriate maternal responsiveness.

Emotion Regulation

In Chapter 6 I argued that the postpartum mother should have the emotional capability to cope with stressful environmental challenges so that she can appropriately care for and protect her infants, and I presented evidence that too much anxiety and fear-related responsiveness is likely to be detrimental to effective maternal behavior. In that chapter, which dealt with nonhuman mammals, I was primarily concerned with the subcortical regulation of anxiety and fearfulness in mothers during the postpartum period, and the relationships between anxiety and maternal aggression (maternal protection of offspring).

In this chapter, I want to focus on an anatomical connection indicated in Figure 8.3, which shows not only that the amygdala projects to the prefrontal cortex to influence various emotional states, but that the mPFC also projects back to the amygdala, allowing it to regulate that output of the amygdala to

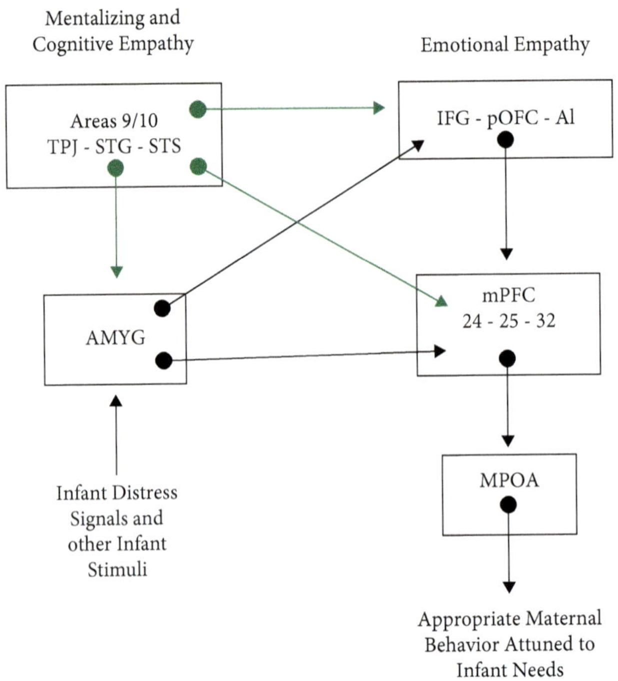

Figure 8.4. Mentalizing and cognitive empathy, in addition to emotional empathy, participate in a mother's appropriate appreciation of her infant's needs so that effective maternal behavior can occur. Anatomical studies have delineated neural pathways through which mentalizing/cognitive empathy cortical neural systems can interact with cortical emotional experience neural networks to allow for effective maternal behavior. The diagram indicates the possible neural circuits that are involved. Mentalizing and cognitive empathy systems (areas 9 and 10 of the dorsomedial prefrontal cortex; temporoparietal junction [TPJ]; superior temporal gyrus [STG]; superior temporal sulcus [STS]) project to the anterior insular region (inferior frontal gyrus [IFG]; posterior orbital frontal cortex [pOFC]; anterior insula [AI]) and to medial prefrontal cortex (mPFC) regions 24, 25, and 32. In addition, mentalizing/cognitive empathy regions also project to the amygdala (Amyg). Therefore, there are several sites where mentalizing/cognitive empathy systems can interact with those neural systems that influence maternal emotional feeling states. Such interactions may result in an appropriate level of activation of areas 24, 25, and 32, which then activate, via projections to the medial preoptic area (MPOA), the subcortical circuits that regulate maternal behavior. Neural connections shown in green emphasize points of interaction between cognitive empathy systems and the other neural systems.

both cortical and subcortical sites. Since negatively valent amygdala neurons (those neurons that respond to aversive or threatening stimuli), through their various neural connections, give rise to fear- and anxiety-related responses (see Figures 6.5 and 6.6), the feedback of the mPFC to the amygdala places the mPFC in a position to either downregulate or upregulate emotional responsiveness. The downregulation of emotional responsiveness is a process that is typically referred to as emotion regulation.

Although many researchers view PFC control mechanisms as downregulating basic aversive responses controlled by the amygdala to dampen overly fearful responses to stressful environmental situations, which then allows an organism to effectively cope with such situations (Heatherton & Wagner, 2011), it is also possible that prefrontal influences on the amygdala output operate to enhance emotional responsiveness. In fact, there is excellent research on rodents that indicates that different parts of the mPFC can exert these differential effects.

Two regions of the mPFC in rodents that project to, and influence the output of, the amygdala, and the fear-related responses that it regulates, are the infralimbic cortex (IL) and the prelimbic cortex (PL). Research has shown that the projections of IL to the amygdala can downregulate basic fear-related responses, while the projections of the PL to the amygdala can upregulate such responses (Milad & Quirk, 2012). Figure 8.5 shows the basic aspects of some of the neural circuitry that underlies these differential effects, and the reader is referred to Numan (2015) for a detailed discussion of the relevant research. This figure shows that threatening or aversive stimuli activate BLA projections to CeA and that the output of the CeA to PAG triggers the occurrence of a variety of basic defensive responses to these external stimuli. It should also be noted that amygdala projections to telencephalic regions (rather than to brainstem regions) can also influence more proactive/voluntary avoidance responses (not shown in Figure 8.5; see Numan, 2015). Excitatory PL input to BLA enhances the basic defensive responses mediated by the amygdala by stimulating the BLA-to-CeA projection. In contrast, excitatory IL input to the amygdala activates inhibitory interneurons that suppress the output of the amygdala to PAG, in this way downregulating fear responses to external stressors. Recall from Chapter 6 that it is CeAm that projects to PAG.

Research on the mechanisms of emotion regulation in humans have distinguished between two types of emotion regulation that are involved in downregulating negative emotional experiences: implicit or automatic emotion regulation, which does not require conscious control, and explicit or cognitive emotion regulation, which involves conscious effort whereby a person reappraises or reinterprets an emotional situation to decrease the intensity of that emotion (Gyurak, Gross, & Etkin, 2011). In the face of stressful situations, coping mechanisms may automatically downregulate emotional reactivity and

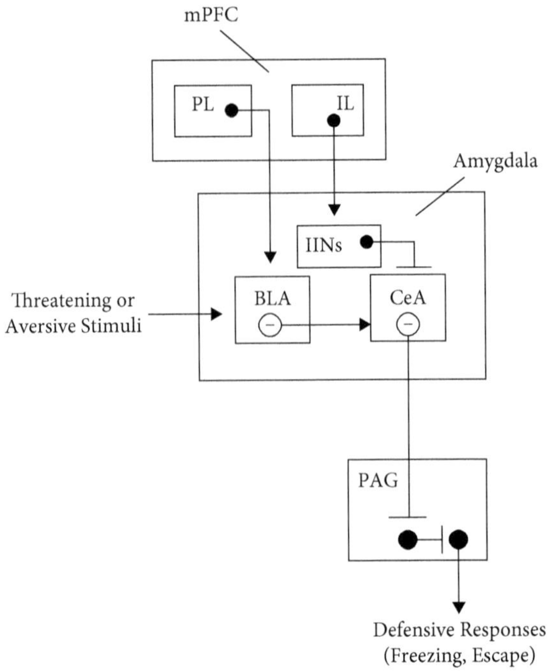

Figure 8.5. Experimental research on rodents has shown that areas of the medial prefrontal cortex (mPFC) can either upregulate or downregulate the amygdala's response to threatening or aversive stimuli. This figure describes the relevant neural circuitry. Threatening stimuli are shown as activating negatively valent neurons (minus sign) in the basolateral amygdala (BLA) which, in turn, activate central nucleus of the amygdala (CeA) projections to the periaqueductal gray (PAG) to give rise to defensive responses (freezing and/or escape responses) through a process of disinhibition within the PAG. The prelimbic part of the mPFC (PL) potentiates such fear responses by activating the BLA. The infralimbic part of the mPFC (IL) dampens fear responses to threatening stimuli, and one way it does this is by exciting inhibitory interneurons (IINs) in the amygdala which, in turn, inhibit the output of CeA to the PAG. Axons ending in an arrow are excitatory and those ending in a bar are inhibitory.

this can be significantly supplemented by an individual's conscious attempt to calm themselves. Using fMRI technology during the acquisition and extinction of a conditioned fear response in humans, Phelps, Delgado, Nearing, and LeDoux (2004) and Delgado, Nearing, LeDoux, and Phelps (2008) explored the neural basis of these two forms of emotion regulation. Subjects were presented with two different stimuli (different colored squares) that served as conditioned stimuli (CS), with one colored square (CS+) signaling the onset of a mild electric shock to the wrist (the unconditioned stimulus [US]), while another colored square indicated that a shock would not be presented (CS–). The conditioned fear response (CFR) in reaction to CS+ was a physiological measure of arousal, the skin conductance response (SCR). To examine neural activations associated with implicit emotion regulation, an extinction procedure was utilized. After the subjects acquired the CFR (an increased SCR to CS+), CS+ was presented over several trials in the absence of wrist shock. Over these extinction trials, the SCR to CS+ decreased, since CS+ no longer predicted the subsequent occurrence of the shock (US). During the initial acquisition trials, when CS+ signaled shock, CS+ presentation was associated with an increase in the SCR and an increased BOLD response in the amygdala. During extinction trials, when CS+ was presented alone, the SCR decreased, the BOLD response in the vmPFC increased, and the BOLD response in the amygdala decreased. In the context of Figure 8.5, it can be proposed that the increased BOLD response in the vmPFC during extinction resulted in neural activations in regions that might be functionally similar to the rodent IL area, with these neural activations resulting in feedback projections to the amygdala to decrease its fear-related responsiveness to CS+.

To study explicit emotion regulation, during acquisition trials subjects were asked to either simply attend to the CS or to think about something calming that was related to the color of the square. The latter task is an example of explicit emotion regulation since conscious cognitive effort is utilized to decrease emotional responsiveness in the face of a threatening stimulus. Although CS+ still activated a SCR during both the attend and explicit regulation conditions, the SCR was lower during the explicit regulation condition, suggesting a decrease in emotional responsiveness to CS+. In comparison to the attend condition, during the explicit emotion regulation condition, the BOLD response increased in both the dlPFC and the vmPFC, and these enhanced responses were associated with a decreased BOLD response in the amygdala.

On the basis of these results, and other analyses, these researchers suggest that implicit emotion regulation involves the activation of vmPFC inputs to the amygdala, which, in turn, inhibit the output of fear-related amygdala neurons. In contrast, explicit emotion regulation involves a pathway where the dlPFC activates the vmPFC, which, in turn, descends to the amygdala to suppress its output. Upon examining Figure 8.3, it is possible that the dlPFC does not project directly to the vmPFC, but instead reaches the vmPFC indirectly through its projections to the heterogeneous region that consists of IFG-pOFC-AI. An interesting speculation is that explicit emotion regulation by dlPFC is mediated by its modification of actual emotional experiences regulated by the IFG-pOFC-AI region (recall that this region is involved in the experience of both positive and negative emotions), and that such a modification affects the input of this latter region to the vmPFC, ultimately activating those particular vmPFC neurons that downregulate the activity of fear-related neurons in the amygdala.

Significantly, the part of the vmPFC that projects to the amygdala to suppress the output of its fear-related neurons appears to be located near area 32 (Delgado et al., 2008). Recall that area 32 also projects strongly to the MPOA. In the context of mother–infant interactions during challenging environmental conditions, it is interesting to speculate that certain area 32 neurons, those that project to MPOA, enhance maternal motivation, while another population of area 32 neurons projects to the amygdala to downregulate anxiety and stress-related responses.

An interesting study by Raij et al. (2018) conforms with the previous results. These authors used noninvasive transcranial magnetic stimulation of the scalp to activate functional connections between the dlPFC and the vmPFC and found that such stimulation reduced conditioned fear responses in humans. One interpretation of these results is that such stimulation substituted for explicit cognitive emotion regulation. The area of the vmPFC that was activated by the stimulation of the dlPFC appeared to the located at the border of the ventral part of area 32 and the anterior part of area 25.

In addition to the role of the dlPFC in explicit emotion regulation, other studies have indicated that the vlPFC, including parts of the IFG, are also involved in this process (Nicholson et al., 2017; Wager, Davidison, Hughes, Linquist, & Ochsner, 2008).

The results reviewed up to this point indicate that the vmPFC provides a critical descending link to the amygdala that is the final common path underlying both explicit and implicit emotion regulation. There is some confusion about which part of the vmPFC in humans in most important in this regard. Wallis et al. (2017) note that on the basis of anatomy many researchers have suggested that area 25 (sgACC) is homologous to the rodent IL, while area 32 (pgACC) is homologous to the rodent PL area. However, the research I have reviewed suggests that the opposite might be the case. Wallis et al. note that BOLD responses in area 32 are correlated with positive affect, while BOLD responses in area 25 are correlated with negative affect. They propose that this functional data indicates that area 32 in humans is likely to be homologous to the rodent IL area, while area 25 (particularly its posterior parts) is probably homologous with the rodent PL area, and they present research findings on marmoset monkeys that are consistent with this proposal (inactivation of area 32 increased a conditioned fear response while inactivation of area 25 decreased a conditioned fear response). Although more research is needed to firmly resolve these issues, in Figure 8.6 I present a tentative (and partial) neural model for emotion regulation in humans. I show that negative emotional stimuli activate negatively valent amygdala neurons. Some of these neurons project downstream to the PAG to give rise to basic

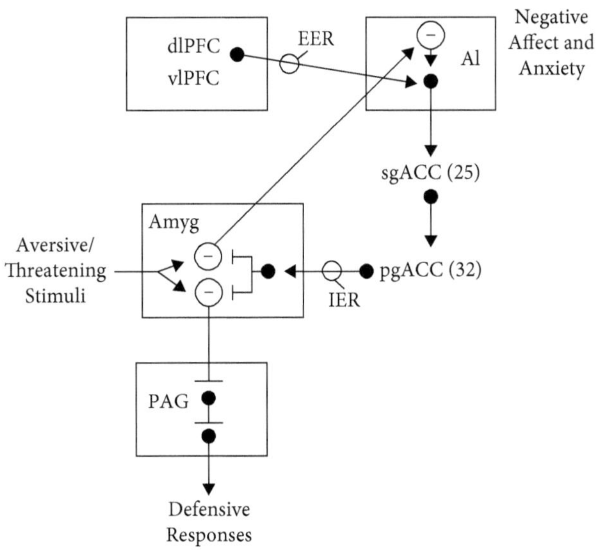

Figure 8.6. A hypothetical neural model for emotion regulation in humans. Negative emotional stimuli, which give rise to anxiety and fearfulness, are shown as activating negatively valent (minus sign) amygdala (Amyg) neurons. Some of these neurons project downstream to the periaqueductal gray (PAG) to stimulate fear-related behavioral responses. Other negatively valent amygdala neurons project to negatively valent neurons in the anterior insular region (AI), which results in negative emotional experiences, such as anxiety. To dampen intense feelings of anxiety and fearful behavioral responses (emotion regulation), certain AI neurons may project to the subgenual anterior cingulate cortex (sgACC= area 25). The sgACC may then activate the pregenual anterior cingulate cortex (pgACC = area 32), whose downstream projections to the amygdala dampen the amygdala's response to threatening stimuli. The pathway from the pgACC to the amygdala has been proposed as a mechanism that may underpin implicit emotion regulation (IER). Explicit emotion regulation (EER) neural systems can supplement the effects of implicit emotion regulation in order to dampen emotional responsiveness. The activation of AI projections to medial prefrontal cortical areas 25 and 32 by the dorsolateral and ventrolateral prefrontal cortex (dlPFC and vlPFC, respectively), may be a mechanism that underlies explicit emotion regulation. According to this model, impairment of proper functional activity within the pgACC would disable both IER and EER. Axons ending in an arrow are excitatory and those ending in a bar are inhibitory.

defensive responses, such as escape behavior. Other negatively valent amygdala neurons project to the AI and give rise to negative emotional experiences such as anxiety. To lower intense levels of anxiety, two neural mechanisms come into action. A projection of certain AI neurons to sgACC (area 25) may allow this region to activate the pgACC (area 32), and the downstream projections of the pgACC to the amygdala act to downregulate amygdala reactivity to stressful and threatening situations. This basic pathway would form the basis of implicit emotion regulation. A pathway from the lateral PFC (dlPFC and vlPFC) to the AI regions could provide a mechanism to allow for explicit emotion regulation. Based on this model, one possibility is that deficits in the reactivity of area 32 (pgACC) to its inputs would disable both forms of emotion regulation, making it difficult for a person to downregulate high levels of anxiety and stress responsiveness (cf. Etkin, Prater, Hoeft, Menon, & Schatzberg, 2010). With respect to mother–infant interactions, one can imagine that deficits in emotion regulation could result in poor maternal responsiveness under challenging and demanding situations because the mother might not be properly attentive to the child's needs, and if a child were unruly or disruptive, maternal abuse might occur.

Possible Sites of OT Action Within the Human Cortex

I have already reviewed the possible subcortical sites where OT might act to influence maternal behavior in women. Is there any evidence that OT may also act on cortical regions to influence human maternal behavior? Using immunohistochemical techniques, Boccia et al. (2013) have localized OTRs in the anterior cingulate cortex of the human brain, and Rogers et al. (2018) have detected OT axon terminals in area 25 of the anterior cingulate cortex. Bethlehem, Lombardo et al. (2017) detected the expression of OTR mRNA in mPFC (areas 24 and 32), dmPFC (area 9), STG, and the lateral PFC (dlPFC, vlPFC). Therefore, OT action within the cortex of humans may modulate the mechanisms that underpin both cognitive and emotional empathy to influence empathic care and may also influence those neural regions involved in explicit and implicit emotion regulation.

Summary

Figure 8.7 presents a summary diagram, in an abridged form, of some of the cortical circuits, and their interactions with subcortical circuits, that may influence maternal responsiveness in women. The left side of the diagram shows that infant stimuli, which can be indicative of either infant happiness or distress, activate positively valent amygdala neurons. The view that infant distress signals activate positively valent neural systems is based on the idea that in normal mothers such stimuli should lead to empathic care and maternal responses meant to relieve the distress of the infant (if infant distress signals such as crying activated negatively valent amygdala neurons, such an abnormal relationship would likely lead to

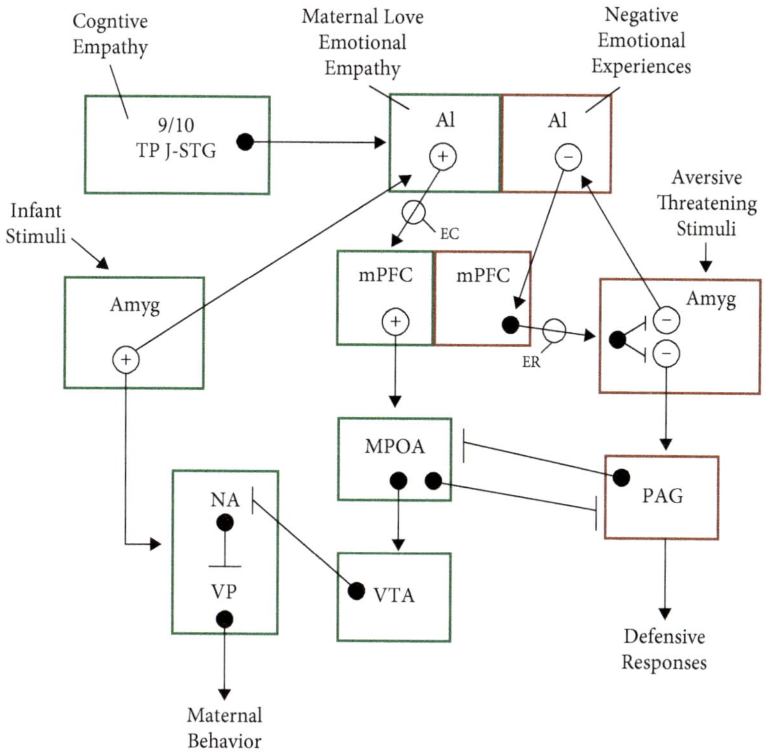

Figure 8.7. A summary diagram (in abbreviated form) that shows how cortical circuits involved in mentalizing/cognitive empathy, emotional empathy/maternal love, and emotion regulation may interact with subcortical neural circuits to promote appropriate and competent maternal responsiveness. The left side of the diagram, outlined in green, describes cortical-subcortical interactions which promote maternal motivation and sensitive maternal caretaking. The right side of the diagram, outlined in red, describes cortical-subcortical interactions involved in negative emotions, defensive responses, and emotion regulation, with the latter allowing a mother to properly care for her infant under challenging/stressful situations. Positively valent neurons (plus sign) represent those neurons that respond to infant cues that evoke positive maternal caretaking. Negatively valent neurons (minus sign) respond to threatening, stressful, or otherwise aversive stimuli and give rise to anxiety, fearfulness, and fear-related responses, but also activate emotion regulation neural systems that dampen these responses. 9 and 10 represent dorsomedial prefrontal cortical regions; AI = anterior insular region; Amyg = amygdala; EC = empathic care; ER = emotion regulation; mPFC = medial prefrontal cortical areas 24, 25, and 32; MPOA = medial preoptic area; NA = nucleus accumbens; PAG = periaqueductal gray; STG = superior temporal gyrus; TPJ = temporoparietal junction; VP = ventral pallidum; VTA = ventral tegmental area. Axons ending in an arrow exert excitatory effects and those ending in a bar exert inhibitory effects. See text for details.

faulty maternal responsiveness). Positively valent amygdala neurons are shown as projecting to both the NA-VP circuit and the AI regions. The projection of positively valent amygdala neurons to positively valent AI neurons is proposed to give rise to feelings of maternal love and emotional empathy. The projections of cognitive empathy regions to AI enable a mother to fully understand the emotions and needs of her infant, and AI projections to the mPFC allow these empathy systems and feelings of maternal love to activate appropriate maternal care through projections to MPOA, which then activates the mesolimbic DA system. If any aspect of these systems were not operating properly, maternal insensitivity and neglect might be the outcome. On the right side of the diagram, threatening, stressful, or otherwise aversive stimuli are shown as activating negatively valent amygdala neurons. Note that for some women, certain infant stimuli, such as unruly infant behavior or persistent crying may be perceived as aversive stimuli. These negatively valent amygdala neurons are shown to project downstream to the PAG (and to other regions) to promote defensive (avoidance/rejection) responses. PAG may also depress maternal behavior by inhibiting the output of MPOA. But note the following reciprocal relationship: An appropriately primed and active MPOA may depress PAG activity and decrease defensiveness (see Chapter 5 of this volume). Negatively valent amygdala neurons also project to negatively valent neurons in AI, giving rise to negative affective states. To deal effectively with such negative emotional states so that a mother can cope with demanding situations, emotion regulation mechanisms are engaged: AI projections to emotion regulatory regions in mPFC stimulate descending circuits that restrain the activity of negatively valent amygdala neurons. Similar to my discussion of the neural circuits that regulate maternal responsiveness in nonhuman animals, this figure shows that for appropriate and sensitive maternal caretaking to occur in women, maternal motivation, maternal love, and maternal empathy should be upregulated, while negative emotions such as anxiety and anger and defensive responses should be downregulated to a level that is appropriately adaptive to the situation—not too high and not too low (see Chapter 6 of this volume). Such emotion regulation may be particularly important under demanding environmental situations, and if this regulation does not effectively occur, it is likely that maternal behavior would be disorganized (the mother may be unable to care for both herself and her infant) and that maternal abuse of an unruly or disruptive infant might also occur.

Finally, it should be realized that current fMRI procedures are usually not able to differentiate positively valent neurons from negatively valent neurons. In the model that I have proposed in Figure 8.7, I hypothesize, for example, that infant cries could activate positively valent neurons in a particular region, which then gives rise to empathy, while in other cases, infant cries might activate negatively valent amygdala neurons that give rise to aversive emotional states. If such neurons are anatomically separated, fMRI techniques with high spatial resolution might be able to segregate the activity of differently valent neurons (cf. Gamer et al., 2010) and compare these responses to measures of maternal sensitivity versus maternal intrusiveness. However, if differently valent neurons are intermixed with a brain region, different procedures would have to be developed to distinguish such neuronal responses from each other. Perhaps the differential functional connectivity of neurons with different valences within certain neural networks would be a method that could tackle this problem (see my discussion of the research of Atzil et al., 2011, in the previous subsection on subcortical neural activations).

Cortical Neural Activations Associated With Maternal Responsiveness in Women

Several pieces of evidence support the model shown in Figure 8.7, which is based, in part, on my understanding of the research I am about to describe. In most of these fMRI studies, cortical BOLD responses are compared when a mother views stimuli of her own infant or those of an unfamiliar infant. Because of the strong mother–infant bond that forms between a mother and her own infant, one would expect to see greater neural activations in regions that mediate maternal love and empathy when a mother views stimuli from her own infant.

In a study by Nitschke et al. (2004), primiparous mothers viewed positive (smiling) facial photographs of their own infant, an unfamiliar infant, or adult faces (familiar and unfamiliar) during the fMRI scanning procedure. Immediately after the scans were obtained, the mothers rated their mood, indicating the pleasantness of their experience while viewing the photos. Mothers rated their mood as being more pleasant after viewing their own infant in comparison to the other stimuli. Importantly, the BOLD response was greater in the IFG-pOFC-AI region when the mothers viewed their own infant in comparison to an unfamiliar infant, and there was a significant positive correlation between the mothers' pleasant mood ratings and the BOLD response in IFG-pOFC-AI (also see Rigo et al., 2019). The IFG-pOFC-AI was not activated by the adult photos. These results are consistent with the proposal that activation of the AI region while viewing positive own infant stimuli gives rise to feelings of maternal love.

In Figures 8.3 and 8.7, I show that positively valent amygdala neurons project to AI and that this connection may underpin feelings of maternal love and emotional empathy. Also recall that Barrett et al. (2012) found that the amygdala BOLD response was greater when mothers viewed positive facial expression of their own infant in comparison to an unfamiliar infant. This study also found that the BOLD response in the amygdala while viewing one's own infant was positively correlated with mothers' positive affect and feelings of attachment to their infant. Importantly, Wonch et al. (2016) found that the functional connectivity between amygdala BOLD responses and BOLD responses in AI was greater when mothers viewed photographs of their own smiling infant in comparison to photos of an unfamiliar smiling infant.

Rocchetti et al. (2014) have proposed that the greater activation of AI, and the greater functional coupling of the amygdala with the AI, when mothers view their own infant may be mediated by the effects of OT action in the brain, since intranasal administration of OT has been found to increase these BOLD responses during a variety of emotional processing tasks in nonmothers, such as rating the emotions of various facial expressions. This proposal receives support from Kim et al. (2011), who found that breastfeeding mothers, in comparison to non-breastfeeding mothers, demonstrated greater BOLD activations in both the amygdala and IFG-AI when listening to the cries of their own infant in comparison to a standard control baby cry. One interpretation of these results is that breastfeeding mothers had greater release of OT into the brain than mothers who did not breastfeed, and that this effect resulted in the greater amygdala and AI responses to own infant cries. Interestingly, breastfeeding mothers tended to have higher levels of maternal sensitivity than non-breastfeeding mothers. Perhaps OT mediated this effect by increasing the mother's empathic response to their infant's distress signals. Realize, however, that these findings are correlational in nature. It is certainly possible that mothers who chose to breastfeed were simply "better" and more sensitive mothers at the start of the postpartum period, even before they began to breastfeed.

Recall that Atzil et al. (2017) found that high synchrony mothers, in comparison to low synchrony mothers, demonstrated stronger resting state functional connectivity between the amygdala, mPFC, MPOA, and NA. Perhaps this connectivity effect is mediated, in part, by an amygdala-to-AI-to-mPFC-to-MPOA-to mesolimbic DA projection (see Figures 8.3 and 8.7). This circuit may be involved in allowing maternal love and empathy to promote sensitive maternal caretaking.

High-quality, synchronous, and sensitive maternal behavior should not only require the engagement of neural systems involved in maternal love and emotional empathy, but should also entail effective cognitive empathy and mentalizing. Several fMRI studies have found that regions involved in cognitive empathy are more active, in terms of their BOLD responses, when mothers are exposed to their own, as opposed to unfamiliar, infant stimuli (Elmadih et al., 2016; Kim, Leckman, Mayes, Newman et al., 2010; Kim et al., 2011). In the study by Elmadih et al. (2016), mothers were characterized as displaying either high or low maternal sensitivity based on observations of mother–infant interactions during play sessions at 4 to 6 months postpartum. Scanning was conducted at 7 to 9 months postpartum while the mothers viewed videos of their own or an unfamiliar infant. Compared to low sensitivity mothers, mothers that had exhibited high maternal sensitivity showed greater BOLD responses in the STG in response to own versus unfamiliar infant stimuli. Similar results, using infant cry stimuli, have been reported by others for activation responses in STG and TPJ (Kim, Leckman, Mayes, Newman et al., 2010; Laurent & Ablow, 2012b).

In an interesting study, Abraham, Raz, Zagoory-Sharon, and Feldman (2018) examined brain responses, at approximately 12 months postpartum, when mothers viewed videos of their interactions with their own infants or videos of unfamiliar mother–infant interactions. They divided brain regions into two networks, an emotional empathy system (IFG, AI, ACC) and a cognitive empathy system (dmPFC, TPJ, STG; see Abraham, Hendler, Zagoory-Sharon, & Feldman, 2016), and they examined the intranetwork and internetwork functional connectivity within and between these two networks, respectively. In comparison to viewing videos of unfamiliar mother–infant interactions, when mothers viewed their interactions with their own infants, there was greater functional connectivity within the emotional empathy and cognitive empathy networks; importantly, there was also greater functional connectivity between the cognitive and emotional empathy systems. This research suggests that the strong bond between a mother and her own infant is associated with a strong functional interaction between the cognitive and emotional empathy systems, which conforms with the neural model shown in Figure 8.4. Clearly, a strong mother–infant bond that results in greater maternal sensitivity requires both feeling and understanding the communicative signals of one's infant.

Finally, there is some evidence that OT neural systems may influence mentalizing and cognitive empathy. Mackinnon et al. (2014; also see Mackinnon et al., 2018), measured blood levels of OT during pregnancy and at 7 to 9 weeks postpartum. Between 7 and 9 weeks postpartum, mothers were also administered a test that measures mentalizing ability—the reading the mind in the eyes test (REMT). For this task, mothers were presented with photographs of the eye region of adults, and they were asked to choose one of four words (one correct, three incorrect) that they felt best described what the person in the photograph was thinking or feeling. This test was followed by observations of a 5-minute mother–infant interaction during a free play session and the degree of maternal sensitivity was measured. Statistical analyses suggested that higher plasma OT levels during the third trimester of pregnancy, but not postpartum OT levels, were associated with higher scores (better mentalizing) on the REMT and that better mentalizing was associated with greater maternal sensitivity. To the extent that plasma levels of OT are indicative of OT release in the brain, one interpretation of these results is that late pregnancy increases in OT release within the brain, which likely interacts with other physiological events associated with late pregnancy, act on the brain to result in long-term increases in cognitive empathy and mentalizing, which, in turn, promotes maternal sensitivity.

Emotion regulation systems should also be activated for mothers to show effective maternal behavior. As indicated by Laurent, Stevens, and Ablow (2011), mothers who can regulate their own stress in the face of their infant's distress are more likely to demonstrate a more sensitive maternal response. In dangerous or stressful situations, highly anxious mothers who exhibit poor emotion regulation may also be less protective parents in that they may be overwhelmed by threatening situations and may freeze or otherwise react inappropriately, rather than effectively protecting their infants from external threats (Bakermans-Kranenburg & van Ijzendoorn, 2017).

As reviewed in Chapter 6, the hypothalamic-pituitary-adrenal physiological stress response, as measured by peripheral corticosterone/cortisol levels, is positively correlated with anxiety and fearfulness, which is probably the result of enhanced corticotropin-releasing factor (CRF) release into the brain that activates central fear/anxiety neural systems. Gordon, Zagoory-Sharon, Leckman, and Feldman (2010) found that maternal salivary cortisol levels were negatively correlated with behavioral measures of mother–infant synchrony, suggesting that more anxious mothers tend to show lower levels of maternal sensitivity. Laurent et al. (2011) exposed mother–infant dyads to a stressful situation that involved a series of mother–infant separations and reunions in a novel laboratory setting. The primiparous mothers could see and hear the distress of their 15- to 18-month-old infants during the separation periods. Salivary cortisol levels were measured prior to and during the stressful series of separations and reunions. As expected, maternal cortisol levels increased in response to this stressor, but there was variability, with some mothers exhibiting a greater cortisol response than other mothers (high-reactive and low-reactive mothers). Increased cortisol reactivity was associated with poorer maternal sensitivity upon the infant's return to its mother. In a scanning session, mothers listened to their own infant's or an unfamiliar infant's cry. The fMRI results showed that mothers who were less physiologically stressed during the mother–infant separations showed increased BOLD responses to their own infant cry in the vlPFC and in the ACC-mPFC (areas 24 and 32) than did mothers who showed higher stress reactivity. Additional supportive evidence that maternal sensitivity is positively related to activation in vlPFC has been reported by others (Kim, Leckman, Mayes, Newman et al., 2010; Laurent & Ablow, 2012b; Musser, Laurent, & Ablow, 2012). These results, taken together, suggest that less stress reactive mothers are more successful in activating both explicit and implicit emotion regulatory regions in response to their infant distress signals, which downregulates their emotional responsiveness and that these processes promote more sensitive maternal caretaking behaviors during stress.

In an important series of studies, Kim and her colleagues have explored the relationships between socioeconomic disadvantage (poverty), maternal behavior, and brain responses to infant stimuli. Kim, Capistrano, and Congleton (2016) interviewed first-time mothers (infant ages were 6 months or less) during a home visit. The mothers' economic status was based on their income-to-needs ratio, and about half of the participants lived in poverty (income-to-needs ratio of 1 or less) or near poverty (income-to-needs ratio of 2 or less). During the home visit, each mother rated the amount of stress they experienced during the past month and their perceptions (positive or negative) of being a parent. Three weeks later, during fMRI scans, the mothers were exposed to the cry of their own or an unfamiliar infant. The hypothesis underlying this study is that the stress associated with poverty may result in decreases in both maternal sensitivity and in the mother's emotional availability to her infant.

The results of this study can be summarized as follows: (a) Mothers who lived in or near poverty experienced more stress in their lives and had more negative perceptions of being a parent than did mothers who were more financially secure; (b) poverty or near poverty was associated with decreased BOLD responses to own infant cries in the dlPFC and mPFC (area 32); (c) increases in perceived stress mediated the relationship between poverty and decreased responsiveness in dlPFC and mPFC; and (d) less positive perceptions of parenting were associated with decreased neural responses to infant cry sounds in the dlPFC. One interpretation of these results in that the stress associated with poverty results in decreases in both explicit and implicit emotion regulation, and the interactions between these two systems (see Figure 8.6), with the result that the mother might perceive infant distress signals as aversive and that this, in turn, gives rise to less positive views of being a parent.

In a follow-up study, Kim, Capistrano, Erhart, Gray-Schiff, and Xu (2017) exposed low- and middle-income primiparous mothers to positive and negative infant faces (smiling vs. crying faces) during an fMRI scanning session. Mothers with lower income-to-needs ratios exhibited decreased amygdala responses to positive infant faces and increased amygdala responses to negative infant faces in comparison to mothers with higher income-to-needs ratios. Outside the scanner, analysis of mother–infant interactions showed that mothers living in poverty exhibited a more intrusive maternal style. With respect to emotion regulation, the fact that amygdala responses to negative infant stimuli were greater in mothers living in poverty suggests that lower levels of explicit and implicit emotion regulation in such mothers, as defined in the study by Kim et al. (2016), may have resulted in infant distress signals activating negatively valent amygdala neurons that give rise to aversive feeling states. The decreased responsiveness of the amygdala to positive infant stimuli may also indicate that poverty may be associated with reductions in the ability of infant stimuli to activate positively valent amygdala neurons, which could lead to deficits in maternal love and empathy.

Kim et al. (2017) make the important point that the increased response of the amygdala to infant distress signals (crying faces), and the higher levels of maternal intrusiveness, in low income mothers may be related to higher levels of maternal vigilance and protectiveness, and that these responses may actually be adaptive in stressful and unpredictable environments, such as living in poverty. However, the way a mother treats her infant has important impacts on the infant's development, and maternal intrusiveness can negatively impact the social and emotional development of her infant. Since many modern societies provide routes for upward mobility, where individuals move out of poverty, the way a mother treats her children under conditions of poverty may have important implications with respect to the success of her children under improved environmental conditions. I will have more to say about these important issues in Chapter 10.

OT neural systems not only enhance maternal motivation, but also exert anxiolytic effects. Is there evidence that OT can influence the neural circuits involved in emotion regulation in humans, in this way reducing negative emotional responses to threatening or stressful stimuli? In healthy human subjects, Sripada et al. (2013) found that intranasal OT treatment increased the resting state functional connectivity between the mPFC (areas 25/32) and the amygdala. Dodhia et al. (2014) found that patients with generalized social anxiety disorder (GSAD) had decreased resting state functional connectivity between the mPFC (area 32) and the amygdala in comparison to healthy controls. Intranasal treatment with OT was found to increase the resting connectivity between these two regions in the GSAD patients, so that this connectivity matched that of healthy controls. Therefore, it is possible that OT action in the brain increases the top–down regulation from area 32 to the amygdala, in this way decreasing the responsiveness of negatively valent amygdala neurons to aversive stimuli. This proposal fits with the findings of Labuschagne et al. (2010), who found that GSAD patients, in comparison to healthy controls, exhibited greater amygdala BOLD responses to fearful/angry faces and that intranasal treatment with OT depressed this amygdala response to the level observed in healthy subjects. Since OT activity is presumably high in the maternal brain, it may therefore promote proper emotion regulation under stressful and challenging environmental conditions, and this effect may enhance adaptive maternal responsiveness. Another implication is that dysfunctions in maternal OT systems may be related to poor emotion regulation.

Upon examining Figure 8.4, one can see that cognitive empathy and mentalizing neural systems, by way of their projections to the mPFC (areas 24, 25, 32) and/or to the amygdala, are positioned to not only influence maternal motivation, but may also affect emotion regulation (see Figure 8.6). Although this relationship is not shown in Figure 8.6, it makes sense that a rational understanding of a stressful and anxiety-provoking situation may play a role in decreasing aversive emotional experiences by way of mentalizing neural regions either directly or indirectly (via projections to mPFC) affecting the amygdala, to downregulate the responsiveness of negatively valent amygdala neurons. There is some recent evidence that appears to support this proposal in the context of the occurrence of positive maternal caretaking in women with subclinical anxiety symptoms (Guo, Moses-Kolko, Phillips, Swain, & Hipwell, 2018). This study emphasized the possibility that interactions between the STG and the underlying STS with the amygdala (presumably via STG/STS projections to mPFC or directly to the amygdala, see Figure 8.4) may play a role in restraining the responsiveness of negatively valent amygdala neurons to infant cry stimuli, which, in turn, fosters more positive maternal caretaking behaviors. It is also possible that mentalizing neural regions, such as STG/STS, increase the responsiveness of positively valent amygdala neurons to infant cry stimuli.

To summarize this section on cortical neural circuits that have been implicated in the maternal behavior of postpartum women, the evidence indicates that maternal love and empathy, combined with the effective regulation of negative emotions, foster positive mother–infant interactions and that the neural connections within and between the cortical regions that mediate these processes affect adaptive maternal responsiveness through their ultimate influences on the subcortical mechanisms involved in the regulation of maternal motivation and behavior (see Figure 8.7). OT neural inputs to various nodes within these cortical and subcortical regions may exert facilitating effects on each of these processes.

Postpartum Depression

Introduction

Postpartum depression is a serious psychiatric disorder that occurs in the early postpartum period, affecting 10% to 15% of mothers, and it may last for more than 7-months postpartum (Brummelte & Galea, 2016; Yim, Tanner Stapleton, Guardino, Hahn-Holbrook, & Dunkel Schetter, 2015). Postpartum depression may be preceded by prepartum or antenatal depression, and therefore some researchers refer to this disorder as peripartum depression, and the occurrence of antenatal depression is a strong risk factor for the subsequent development of postpartum depression (Brummelte & Galea, 2016). Postpartum depression is composed of a heterogeneous group of symptoms that can include depressed mood (sadness), anhedonia (inability to experience pleasure), and increases in aversive mood states such as distress, irritability, and anger, which can generally be described as increases in stress reactivity (Lovejoy, Graczyk, O'Hare, & Neuman, 2000). This last characteristic is most likely the result of the fact that postpartum depression is typically associated with anxiety (Brummelte & Galea, 2016; Yim et al., 2015; Pawluski, Lonstein, & Fleming, 2017). Significantly, pure prenatal anxiety is a strong risk factor for the subsequent development of full postpartum depression (Anniverno, Bramante, Mencacci, & Durbano, 2013).

One of the most common measures used to diagnose postpartum depression is the Edinburgh Postnatal Depression Scale (EPDS), which is a 10-item self-report questionnaire that is administered to the mother. This scale includes measures that detect sad mood, anhedonia, and anxiety. Scores on this test can range from zero to 30, and a cutoff score of 13 or higher is usually used as a diagnosis of postpartum depression, while scores of between 10 and 12 are indicative of minor depression (Glynn & Sandman, 2014). Based on the scoring system of this scale, a mother who is predominantly anxious, with relatively low scores (but higher than zero) on measures of sadness and anhedonia, would be classified as depressed, as would mothers with high sadness and anhedonia, but low anxiety. Therefore, the use of total scores on this scale makes it difficult to determine the individual roles of each pathology to observed behavioral outcomes (Pawluski et al., 2017), such as poor maternal behavior.

It is important to discuss postpartum depression in the context of understanding the parental brain because women with postpartum depression exhibit faulty maternal behavior, which, in turn, influences the social, emotional, and cognitive development of their children (Bernard, Nissim, Vaccaro, Harris, & Lindheim, 2018; Brummelte & Galea, 2016; Lovejoy et al., 2000; Pawluski et al., 2017). Mothers with postpartum depression engage in fewer positive interactions with their child, display lower levels of maternal sensitivity/synchrony, and higher levels of intrusiveness and anger. The latter is associated with decreases in tolerance toward a distressed child and the use of coercive behaviors to influence the behavior of the child.

It is interesting to speculate that mothers who present with high levels of anhedonia and sadness may withdraw from their child, which could lead to maternal neglect, while mothers that present with high levels of stress reactivity and anxiety and are easily disturbed and irritable may interact more negatively with their child, which could lead to maternal abuse (Numan & Insel, 2003). Note, however, that both anhedonia and anxiety may co-occur during postpartum depression (Putnam et al., 2017), which might lead to both neglect and abuse.

Taylor, Atkins, Kumar, Adams, and Glover (2005) developed a Mother-to-Infant Bonding Scale, which is an 8-item maternal self-report scale that measures a mother's positive (affection and love) and negative emotions (no positive feelings or dislike) toward her child. High scores on this scale are indicative of poor mother–infant bonding. Not surprisingly, high scores on this bonding scale were positively correlated with high scores on the EPDS. These results suggest that postpartum depression may disrupt the development of a strong mother-to-infant bond. (The reverse relationship is also possible: Poor bonding may influence the development of postpartum depression.) O'Higgins, Roberts, Glover, and Taylor (2013) emphasize that not all women with postpartum depression, as measured with the EPDS, experience problems with bonding to their infant. Perhaps it is an inability to experience pleasure while interacting with one's infant (anhedonia) that is the primary factor that leads to poor maternal bonding. Mothers that are primarily anxious may bond with their infants, but may be less tolerant toward a disruptive or distressed infant.

The causes of postpartum depression are not fully understood, and multiple factors are involved. Research has indicated that certain abnormal maternal reactions to the physiological events associated with late pregnancy might be involved (Brummelte & Galea, 2016; Numan & Insel, 2003; Payne & Maguire, 2019; Sherer, Posillico, & Schwarz, 2018; Yim et al., 2015), and I will discuss certain aspects of this research later in this chapter in the subsection on corticotropin-releasing factor, oxytocin, and postpartum depression. I have already indicated that prepartum depression or anxiety are strong predictors of the subsequent development of postpartum depression. Early life stress (childhood maltreatment) also increases the likelihood that the affected children will develop postpartum depression when they become mothers, and I will discuss this relationship in Chapter 10, which deals with the development of the human parental brain. Finally, current life stressors and low social support are risk factors for the development of postpartum depression (Yim et al., 2015). In the context of the role of low social support as a risk factor for the development of postpartum depression, Hagen (1999) has offered a very speculative hypothesis. He suggests that postpartum depression may actually be an evolutionary adaptive response to the lack of social support, whereby a mother decreases her investment in offspring that are not likely to survive. Recall that Hrdy (2016) has presented evidence that cooperatively breeding New World monkey mothers may abandon their young if sufficient alloparental support is not available. Also recall that alloparental support may have been particularly important for successful maternal reproduction in ancestral humans (Hrdy, 2009). Therefore, perhaps a mother's perception of low levels of social support in modern societies promotes anhedonia and disrupts the mother–infant bond. Although this proposal makes some sense, it is difficult to understand why anhedonia and other aspects of postpartum depression are global and are not specifically related to emotions that are selectively displayed toward one's infant. It is also worth emphasizing that in modern Western societies, many children who are neglected or abused by their mothers survive and may be taken away from the mother and cared for by others. Because of the negative impacts of postpartum depression on the mother's interaction with her social world, and on the socioemotional and cognitive development of the poorly cared for child, it is important to understand the proximate causes of postpartum depression in the hope of developing effective therapies to prevent the occurrence of postpartum depression and/or to reduce its symptoms. Clearly, increasing social support for mothers in need might be one remedy for a certain subgroup of mothers with a risk for postpartum depression.

Postpartum Depression and the Maternal Brain

Because postpartum depression influences the nature of mother–infant interactions, investigators have used fMRI procedures to determine whether brain responses to infant stimuli differ between mothers with postpartum depression and healthy control mothers. Wonch et al. (2016) presented positive infant faces (smiling infants) from own or unfamiliar infants to these two groups of mothers at 2 to 5 months postpartum. The postpartum depressed mothers had higher depressive symptoms and anxiety than did the healthy control mothers. When presented with positive infant stimuli while in the scanner, the nondepressed mothers exhibited a greater BOLD response in the bilateral amygdala to their own infant's smiling face in comparison to an unfamiliar infant's smiling face. For mothers with postpartum depression, the BOLD response when viewing their own infant, in comparison to an unfamiliar infant, was only greater in the right amygdala. Importantly, a functional connectivity analysis indicated a correlation between BOLD responses in the amygdala and insular cortex (strong functional connectivity) when nondepressed mothers viewed photos of their own infant, while such connectivity was nonexistent in the depressed mothers. Perhaps these differences are related to lower levels of joy, happiness, and maternal love in depressed mothers while viewing positive infant faces (see Figure 8.3) because the amygdala was not capable of effectively influencing positive maternal emotional experiences via projections to the AI. It is also possible that the increased activity in the right amygdala of depressed mothers was the result of own-infant faces activating negatively valent amygdala neurons that might give rise to withdrawal/avoidance type responses (also see Lenzi et al., 2016). Since the entire amygdala was analyzed in this study, it cannot be determined whether infant stimuli might have affected different regions of the amygdala in the two groups of subjects. It would certainly be interesting to know whether own-infant faces activated the CeA in depressed mothers, giving rise to escape/avoidance-like covert responses, while the BLA/BMA was activated by own-infant faces in the nondepressed mothers, giving rise to positive feeling states and covert approach responses. Increased activation of negatively valent amygdala neurons would also be suggestive of deficits in emotion regulation during postpartum depression.

With respect to feeling states, Wonch et al. (2016) found that both groups of mothers reported feeling more positive when viewing their own infant in comparison to a strange infant, and the two groups did not differ in the intensity of their positive affect. These results do not seem to conform with my interpretation of the fMRI data. However, it is possible that the depressed mothers did not report their actual affective state, but instead reported a socially acceptable response. They may have responded in a way that they thought was expected. Additionally, perhaps potential negative feeling states were unconscious and automatic under these testing conditions, and normal mentalizing functions allowed the depressed mothers to give a rational and logical response to conform to the situation.

Laurent and Ablow (2013) reported that depressed mothers, in comparison to heathy controls, exhibited decreased BOLD responses in the insula when they viewed their own infants' joy faces. These findings would fit with the aforementioned findings of a decreased functional connectivity between the amygdala and insula in depressed mothers. Laurent and Ablow also reported that depressed mothers showed reduced activation of the ACC (areas 24 and 32) when viewing crying/distressed faces of their own infants (also see Lenzi et al., 2016). This finding may be related to decreased emotion regulation capability in mothers with postpartum depression (see Figure 8.6). A deficit in emotion regulation may lead to enhanced anxiety and irritability in mothers with postpartum depression, and such a deficit could result in increases in maternal intrusiveness and coercive maternal responses.

Laurent and Ablow (2012a) reported on the neural responses that occurred in various brain regions when depressed and nondepressed mothers listened to recordings of their own infant cry sounds while in a scanner. In comparison to depressed mothers, nondepressed mothers showed greater BOLD responses to their own infant's cry in the nucleus accumbens (cf. Witteman et al., 2019). Own-infant cry sounds also activated the AI and mPFC in nondepressed mothers, but these regions were not activated in depressed mothers. Upon examining Figure 1 in the Laurent and Ablow study, it also appears that mothers with higher levels of depression showed decreased BOLD responses in the preoptic region when listening to their own infant cry. These researchers did not mention this possible effect and they assumed that the decreased response that they observed occurred in the thalamus rather than in the preoptic region. In a related study, Ho and

Swain (2017) have reported that depressed mothers, in comparison to healthy controls, show decreased functional connectivity between the amygdala and the nucleus accumbens when they listened to own-infant cry sounds in comparison to a generic infant cry. These results, taken together, suggest that postpartum depression may be associated with deficits in empathy and a desire to help one's distressed infant because of decreased neural activity across the circuits shown in Figure 8.3: AI-to-mPFC-to-MPOA-to-mesolimbic DA system (see Post & Leuner, 2019, for a review of the role of dysfunctions in the mesolimbic DA system in women with postpartum depression).

In summary, the maternal behavior deficits that are associated with postpartum depression may result from underactivity in neural circuits that underpin maternal love, maternal empathy, maternal motivation, and emotion regulation in response to various infant stimuli. Decreases in maternal love and empathy, which could promote child neglect, may be related to anhedonia, while the increased anxiety and stress reactivity that are associated with postpartum depression may be the result of deficits in emotion regulation, which, in extreme cases, could lead to child abuse. Decreases in maternal motivation may be related to both anhedonia and increased anxiety and stress reactivity (see Figure 8.7).

Corticotropin-Releasing Factor, Oxytocin, and Postpartum Depression

In Chapter 6, I described the opposing roles of OT and CRF with respect to anxiety-related behaviors in animals, with OT neural systems exerting anxiolytic effects and CRF neural systems exerting anxiogenic effects. There is some evidence that lower levels of OT and higher levels of CRF may promote postpartum depression, particularly with respect to the high distress and anxiety components that are typically associated with postpartum depression.

Glynn and her colleagues have presented evidence that high levels of CRF in blood plasma during late pregnancy are good predictors of the subsequent development of postpartum depression in women (Glynn, Davis, & Sandman, 2013; Glynn & Sandman, 2014; cf. Meltzer-Brody et al., 2011). Glynn et al. note that CRF is synthesized by the placenta of women beginning by the seventh week of gestation. Placental CRF (pCRF) production increases throughout the remainder of gestation and reaches very high levels in maternal blood plasma during the last trimester of pregnancy. One reason for this heightened production of pCRF is that cortisol stimulates pCRF production. During pregnancy, increased pCRF in blood plasma activates adrenocorticotropic hormone (ACTH) release from the anterior pituitary, and ACTH stimulates cortisol release from the adrenal cortex.

Cortisol then causes further increases in pCRF production and release. This positive feedback loop results in high levels of plasma ACTH, cortisol, and pCRF throughout the later stages of pregnancy in all women. But there are interindividual differences in the levels of these hormones that are produced during pregnancy in women. Glynn and Sandman (2014) found that women who exhibited higher levels of pCRF throughout the second half of pregnancy were at risk for the subsequent development of postpartum depression: Higher levels of plasma pCRF at 25, 31, and 36 weeks of pregnancy predicted postpartum depression symptoms at 3 months postpartum.

I would like to offer a proposal that might explain the increased stress reactivity and anxiety that are associated with postpartum depression. pCRF, because it is a peptide, is not likely to enter the brain from maternal plasma, but cortisol does have access to the brain. Perhaps the higher than normal levels of cortisol that occur during pregnancy in some women enter the brain and cause a higher than normal increase in the synthesis of brain CRF within the central nucleus of the amygdala (CeA), as I described in Chapter 6. Increased CRF within CeA neurons, coupled with current life stressors, such as poverty or lack of social support, may precipitate the high levels of anxiety that are associated with postpartum depression. Although I find this proposal to be attractive, Glynn and Sandman (2014) have presented some evidence that opposes my hypothesis: While prepartum pCRF was predictive of postpartum depression at 3 months postpartum, neither prepartum cortisol nor ACTH levels were predictive. Clearly, more research needs to be done to explore these issues. Since cortisol exerts positive feedback effects on CRF production in CeA, it is worth considering the possibility that CeA glucocorticoid receptors are more responsive to the positive feedback effects of cortisol in women who will develop postpartum depression. Such a relationship might allow CRF, but not cortisol, levels to be predictive of postpartum depression.

Only a few studies have examined the relationship between OT and postpartum depression. Skrundz, Bolten, Nast, Hellhammer, and Meinlschmidt (2011) measured plasma levels of OT in women during the third trimester of pregnancy. At 2 weeks postpartum, the women were administered the EPDS. The results indicated a negative relationship between plasma OT and postpartum depression scores. Based on these findings, and on the assumption that plasma OT levels reflected the release of OT within the brain, these researchers suggested that enhancing OT release during pregnancy might be a potential therapy to prevent the development of postpartum depression. With respect to postpartum period, Stuebe, Grewen, and Meltzer-Brody (2013) have reported a negative correlation between plasma OT levels and postpartum anxiety and depression scores.

These results are interesting with respect to the antagonistic relationship between OT and CRF on the expression of anxiety, as described in Chapter 6. Perhaps deficiencies in OT enhance CRF release from CeA because OT is less effective in inhibiting such release. In the context of the findings indicating a positive relationship between prepartum CRF and postpartum depression, perhaps women who exhibit higher than normal levels of CRF synthesis and release within the brain and lower than normal levels of OT release within the brain during late pregnancy are more likely to develop postpartum depression, particularly with respect to the stress reactivity and anxiety components of this disorder.

In light of the findings of Skrundz et al. (2011), a preliminary study by Mah, van Ijzendoorn, Smith, and Bakermans-Kranenburg (2013) examined whether the administration of intranasal OT could ameliorate the symptoms of postpartum depression. Using a within-subjects design with women who were diagnosed with postpartum depression, at 3 to 12 months postpartum, women received intranasal OT or placebo, one week apart and in random order. They found that the administration of OT did not improve EPDS scores. Based on the findings of Skrundz et al., perhaps OT treatment during late pregnancy, rather than during the postpartum period, would have been found to be effective in preventing the subsequent onset of postpartum depression (women who demonstrate prepartum depression and/or anxiety could be administered OT or control treatments, and the effects of such prepartum manipulations on the subsequent development of postpartum depression could be evaluated). Also, acute treatments at any time point may not be effective, and a more prolonged enhancement of OT action within the brain might be needed to ameliorate the stress reactivity and anxiety symptoms associated with postpartum depression. It is also possible that some women with postpartum depression may not have deficits in brain OT per se, but instead may have deficiencies in the density of OTRs or in the sensitivity of the OTR to OT (Kim et al., 2014). Since OT exerts a positive feedback effect on its own release by stimulating PVN OTRs, as described in Chapter 5, perhaps the lower levels of OT release associated with postpartum depression are related to the decreased responsiveness of OTRs to OT. Such a decreased responsiveness to OT may not only involve the PVN, but may also involve the responsiveness of other brain regions to OT. Exogenous administration of OT may not be capable of overcoming such deficiencies in OTR responsiveness to OT. Finally, it is also possible that even if OT effectively activates OTRs, the downstream targets of the OT receptive neurons may be dysregulated in some way (see last paragraph in this section).

There is one interesting study (Mah, Bakermans-Kranenburg, van Ijzendoorn, & Smith, 2015), which suggests that intranasal administration of OT may enhance maternal aggression, or maternal protection of their infants, in women with postpartum depression. Women diagnosed with postpartum depression were subjected to the Enthusiastic Stranger Paradigm. In this test, women and their 3- to 11-month-old infants are placed in a laboratory waiting room and an unknown adult enters the room and approaches the infant and eventually touches the infant if such behavior is not prevented by the mother. Maternal protective responses were measured on a scale of 1 to 5, from no interference with the stranger's approach to verbal or overt motor responses to prevent the stranger from approaching and touching the infant. Depressed mothers that received intranasal OT 55 min prior to the Enthusiastic Stranger Paradigm were significantly more protective of their infants than were mothers that received the placebo treatment. In a manner similar to my discussion of maternal aggression in animals, Bakermans-Kranenburg and van Ijzendoorn (2017) suggest that highly anxious parents with postpartum depression may be overwhelmed by threatening situations, which prevents them from protecting their infants. These researchers suggest that OT administration, by decreasing anxiety, may have promoted maternal protective responses. This interpretation fits with the research I described in the section on cortical neural activations associated with maternal responsiveness in women, which indicated that OT may improve emotion regulation. OT may not only depress anxiety by directly suppressing the release of CRF from CeA, but may also promote a top–down mPFC inhibition of amygdala reactivity to threatening and stressful stimuli. It should be noted, however, in this study that employed the Enthusiastic Stranger Paradigm, no specific measures of maternal anxiety under the two conditions were taken. Therefore, it is also possible that OT increased maternal protective responses because it enhanced maternal motivation rather than because it decreased maternal anxiety.

Much more research is needed to explore the relationships between OT, CRF, and the symptoms of postpartum depression. The work that I have reviewed is certainly suggestive of the possibility that a dysregulation of central CRF and OT systems contributes to the heightened anxiety and stress reactivity that can be associated with postpartum depression. Such a dysregulation could lead to lower maternal protective responses and increases in maternal intrusive and coercive behaviors directed toward a distressed or unruly child. These relationships are worthy of receiving further research efforts. In this regard, a recent drug treatment has been developed that appears effective in reducing the symptoms of postpartum depression when administered to women in the early postpartum period (Kanes et al., 2017). The drug, brexanolone, is a synthetic form of allopregnanolone. Allopregnanolone is a metabolite of progesterone, and it acts in the brain by enhancing the activity GABA at GABA-A receptors (Numan & Insel, 2003). Because progesterone and its metabolites are secreted at high levels during pregnancy and then these hormones decline precipitously at parturition, it has been proposed that for a certain susceptible population of women who develop postpartum depression, GABA receptors in the brain may become hyposensitive to GABA as a result of the prolonged exposure to allopregnanolone followed by its decline, leading to an enhanced neural activity that results from decreased GABAergic inhibition (for a review of this phenomenon, see Numan & Insel, 2003). The postpartum administration of brexanolone may ameliorate this effect by stimulating GABA receptors. In this regard, please refer to Figure 6.5, where it was suggested that OT may activate GABAergic inhibitory interneurons in the amygdala, which, in turn, suppress the output of CeAm neurons to PAG and CeA-CRF neurons to dlBST. For certain women who are susceptible to postpartum depression, the functional effects of OT activation of amygdala GABAergic interneurons may be less effective, because CeA neurons are less responsive to the GABA that is released. Such a result would lead to enhanced amygdala projections to PAG and enhanced CRF release into the brain, leading to heightened anxiety. The administration of brexanolone may counteract some of these effects. Of course, brexanolone may act at other sites to enhance GABAergic inhibitory effects in the brain.

The Paternal Brain in Men

Compared to research on the neural activations associated with maternal behavior in women, much less research has been devoted to the paternal brain (Glasper, Kenkel, Bick, & Rilling, 2019; Rajhans, Goin-Kochel, Strathearn, & Kim, 2019; Rilling & Mascaro, 2017). One possible reason for this research bias is that in many human societies, the mother is the primary caregiver, while the father plays a secondary role in the direct care of offspring (Fernandez-Duque, Valeggia, & Mendoza, 2009; Hrdy, 2009; Kramer & Veile, 2018). In comparing the nature of paternal behavior in humans with other mammals, in most nonhuman mammalian species, when paternal behavior occurs, it typically occurs in all males (California mouse, dwarf hamster, titi monkey). Therefore, paternal behavior in the typical biparental nonhuman mammal is considered to be obligate (necessary). In contrast, in human societies there is high variability in the degree to which direct care of infants is exhibited by the father, and, therefore, paternal behavior in humans can be considered facultative, that is, conditional and dependent upon specific ecological, social, and cultural circumstances. One can propose, therefore, that paternal experience with infants might play a primary role in boosting parental motivation and the desire to care for infants in men (Storey & Zeigler, 2016).

In relation to the issue of the importance of father–infant interactions for the development of paternal responsiveness in men, one can compare brain responses to infant stimuli in fathers and nonfathers. In an fMRI study, Seifritz et al. (2003) compared BOLD responses to infant cry or infant laughing sounds (from unknown infants) in men without children and in fathers whose infants were 1 to 3 years old. Interestingly, fathers showed a greater neural response in the amygdala and insula to infant crying when compared to laughing, while nonfathers showed the opposite response. One interpretation of these results is that infant laughing sounds result in positive emotional experiences in both groups of men. However, in fathers, as a result of paternal experience, infant cries also evoke strong emotional empathic responses (amygdala-to-AI), which would presumably be related to a desire to help the distressed infant.

In a related study (Mascaro, Hackett, Gouzoules, Lori, & Rilling, 2014), fathers of 1- to 2-year-old children were administered a parental responsibility scale, which assessed the degree of paternal involvement in infant care. Subsequently, while in a scanner, the fathers were exposed to unfamiliar infant cry stimuli or control neutral baby vocalizations. A curvilinear relationship between paternal responsibility and AI BOLD response to infant cry stimuli was detected. Fathers with low or high levels of AI BOLD responses engaged less in childcare, while fathers with moderate AI BOLD responses exhibited the highest levels of paternal caretaking. One way to interpret these findings is that low levels of AI activity signify low emotional empathy that is associated with lower levels of paternal experience. High levels of AI activity may be related to the recruitment of negatively valent AI neurons associated with negative emotions (the cry is aversive), which, in turn, is associated with lower paternal involvement in childcare. Moderate levels of AI activity, which are related to higher levels of paternal experience, may be associated with a robust and selective emotional empathic response (Mascaro et al., 2014).

Mascaro, Hackett, and Rilling (2013) also assessed the degree of paternal involvement in direct child-care activities in a group of fathers who had 1- to 2-year-old children. fMRI scans indicated that pictures of one's own child, in comparison to an unknown adult, were associated with increased BOLD responses in the VTA. Most important, there was a positive correlation between the degree of paternal caregiving experience and the intensity of the BOLD response in VTA. Mascaro et al. offer two potential explanations for these results. First, fathers who find their child's face to be more attractive (due to higher activity in the mesolimbic DA system), may as a result, engage in more paternal behavior. Alternatively, fathers who engage in more paternal care may develop of stronger father–infant bond, which may be reflected in a greater BOLD response in VTA and improved paternal behavior over time. The question, therefore, is whether fathers who engage in higher levels of paternal caretaking have higher VTA BOLD responses as soon as their infants are born, or does the enhanced VTA response develop along with paternal experience. Based on the premise that paternal experience with infants increases paternal motivation in men

(Storey & Zeigler, 2016), I would predict that the latter mechanism is more likely. However, as the authors indicate, longitudinal studies are needed to gain insight into the possible direction of causality that links the VTA BOLD response with the degree of paternal behavior.

Abraham et al. (2014) employed a novel approach toward investigating how the degree of father–infant interactions and experience influence brain responses to videos of parent–infant interactions. This study recruited families composed of either heterosexual couples or male homosexual couples who were caring for infants of about 1 year of age. For the heterosexual couples that were accepted into the study, the mother was always the primary caretaker, and the father was the secondary caretaker. For the homosexual male parents, both parents were considered primary caretakers in that they devoted equal amounts of time in caring for their child. Therefore, the primary caretaker fathers demonstrated more daily care of their child than did the secondary caretaker fathers. What is not clear from the reported data is whether primary caretaker fathers devoted the same amount of paternal care as the maternal care engaged in by the primary caretaker mothers.

In behavioral observations, it was found that primary caretaker mothers and fathers exhibited greater parent–infant synchrony than did secondary caretaker fathers. While in a scanner, parents were exposed to videos of themselves interacting with their child or to videos of an unknown parent–child interaction. Primary caretaker mothers showed greater amygdala BOLD signals in response to viewing self–infant interactions than did secondary caretaker fathers. Perhaps this is related to greater parental motivation, parental emotional empathy, and parental love in primary caretaker mothers, which results in more synchronous parenting style. Significantly, primary caretaker fathers exhibited an amygdala response equal to that of primary caretaker mothers. This paternal amygdala response was associated with a higher BOLD response in the STS (a mentalizing/cognitive empathy region) when primary caretaker fathers were compared to primary caretaker mothers. Further, the functional connectivity between the amygdala and STS during self–infant video viewing was greater in primary caretaker fathers than in primary caretaker mothers and secondary caretaker fathers; when all fathers were analyzed together, there was a positive correlation between reported time spent in infant care and amygdala-STS functional connectivity.

The authors conclude from their data that as a result of the high levels of paternal experience associated with a primary caretaker role in fathers, mentalizing regions are engaged in such fathers to activate the amygdala, and amygdala activation would then presumably enhance emotional empathy and love along with parental motivation. In contrast, such mentalizing is not necessary for amygdala activation in mothers, where maternal behavior in presumed to occur more "naturally."

Although this interpretation is intriguing, several caveats are worth pointing out. First, perhaps primary caretaker mothers had more parental experience than did the primary caretaker fathers. Second, homosexuality may have influenced the results. Finally, Guo, Moses-Kolko, Phillips, Swain, and Hipwell (2018) have reported that among mothers with high anxiety, those mothers that demonstrated stronger amygdala-STS functional connectivity while listening to their own infant cry sounds were observed to have higher levels of warm and involved caretaking of their own infants. A similar level of amygdala-STS connectivity was not observed in those mothers with lower anxiety scores.

Given these qualifications, it is possible that the primary caretaker fathers in the Abraham et al. (2014) study were more anxious than the primary caretaker mothers, perhaps because they had less parental experience, and that mentalizing STS interactions with the amygdala were needed to down-regulate their anxiety and promote the output of positively valent amygdala neurons. If this is the case, then increases in parental experience in men may both promote parental motivation and decrease anxiety.

Finally, there is evidence that intranasal OT administration improves father–child interactions during play sessions (Naber, van Ijzendoorn, Deschamps, van Engeland, & Bakermans-Kranenburg, 2010). Further, Li, Chen, Mascaro, Haroon, and Rilling (2017) reported that intranasal OT treatment increased BOLD responses in several parental brain regions while fathers viewed pictures of their 1- to 2-year-old children. On the assumption that the fathers in these studies were likely to be secondary caregivers, it is possible that when fathers assume a primary role in childcare, experience-induced increases in endogenous OT improve their paternal behavior, perhaps by increasing their parental motivation and confidence in their role as a primary caretaker.

It can be proposed that extensive father–infant interactions stimulate paternal motivation in men. This is based on the view, as developed in Chapter 7, that parental circuits are likely to be present in the brains of males of all mammalian species in a latent form, and that certain external and internal triggers are necessary for these circuits to become active. The primary trigger in human males may be those sociocultural factors that permit extensive father–child interactions.

This brief review of the paternal brain has emphasized that, in a manner similar to the maternal brain, the paternal brain involves activity in subcortical neural circuits that influence motivation (amygdala and VTA), and in cortical circuits involved in emotional empathy and love (AI). Mentalizing brain regions (STS) and brain areas involved in emotion regulation also appear to be important. In modern societies, fathers are assuming a larger role in child care. This change should foster increased research into the paternal brain in men (see Li et al., 2018).

General Conclusions

Most research on the parental brain in humans has been performed on mothers. The available research indicates that the subcortical circuits that control maternal motivation in nonhuman mammals are also operative during human maternal behavior. Cortical neural circuits interact with these subcortical circuits in human mothers, and this interaction is presumed to regulate the translation of feelings of maternal love and empathy into overt maternal responses. Cortical–subcortical interactions also promote emotion regulation, which allows for effective and competent maternal behavior under demanding environmental conditions. Deficiencies in these neural processes have been associated with less than adequate maternal responsiveness. In such cases, it is possible that intervention strategies that are designed to promote more effective parenting through the use of various types of training procedures might be able to modify brain activity and improve maternal responsiveness to an infant's needs (see Giuliani, Beauchamp, Noll, and Fisher, 2019).

9

Development of the Parental Brain in Nonhuman Mammals

Introduction

There is a large literature on the intergenerational continuity of faulty or abnormal maternal behavior in humans (Assink et al., 2018; Berlin, Appleyard, & Dodge, 2011; Lomanowska, Boivin, Hertzman, & Fleming, 2017; Numan & Insel, 2003): Children who have been abused (emotionally or physically) or neglected (not cared for and protected) by their parents are more likely to become neglectful or abusive parents themselves in comparison to children who have not been maltreated. In a recent meta-analytic review, Assink et al. (2018) concluded that parents who experienced abuse or neglect in their own childhood were three times more likely to abuse or neglect their children in comparison to parents who did not experience abuse or neglect in their childhood. Three aspects of the review by Assink et al. are worth emphasizing. First, these human studies do not resolve the causal mechanisms that may contribute to the intergenerational continuity of faulty maternal behavior. For example, such intergenerational transmission could be due to genetic inheritance, the early adverse effects of poor parenting on the child's brain development, or both. Second, being abused or neglected does not destine one to become an abusive or neglectful parent. Third, parents without a history of being maltreated may still maltreat their own children, indicating that factors in addition to a history of being maltreated as a child can influence the occurrence of faulty parental behavior.

What might be the neural underpinnings that lead a mother to abuse and/or neglect her offspring? Figure 9.1 presents two possibilities. First, the neural circuits that regulate maternal motivation may not develop properly so that a mother is not attracted to her infant(s) and does not find her interactions with them to be rewarding. Second, deficient development of the neural circuits involved in emotion regulation that restrain stress reactivity, anxiety, and fearfulness may result in a mother who cannot care for her infant(s) properly, particularly under challenging and demanding environmental situations or when her infant is being disruptive and is difficult to control. These two possibilities of abnormal brain development are not mutually exclusive in that both processes may be evident in the same mother. It is also likely that other aspects of

The Parental Brain. Michael Numan, Oxford University Press (2020). © Oxford University Press.
DOI: 10.1093/oso/9780190848675.001.0001

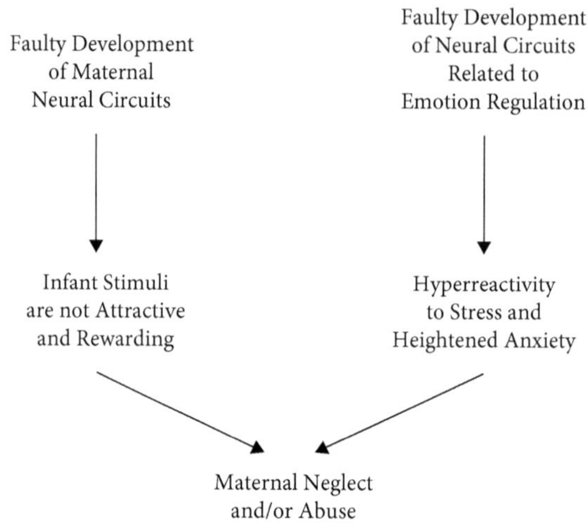

Figure 9.1. Two potential routes through which central neural dysfunctions may lead to abnormal maternal behavior: maternal neglect and/or abuse of her infant(s). First, the neural circuits that underpin maternal motivation may develop abnormally, resulting in a mother who is not attracted to her infant(s) and does not find interactions with them to be rewarding. Second, deficient development of the neural circuits involved in emotion regulation may result in a mother who does not appropriately care for her infant(s) under stressful environmental conditions.

atypical brain development influence the occurrence of faulty maternal responsiveness, such as deficits in the neural systems that regulate attentional processes, so that a mother is easily distracted by irrelevant stimuli that then interferes with caretaking activities (Lovic & Fleming, 2004). The focus of the present chapter, however, will be on the two processes represented in Figure 9.1, since that has been the concern of most of the animal research on this topic.

Figure 9.1 does not indicate whether the proposed alterations in the maternal brain could be due to genetic or experiential influences. In this chapter, which emphasizes research on nonhuman animals, I want to present a significant body of research that emphasizes the role of experience in the intergenerational continuity of maternal responsiveness. Specifically, I will show that the way a mother treats her female offspring influences the daughters' brain development and her subsequent maternal behavior once she becomes an adult and has her own offspring.

Important behavioral findings in nonhuman primates and in rats clearly show that the intergenerational continuity of maternal responsiveness is influenced by the way a mother treats her young. Maestripieri (2005) found that in captive social groups of rhesus monkeys, about 5% to 10% of mothers physically abuse their infants. Maternal abuse, which includes dragging, crushing, throwing, and biting the infant, occurs in short bouts during the first months of the infant's life and alternates with otherwise normal maternal behavior. In some cases, such abuse is severe enough to result in the death of the infant. Also, abusive mothers remain abusive across multiple births, demonstrating that it is a stable characteristic. Maestripieri performed a cross-fostering study utilizing multiparous females, which allowed him to know the maternal abuse status of each mother ahead of time. Female infant rhesus monkeys were raised with a known abusive foster mother (the occurrence of infant abuse was verified) or with a known nonabusive control foster mother. When the cross-fostered infants grew into adulthood, their maternal behavior toward their own offspring was examined. Infants that were born to abusive mothers but were cross-fostered at birth to nonabusive mothers did not abuse their own offspring. In contrast, 50% of infants that were born to nonabusive mothers but were cross-fostered to abusive mothers were observed to abuse their own offspring. This research clearly supports the primacy of an experiential mode of intergenerational transmission of faulty maternal behavior in this population of rhesus monkeys. However, note that 50% of infants raised by abusive foster mothers did not develop an abusive phenotype. In fact, about 50% of rhesus infant females raised by an abusive biological mother also do not develop an abusive phenotype (Maestripieri, 2005). Gene by environment interactions (G × E) might be involved in these outcomes: Infants with particular genotypes may be more or less susceptible to the negative impact of abuse on the development of their maternal behavior (Kinnally, Ceniceros, & Martinez, 2018).

In a related study, Maestripieri, Lindell, and Higley (2007) indicate that the maternal behavior of rhesus monkeys also differs on the dimension of rejection, which measures variation in the extent to which the mother permits infant contact through suckling, carrying, and holding. Mothers showing high levels of rejection are low-contact mothers. In a cross-fostering study, it was found that the rates of maternal rejection of adult cross-fostered females toward their own infants was positively correlated with the rate of rejection that they received from their foster mother. Similar findings have been reported by Kinnally et al. (2018) both for the intergenerational transmission of maternal rejection and for the intergenerational transmission of maternal protectiveness (which is another dimension of maternal behavior in rhesus monkeys—the mother restricts the movement of her infant away from her).

The other major behavioral finding supporting an experiential basis for the intergenerational continuity of maternal responsiveness comes from research on normal variations in the maternal behavior of rats. Maternal rats demonstrate individual differences in the duration in which they lick/groom (LG) their pups (Champagne, Francis, Mar, & Meaney, 2003). High LG mothers spend more time engaging in this pup-directed maternal response during the first post-partum week than do their low LG counterparts. Note that unlike the previously described research on abusive rhesus monkeys, this research on rodent maternal behavior does not involve dysfunctional parenting. All these females, irrespective of their LG phenotype, remain in contact with and nurse their young for similar amounts of time and raise their young to weaning. High LG mothers, however, might be viewed as being more attentive to their infants than low LG mothers.

Significantly, there is an intergenerational continuity of these behavioral differences in LG, since the daughters of high LG mothers display high LG toward their own offspring in adulthood, and the daughters of low LG dams become low LG mothers (Champagne et al., 2003). Importantly, cross-fostering studies demonstrate that this intergenerational continuity is experientially transmitted. Female rat pups born to high LG mothers but cross-fostered to low LG mothers become low LG mothers in adulthood, and females born to low LG dams but cross-fostered to high LG dams demonstrate the high LG phenotype during their adult postpartum period (Champagne et al., 2003; Francis, Diorio, Liu, & Meaney, 1999).

Although one might view these differences in rodent maternal responsiveness as subtle, I will show that they influence the socioemotional development of the affected offspring. Also consider that if relatively nuanced differences in the display of normal maternal behavior can impact the socioemotional development of offspring, then more drastic maternal treatment effects involving abuse and neglect should have an even greater impact.

Normal Variations in Maternal Licking/Grooming of Pups Affect the Development of the MPOA-to-VTA-to-NA Circuit in Rodent Offspring

Figure 9.1 proposes that if maternal brain circuits were to be altered during development, then this would affect maternal motivation by influencing the attractive and rewarding value of infant stimuli. In the previous section, I presented evidence that the way a mother treats her offspring can influence the development of maternal behavior in her young. In this section, I review evidence that the way a mother treats her offspring can influence the development of the medial preoptic area (MPOA)-to-ventral tegmental area (VTA)-to-nucleus accumbens (NA) circuit, which would be an avenue through which maternal motivation could be affected. This research will provide evidence that high LG mothers are more attentive to their young than low LG mothers because pup stimuli more easily activate MPOA-to-VTA-to-NA circuits in high LG mothers compared to their low LG counterparts. Importantly, this neural phenotype is experientially transmitted to the mother's offspring, which allows female pups that are raised by high LG mothers to be more attentive to their own offspring.

In Chapter 5, I reviewed the evidence that MPOA activation of DA release into NA is important for the appetitive pup-directed aspects of maternal behavior. Champagne et al. (2004) found that high LG lactating rats have a greater release of dopamine (DA) into NA than do low LG postpartum females during mother–infant interactions and that this DA release precedes and predicts a bout of licking and grooming. When a DA reuptake inhibitor was systemically administered to low LG dams, DA release into NA was increased (due to decreased reuptake), which increased the level of licking and grooming of pups in these females to levels comparable to that of naturally high LG mothers. This research supports the view that the normally observed differences in LG among postpartum rats may represent natural variations in appetitive maternal motivation as reflected by DA release into NA. Furthermore, high LG postpartum mothers also have more estrogen receptors (ER-alpha; Champagne, Weaver, Diorio, Sharma, & Meaney, 2003) and oxytocin receptors (OTRs; Champagne, Diorio, Sharma, & Meaney, 2001; Francis, Champagne & Meaney, 2000) in the MPOA, and D1 DA receptors in NA (Champagne et al., 2004) than do low LG mothers. Since estradiol binding to ER-alpha ultimately increases the expression of OTRs in MPOA (see Chapter 5, Figure 5.4, of this volume), the following sequence of events is possible: At parturition, increasing levels of estradiol activate the expression of OTRs in MPOA of high LG mothers to a greater extent than they do in low LG mothers (Champagne et al., 2001). This increased expression of OTRs in MPOA during the postpartum period of high LG mothers allows pup stimuli, which presumably cause oxytocin (OT) to be released into MPOA, to more effectively activate the MPOA-to-VTA-DA connection, resulting in a greater release of DA onto D1 receptors in NAs, which then stimulates enhanced maternal motivation as evidenced by increased levels of maternal licking and grooming. There is some experimental evidence that supports this proposal: intracerebroventricular (ICV) administration of an OTR antagonist (OTA) depresses the licking and grooming levels of day 3 postpartum high LG dams to levels shown by low LG mothers (Champagne et al., 2001). However, since OTA was injected ICV, the involvement of OT action on OTRs in MPOA is not conclusive. It would be important to demonstrate that direct application of OTA to MPOA would decrease licking and grooming in high LG dams.

Since adequate maternal behavior occurs in low LG mothers, the proposed action of OT on OTRs in MPOA, which potentially enhances the functional connectivity between MPOA and VTA-DA neurons, is modulatory in nature and is not essential for the maintenance of maternal motivation. Therefore, these research findings are not related to the presence versus the absence of maternal motivation, but instead are related to variations in maternal motivation within the normal range, as indicated by the observed levels of maternal licking and grooming of pups.

With respect to the intergenerational continuity of maternal responsiveness, I have already reviewed that pups raised by biological or foster high or low LG females grow up to express the behavioral phenotype of the mother that raised them. The question posed here is whether the neural differences expressed by high and low LG mothers are also transmitted to their offspring. Table 9.1 shows findings that answer this important question in the affirmative.

The most definitive findings are with respect to the expression of ER-alpha in MPOA (Champagne et al., 2006; Pena, Neugut, & Champagne, 2013). Female rat pups raised by biological or foster mothers that provide them with high levels of LG during the early postpartum period grow up to show high LG toward their own offspring and such females have higher levels of ER-alpha in MPOA than do females that were reared by low LG mothers, with the latter females

Table 9.1 Maternal Treatment Effects on the Development of Medial Preoptic Area-to-Ventral Tegmental Area-Dopamine Neural System in the Affected Female Rat Offspring

Phenotype of adult female rat offspring	Type of maternal rearing condition			
	Raised by biological mother		Cross fostered condition	
	HLG	LLG	HLG-to-LLG	LLG-to-HLG
Maternal behavior	HLG	LLG	LLG	HLG
ER-alpha expression in MPOA	High	Low	Low	High
OTR expression in MPOA	High	Low	—	—
TH neurons in VTA	High	Low	—	—

A dash under a particular condition means that no data is available for that condition.
Abbreviations: ER = estrogen receptor; HLG = high licking/grooming of neonates by mother; LLG = low licking/grooming of neonates by mother; HLG-to-LLG = offspring born to HLG mothers but cross-fostered to LLG mothers; LLG-to-HLG = offspring born to LLG mothers cross-fostered to HLG mothers; OTR = oxytocin receptor; TH = tyrosine hydroxylase.

The data in this table were derived from multiple research reports from Champagne and her colleagues. See text for references.

displaying lower levels of LG toward their offspring. Other supportive evidence, that did not include cross-fostered control groups, shows that female offspring raised by biological high LG mothers have more OTRs in MPOA (Champagne & Meaney, 2007) and tyrosine hydroxylase immunoreactive (TH-ir) neurons in VTA (Pena, Neugut, Calarco, & Champagne, 2014) than do females raised by biological low LG mothers. Tyrosine hydroxylase is an enzyme involved in DA synthesis and therefore the number of TH-ir neurons in VTA is representative of the number of VTA-DA neurons. These results strongly suggest that the way a mother, whether biological or foster, treats her young pups affects the development of ER-alpha in MPOA. This basic effect then appears to drive the other observed neural phenotypes. High levels of LG result in higher levels of MPOA ER-alpha, which then allows estradiol to induce a higher expression of OTRs within MPOA. These dual outcomes presumably promote enhanced functional, and perhaps also structural, connectivity between MPOA and VTA-DA neurons, which may, in turn, promote an enhanced development of the number of VTA-DA neurons (Pena & Champagne, 2015). These combined effects are proposed to enhance aspects of maternal motivation in the affected offspring by promoting the development of a more effective MPOA-to-VTA-DA-to-NA circuit.

In addition to differences in postpartum LG behavior, is there any other evidence that suggests that maternal motivation is enhanced in female rats that are raised by high LG mothers? Champagne et al. (2001) have reported that virgin female rat sensitization latencies are shorter in females that were raised by high LG mothers than in virgins that were raised by low LG mothers (also see Pena & Champagne, 2015). These results suggest that enhanced functional connectivity between MPOA and VTA may promote aspects of maternal motivation even in nulliparous females and therefore outside the postpartum period.

What are the mechanisms through which maternal licking and grooming influences the development of ER-alpha containing MPOA neurons? Champagne (2008) has reviewed the research that demonstrates the involvement of epigenetic mechanisms. To understand epigenetic processes, one needs to understand the mechanisms that regulate gene transcription. (For a detailed primer of basic genetic and epigenetic molecular mechanisms, see Numan, 2015; Stolzenberg, Grant, & Bekiranov, 2011.) Briefly, the specific nucleotide sequence contained within the DNA of a particular gene controls the synthesis of a particular protein within a cell. For example, the oxytocin receptor gene (OXTR) controls the synthesis of OTRs. A gene can be divided into two major parts, a regulatory region (promoter region) and a coding region. Various transcription factors can bind to regulatory regions within a gene, and these factors can then stimulate gene transcription through activation of the gene's coding region. For example, Figure 5.4 in Chapter 5 shows that when estradiol binds to ER-alpha, the E–ER complex serves as a transcription factor that stimulates the transcription of the OXTR gene. Transcription involves the activation of the DNA coding region, which then transcribes a specific messenger ribonucleic acid (mRNA). The nucleotide sequence of mRNA is then translated, at the level of the cell's ribosomes, into the amino acid sequence that makes up a particular protein. Figure 9.2 depicts

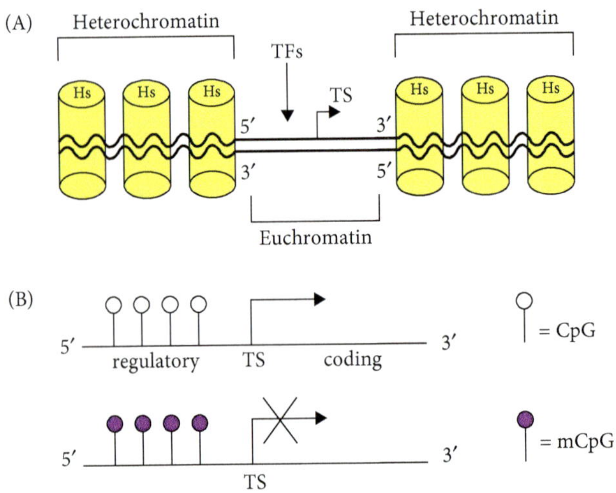

Figure 9.2. Basic epigenetic processes. (A) The distinction between heterochromatin and euchromatin. Chromatin refers to the DNA and proteins that make up chromosomes. In the heterochromatin condition, DNA is tightly wound around a group of histone proteins (Hs), which prevents the ability of transcription factors (TFs) and RNA polymerase to bind to regulatory/promoter regions located within the genes of the heterochromatin segment. Heterochromatin, therefore, contains genes that are inactive, in that their transcription to mRNA is suppressed. Euchromatin represents the areas of a chromosome that contain genes that are in an active or open state: DNA is separated from Hs, which allows the exposed genes to be transcribed. (B) The inhibitory effects of DNA methylation on gene transcription. DNA methylation refers to the addition of a methyl group onto cytosine bases within DNA. Methylation is facilitated when a cytosine base is followed by guanine (CpG site) in a particular nucleotide sequence. The figure distinguishes unmethylated (CpG) from methylated (mCpG) cytosine bases. The *p* in CpG refers to the phosphate linkage between the two nucleotides within a DNA strand. DNA methylation with the regulatory/promoter region of a gene can suppress gene transcription in two ways: (a) by preventing the binding of TFs to these sites and (b) by attracting histone deacetylases that remove acetyl groups from histones, which favors a closed heterochromatin state. The X across the arrow at the transcription start site (TS) indicates that transcription at the coding region of a particular gene is suppressed.

Source: Reprinted from Figure 2.6 in Numan (2015) with permission from Elsevier.

a chromosome segment, within a nucleus of a cell, made up of DNA and associated histone proteins. A double-stranded DNA segment can be composed of heterochromatin and euchromatin. When DNA is tightly wound around histone proteins, it is in a closed heterochromatin state that makes it difficult for transcription factors to stimulate gene transcription. When DNA is in a euchromatin state, it separates from histones and is in an open state that allows transcription factors to stimulate gene expression (transcription). Epigenetic processes refer to those molecular mechanisms that shift a gene between these closed and open states, which would therefore either suppress or promote the synthesis of particular proteins. Several factors are involved in shifting DNA strands between open and closed states. One factor is the degree to which cytosine bases within a DNA segment are methylated. DNA methylation tends to promote a closed heterochromatin state, in this way inhibiting or depressing gene transcription and protein synthesis.

With respect to epigenetic processes, Champagne et al. (2006) found that adult female rats that were raised by either biological or foster low LG dams had lower expression of ER-alpha in MPOA compared to females raised by high LG mothers. This difference in ER-alpha expression was correlated with differences in DNA methylation within the promoter region of the ER-alpha gene (referred to as the Esr1 gene) within cells from the MPOA. Females raised by low LG mothers exhibited greater DNA methylation, which would account for decreases in the synthesis of ER-alpha within MPOA.

Putting all of these findings together, the likely sequence of events underlying the experience-based intergenerational continuity of normal variations in the maternal behavior of rats can be described as follows. Female neonates that receive more maternal attention (as measured by LG level) during the first week postpartum develop low levels of DNA methylation within the promoter region of the Esr1 gene in MPOA neurons (Pena et al., 2013), which results in the greater expression of ER-alpha within MPOA neurons in comparison to females that receive lower levels of maternal attention. This epigenetic process appears to be the first developmental step in the previously described cascade of events that ultimately leads to enhanced connectivity across the MPOA-to-VTA-to-NA circuit, which, in turn, allows maternal treatment effects to modulate the maternal motivation of the affected offspring.

Although I have been describing maternal treatment effects on the development of maternal circuits in female rats that fall within the normal range of maternal motivation, the implications of these findings are immense. In particular, it is easy to conceive that extreme forms of maternal abuse and/or neglect might completely disrupt the development of maternal motivational neural circuits in the affected infants so that when such infants grow up and have their young, their maternal responsiveness might be severely compromised.

Normal Variations in Maternal Licking/Grooming of Pups Affect the Development of Neural Circuits That Regulate Stress Reactivity, Fearfulness, and Anxiety in Rodent Offspring

In relation to Figure 9.1, there is also excellent evidence that the manner in which a mother treats her offspring can influence their emotional development by affecting the way neural circuits that regulate stress reactivity, fearfulness, and anxiety develop. In this section I will focus on research that shows that the level of maternal licking/grooming of her offspring influences the development of such circuits in the affected infants.

Several studies have shown that the adult female offspring of high LG rat mothers are less fearful/anxious than are females raised by low LG mothers, and cross-fostering studies show that these differences in fearfulness are mediated by the rearing environment (maternal treatment effects during the early postnatal period) rather than inherited genetically from the biological mother (Francis, Diorio, Liu, & Meaney, 1999; Pan, Fleming, Lawson, Jenkins, & McGowan, 2014; Uriarte, Breigeiron, Benetti, Rosa, & Lucion, 2007). In these studies, anxiety/fear-related behavior was tested during adulthood with the open field test. In this test, females are placed in a novel arena, and the latency to enter the center (less protected part) of the arena and the amount of time spent in the central part are used as measures of anxiety. Adult virgin female rats raised by high LG mothers enter the center sooner and spend more time in the center of the field, indicating lower levels of anxiety. For example, during a 5-minute test, females raised by high LG mothers spend about 35 to 40 seconds in the center of the field, while females raised by low LG mothers only spend about 10 to 15 seconds in the center of the field (Francis et al., 1999).

The purpose of my analysis in this section is to examine whether such emotionality differences not only occur in adult nonlactating females (as previously shown), but also whether these differences in fearfulness persist into the postpartum period. Such a developmental process would create a mechanism through which maternal treatment effects could influence the emotionality of the mother's adult offspring during their postpartum period, which would be a potential mechanism for an experience-based intergenerational continuity of aspects of maternal behavior. In this regard, Francis, Champagne, and Meaney (2000) have shown that the adult postpartum offspring of low LG rat mothers are more anxious, as measured in the open field test, than are the adult postpartum offspring of high LG mothers.

I would like to offer the following proposal: The amount of LG that a female neonatal rat receives from its mother during the early postpartum period affects the development of two neural systems. First, as described in the previous section, maternal LG affects the development of the MPOA-to-VTA-to-NA circuit, and the evidence reviewed indicates that variation in the development of this maternal motivational system is the likely mechanism through which the LG phenotype is transferred across generations from the mother to her female offspring. Second, as the behavioral evidence just described indicates, the level of maternal licking and grooming received by a female neonate also affects the development of emotionality in the affected infants. Fearful mothers that show low LG produce fearful offspring. Such an intergenerational transfer of emotionality may affect the maternal behavior of mothers under stressful or demanding environmental conditions, with high stress reactive mothers showing less competent maternal behavior than their less fearful counterparts (see Chapter 6 of this volume). This proposal is summarized in Figure 9.3.

What are the underlying neural mechanisms that are influenced by the level of LG that a neonatal rat receives from its mother to affect the development of fear/anxiety/stress reactivity? First, adult offspring that have received low levels of LG from their mothers during the early postnatal period show an increased hypothalamic-pituitary-adrenal (HPA) physiological stress response (see Chapter 6 of this volume) in comparison to their high LG counterparts: In response to restraint stress, more corticosterone is released from the adrenal cortex into the blood of adults that received low LG in comparison to those that received high LG (Liu et al., 1997). These results were obtained from adult male offspring, and the extent to which these physiological stress reactivity effects also apply to females is an open question (see Pan et al., 2014). Importantly, adult offspring that received less licking and grooming also have lower levels of glucocorticoid receptors (GRs) in their hippocampus. The action of corticosterone on GRs in the hippocampus is part of a negative feedback loop that activates neural systems that inhibit corticotropin-releasing factor (CRF) release from the paraventricular nucleus (PVN; Ulrich-Lai & Herman, 2009). Therefore, when a stressful stimulus activates the HPA axis, the increased corticosterone levels act on the hippocampus to suppress further CRF release into the anterior pituitary, which suppresses further adrenocorticotropic hormone release, and this, in turn, depresses the continued release of corticosterone (see Chapter 6 of this volume). If low levels of licking and grooming suppress the development of hippocampal GRs, this would result in a greater physiological stress response because of decreased negative feedback.

How does the level of LG that a neonatal rat receives affect the development of hippocampal GRs? Evidence indicates that epigenetic processes are involved (Weaver et al., 2004). Low levels of LG, whether from a biological or foster mother, are associated with increased levels of DNA cytosine methylation in the promoter region of the GR gene within hippocampal cells of offspring, while the opposite is true for neonates that receive high LG. Interestingly, on the day after

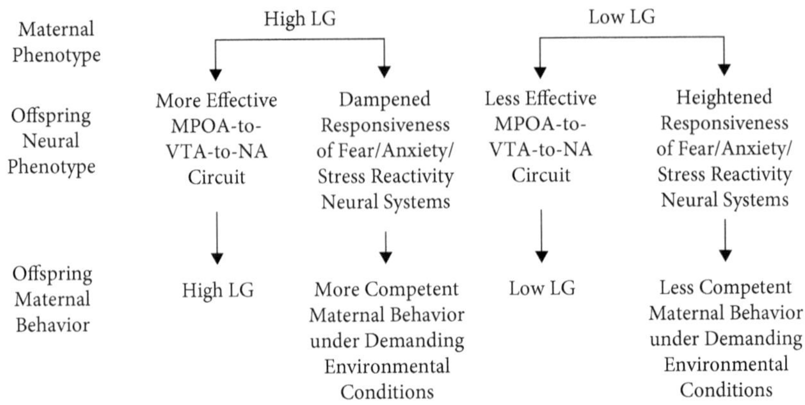

Figure 9.3. The amount of licking/grooming (LG) that a female neonatal rat receives from its mother has been proposed to affect the development of two neural systems which, in turn, influences the maternal behavior of the affected infants when they become adults and have their own offspring. First, exposure of neonatal females to high levels of LG results in the development of a more effective medial preoptic area (MPOA)-to-ventral tegmental area (VTA)-to nucleus accumbens (NA) circuit in these females. The enhanced development of this circuit results in females that also direct high levels of LG to their own offspring when they become adults. Exposure of neonatal females to low levels of LG produces the opposite effect: a less effective MPOA-to-VTA-to-NA circuit and the expression of lower levels of LG in the affected offspring when they become adults and have their own offspring. Second, exposure of neonatal females to high LG results in the development of a dampened responsiveness of fear, anxiety, and stress reactivity neural systems to threatening/stressful stimuli in the affected offspring. In adulthood, such females are likely to show more competent maternal behavior toward their offspring under challenging/demanding environmental conditions. In contrast, exposure of neonatal females to low LG results in the development of heightened responsiveness of fear, anxiety, and stress reactivity neural systems to threatening/stressful stimuli. In adulthood, such females are predicted to display less competent maternal behavior toward their own offspring under challenging/demanding environmental conditions.

birth (postnatal day 1), the degree of DNA cytosine methylation is high in all offspring. However, on postnatal days 6, 21, and 90, the degree of methylation of the GR receptor gene within hippocampal cells is decreased in the offspring of high LG mothers, but not in the offspring of low LG mothers (Weaver et al., 2004). These results suggest that increased levels of LG during the first postnatal week demethylate the infant's GR gene within the hippocampus, which increases gene transcription and the subsequent expression of GRs within hippocampal neurons. This epigenetic effect occurs early in life and is maintained into adulthood, which results in the adult offspring of high LG mothers having a decreased physiological stress response, while the offspring of low LG mothers have an enhanced stress response. In further support, Weaver et al. (2004) treated adult rats that were raised by low LG mothers with a drug that causes demethylation of DNA. This drug treatment, administered intracerebroventricularly, decreased DNA cytosine methylation within the promoter region of the GR gene, which was then associated with increased hippocampal GR expression and a decreased corticosterone response to restraint stress. These rats expressed characteristics similar to those of rats raised by high LG mothers.

The research I have just described involves a neuroendocrine stress response, but how can we relate these findings to the increased behavioral stress reactivity and fearfulness of adult rats that were reared by low LG mothers? A review of the research presented in Chapter 6 will help the reader understand what I am about to propose. First, recall not only that CRF activates adrenocorticotropic hormone release from the anterior pituitary, but that its release into the brain as a neurotransmitter/neuromodulator also exerts anxiogenic effects. Some of these effects could be due to the central (as opposed to the anterior pituitary) projections of CRF-containing PVN neurons (see Chapter 6 of this volume). Another mechanism that relates the enhanced physiological stress response of rats that were raised by low LG mothers to the enhanced anxiety-related behavior of these offspring involves the anxiety-producing effects of CRF neurons within the central nucleus of the amygdala (CeA). Recall that corticosterone acts on GRs in CeA to stimulate the synthesis of CRF. This effect increases anxiety, as measured in the elevated plus maze, as shown in adult rats with implants of corticosterone in CeA (Shepard et al., 2000). Interestingly, there is some evidence that postnatal LG may affect the development of GRs in the hippocampus but not in the CeA. Epigenetic influences on GR expression may, in certain instances, be tissue-specific (Meaney et al., 1985). These results suggest that low levels of LG may decrease the development of GRs in the hippocampus of offspring without affecting GR expression in CeA. Given this possibility, the enhanced corticosterone release in response to stress in adults born to low LG mothers may stimulate greater synthesis and release of CRF from CeA neurons. Higher than normal levels of CRF in CeA may result in increased levels of anxiety-related behaviors in response to stressful/threatening stimuli in these offspring in comparison to their counterparts that received high LG during their early postnatal development. Based on the data reviewed in Chapter 6, perhaps CeA-CRF projections to the dorsal BST are involved in this effect (see Figure 6.5). With respect to this proposal, the following findings are supportive. Neonatal rats that receive lower levels of LG during the early postnatal period have increased expression of CRF in PVN and CeA in adulthood than do their counterparts that received higher levels of LG (Francis et al., 1999; Francis & Meaney, 1999). CRF anxiogenic neural systems are upregulated in rats that received less maternal attention during their early development.

In Chapter 6, I described the opposing roles of OT and CRF neural systems with respect to the regulation of fearfulness and anxiety, and I reviewed the evidence that OT action on OTRs in CeA exerts anxiolytic effects. Significantly, adult female rats that have been raised by low LG mothers exhibit lower OTR binding sites in CeA than do females that were raised by high LG mothers, and these differences are observed in both nonlactating and lactating postpartum rats (Champagne & Meaney, 2006; Francis et al., 2000). Since OTR expression in CeA is not regulated by estradiol (see Chapter 4 of this volume), the mechanisms through which maternal treatment effects influence the development of OTRs in CeA of rats remains to be determined. Importantly, recent research on prairie voles indicates that lower levels of normal maternal care are associated with increased methylation of the OXTR gene in the affected offspring, which is associated with decreased OTRs in the nucleus accumbens (Perkeybile et al., 2019). Perhaps this mechanism also applies to the effects of maternal care on the development of OTRs in the CeA of the mother's offspring in rats. In partial support of this idea, McCoy et al. (2019) have recently reported that the degree of overall DNA methylation within amygdala neurons of rats is positively correlated with their anxiety-related behavior. (Interestingly, decreased expression of OTRs in NA of prairie vole offspring that receive lower levels of maternal care is likely to influence the development of the neural circuits that regulate parental motivation, rather than those that regulate fearfulness, anxiety, and stress reactivity.)

These results suggest that the increased anxiety/stress reactivity phenotype of adult rats that received low LG during their early development are the result, in part, of a downregulation of the anxiolytic properties of OT and an upregulation of the anxiogenic effects of central CRF systems.

How might the upregulation of anxiety affect the maternal behavior of females that received low levels of LG during their early development? Figure 9.3 proposes that such increases in stress reactivity and anxiety might affect maternal competence under challenging or stressful environmental conditions. To the best of my knowledge, this idea has not been fully examined, although there is some supportive evidence. First, maternal aggression is slightly decreased in low LG postpartum female rats in comparison to their high LG counterparts (Ruthschilling et al., 2012). Further supportive evidence comes from a study by Padoin, Cadore, Gomes, Barros, and Lucion (2001). These researchers exposed neonatal rat pups of both sexes to a brief handling procedure (neonatal handling) from the first to the 10th day after parturition. Pups were removed from their mother and handled by the experimenter for a few minutes and then returned to their mother on each day. Research by others has shown that similar brief neonatal handling procedures result in the mother showing enhanced LG of her pups upon their return (Liu et al., 1997; Numan & Insel, 2003), although Padoin et al. did not measure maternal LG behavior after the pups were returned to their dams. In adulthood, Padoin et al. found that pups of both sexes exposed to brief neonatal handling, in comparison to pups that were not handled, displayed decreased anxiety-related behavior in an open field test. The females that were tested in the open field test were nonlactating females. In a separate experiment, additional groups of such briefly handled and nonhandled neonatal females were mated in adulthood and tested for maternal aggression toward a male intruder on day 7 postpartum. The briefly handled females displayed more maternal aggression than did the nonhandled control females. These results imply that exposure of pups to higher levels of LG decreases their adult anxiety-related behaviors and stress reactivity, which then enhances their maternal protective responses during a challenging test with a male intruder. Note, however, that this study found modest differences in maternal aggression and that effective maternal aggressive responses occurred in all females.

This general issue is an important area for future investigation. However, it may be the case that the differences in anxiety/stress reactivity development that result from normal variations in maternal LG may not be extreme enough to severely disrupt the subsequent development of maternal competence under challenging environmental conditions. Normal variations in maternal behavior may produce offspring with variations in emotional temperament that fall within a normal range, allowing for generally adaptive maternal responsiveness. Perhaps a more extreme upregulation of anxiety-related neural systems, such as might arise from maternal abuse and/or neglect, would be needed to seriously compromise maternal behavior under such conditions. In support, the evidence reviewed in Chapter 6 showed that experimentally induced hyperactivation of central CRF systems not only disrupts maternal aggression but also disrupts infant-directed maternal behavior.

The Effects of Various Forms of Maternal Neglect of a Young Infant on the Subsequent Development of Maternal Behavior and Its Associated Neural Systems in the Affected Infant

Research on Macaque Monkeys

It is likely that severe forms of abnormal maternal care that fall outside the range of normal variations in maternal behavior would have more drastic effects on the development of maternal behavior in the affected female infants. The early and important research of Harlow and his colleagues (Ruppenthal, Arling, Harlow, Sackett, & Suomi, 1976) on the effects of various forms of maternal deprivation on the development of maternal behavior in the affected female infants is instructive in this regard. In this analysis, I will compare the adult maternal behavior of female rhesus monkeys that have been either mother-reared (MR) or peer-reared (PR). MR monkeys are raised in a laboratory setting within social groups composed of several mother–infant pairs. These offspring interact with both their mother and their peers during their development. PR monkeys, in contrast, are separated from their mothers near the time of birth and reared by humans in a nursery for several weeks, after which they are placed in peer groups with other PR monkeys until 6 months of age. Following this, they are placed in larger social groups. A comparison of the adult maternal behavior of females reared under these two conditions provides some information on the importance of the infant's mother for the development of the infant's maternal responsiveness in adulthood, which could shed light on the involvement of maternal deprivation (neglect) on the experience-based intergenerational transmission of faulty maternal behavior. The research reviewed by Ruppenthal et al. (1976) indicated that greater than 95% of MR female rhesus monkeys showed adequate maternal behavior toward their own offspring when they became mothers in adulthood. In contrast, 25% of PR female rhesus monkeys were abusive and showed high rates of maternal rejection. Therefore, under laboratory conditions, although the maternal behavior of PR monkeys is not eliminated, a significant proportion of such females show abnormal maternal behavior as evidenced by increases in abusive and rejection responses toward their infants.

Research has also shown that various forms of maternal neglect in macaque monkeys are associated with decreased levels of OT and increased levels of CRF in the cerebrospinal fluid (CSF), and decreased expression of OTRs in the brains of the affected offspring (Baker et al., 2017; Coplan et al., 1996; Winslow, Noble, Lyons, Sterk, & Insel, 2003). The disruption of maternal behavior in PR monkeys could be due to increased anxiety and stress-reactivity resulting from enhanced CRF release; deficient development of OT neural systems might disrupt maternal neural circuits and enhance neural activity within fear/anxiety neural circuits. There is additional behavioral evidence that macaque monkeys that are subjected to maternal deprivation, neglect, or abuse during early development are more stress reactive and fearful and more subordinate and solitary and display lower levels of prosocial interactions with other monkeys (Baker et al., 2017; Coplan et al., 1996; Corcoran et al., 2012; Feng et al., 2011; Sanchez, McCormack, & Howell, 2015; Winslow et al., 2003).

Rodent Models on the Effects of Abnormal Maternal Caretaking on the Development of Maternal Behavior in the Affected Offspring

Fleming and her colleagues (Gonzalez, Lovic, Ward, Wainright, & Fleming, 2001) have developed a model of artificial rearing (AR) of rat pups that is analogous to the PR condition utilized in macaque monkeys. AR female rat pups were separated from their mothers on day 4 of life and raised in complete social isolation until 21 days of age, being fed through a gastric tube during this isolation period. Control MR female pups were raised normally by their mothers until 21 days of age (MR pups received a sham gastric tube implant). At 21 days of age, all females were grouped with a same-sex conspecific until adulthood. AR and MR females were then mated and their maternal behavior toward their own pups was examined. A significant disruption of maternal behavior in AR females, in comparison to MR controls, was observed. AR mothers showed lower levels of nursing behavior, retrieving, and LG of their pups compared to MR pups. Although AR did not eliminate maternal responsiveness, it significantly decreased several aspects of maternal behavior. It is interesting to compare the maternal behavior of AR females with females that receive low LG during development. Maternal behavior is decreased to a greater extent in AR females, indicating that maternal deprivation causes a greater interference with aspects of maternal behavior than does exposure to low levels of licking and grooming.

To provide evidence for an intergenerational continuity of the observed decreases in maternal behavior in AR mothers, Gonzalez et al. (2001) also found that the daughters of AR mothers exhibited lower levels of nursing behavior and licking/grooming of their pups in comparison to the daughters of MR pups.

Gonzalez et al. (2001) examined whether differences in emotionality might have influenced the differences in maternal behavior observed in AR and MR females. MR and AR females were tested in an open field test after their pups were weaned. Surprisingly, the data showed that the AR-reared females were more active and less emotional in the open field. Although it would have been more appropriate to perform this test during the postpartum period, this finding suggests that increases in anxiety did not contribute to the maternal deficits in the AR females. However, these authors did provide evidence that AR postpartum mothers were more distracted by extraneous stimuli, leaving their nest often to explore other parts of the cage. Perhaps this finding is indicative of increased reactivity and vigilance to relatively innocuous stimuli.

If increased fear and anxiety did not contribute to the maternal behavior deficits in AR mothers, then perhaps AR mothers did not have a fully developed maternal circuit. There is some evidence that supports this view. Afonso, King, Novalov, Burton, and Fleming (2011) found that AR-reared mothers had reduced pup-stimulated dopamine elevations in the nucleus accumbens than did MR dams. One interpretation of these results is that as a result of maternal deprivation, the neural connectivity between the MPOA and the mesolimbic DA system is not fully developed. Levy, Melo, Galef, Madden, and Fleming (2003) found that the formation of maternal memory (see Chapter 5 of this volume) is also disrupted in AR mothers. AR and MR mothers were allowed to interact with their pups for 24 hours after birth and then were separated from their pups. There were no differences in maternal behavior between the two groups during this initial 24-hour period (major differences in maternal behavior between AR and MR mothers typically emerge between days 4 to 10 postpartum, presumably after the effects of hormonal stimulation have waned; cf. Novakov & Fleming, 2005). Two weeks later, during a retention test for maternal memory, all females were tested with foster pups, and their sensitization latencies were measured. Pup-stimulated maternal behavior occurred after 1 day of pup exposure in the MR females, but the latency to onset of maternal behavior in AR females was 5 days, which approaches the latency typically observed in naïve virgin females.

Another rodent model that has been used to explore the effects of maternal deprivation/neglect on the development of maternal behavior in the affected offspring is the use of prolonged maternal separation (PMS). The PMS model typically involves separating pups from their mother for 3 to 6 hours per day for about the first 2 weeks postpartum (the pups are placed in an incubator to keep them warm during the separation period). The PMS procedure is less severe that the AR procedure, since the latter, but not the former, results in complete maternal deprivation. The pups subjected to PMS are typically weaned at 21 days of age and behavioral tests are performed on them in adulthood.

Lovic, Gonzalez, and Fleming (2001) found that such a PMS procedure (5 hours/day) disrupted the subsequent maternal behavior of the affected female offspring: Their nursing behavior and LG of their pups occurred at lower levels than that observed in control females. There is some evidence that the disruption of maternal behavior by PMS might be due to enhanced anxiety and stress reactivity of the affected offspring in adulthood. Caldji, Francis, Sharma, Plotsky, and Meaney (2000) found that PMS results in adult rats that show increased anxiety-related behaviors on a variety of tests. Plotsky et al. (2005) reported that exposure to PMS increased CRF levels in the CeA, PVN, and the dorsal part of the bed nucleus of the stria terminalis of adult rats in comparison to controls. A cautionary note, however, is that this research was performed on male, but not female, rats. However, Chen et al. (2012) found that PMS resulted in adult female rats that have an enhanced corticosterone response to restraint stress in comparison to control females, and Uchida et al. (2010) reported both an enhanced corticosterone response to restraint stress and an increase in anxiety-related behavior in adult male and female rats that were subjected to PMS during the neonatal period. Another relevant study on female rats was performed by Boccia and Pedersen (2001). These researchers subjected neonatal female rats to either PMS (3 hours/day) or to brief 15-minutes separations on each day over the first 2 postnatal weeks. In adulthood, the females were mated and their pup-directed maternal behavior, maternal aggression toward a male intruder, and anxiety-related behavior in the elevated plus maze was examined. The females exposed to PMS during the neonatal period showed lower levels of LG toward their own pups in adulthood, lower levels of maternal aggression, and increased anxiety in the elevated plus maze. The only problem with interpreting this study is that it is not clear whether the observed effects are related to the anxiety-promoting effects of PMS, the anxiety-reducing effects of brief maternal separations (see discussion of Padoin et al., 2001, in the previous section on normal variations in maternal licking/grooming of pups affect the development of neural circuits that regulate stress reactivity, fearfulness, and anxiety in rodent offspring), or both.

Surprisingly, with respect to anxiety, only a few studies have examined whether PMS results in epigenetic effects that influence the methylation status within the promoter region of the GR gene in the hippocampus of the affected rodent offspring, and the research that does exist has presented conflicting results (Daniels et al., 2009; Zhu et al., 2017). There were several methodological differences between these two studies. Relevantly, the duration of each daily maternal separation period and the number of days on which these separations occurred were longer in the Zhu et al. study, and it was this study that found that PMS decreased the expression of GRs in the hippocampus and this decrease was associated with increased levels of methylation within the promoter region of the GR gene.

Is there any evidence that PMS in rodents might interfere with the development of maternal behavior because it affects the development of maternal circuits? Some support is provided by the findings of Zhu et al. (2010). PMS resulted in adult female and male rats expressing lower levels of D1 and D2 DA receptor mRNA levels in the nucleus accumbens. Veenema, Bredewold, and Neumann (2007) have found in mice that neonatal PMS is associated with lower levels of OT immunoreactivity in the PVN of adult lactating females in comparison to controls that were not subjected to PMS. Clearly, more research needs to be done on this important question.

Nephew and his colleagues have developed a novel, and perhaps more ecologically valid, model of the effects of maternal neglect on the development of maternal behavior in the neglectful mother's female offspring (Carini & Nephew, 2013). In this model, maternal postpartum rats are subjected to chronic social stress (CSS) by exposing them to an adult male intruder 1 hour per day on days 2 to 16 postpartum. Because of the frequent engagement in maternal aggression, the pups of the mothers that are exposed to CSS are left unattended for prolonged periods during each daily exposure to the intruder male. In addition, during the periods when the male is not present in the cage, the maternal behavior of the postpartum females exposed to CSS is decreased, presumably because of their exposure to CSS and their expectation of future incursions from a threatening intruder male. Such females nurse and LG their pups significantly less that control females not exposed to CSS (Nephew & Bridges, 2011).

Carini and Nephew (2013) mated the adult female offspring of mothers that were either exposed to CSS or that were not chronically exposed to this stressful event. The females that were "neglected" during their early postnatal development exhibited disruptions in their adult maternal behavior toward their own offspring. The female offspring of CSS mothers showed deficits in all aspects of maternal behavior in comparison to control females, which included less efficient retrieval of scattered pups and decreased durations of nursing behavior and LG of pups. This decreased maternal behavior was associated with increases in nonmaternal behaviors (increased locomotion outside the nest area) that the researchers suggested was indicative of restlessness and anxiety. Importantly, the postpartum female offspring of mothers that were exposed to CSS had higher basal levels of plasma corticosterone and exhibited lower levels of maternal aggression than did control females. These results suggest that exposure to maternal neglect interferes with the subsequent development of maternal behavior in the affected female offspring. Increases in behavioral and physiological stress reactivity may have contributed to these reductions in both maternal responsiveness and maternal aggression.

Nephew et al. (2018) applied fMRI techniques to examine the brain function of postpartum female rats whose mothers were exposed to CSS. They provided evidence that exposure to this form of maternal neglect resulted in the aberrant development of the medial regions of the prefrontal cortex (mPFC; also see Roth, Lubin, Funk, and Sweatt, 2009), which included reduced resting state functional connectivity between the mPFC and MPOA. It is certainly interesting to speculate that the CSS model of maternal neglect interferes with the subsequent development of competent maternal behavior in the affected female offspring via a dual mechanism. Dysfunctions in mPFC-to-MPOA connectivity may interfere with the function of neural circuits controlling maternal motivation (see Chapter 5 of this volume), while interference with the connectivity between the mPFC and the amygdala may disrupt the neural processes involved in emotion regulation (see Figure 8.5), resulting in enhanced behavioral stress reactivity and anxiety. There is recent evidence that deficits in OT action on OTRs in mPFC (see Table 4.1) may be involved in these potential deficits in emotion regulation that result from early parental deprivation in rodents (He et al., 2019), which conforms with the known anxiolytic effects of OT and also with the fact that parental treatment effects can influence the development of OTRs in offspring. Later in this chapter in the subsection on the effects of maternal neglect/deprivation in rodents on the development of mPFC-to-amygdala neural circuits that regulate fear-related behaviors, I will present evidence from rodent models that maternal neglect does indeed disrupt mPFC interactions with the amygdala, which interferes with the normal development of emotion regulation neural systems in the affected offspring.

The Effects of Maternal Neglect/Deprivation in Rodents on the Development of Amygdala Neural Circuits That Regulate Fear-Related Behaviors

In a series of important studies, Sullivan and her colleagues have shown that the acquisition of conditioned avoidance responses to neutral stimuli that have been paired with a noxious stimulus typically does not occur in male and female neonatal rats prior to day 10 of postnatal life (Rincon-Cortes & Sullivan, 2014; Tallot, Doyere, & Sullivan, 2016). To examine this process, postnatal rats are briefly removed from their mother on postnatal day 8 and exposed to a novel odor that is paired with shock. On postnatal day 9, these infant pups are placed in a Y-maze with the shock-paired odor in one arm of the maze and another odor that was not paired with shock in the other arm. These rats do not avoid the arm that contained the odor that was paired with shock. However, after postnatal day 10, rats will learn to avoid the odor that was paired with shock. Sullivan and her colleagues have related these findings to the occurrence of a stress hyporesponsive period in neonatal rats between postnatal days 1 and 9. During this period, brief electrical shocks do not activate the HPA axis and do not increase corticosterone levels. However, beginning on day 10 of life, a physiological stress response is observed. The implication is that stress (shock)-induced corticosterone release is required for avoidance learning to occur in young rats. In support of this view, when postnatal day 8 rats were administered corticosterone systemically prior to odor–shock pairings, they demonstrated precocious odor-aversion learning (Moriceau, Wilson, Levine, & Sullivan, 2006). More important, these researchers observed similar precocious odor-aversion learning when corticosterone was directly applied to the amygdala of rats on postnatal day 8. Conversely, although postnatal day 12 rats learn a conditioned odor aversion, this learning is blocked by interfering with the action of corticosterone on the amygdala. One interpretation of these results is that amygdala fear-learning neural circuits are not fully functional in neonatal rats prior to day 10 of life, but that a premature activation of corticosterone release will result in a precocious activation of such circuits.

How can we relate these findings to the effects of maternal deprivation/neglect on the development of enhanced anxiety and fearfulness in the affected offspring? First, it should be noted the PMS during the stress hyporesponsive period is a severe enough stress to activate increases in plasma corticosterone levels in neonatal rats prior to day 10 of life (Suchecki, Rosenfeld, & Levine, 1993). Further, Moriceau, Shionoya, Jakubs, and Sullivan (2009) have shown that exposing neonatal rats to maternal neglect (decreased maternal caretaking behaviors by their mothers) increases corticosterone levels in rats on postnatal day 7 and that such neglected neonates are able to learn a conditioned odor aversion, while their control counterparts that were not subjected to maternal neglect did not demonstrate odor-aversion learning.

These results, taken together, indicate that maternal deprivation/neglect during the early postnatal period can prematurely activate amygdala neural circuits that underpin fear-related behaviors and fear learning in rats. Interesting research on adult mice has provided evidence on the underlying mechanisms through which corticosterone action on the amygdala might activate fear-related neural circuits (Kolber et al., 2008). During an acquisition trial, mice were exposed to an auditory stimulus paired with shock. Several days later, the auditory stimulus was presented alone, and a conditioned fear response (freezing/behavioral immobility) was observed. However, if GRs in the CeA were experimentally blocked during the acquisition phase, then the learning of the conditioned fear response was disrupted. Significantly, 60 minutes after the initial pairing of the tone with the shock, CRF mRNA levels were increased in the CeA of control mice, but were not increased in the CeA of mice in which GRs were blocked. Importantly, ICV injections of CRF prior to training were able to reverse the disruption of conditioned fear learning in mice with blocked GRs in CeA. Kolber et al. proposed that increases in corticosterone release during fear conditioning activate GRs in CeA, which results in the increased synthesis and release of CRF. The neural effects of CRF on downstream targets are then proposed to facilitate fear conditioning. More simply stated, shock-induced increases in CRF may induce increased fearfulness, which is necessary for associating a novel CS with the noxious US (shock).

Important behavioral research by Callaghan and Richardson (2011, 2013) buttresses the view that maternal deprivation/neglect enhances the development of fear and anxiety in rodents or, alternatively, that adequate maternal care delays the development of anxiety and fearfulness. These researchers used fear-learning paradigms to examine the development of fear systems in young rats and adult rats. In one paradigm, rats are shocked in one compartment of a two-compartment cage, with one compartment being black and the other white. If the rats are shocked in the black compartment and then tested 24 hours later in the two-compartment cage without shock, they avoid the black compartment. This outcome is true for rats trained on postnatal day 18 (preweanling rats) and for adults tested at 100 days of age. However, if the rats are tested at 1 week after training, the adult rats avoid the black compartment, while the rats that were trained during the neonatal period forget and do not avoid the black compartment. This process of the forgetting of a fearful experience in preweanling rats has been referred to as infantile amnesia and suggests that fear-related neural systems are not fully developed in rats that are still being cared for by their mother. However, if neonatal rats are subjected to PMS (3 hours/day) over postnatal days 2 to 14, then when they are fear-trained on postnatal day 18 and tested for retention 1 week later, they behave like adult rats and avoid the shock-associated chamber: They do not show infantile amnesia. These results, along with those from Sullivan's group, indicate that various models of maternal deprivation/neglect enhance, or prematurely activate, the development of fear- and anxiety-related behaviors and their associated neural systems in the affected offspring. It has been hypothesized that the accelerated development of fear and anxiety neural systems controlled by the amygdala may interfere with the normal development of top–down regulation of the amygdala by the mPFC (Callaghan & Richardson, 2013; Callaghan, Sullivan, Howell, & Tottenham, 2014). More specifically, early life stress as represented by maternal neglect and/or abuse may disrupt the effective development of neural systems involved in emotion regulation in the affected offspring, which would contribute to the adult phenotype of enhanced stress reactivity, anxiety, and fearfulness in these rats. I will elaborate on this important proposal later in this chapter in the subsection on the effects of maternal neglect/deprivation in rodents on the development of mPFC-to-amygdala neural circuits that regulate fear-related behaviors.

In considering the effects of PMS on the development of unlearned fearfulness in the affected offspring (e.g., as tested in the open field) and the development of fear learning systems, it is possible that partially overlapping and partially distinct underlying mechanisms influence the development of these two types of fearfulness (cf. Kan, Callaghan, & Richardson, 2016).

It certainly would have been interesting for Sullivan and her colleagues and Callaghan and Richardson to have tested the adult maternal behavior of those neonatal females that were subjected to early life stress in the form of maternal neglect and that also demonstrated an accelerated development of fearfulness. This is an important area for future research and would inform us about some of the neural processes that may be involved in the intergenerational transmission of faulty maternal behavior.

The Effects of Maternal Neglect/Deprivation in Rodents on the Development of mPFC-to-Amygdala Neural Circuits That Regulate Fear-Related Behaviors

In Figure 8.5, I described the relationships between the mPFC and the amygdala in regulating fear-related responses in adult rodents. Numan (2015) has reviewed the evidence that input from the infralimbic (IL) part of the mPFC to the amygdala depresses fearfulness while input to the amygdala from the prelimbic (PL) part of the mPFC enhances fear-related responses. Important information on the development of PL involvement in fear learning has been presented by Li, Kim, and Richardson (2012). Infant and juvenile rats were trained to acquire a conditioned fear response to an auditory stimulus that was paired with shock. This training occurred on either postnatal day 16 (infants) or postnatal day 23 (juveniles that have been weaned). Two days later a retention test was given where the auditory stimulus was presented alone. Rats in both groups displayed a conditioned fear response (immobility/freezing in response to the tone). Please note that a longer training-test interval would have been necessary to demonstrate infantile amnesia in the infant rats. Importantly, when the PL area was temporarily inactivated on the test day in the juvenile rats, the conditioned fear response was suppressed, but this was not the case for the infant rats. These results suggest that the expression of a conditioned fear response is not influenced by the PL cortex in infant rats, while an adult form of PL-to-amygdala interactions has developed in juvenile rats. In relating these findings to those of Callaghan and Richardson (2011, 2013), given that PMS enhances the development of fear learning, perhaps it also enhances the ability of the PL area to positively regulate amygdala output to a threatening stimulus. To the best of my knowledge, this idea has not been tested in rodents. The important point I want to make, however, is that if PMS early in life does enhance the development of mPFC to amygdala connectivity, perhaps this premature development results in suboptimal connectivity between mPFC and amygdala, with the result that emotion regulation neural systems might be compromised in adulthood. An interesting, but speculative, proposal, is that an enhanced development of PL-to-amygdala connectivity by maternal deprivation, which would enhance fearfulness, might interfere with the full development of IL-to-amygdala connectivity, which would typically decrease fearfulness. In support of the general view being presented here, several anatomical studies have provided evidence that PMS or other forms of maternal neglect disrupt the normal development of the mPFC and its connection to the amygdala in the affected offspring (Ishikawa, Nishimura, & Ishikawa, 2015; Uchida et al., 2010; Yan et al., 2017). In reference to Figure 8.5, perhaps various forms of maternal neglect interfere with the development of neural circuits in the affected offspring so that the IL region has decreased efficacy in activating the inhibitory interneurons in the amygdala, which typically act to suppress the output of CeA to PAG (see Santiago, Lim, Opendak, Sullivan, & Aoki, 2018).

What might be the mechanism through which maternal deprivation/neglect/abuse affects the development of emotion regulation neural systems in infants? I will offer a hypothetical model and then present the evidence that supports it. I propose that the profound stress to the young infant caused by PMS or other forms of maternal neglect during the early postnatal period in rodents activates a supernormal release of CRF within the brain. This enhanced CRF release then activates serotonin (5-hydroxytryptamine [5-HT]) neurons in the dorsal raphe nucleus (DR) in the brainstem that project to mPFC. Abnormally high 5-HT release into the mPFC during the neonatal period then affects the development of the mPFC, and, in particular, downregulates the ability of the IL part of the mPFC to suppress the amygdala's responsiveness to threatening stimuli, in this way heightening anxiety and behavioral stress reactivity in adulthood.

The two major types of CRF receptors, CRFR1 and CRFR2, are both located in the DR (Waselus, Nazzaro, Valentino, & Van Bockstaele, 2009). In rodents, low levels of CRF activate CRFR1, while higher (stress-induced) levels activate CRFR2, because CRF has a higher affinity for the former receptor (Lukkes, Forster, Renner, & Summers, 2008). Significantly, activation of DR-CRFR2 by CRF depolarizes DR-5-HT neurons, resulting in the release of 5-HT into the forebrain, including the mPFC (Forster et al., 2008; Waselus et al., 2009). The sources of CRF input to DR 5-HT neurons probably include the CeA, the dorsal part of the bed nucleus of the stria terminalis, and PVN (Valentino, Liouterman, & Van Bockstaele, 2001). To increase CRF synthesis, and presumably release, in neonatal rodents, Kolber et al. (2010) produced a transgenic mouse line that resulted in a temporally restricted enhancement of CRF synthesis within the forebrain, which included CeA, between embryonic day 15 and postnatal day 21 (the time of weaning). In adulthood, these mice exhibited heightened anxiety-related behaviors. Given the findings of Kolber et al., it is certainly possible that increased corticosterone levels in neonates induced by PMS or other forms of maternal neglect activate the synthesis and release of CRF from CeA (see Chapter 6 of this volume), with the result that an adult anxiety-prone phenotype develops. Note how this proposal aligns with the findings of Sullivan's group described in the subsection on the effects of maternal neglect/deprivation in rodents on the development of amygdala neural circuits that regulate fear-related behaviors. Is there evidence that the effects of such enhanced CRF activity might be due, in part, to CRF-induced increases in 5-HT release into the mPFC of the neonatal rodent brain?

Early in life, 5-HT has important effects on brain development (for reviews, see Brummelte, Glanaghy, Bonnin & Oberlander, 2017; Numan, 2015; Yu et al., 2014). A variety of studies have shown that experimentally induced increases in 5-HT action in the brain of neonatal rodents results in the development of increased anxiety-related behavior in adulthood. Many of these studies have enhanced the postsynaptic effects of 5-HT by systemically treating neonatal rodents with selective serotonin reuptake inhibitors, such as fluoxetine, which block the reuptake of 5-HT after its release from axon terminals and thus prolong the postsynaptic action of 5-HT. In an important study, Rebello et al. (2014)

injected male and female mice with fluoxetine daily between postnatal days 2 and 11. This treatment resulted in the expression of enhanced anxiety-related behaviors in adulthood (postnatal day 90) as measured on a variety of tests. Similar daily injections on postnatal days 12 to 22 or 22 to 41 did not enhance adult anxiety. Interestingly, this sensitive period between postnatal days 2 to 11 overlaps with the postnatal period where maternal LG of pups affects the development of emotionality in the mother's offspring. Strikingly, Rebello et al. found that treatment with fluoxetine over postnatal days 2 to 11 affected the anatomy and function of the adult mPFC. Dendritic atrophy occurred in the pyramidal cells of the IL mPFC, and the neuronal excitability of the IL area decreased while that of the PL cortex increased. These results suggest that supernormal early postnatal action of 5-HT in the brain alters the development of the mPFC function, upregulating PL and downregulating IL. These functional and anatomical changes would conform with an adult phenotype of enhanced anxiety.

Two types of 5-HT postsynaptic receptors are the 5-HT1A and 5-HT2A receptor. 5-HT exerts inhibitory postsynaptic effects when it binds to the 5-HT1A receptor and excitatory effects when it binds to the 5-HT2A receptor (Numan, 2015). Sarkar, Chachra, and Vaidya (2014) found that if neonatal mice were treated with fluoxetine daily between postnatal days 2 and 21, anxiety-related behaviors were increased in adulthood. However, co-treatment of the neonatal mice with both fluoxetine and a 5-HT2A receptor antagonist blocked the development of anxiety in adult mice. These results suggest that the supernormal activation of 5-HT2A receptors (perhaps in mPFC) during the early postnatal period contributes to the development of adult anxiety in mice. This conclusion was supported by the additional finding that treating neonatal mice with only a 5-HT2A receptor agonist, which would activate these receptors, also resulted in increased anxiety in adulthood.

Can any of these effects of enhanced 5-HT function during the neonatal period on the development of increased emotionality be related to the effects of maternal deprivation on the development of emotionality? First, Xue, Shao, Li, Shao, and Wang (2013) reported that when neonatal rats were subjected to PMS (4 hours/day between postnatal days 1 and 21), 5-HT levels were increased in the mPFC on postnatal day 21 in comparison to nonseparated controls. These results indicate that PMS enhances the release of 5-HT in mPFC. Importantly, Benekareddy, Vadodaria, Nair, and Vaidya (2011) found that the adult anxiety phenotype induced by PMS in rats (3 hours/day between postnatal days 2 and 14) was prevented in the rats that were subjected to PMS but were also treated daily with a 5-HT2A receptor antagonist. Finally, and importantly, Benekareddy, Goodfellow, Lambe, and Vaidya (2010) compared the neural responses of the PL mPFC to 5-HT in male and female adult rats that were subjected to either PMS during postnatal days 2 to 14 or were not separated from their mothers during this sensitive postnatal period. In control rats, 5-HT exerted inhibitory effects on PL neurons, and this effect was mediated by an action on 5-HT1A receptors. In contrast, in rats exposed to PMS, in adulthood 5-HT exerted excitatory effects on PL neurons that were mediated by an action of 5-HT on the 5-HT2A receptor. These results, taken together, indicate that maternal deprivation early in life alters the development of the mPFC. In particular, PMS alters the functional balance of 5-HT receptors in the PL cortex, biasing responses toward excitatory effects mediated by the 5-HT2A receptor. The enhanced output of the PL cortex to the amygdala, caused by this modified mPFC developmental pattern, can be conceived as contributing to the adult expression of enhanced anxiety and behavioral stress reactivity to threatening stimuli (see Figure 8.5). The modified development of the PL cortex may, in turn, suppress the development of the typical fear-reducing effects of IL cortex projections to the amygdala.

The data presented herein support the proposal that the severe stress to young infants caused by PMS or other forms of maternal neglect during the early postnatal period in rodents activates a supernormal release of CRF within the brain and that this enhanced CRF release then activates 5-HT neurons in the DR that project to mPFC. Abnormally high 5-HT release into the mPFC during the neonatal period then affects the development of the mPFC, and, in particular, downregulates the ability of the IL part of the mPFC to suppress the amygdala's responsiveness to threatening stimuli while also upregulating the ability of the PL-to-amygdala pathway to enhance fearfulness, in this way heightening anxiety and behavioral stress reactivity in adulthood. Although the evidence that I have presented is persuasive, it is not conclusive, and some links in the hypothesized causal chain are in need of experimental validation. For example, it would be important to show that PMS causes the release of CRF into DR of neonatal rodents and that blocking the action of CRF on CRFR2 at the level of DR would interfere with the anxiety promoting effects of early maternal deprivation.

Research on Nonhuman Primates: 5-HT and G × E Interactions

In the introduction to this chapter I indicated that being abused or neglected by one's parent(s) does not destine one to become a neglectful or abusive parent; other factors may influence the vulnerability of young infants to the potentially negative impact of early life parental neglect/abuse on the development their maternal behavior and emotionality. As I previously indicated, one of these additional factors may involve gene by environment (G × E) interactions: Infants with particular genotypes may be more or less susceptible to the negative impact of neglect/abuse on the development of their subsequent parental behavior. In light of the research discussed in the previous section, one possibility is that infants with a genetic make-up that results in enhanced 5-HT action early in life would be more susceptible to the negative impacts of parental neglect/abuse than would infants with a genotype that restrained 5-HT action early in life (Caspi et al., 2002).

In human and nonhuman primates, there is a genetic polymorphism within the regulatory (promoter) region of the 5-HT transporter (5-HTT) gene (Bennett et al., 2002; Homberg & Lesch, 2011). Note that the 5-HTT protein is involved in the reuptake of 5-HT, which terminates the postsynaptic effects of 5-HT. If one allele of the 5-HTT gene were less effective in producing the 5-HTT protein than an alternate allele, then the former allele would be associated with heightened 5-HT action. The genetic polymorphism within the promoter of the 5-HTT gene is referred to as the serotonin transporter-linked polymorphic region (5-HTTLPR), and this polymorphism is the result of a variation in the number of nucleotides within this region, giving rise to either a short (s) allele or a long (l) allele. Importantly, the s allele is associated with decreased transcription of 5-HTT mRNA and decreased expression of 5-HTT protein. In line with the issues I have been discussing, these genotype differences should result in carriers of one (s/l) or two (s/s) copies of the short allele to have a longer duration of stress-induced activation of 5-HT in the brain upon exposure to maternal neglect/abuse early in life.

Coplan et al. (1996) initially reported that adult macaque monkeys that were exposed to maternal neglect early in life had higher levels of CRF in CSF and also developed heightened anxiety in comparison to their nonneglected control counterparts. In a subsequent study, Coplan et al. (2011) reported that CSF levels of CRF were elevated in s/l and s/s monkeys, but not in l/l monkeys, that were exposed to maternal neglect early in life. Relevantly, McCormack, Newman, Higley, Meaestripieri, and Sanchez (2009) have reported that the proportion of mothers carrying the s allele of the 5-HTTLPR was higher in abusive than in nonabusive rhesus monkey mothers. Since Maestripieri (2005) found that only about 50% of rhesus monkey females that experience abuse as infants grow up to become abusive mothers, these results suggest that a gene by environment interaction may be involved in the intergenerational continuity of abusive maternal behavior in rhesus monkeys.

Given these findings, which relate to the development of emotionality, it is possible that when carriers of the short allele of the 5-HTTLPR are exposed to maternal neglect/abuse early in life they have an enhanced action of 5-HT in the brain in response to CRF stimulation in comparison to individuals with the l/l genotype. This enhanced 5-HT action might then result in an atypical development of emotion regulation neural systems, resulting in heightened anxiety and behavioral stress reactivity in the affected offspring in adulthood (which would include higher levels of CRF in CSF, perhaps resulting from the heightened activity of CeA-CRF neurons). The development of heightened emotionality may then affect the subsequent maternal behavior of the neglected/abused infants, particularly when they are caring for their own infants under demanding or challenging conditions. Maestripieri (2011) has noted that highly anxious macaque mothers tend to be overprotective of their infants and may abuse their infants when they do not respond to maternal retrieval cues and, therefore, do not return to their mothers promptly. These highly anxious mothers appear to have difficulty in coping with stressful situations. Maestripieri presents evidence that a dysregulated 5-HT system may contribute to the development of high anxiety and maternal abuse of infants.

Conclusions

In this chapter, I have tried to explain the mechanisms that might regulate the intergenerational continuity of faulty maternal behavior, such as the transmission of maternal neglect and abuse across generations. This chapter dealt with animal models or preclinical research, which tends to be experimentally based and can therefore provide information on the relevant causal mechanisms.

I have emphasized the importance of experiential factors over genetic inheritance in the intergenerational continuity of maternal responsiveness, although the importance of G × E interactions was considered. The basic thrust of my presentation was that the way a mother treats her offspring affects the development of neural systems in the infant, which, in turn, affects the development of the offspring's maternal behavior in adulthood. I emphasized two important processes: (a) The way a mother treats her infants can affect the development of those neural circuits that regulate maternal motivation; less than adequate maternal care can disrupt the full development of maternal neural circuits so that in adulthood the offspring show lower levels of maternal motivation, and (b) the way a mother treats her infants can affect the development of those neural circuits involved in emotionality and emotion regulation; less than adequate maternal care can alter the development of these neural circuits so that in adulthood the offspring show less than adequate maternal behavior, particularly under stressful/demanding/challenging conditions.

The best evidence for these two types of maternal treatment effects was derived from research on the effects of normal variations in maternal behavior on the development of maternal neural circuits and on fear/anxiety-related circuits in rodent offspring. With respect to the effects of severe forms of maternal neglect/deprivation on the development of the affected offspring, most research has dealt with the effects of PMS on the development of emotionality and emotion regulation neural systems. Clearly, more research needs to explore whether severe (abnormal) forms of maternal neglect and abuse can affect the development of those neural circuits that regulate maternal motivation in the affected offspring.

Although the research on the effects of maternal neglect on offspring development in animals has emphasized the effects of maternal maltreatment on the development of emotional control systems, most of this research has not focused on the second link in the chain that I have been concerned with: That alterations in an offspring's emotionality and emotion regulation due to maternal maltreatment can influence the offspring's subsequent maternal behavior. However, wherever possible, I have emphasized those research findings that do indeed show that enhanced anxiety and deficits in emotion regulation caused by various models of maternal neglect in animals are associated with deficits in maternal behavior. This connection has been best demonstrated in the research of Boccia and Pedersen (2001) and Nephew and his colleagues (Carini & Nephew, 2013; Nephew et al., 2018), but more research on this important issue is needed. If maternal neglect results in the development of emotional disturbances in the mother's offspring, which then negatively impacts the maternal behavior of the offspring, then the intergenerational continuity of faulty maternal behavior is a likely contributor to the intergenerational continuity of mood and anxiety disorders.

In Chapter 8, I reviewed the evidence for humans that postpartum depression and anxiety disrupt the normal functioning of the human parental brain and parental behavior. Therefore, if maternal maltreatment of infants and children results in the development of mood and anxiety disorders throughout the lives of the affected offspring, then their maternal behavior should also be negatively affected. I will explore this issue in Chapter 10.

It has been proposed that the enhanced anxiety and fearfulness that develop in offspring that have received less than adequate maternal care may actually be an adaptive mechanism that prepares an organism for an adult life in an adverse environment (Cameron et al., 2005). As an example, a maternal rodent that is socially subordinate, with the result that she is only able to obtain an inferior nest site that contains dispersed food resources and increased exposure to predators, may spend more time off the nest, not attending to her infants. The decreased maternal care that the offspring receive, which increases the offspring's anxiety, fearfulness, and heightened reactivity to potential threats may, in this way, prepare these offspring for life under such adverse environmental conditions in adulthood. To develop this idea further, note that in adult rodents, CRF-induced release of 5-HT into the mPFC typically exerts anxiolytic effects (Forster et al., 2008; Ren et al., 2018), which helps the adult cope with a stressful environment. However, due to the developmental effects of stress (CRF)-induced release of 5-HT into mPFC that I previously described in this chapter, such release in young rodents, due to decreased maternal care, actually results in the development of enhanced anxiety in the adult. It could be proposed that such a developmental effect prepares that organism for an adult life in an adverse environment.

Although this proposed adaptive mechanism may be appropriate for the effects of relatively normal variations in maternal behavior, I do not think it makes sense when considering severe forms of maternal neglect/abuse on infant development, since, for most mammals, the young might not survive under such extreme treatment conditions in nature. Also, and significantly, the proposal by Cameron et al. (2005), even when applied to normal variations in maternal behavior, would only be adaptive if the environment of the adult offspring matched the environmental conditions under which they were born (Schmidt, 2011). Although there may be environmental consistency across generations in most animal societies, the development of heightened anxiety/fearfulness in human children as the result of maternal maltreatment is likely to be maladaptive, particularly in modern societies. Modern societies support upward mobility, and such upward mobility is a cornerstone of social advancement. Therefore, the negative impacts of maternal maltreatment on the socioemotional development of children in modern societies is likely to be a maladaptive trait that would tend to hinder or prevent upward mobility.

Finally, this chapter has focused on maternal treatment effects on the development of female offspring and the intergenerational transmission of maternal phenotypes. In biparental species, the behavior of the father toward his offspring is also expected to have effects on their development, and the reader is referred to a review by Feldman, Braun, and Champagne (2019) for important information on this topic. Further, while I have emphasized the effects of the mother's behavior on the development of her young, there is also evidence that the behavior of an infant, influenced by its genotype, also affects the treatment it receives from its mother (see Pan et al., 2018; Potter, Ashbrook, and Hager, 2019).

10
Development of the Parental Brain in Humans

Introduction

This chapter will analyze the research on the intergenerational continuity of maternal responsiveness in humans. Most of this research is correlational in nature, and, therefore, firm conclusions about causality cannot be reached. However, against the background of the experimental animal research from Chapter 9, I will present the human research that indicates that the way human parents treat their children is an important factor that can affect the development of emotion neural circuits and parental neural circuits in their offspring. The development of these circuits, in turn, influences the parental responsiveness of these offspring when they have their own children.

Many factors influence the way a human mother interacts with her children both within and outside the normal range, with the latter leading to child maltreatment (abuse or neglect). As one example, living in poverty and its associated stresses can have negative impacts on mother–child interactions. However, in this chapter, as in the previous one, I will focus on how maternal treatment effects, which are experiential in nature, influence the development of parenting in the affected offspring.

The Intergenerational Transmission of Maternal Responsiveness: A Behavioral Analysis

Intergenerational Transmission of Child Maltreatment

Parental maltreatment of their child (child maltreatment) includes sexual, physical, and emotional abuse and physical and emotional neglect. Based on self-reports, nonsexual forms of child maltreatment account for about 90% of abuse and neglect cases (Assink et al., 2018). In a meta-analytic review of 84 primary studies that examined whether an association existed between parents' experience of maltreatment in their childhood and the maltreating behaviors of such parents toward their own children, Assink et al. (2018) concluded that a parental

The Parental Brain. Michael Numan, Oxford University Press (2020). © Oxford University Press.
DOI: 10.1093/oso/9780190848675.001.0001

history of experiencing child maltreatment was an important risk factor for these parents to engage in abuse and/or neglect of their own children. Similar results have been reported by others (Bartlett, Kotake, Fauth, & Easterbrooks, 2017; Babcock Fenerci & Allen, 2018; Schofield, Conger, & Conger, 2017). Significantly, Bartlett et al. reported that this association was particularly strong when considering the mother as both a victim in her childhood and the perpetuator of her child's maltreatment: 50% of mothers who were maltreated by their mothers maltreated their child, while only 29% of non-maltreated mothers abused or neglected their children.

As indicated by Assink et al. (2018), these results also indicate that not all parents with a history of maltreatment abuse their own children, suggesting that there are protective factors that restrain the intergenerational transmission of child maltreatment. Gene by environment (G × E) interactions might be involved in such protective effects, as might the occurrence of social support by individuals other than parents (Schofield, Lee, & Merrick, 2013). Finally, these results also show that child maltreatment can occur in parents who have not been maltreated in their childhood, which indicates that a parental history of abuse and/or neglect is not the only factor that might cause a mother to maltreat her children.

Assink et al. (2018) have proposed several possible pathways that might mediate the intergenerational transmission of child maltreatment, and most of these mechanisms are presumed to be experiential in nature. The first mediating mechanism is that the child's experience of maltreatment may result in the development of impaired emotion regulation (Poole, Dobson, & Pusch, 2018) and that a severe disruption of emotion regulation might lead to high levels of irritability that promote child abusive behaviors in adulthood once the victim becomes a parent. There is also evidence that the original maltreating parent may have problems with emotion regulation, which then creates a similar emotional dysregulation in their offspring (Bridgett, Burt, Edwards, & Deater-Dekard, 2015). Second, child maltreatment may hinder the ability of the maltreated child to form normal social bonds in adulthood (also see Cyr, Euser, Bakermans-Kranenburgh, & van Ijzendoorn, 2010; Raby, Labella, Martin, Carlson, & Roisman, 2017). Such an impairment, of course, could disrupt mother-to-infant bonds. Note how these two proposed mediating pathways match the models that I presented in Chapter 9 (see Figure 9.1).

Another mediating pathway proposed by Assink et al. (2018) is the transfer of psychopathology from one generation to the next, through experiential and/or genetic processes. In Chapter 8, I reviewed research that showed that postpartum depression is associated with faulty maternal behavior, and I proposed that depressed mothers that express high levels of anhedonia may be at risk for showing maternal neglect, while highly anxious depressed mothers may be at risk for displaying maternal abuse. Importantly, and with respect to experiential influences, childhood maltreatment increases the risk for the development of postpartum depression (Li, Long, Cao, & Cao, 2017; Muzik et al., 2013) as well as lifetime adult depression and anxiety (Gallo, Munhoz, Loret de Mola, & Murray, 2018; Spertus, Yehuda, Wong, Halligan, & Seremetis, 2003), and this latter relationship is particularly prominent in women. Additionally, the involvement of prepartum (antenatal) depression is worth considering. Plant, Barker, Waters, Pawlby, and Pariante (2013) reported that mothers who experienced maltreatment during their own childhoods were more likely to develop prepartum depression than mothers without childhood maltreatment. Further, mothers that experienced childhood maltreatment and also developed prepartum depression were likely to maltreat their own children (cf. Bouvette-Turcot et al., 2019).

With respect to genetic inheritance, most studies on the intergenerational transmission of child maltreatment are correlational, so genetic inheritance not only could contribute to psychopathology, but could also influence emotion regulation and social bond formation across generations.

Whether through experiential or genetic pathways, the transfer across generations of psychopathological disorders, such as anxiety, may reflect deficits in emotion regulation, and depression-associated anhedonia may result in poor social bond formation.

Finally, the intergenerational transmission of child maltreatment might result from the transfer of general environmental risk factors, such as living in poverty, across generations (Assink et al., 2018). In this case, maternal treatment effects would not be involved. Instead, particular environmental stressors may promote child maltreatment in parents and their adult offspring based on the assumption that these environmental risk factors do not change significantly across generations.

Although general environmental risk factors may act alone to promote the development of faulty maternal behavior, it is likely that interactions between experiencing child maltreatment and being exposed to a stressful environment, both in childhood and adulthood, co-act to promote the development of abnormal maternal behavior. Such abnormal development may also be influenced by G × E interactions.

Intergenerational Transmission of Normal Variations in Maternal Behavior

Intergenerational Transmission of Attachment Styles Influences Parental Behavior

Bowlby (1973, 1980, 1983) and Ainsworth, Blehar, Waters, and Wall (1978) have made outstanding theoretical and empirical contributions to our understanding of the development of social attachments in children and adults. With respect to a child's attachment to its mother, mothers who respond sensitively, synchronously, and nonintrusively to their infant's signals (high maternal sensitivity) have infants that form a secure attachment to them. In contrast, mothers who are consistently insensitive to their infant's signals of both distress and desires for engagement/interaction (consistently low maternal sensitivity), and mothers who are inconsistent in their responses to their infant's signals (sometimes responding appropriately, sometimes ignoring their infants and sometimes being overly intrusive) have infants that form insecure attachments to their mothers. Mothers who consistently ignore their infants have infants that develop an insecure-avoidant phenotype, while mothers who are inconsistently sensitive to their infants have infants that develop an insecure-anxious phenotype (Benoit, 2004; Nelson-Coffey, Borelli, & River, 2017).

The Strange Situation Procedure (Ainsworth et al., 1978) has been used to classify these three major types of infant-to-mother attachments. In this procedure, a mother and infant are placed in a novel setting that contains toys for the child to play with. Afterwards, a stranger enters the room, and this is followed by the mother leaving and then returning to the room. Secure infants play with the toys in the mother's presence and are not upset when a stranger enters the room. Secure infants show mild distress when the mother leaves the room, but they calm down upon her return. Insecure-avoidant infants basically show low emotional responses under all conditions and do not rely on their mothers for support: They do not show distress when their mother leaves the room, and they do not seek contact when the mother returns. They are emotionally detached. Finally, insecure-anxious infants are behaviorally inhibited in the strange room, frightened by the stranger even in the mother's presence, and respond with anger and clinging upon the mother's return after the separation period.

Bowlby's theoretical formulations and Ainsworth's empirical findings have emphasized an experiential, rather than a genetic, basis to the development of infant-to-mother attachments: The way a mother responds to her infant affects the development of the child's attachment type. Support for an experiential basis of infant-to-mother attachments has been provided by Leonetti (2014). Infants adopted by secure mothers (see the following discussion) at an early developmental stage (at about 5 months of age) were more likely to develop a secure attachment to their mothers than were infants adopted by insecure mothers (also see Pace, Di Folco, Guerriero, & Muzi, 2019).

Importantly, Bowlby proposed that the nature of an infant's attachment to its mother influences its adult social behavior and its ability to form sensitive and close social bonds with others, including its parental responsiveness toward its own infants in adulthood. Recent research has confirmed that there is a moderate stability of attachment styles across a person's lifespan (Lee & Hankin, 2009; Sroufe, 2005; Verhage et al., 2016). This moderate, rather than strong, stability, is in part due to protective factors, such as the occurrence of strong social support networks later in life, that might compensate for an insecure infant-to-mother attachment. Also, children with secure attachments to their mother may develop insecure attachments in adulthood due to other high-risk conditions, such as a lack of adult social support. Importantly, in a 30-year longitudinal study, Sroufe (2005) found that insecure attachments in early childhood were related to deficiencies in both emotion regulation and prosociality (social competence, social bonding, and the formation of close friendships) later in life.

Adult attachment relationships are typically conceived as varying along the dimensions of avoidance and anxiety (similar to infant-to-mother attachments; Nelson-Coffey et al., 2017). Secure adult attachments are represented by low levels of both anxiety and avoidance. Insecure-avoidant adults tend to rely on themselves, and they do not depend on strong social relationships with others. Insecure-anxious adults tend to worry about the permanence of their attachments to others and are overly sensitive to rejection and the possibility that a social relationship will be terminated by a partner. In relation to parenting, Nelson-Coffey et al. found that parents who scored high on measures of attachment avoidance had fewer positive and fewer negative emotions during episodes of caring for their own child. They suggested that such parents may have been more emotionally detached from their children than secure parents. In contrast, parents that scored high on anxious attachment tended to show higher levels of negative emotions throughout the day, both during child care and during nonchild care episodes. They hypothesized that parents who were high in attachment anxiety may have experienced feelings of frustration and anger during childcare and feelings of anxiety and worry when apart from their child. Although the effect sizes in this study were small, it is interesting to consider the possibility that adults with high attachment avoidance may have lower levels of parental motivation and empathy as a result of their poorer bonding to their infants, while adults with attachment anxiety may have deficiencies in emotion regulation. This proposal conforms with Sroufe's (2005) findings that insecure child-to-mother attachments are associated with deficiencies in emotion regulation and social competence later in life. The occurrence of such developmental trajectories may be moderated by other risk factors, such as socioeconomic status and social support, experienced in adulthood (cf. Belsky & Pasco Fearon, 2002).

In support of these views, Jones, Cassidy, and Shaver (2015) noted that adult attachment avoidance is associated with less sensitive and supportive parental behavior. Strong support for an intergenerational link in such attachment relationships has been provided by Stern, Stern, Borelli, and Smiley (2015). In a correlational analysis, these researchers reported that the level of a parent's empathy (both emotional and cognitive empathy) mediated a negative correlation between measures of the parent's attachment avoidance and their child's attachment security. Parents that displayed an avoidant attachment style in adulthood were less empathic, and their children were more insecure. In relation to attachment anxiety, Stevenson-Hinde, Chicot, Shouldice, and Hinde (2013) have reported that maternal anxiety is associated with inconsistent maternal sensitivity, which, in turn, was associated with an insecure infant-to-mother attachment in their children.

Finally, in a study by Leerkes and Siepak (2006), college women were shown videos of facial expressions of infants showing either fear or anger. The students were asked to identify the emotions depicted by the expressions. The accurate identification and discrimination of infant distress signals by these women was found to be negatively correlated with the women's history of parental emotional rejection and their adult insecure attachment scores (either avoidant or anxious). These results imply that a history of lower levels of experienced maternal sensitivity in childhood promotes insecure adult attachments in the affected offspring, which, in turn, affects their empathic responsiveness toward infant distress signals. Although this study was conducted on a population of primarily nulliparous women, it is probably also relevant to maternal sensitivity and empathy in postpartum women.

The purpose of this section was to provide evidence for the following relationships. The way a mother treats her child is associated with the development of a secure or insecure attachment style in the child, and these attachment styles are moderately stable throughout life. In adulthood, these variations in attachment security are associated with variations in maternal sensitivity and empathy, which, in turn, influence the attachment styles of infants in the next generation. This research is primarily correlational, but a causal experiential basis for these relationships is likely to be significant. It appears that the development of avoidant attachment may lower maternal sensitivity and empathy and may be associated with lower levels of mother-to-infant bonding. The development of anxious attachment may result in inconsistent maternal sensitivity that may be due to lower levels of emotion regulation. Although the variations that I have described fall within the normal range, it can be proposed that extreme variants could result in child maltreatment. A mother with abnormally high levels of avoidant attachment may be likely to physically or emotionally neglect her child, while mothers with abnormally high levels of anxious attachment may be prone to physically or emotionally abuse their children, particularly under stressful and challenging environmental conditions.

The Relationship Between Normal Variations in Maternal Behavior and the Development of Emotionality and Sociality in the Affected Children

The research reviewed in the previous section suggests that normal variations in the way a mother treats her child can influence the development of emotionality and/or social competence (the ability to form strong social bonds) in her offspring, which would then affect the child's parental behavior in adulthood. In this section, I will describe several additional studies relevant to the development of emotionality and social competence with respect to the first link in this intergenerational chain: Normal variations in the way a mother treats her child are related to the emotional reactivity and social development of the child. In a laboratory study, Laurent, Harold, Leve, Shelton, and van Goozen (2016) studied mother–infant interactions and the child's emotional reactivity in the following way. First, mothers' interactions with their 1- to-3-year-old child were examined during a free play session. Then the mother left the room, and the child was exposed to a fearful condition (the approach of a novel mechanical moving toy). Mother–infant interactions were then re-examined upon the return of the mother after the separation period. The children of less sensitive and more intrusive mothers, particularly when this mother–infant interaction style was measured upon the mother's return and after the child was exposed to the fear condition, showed higher levels of behavioral distress (facial expressions of fear, startle and escape responses, and distress vocalizations) during the toy-stress condition and they also exhibited a prolonged hypothalamic-pituitary-adrenal (HPA) stress response as measured by levels of cortisol in the child's saliva during the poststress interval (also see Blankenship, Chad-Friedman, Riggins, & Dougherty, 2019). One interpretation of these results is that the less sensitive and more intrusive mothers typically do not respond appropriately to a distressed child and do not properly soothe and comfort their child during stressful occasions. Specifically, they do not serve as a secure base for the child. This experiential effect then leads to enhanced stress reactivity and fearfulness during future episodes that expose the child to a fearful situation. However, because these results are correlational, it is possible that more intrusive mothers and their more fearful children both exhibit heightened emotionality and that genetic transmission could be involved. In relation to the previous section, it would be interesting to know whether mothers who were more intrusive and less sensitive exhibited either an anxious attachment style or an avoidant attachment style. Based on the high level of intrusiveness, I would assume that these mothers had an adult anxious attachment style.

Abraham, Hendler, Zagoory-Sharon, and Feldman (2016) assessed parents' interactions with their 1-year-old infants. Some of the primary parental caregivers were the child's biological mothers while others were the child's biological fathers. Subsequently, the child's emotion regulation capability was assessed when the child was about 3.5 years old. They present evidence for the following associations: Increased parental empathy is associated with higher levels of primary caregiver–infant synchrony and the primary caregiver's sensitivity to the infant's needs, which, in turn, is associated with greater emotion regulation capability in preschool children when confronted with a fear-eliciting situation (also see Abraham, Raz, Zagoory-Sharon, & Feldman, 2018).

Finally, Feldman, Gordon, Influs, Gutbir, and Epstein (2013) examined mother–child synchrony and maternal sensitivity throughout the first 3 years of a child's life. At 3 years of age, they assessed the social behavior of the child while the child was playing with a close friend. Greater mother–infant synchrony and maternal sensitivity was associated with greater cooperative behavior when their 3-year-old child interacted with a friend. Similar findings have been reported by Feldman, Bamberger, and Kanat-Maymon (2013).

On the assumption that these effects of parental responsiveness on the development of emotionality and sociality in children are maintained into adulthood, they should, in turn, influence the parental responsiveness of these children when they have their own children in adulthood.

Normal Variations in Parental Behavior Are Related to the Development of the Parental Brain in Offspring

Introduction

In Chapter 9, I reviewed the evidence that normal variations in the way a rodent mother treats her offspring can influence the development of subcortical neural circuits that influence maternal responsiveness. In Chapter 8, I reviewed research on the human parental brain, which suggested that this core subcortical circuitry is overlaid by, and interacts with, cortical circuits regulating empathy and love so that maternal feeling states and cognitive empathy can affect maternal motivation and caregiving behavior directed toward infants. Cortical circuits involved in emotion regulation also interact with the subcortical circuits that underpin maternal motivation, in this way influencing maternal caregiving responses under challenging environmental conditions. Later in this chapter, I will present evidence that suggests that normal variations in human maternal behavior may influence the development of some of these cortical and subcortical circuits in the mother's young, which would be expected to result in individual variations in the maternal responsiveness of the affected offspring when they have their own children in adulthood.

With respect to attachment theory, the development of secure and insecure infant-to-mother attachments has been proposed to affect the attachment style of the adult, which, in turn, could influence the adult's parental behavior. This proposal has important implications. I, as well as others, have proposed that the maternal caregiving neural circuitry in mammals may provide a foundation upon which other strong affiliative social bonds are built (Numan, 2015; Numan

& Young, 2016). What this idea suggests is that aspects of the neural circuitry that underpins the mother-to-infant bond may also contribute to the infant's infant-to-mother bond.

With regard to the view that the neural circuitry that underpins the mother-to-infant bond may contribute to the infant's infant-to-mother bond and to the infant's other social relationships in adulthood, how would this process work in a psychological and neural sense? If a child is raised by an insensitive parent or by an inconsistently sensitive parent, the child may learn that it cannot reliably depend on its primary caregiver for protection from frightening situations and for consistent positive social interactions. Depending of the type of parental responsiveness, this might cause the child to develop either an avoidant (resulting from consistently insensitive parental caregiving) or anxious (resulting from inconsistently sensitive parental caregiving) attachment, which would then affect the ability of the child to form appropriate and adaptive social attachments with others throughout its lifespan. It is my proposal that these deficiencies in attachment processes are likely to include elements of the neural circuitry that contribute to adult parental behavior. Avoidant attachment may, in part, be the result of deficiencies in the parental neural circuitry outlined in Figures 8.3 and 8.4. Anxious attachment may result, in part, from emotional dysregulation (problems with emotion regulation and hyperactivity of subcortical mechanisms regulating anxiety and fear), which then has a negative impact on social bond formation (see Figure 8.7). These hypothetical proposals are based on the view that the parental neural circuity is present in the infant's brain, can be modified by parent–infant interactions (see Chapter 9) and that the parental neural circuitry is the primordial social bonding system that forms the foundation upon which other types of social bonds are built.

With regard to these proposals, and particularly with respect to adult avoidant attachment, Cittern and Edalat (2017) have described an attachment-based psychotherapy whose goal is to help convert adult insecure attachment styles to secure adult attachment styles. This psychotherapeutic approach is based on methods meant to increase emotional empathy, which is conceived as being low in insecure adults, but high in adults with secure attachments. The type of adult attachment, of course, is conceived as being based on the type of infant-to-mother bond that developed in childhood. Importantly, Cittern and Edalat have presented a theoretical computational neural model to explain how these therapeutic effects might work. Their neural model is based on my neural models that show how emotional empathy may drive maternal caregiving responses toward a distressed infant (see Figure 8.3 in this book and Figure 7.2 in Numan, 2015). In their computational neural model, effective attachment-based psychotherapy in adults, which is meant to promote secure adult attachment styles, increases neural activity in the anterior insular (AI) region, which, in turn, increases activity across a medial prefrontal cortex (mPFC)-to-medial preoptic area (MPOA)-to-ventral tegmental area (VTA)-to-nucleus accumbens (NA)-ventral pallidum (VP) circuit. This allows emotional empathy to promote empathic care, which results in prosocial aid-giving behaviors toward another individual in need of aid. Note how such a model would not only apply to parent–infant interactions, but also to interactions between adults and the ability of such adults to form close social bonds with others (see Chapter 11 of this volume).

In summary, if a child is raised by an insensitive or inconsistently sensitive mother, the child may learn that it cannot depend on appropriate and timely aid from others. This may then affect the ability of the child to form strong social bonds with others, and aspects of the neural circuitry that underlies such social bond formation may include parts of the parental brain circuitry.

Evidence That Normal Variations in Parental Behavior Are Related to the Development of the Parental Brain in Offspring as Measured in Adulthood

Strathearn, Fonagy, Amico, and Montague (2009) related a mother's attachment style to both her blood levels of oxytocin (OT) and her brain responses to images of her infant's face. Initially, primigravid women were interviewed during pregnancy and their attachment relationship with their own parents was assessed. The subsequent analyses included mothers with either a secure attachment style or an insecure-avoidant attachment style (women with an insecure-anxious style were not examined). This aspect of the study relates to the potential experience-based intergenerational transmission of maternal behavior and is based on the possibility that if a female had a less than adequate attachment relationship with her own mother, then such an experience might influence the development of her parental neural system and her maternal behavior. In the second part of the study, which occurred at 7 months postpartum, each mother's plasma OT levels were measured after a 5-minute mother–infant interaction. In the third part of the study, at 11 months postpartum, brain images using functional magnetic resonance imaging (fMRI) were collected while mothers viewed faces of their own or unfamiliar infants. The most important findings were as follows: (a) During the free play mother–infant interaction period, blood levels of OT increased above baseline in the secure mothers but not in the insecure-avoidant mothers; (b) the NA BOLD response was greater in secure mothers than in insecure-avoidant mothers upon viewing images of their infant displaying a happy facial expression. This BOLD response was likely due to the level of ventral tegmental area

(VTA)-dopamine (DA) axon terminal activity within NA; (c) with all females included, there was a significant positive correlation between blood OT levels measured during the play session at 7 months postpartum and the NA BOLD response measured at 11 months postpartum while mothers viewed images of their infant's face. To the extent that blood levels of OT accurately represented the central release of OT within the brain, these results imply greater OT neural system activation and mesolimbic DA system activation in secure, compared to insecure, mothers during mother–infant interactions, whether overt (during play session) or covert (during fMRI scanning).

Note how these results correspond to aspects of the rodent research reviewed in Chapter 9. In rodents, normal variations in maternal attentiveness influence the development of the MPOA-to-VTA-DA-to-NA pathway in offspring, which then affects the adult maternal responsiveness of the affected offspring. OT is primarily involved in the onset of maternal behavior in animals, but it modulates the level of maternal attentiveness during the postpartum period, and it also boosts maternal motivation during the postpartum period under challenging conditions.

These results indicate that human mothers with an insecure-avoidant attachment style may form a weaker social bond with their infants during the postpartum period and that this, in turn, is correlated with a lower level of OT and mesolimbic DA neural system activation during mother-infant interactions, perhaps due, in part, to a decreased ability of infant stimuli to activate MPOA projections to VTA-DA neurons and to paraventricular nucleus (PVN)-OT neurons.

The contribution of OT receptor (OTR) expression in the MPOA to the findings of Strathearn et al. (2009) is not known. Since experimental animal studies have shown that maternal treatment effects influence OTR expression in MPOA of her offspring, which in turn affects activity across MPOA-to-mesolimbic DA subcortical circuits that regulate maternal motivation, similar effects may also occur in humans. Furthermore, in humans OTRs are not only located subcortically, but are also present in cortical sites that are involved in empathy. Therefore, normal variations in maternal behavior could also influence the expression of OTRs in these regions, which could affect the development of emotional and cognitive empathy in their children (also refer to the subsection on the relationships between childhood maltreatment and OT neural systems later in this chapter).

A major deficiency in the research of Strathearn et al. (2009), particularly in relation to the OT findings, is that the quality and quantity of mother–infant interactions during the free play period were not recorded. It would have been interesting to know whether the insecure-avoidant mothers showed lower levels of positive affective interactions with their infants than did the secure mothers. Also note the correlational nature of these findings. If secure mothers were found to engage in more positive physical interactions with their infants, one would not be able to determine whether increases in OT caused increases in positive maternal responses or whether increased contact with infants induced the rise in OT levels (cf. Feldman, Gordon, Schneiderman, Weisman, & Zagoory-Sharon, 2010; Feldman et al., 2012).

Some of these issues have been partly resolved in a recent study by Kohlhoff et al. (2017). In comparison to mothers with a secure attachment style, mothers with an insecure-avoidant attachment style showed lower levels of maternal sensitivity when interacting with their 3- to 4-month-old infants and also had lower blood levels of OT. They suggested that lower levels of endogenous OT were related to both higher levels of attachment avoidance and lower levels of maternal sensitivity. They proposed that the way a child is treated by its mother affects the development of its OT neural system, which, in turn, affects both the attachment style and maternal sensitivity of the child in its adulthood. It has also been found that adult women with an insecure attachment style, when compared to securely attached women, have lower blood levels of OT when measured at about 24 weeks of pregnancy (Samuel et al., 2015).

There is also evidence that amygdala responsiveness to infant stimuli in mothers can be affected by the early relationships of these mothers with their parents. Kim, Fonagy, Allen, and Strathearn (2014) studied a group of mothers that they classified as having, or not having, unresolved traumatic relationships, such as insecure attachments to their primary caregiver in childhood. This concept of unresolved trauma involves less severe types of early trauma. According to Kim and Strathearn (2017), such early attachment trauma undermines an individual's capacity to develop and maintain future attachment relationships. Unfortunately, the adult attachment style of the mothers with unresolved trauma was not further categorized with measures that could classify such women as either insecure-avoidant or insecure-anxious. At 7 months postpartum, while in an fMRI scanner, mothers were presented with facial images of their own or unfamiliar infants. Compared to normal mothers, mothers with unresolved trauma showed blunted amygdala BOLD responses to the sad facial expressions of their own infants. The maternal behavior of these women was not examined, but these results suggest that empathic responding to infant distress signals may be compromised in women with unresolved trauma. I would predict that such women probably had an insecure-avoidant attachment style with the result that infant distress signals were less effective in activating positively valent amygdala neurons.

Evidence also indicates that normal variations in maternal behavior can affect the development of cortical neural circuits that influence maternal responsiveness in mothers. Kim, Leckman, Mayes, Newman et al. (2010) studied healthy mothers during the first postpartum month. These mothers were administered the Parental Bonding Instrument questionnaire, which measured their recollections of the maternal care they received in their childhood. Recollections of maternal care ranged from parental closeness, emotional warmth, and affection to maternal indifference and insensitivity to the child's needs. These women were then divided into two groups, the high perceived maternal care (HPMC) and low perceived maternal care (LPMC) groups. Subsequently, while in an fMRI scanner, these women listened to a standard infant cry sound or to white noise. The results showed that mothers in the HPMC group showed larger BOLD responses in the lateral PFC and in the superior temporal gyrus (STG) in comparison to mothers in the LPMC group. These results suggest that mothers that recalled higher levels of positive parental care in childhood had neural responses indicative of higher levels of both explicit emotional regulation and cognitive empathy (see Chapter 8 of this volume). The actual maternal behavior of these mothers toward their own infants was not examined in this study.

Several studies have related the adult attachment styles of nulliparous women to their brain responses to infant stimuli. Lenzi et al. (2013) compared women with a secure attachment style to those with an insecure-avoidant attachment style. First, in interviews with the women, it was noted that secure women looked forward to motherhood while the insecure-avoidant women had difficulty in imagining themselves as mothers in the future. These women then viewed facial expressions of young infants while in a scanner, and they were asked to try to empathize with the child's expressions. The most important finding was that when secure women viewed infant emotional expressions, the ventromedial prefrontal cortex (vmPFC), including the pregenual anterior cingulate cortex (pgACC; presumably involving area 32), showed increased BOLD responses, while this region was not activated in the insecure-avoidant women. Because the vmPFC, particularly area 32, projects to MPOA, I suggested in Chapter 8 that this projection might be part of a neural route through which maternal feelings of love and empathy are translated into infant-directed caregiving responses. To the extent that the adult attachment style of these women was related to their early experiences with their own mother, these results suggest that normal variations in experienced maternal care during childhood can influence the full development of neural circuits underlying maternal responsiveness in these adult nulliparous women.

In Chapter 8, I reviewed the research of Riem et al. (2011) and Bakermans-Kranenburg et al. (2012), which showed that when nulliparous women listened to infant cry sounds in comparison to control sounds, that infant cry sounds activated the amygdala and that such activation was decreased by intranasal application of OT. In the analysis of that research, it was suggested that for some nulliparous women infant cry sounds may be aversive as a result of the activation of negatively valent amygdala neurons. Also recall that in postpartum women, infant cry sounds also activate the amygdala but that the extent of amygdala activation is positively correlated with maternal sensitivity (Kim et al., 2011). It was suggested that in postpartum women, who have been physiologically primed for motherhood, infant cry sounds are likely to activate positively valent amygdala neurons that give rise to maternal empathic responses.

In subsequent research, Riem and her colleagues examined the amygdala response to infant cries in nulliparous women as it related to their adult attachment style. Riem, Bakermans-Kranenburg, van Ijzendoorn, Out, and Rombouts (2012) reported that nulliparous women with an insecure-anxious attachment style, in comparison to women with secure attachment, showed greater amygdala activation to infant cry sounds and that this was associated with the use of excessive force on a handgrip gauge while listening to such sounds. Riem, Bakermans-Kranenburg, and van Ijzendoorn (2016) reported that intranasal application of OT reduced both the amygdala activation and the use of excessive handgrip force while listening to infant cries in women with an insecure-anxious adult attachment style. These results suggest that for a certain group of nulliparous women, particularly those with an insecure-anxious attachment style, infant cries are aversive. This aversive response could be the combined result of a direct stimulatory effect of infant cry sounds on negatively valent amygdala neurons coupled with deficits in emotion regulation in women with an insecure-anxious attachment style. Perhaps this attachment style, which was likely affected by the women's early experiences with their own mothers, would affect the degree of allomaternal behavior of these women under more natural conditions. Although intranasal OT ameliorated the effects of infant cry sounds on amygdala reactivity and the use of excessive handgrip force, it is also possible that motherhood in these insecure-anxious women would be associated with higher levels of maternal intrusiveness and lower levels of maternal sensitivity, as suggested by the research reviewed earlier in this chapter in the subsection on intergenerational transmission of attachment styles influences parental behavior. Perhaps insecure-anxious mothers have lower levels of endogenous OT release within the brain, as do insecure-avoidant mothers (Strathearn et al., 2009). Since OT not only promotes mother–infant bonding but also exerts anxiolytic effects, the following findings are relevant: Eapen et al. (2014) found that in postpartum women, there was a significant negative correlation between plasma OT levels and the degree of the mother's anxious attachment (a similar negative correlation was also found between OT and avoidant attachment).

Evidence That Normal Variations in Parental Behavior Are Related to the Development of the Parental Brain in Offspring as Measured in Childhood

Previously in this chapter, I presented research from Feldman and her colleagues (Abraham et al., 2016; Feldman et al., 2013) that maternal empathy, maternal–infant synchrony, and maternal sensitivity were positively related to the development of cooperative prosocial behaviors and effective emotion regulation abilities in their preschool children. These researchers have also presented evidence that the intergenerational continuity of OT neural systems between the parent and child may underpin these relationships. Higher maternal plasma OT levels were correlated with more effective maternal behavior and with the child's salivary OT levels, with the latter positively correlating with the child's emotion regulation and with the child's cooperative social behavior with a best friend. Their proposed interpretation is based on the assumption that peripheral levels of OT are positively correlated with activity within the brain's OT neural system.

Conradt et al. (2016) have reported a negative correlation between maternal sensitivity and the degree of methylation within the promoter region of their 4-month-old infant's glucocorticoid receptor (GR) gene. These researchers also reported that an increased infant cortisol stress response was associated with higher levels of GR gene methylation, which was presumably the result of the decreased expression of GRs within the hippocampus. Therefore, lower levels of maternal sensitivity may increase physiological, and presumably behavioral, stress reactivity in their infants. However, since the degree of methylation of the GR gene was measured from buccal epithelial cells, it is not necessarily the case that this degree of methylation would apply to brain cells, such as neurons in the hippocampus.

Note how these studies map on to the results of the animal studies reviewed in Chapter 9. The way a mother treats her infants may affect the development of both OT and corticotropin-releasing factor (CRF) neural systems in the infant's/child's brain, which would then influence the parenting style of the affected offspring in adulthood.

The Relationships Between Abnormal Parental Care and the Development of Neural Circuits That Could Impact Parental Behavior in the Affected Offspring

Introduction

In the previous section on normal variations in parental behavior are related to the development of the parental brain in offspring, I explored the relationships between normal variations in parental behavior and the development of the parental brain in the affected offspring. In this section, I will review research that deals with the effects of severe forms of parental abuse and/or neglect on the development of neural circuits that could influence the adult maternal responsiveness of children that have been exposed to such atypical parental caregiving. In certain instances, to present a coherent and cohesive analysis of complex data, I will buttress these findings with supportive data derived from normal variations in maternal behavior.

The Development of Emotion Regulating Neural Systems in Children That Have Been Raised in Orphanages

Tottenham and her colleagues have engaged in a long-term research project that has examined the emotional development, from both a behavioral and neural perspective, of children that have been raised in orphanages during infancy and then subsequently adopted into families (for a review, see Tottenham, 2015). Children raised in orphanages throughout their first year of life, prior to adoption, are typically exposed to low levels of quality parental care because one caregiver may be responsible for many children, and each institutionalized child is usually cared for by many different individuals. Such conditions are likely to be associated with high levels of inconsistent care and emotional neglect. Tottenham's group has presented convincing evidence that previously institutionalized children develop deficits in emotion regulation, particularly with respect to mPFC regulation of amygdala reactivity to fearful stimuli, which results in an anxious behavioral phenotype. This research is importantly related to the animal research showing that prolonged daily maternal separations of neonatal rodents from their mother (a rodent model of maternal neglect) affects the development of the mPFC in the affected offspring so that in adulthood the ability of the mPFC to downregulate amygdala reactivity to threatening stimuli is decreased.

A series of studies have compared previously institutionalized children who were adopted after about 15 months in an orphanage with a control comparison group of children that were never institutionalized. Behavioral and neural analyses were conducted when these children (both girls and boys) were about 9 years old. The previously institutionalized children, in comparison to controls, had larger amygdala volumes, increased anxiety and fear-related responses, and increased amygdala BOLD responses to photos of fearful faces (Malter Cohen et al., 2013; Tottenham et al., 2010).

Gee, Humphreys et al. (2013) have explored the normative development of the functional connectivity between the mPFC (comprising areas 24, 25, and 32) and the amygdala in healthy children that were raised from birth in their biological families. The children were separated into two age groups, those younger than 10 years of age and those older than 10 years of age. While in an fMRI scanner, these children viewed photos of neutral or fearful facial expressions. For the younger children, there was a positive correlation between the amygdala BOLD response to fearful faces and the BOLD response in the mPFC. For the older children, a negative correlation existed: When presented with fearful faces, increased BOLD responses in the mPFC were associated with decreased BOLD responses in the amygdala. One interpretation of these results, as suggested by Gee et al., is that in young children, amygdala stimulation by fearful stimuli activates bottom–up excitatory connections to the mPFC, but that top–down inhibitory control of the amygdala by the mPFC has not yet developed. In older children, however, such top–down inhibitory control has developed, resulting in a negative task-related functional connectivity between the mPFC and amygdala.

In a subsequent study (Gee, Gabard-Durnam et al., 2013), previously institutionalized children and comparison control children were each divided into a younger (6–10 years) and an older (11–18 years) age group. When exposed to fearful facial expressions while in a scanner, the control children demonstrated the previously described normative development of mPFC-amygdala functional connectivity: For the younger children, the functional connectivity was positive, but this switched to a negative functional connectivity for the older children. In contrast, for both age groups of the previously institutionalized children, a negative functional connectivity between the mPFC and amygdala was observed. These results suggest that top–down inhibition of amygdala responsiveness to fearful stimuli by the mPFC develops earlier in previously institutionalized children than in the comparison control group. One way to interpret these results is that for younger control children, their mothers typically serve as a secure base to buffer their fear responses, which delays the development of mPFC inhibitory control of amygdala reactivity to fear (Gee et al., 2014). For previously institutionalized children, however, a lack of consistent care from a reliable caregiver who would serve as a secure base during infancy may have hastened the development of top–down inhibitory control of amygdala fear circuits by the mPFC as an adaptive response to deal with dangerous and/or threatening situations. Similar conclusions have been reached by Thijssen et al. (2017).

Other important findings from the Gee, Gabard-Durnam et al. (2013) study were that measures of anxiety indicated that despite an early development of mPFC-to-amygdala inhibitory control, previously institutionalized children were more anxious than the children in the comparison control group, and their amygdala BOLD responses to fearful faces were also greater than those observed in the comparison group. Interestingly, not all children who were raised in orphanages demonstrated an early development of mPFC-to-amygdala negative functional connectivity, and those that showed such early development were less anxious than those that did not. However, the former children were still more anxious than the comparison control children. Therefore, the early development of the presumed top–down inhibitory control of amygdala fear circuits by mPFC may have only partially ameliorated anxiety and fearfulness in the previously institutionalized children because their emotion regulation neural capabilities were not fully realized. Perhaps the early development of this negative connectivity, because it occurred at an atypical and premature developmental stage, was not fully complete. Indeed, in a behavioral test, previously institutionalized children showed deficits in emotion regulation, as measured by their decreased ability to perform effectively on a task under conditions where they were anticipating the possible occurrence of a frightening stimulus (Malter Cohen et al., 2013). It is interesting to speculate that an early but partial development of emotion regulation neural systems (refer to Figure 8.6) in children while they are in orphanages and are receiving suboptimal care may be an adaptive mechanism. The increased anxiety and hypervigilance that is evident in such children (Bowlby, 1973; Tottenham, 2015) may allow them to function effectively in a potentially high-risk environment. However, and this is important, the persistence of heightened anxiety and deficits in emotion regulation would not be adaptive after these children are adopted into a normal family setting. I have emphasized this point before: What is adaptive under one set of environmental conditions may result in maladaptive responses under another set of environmental conditions. The problem, therefore, is that the heightened anxiety that develops in institutionalized children persists during later periods of their lives when they are no longer living in a high-risk environment.

It is interesting to compare these findings from Tottenham and her colleagues with the animal research reviewed in Chapter 9. The animal research suggested that maternal neglect (prolonged maternal separation) causes the early development of amygdala fear circuits and enhanced fearfulness and that this early development might be related to increased mPFC stimulation of amygdala fear circuits. In contrast, the human research on previously institutionalized children suggests that a history of suboptimal care from a caregiver is associated with an early development of the ability of the mPFC to downregulate amygdala fear circuits, but this early development is only partially successful. In both case, however, the outcomes are similar: Lower levels of parental care are associated with the development of increased anxiety and deficits in emotion regulation in the affected offspring.

It would certainly be interesting to know what the parental behavior of previously institutionalized individuals toward their own children would be like. It would also be important to explore their brain responses to various infant stimuli. I am not aware of any research on these important issues. However, since postpartum anxiety and depression are associated with a disruption of mother–infant interactions and parental brain responses, I would predict that deficits in parental behavior would be detected in at least some mothers who were raised in orphanages for long durations during their childhood, particularly when such mothers are interacting with a distressed infant under demanding environmental conditions.

The lack of research on an intergenerational analysis of parenting in previously institutionalized individuals is likely related to the fact that many preclinical and clinical researchers have a primary focus on how poor or suboptimal parenting is related to the development of psychopathology, such as anxiety, depression, and other disorders in the affected children (Herpertz & Bertsch, 2015; Nemeroff, 2016). But the emotional, motivational, and cognitive characteristics of many of these disorders are sure to influence parenting. Therefore, the further exploration of parenting behavior in such individuals would inform us about the intergenerational link through which faulty socioemotional characteristics are transmitted across generations.

Finally, and interestingly, Fareri and Tottenham (2016) have speculated that the early development of mPFC-to-amygdala functional connectivity in previously institutionalized children may interfere with the normal development of amygdala connectivity with the NA-VP circuit. In support, Goff et al. (2013) have reported that adolescents that had been previously institutionalized in orphanages during early childhood demonstrated a hypoactive BOLD response in the NA-VP region while viewing happy faces in a scanner when compared to a control group of adolescents that were raised within their families from birth. If similar results could be shown to be the case for such groups of postpartum women, particularly while viewing infant faces, then the suboptimal care received by infants raised in orphanages might also interfere with the normal development of subcortical circuits that are known to regulate maternal motivation.

The Relationships Between Child Maltreatment and the Development of the Parental Brain

Introduction

My use of the term *child maltreatment* (CMT) refers to cases where a child has been abused and/or neglected by its parents. Such cases of CMT include physical and emotional neglect and physical, emotional and sexual abuse. As reviewed by McCrory, Gerin, and Viding (2017), children who experience maltreatment are typically exposed to more than one type of abuse or neglect. Depending on the study, instances of CMT are sometimes documented in welfare department state records, or they are recalled in retrospective reports by either the child or the parents. In many studies, CMT scores represent a cumulative measure of all forms of neglect and abuse, and the behavioral and neural outcomes associated with different types of abuse and/or neglect are not analyzed. I think it is important to differentiate the potential influences of different forms of abuse and neglect, and I will note those studies that focus on relationships between particular forms of CMT and their associated outcomes.

Not surprisingly, CMT is associated with the development of psychopathology in the maltreated child, although not all maltreated children develop clinical levels of psychopathology (McCrory et al., 2017). However, even when a psychiatric disorder, such as severe anxiety and depression, has not developed, CMT can be associated with brain changes which might represent latent risk factors that could promote the future development of psychopathology in certain individuals, particularly when such individuals are exposed to additional stressors later in life (McCrory et al., 2017).

As I will show, CMT has been associated with increased amygdala reactivity to threatening stimuli, dysregulated emotion regulation, and decreased responsiveness of the NA-VP circuit to rewarding stimuli. Clearly, such alterations would be expected to promote anxiety, fearfulness, and depression (particularly anhedonia). Interestingly, the results I reviewed with respect to children raised in orphanages presents a similar picture. It seems obvious that increased fearfulness, anxiety, and threat detection should develop in children that have been abused. These outcomes might seem less obvious in cases of extreme neglect (abnormal parental neglect should not be equated with the parental insensitivity that can occur in normal parents). On closer scrutiny, however, one can understand that in the absence of parental support and comforting under threatening or strange situations, it is highly likely that increased fearfulness and anxiety, which would be associated with enhanced vigilance for potential threats, would develop in children exposed to severe forms of physical and/or emotional neglect (Puetz et al., 2019).

Although CMT is typically associated with the development of increased anxiety and fearfulness, there is some recent evidence that indicates that when children are both abused and neglected by their parents they may develop hypoactive emotional responses or emotional detachment (Puetz et al., 2019).

Unfortunately, as in the case of previously institutionalized children, studies on the neural alterations associated with CMT have not been extensively examined in postpartum mothers (and fathers). This fact presents a serious difficulty with respect to making firm conclusions about the neural underpinning of the intergenerational continuity of abnormal parental behavior. The studies I will review primarily deal with the relationships between CMT and behavioral and neural outcomes in nonparents. With respect to the intergenerational sequence of (a) exposure to child maltreatment, (b) associated behavioral and brain changes in nonparents, and (c) associations between these brain changes and the parental behavior of the child who was abused/neglected, most of the research only deals with the first two steps in the sequence. Despite this problem, it is expected that the enhanced anxiety, fearfulness, and emotional dysregulation associated with CMT would negatively impact the subsequent parental behavior of an individual who had been maltreated as a child. Such emotional changes might promote intrusive, abusive, and coercive parental behaviors and a lack of appropriate parental protective responses. Further, the disruption of reward-processing neural circuits would likely contribute to deficits in parental motivation and give rise to various forms of neglectful parenting.

Evidence That Child Maltreatment Is Associated With Alterations in Neural Circuits Relevant to Parental Behavior

I am only aware of two studies that examined the brain and maternal behavior outcomes observed in postpartum mothers who were exposed to CMT during their childhood. Both of these studies used magnetic resonance imaging anatomical procedures to measure either the gray matter volume (GMV) within certain cortical neural regions (Mielke et al., 2016) or cortical white matter neural tract integrity using diffusion tensor imaging (Rodrigo et al., 2016). Based on direct observations of mother–infant interactions, both studies found that mothers who were maltreated during their childhood were poorer mothers (decreased maternal sensitivity) than were control mothers who were not maltreated. The maltreated mothers were also at risk for maltreating their children. A preliminary interpretation of the combined anatomical results from both studies suggests that the association between experiencing maltreatment and the occurrence of lower levels of maternal behavior in the affected mothers may be mediated, in part, by a reduction in emotional empathy (decreases in GMV within the AI). However, maternal sensitivity and emotional availability between a mother and her child may sometimes improve in mothers who have experienced maltreatment, and such improvement is associated with anatomical evidence for an increase in cognitive empathy (increased GMV within STS/STG, and an increased volume of a white matter tract connecting facial regions of the occipital [visual] lobe with the STG). If mothers who were maltreated as children can compensate for deficits in emotional empathy by using cognitive empathy networks, then such a compensatory mechanism may ameliorate some of their maternal behavior deficits.

In a related fMRI study, Wright, Laurent, and Ablow (2017) studied mothers who received different degrees of emotional and physical neglect in their childhood. Mothers who reported higher neglect in their own childhood, when compared to those who were exposed to less neglect, showed greater activation to their own infant's cry sound in the insular cortex and in the ACC. Although the maternal behavior of the mothers in this study was not reported, Wright et al. suggest that this increased activation of the insula-ACC network in the mothers that were exposed to high levels of neglect was probably due to the enhanced activity of negatively valent neurons in these regions that mediated an aversive emotional state (refer to Figures 8.6 and 8.7). If that was indeed the case, one would predict that such mothers would show less sensitive maternal behavior toward their own child due to a lack of empathy coupled with an increase in avoidance/rejection responses, perhaps influenced by deficits in emotion regulation. It is interesting to speculate that the decreased GMV in the AI that was reported in the Mielke et al. (2016) was due to atrophy of those positively valent AI neurons that would give rise to a positive affective state in response to infant stimuli.

Due to the correlational nature of these studies, note how certain findings are interpreted within the context of expected outcomes. Wright et al. (2017) suggested that CMT may have resulted in increased activity within negatively valent AI neurons when a mother listened to her infant's cries. But if these mothers would have been warm and sensitive mothers who were not exposed to CMT, an opposite proposal might have been put forward if infant cries, in comparison to control sounds, were associated with an enhanced BOLD response in AI. In this case, it would have probably been proposed that infant cries activated positively valent AI neurons related to empathy. Clearly, more detailed research needs to be employed, including functional connectivity and neurochemical analyses, to add strength to many of these proposals and to differentiate positively valent from negatively valent neurons(see the study by Lutz et al., 2018, which is described later in this section). This general problem with respect to interpreting correlational data is also relevant to the research findings reviewed in Chapter 8 on the human parental brain, which was alluded to in the summary of the section on the cortical neural regions and circuits relevant to the maternal behavior in women in that chapter.

It is surprising and unfortunate that research has not focused on the possibility that exposure to child maltreatment might affect the subsequent maternal behavior of women by influencing the development of functional activity and connectivity within neural circuits along a mPFC-to-rostral hypothalamus (MPOA)-to-mesolimbic DA system route. This is a fertile area for future research.

Most of the other studies that have explored the potential developmental outcomes of a history of CMT have studied nonparents (usually children and adolescents) and have primarily, but not exclusively, focused on the relationship between CMT and the development of emotional dysregulation.

Several studies have found that male and female adolescents with a history of CMT (typically determined through self-report questionnaires), when compared to control subjects without such a history, demonstrate increases in internalizing (anxiety, depression) and externalizing (impulsive aggression) disorders. During the passive viewing of negative emotional stimuli, such as angry or fearful faces, while in a scanner, the CMT group also demonstrates greater BOLD activation within the amygdala and AI in comparison to the control subjects (Dannlowski et al., 2012; Hein & Monk, 2017; McCrory et al., 2017; McLaughlin, Peverill, Gold, Alves, & Sheridan, 2015). One interpretation of these results is that negative emotional stimuli cause a greater activation across amygdala-to-AI neural circuits that underpin aversive states in adolescents with a history of being abused or neglected than in their non-maltreated counterparts. It is highly likely that an early history of abuse and/or neglect results in a hypervigilant state with respect to potentially threatening stimuli. Significantly, Lutz et al. (2018) examined postmortem brain tissue of individuals with and without a history of prior CMT and they found that the kappa opioid receptor (KOR) was decreased in AI tissue obtained from the CMT group in comparison to the control group. They also provided evidence that epigenetic mechanisms may be involved in this effect. Just how the downregulation of KORs might alter the function of the AI in individuals with a history of CMT remains unknown. Perhaps this change in KORs mediates the experience of heighted negative emotions in response to aversive stimuli, although deficits in emotional empathy or the experience of positive emotional states are alternative possibilities.

Interestingly, in the previously cited study by McLaughlin et al. (2015), it was reported that participants in both the CMT and comparison groups were able to decrease the amygdala activation that resulted from viewing negative stimuli through the use of cognitive reappraisal methods, which would be an example of explicit emotion regulation. However, during such explicit emotion regulation, the CMT group exhibited greater activation of the lateral PFC (a region involved in explicit emotion regulation) than did the control group. One interpretation of these results is that the maltreated adolescents, in comparison to the controls, had deficits in implicit emotion regulation and also required more cognitive effort to decrease emotion-related neural responses through the use of explicit emotion regulation mechanisms. McLaughlin et al. suggest that because maltreated adolescents appear to require greater neural resources to regulate their emotions effectively, emotion regulation processes may break down under conditions of high cognitive load and ongoing stress. One should be able to see that if similar processes occurred in mothers with a history of CMT, their maternal caretaking behaviors might be disrupted under complex and stressful environmental conditions.

Two important studies have supported aspects of the analysis of the McLaughlin et al. (2015) study. Herringa et al. (2013) studied a group of 18-year-old females who filled out the Childhood Trauma Questionnaire (CTQ; a retrospective self-report of experienced abuse and/or neglect during childhood). The variance of CTQ scores among these females was driven most strongly by a childhood history of emotional abuse and emotional neglect. Higher CTQ scores predicted more internalizing symptoms. While in a scanner, the resting state functional connectivity (rsFC) between the vmPFC and the amygdala was measured. Variations in rsFC are considered as measuring differences in the strength of connectivity between brain regions in the absence of a specific task (the subject rests quietly in the scanner). Significantly, higher CTQ scores were associated with decreased rsFC between the vmPFC and the amygdala. Using a statistical mediation model, Herringa et al. (2013) propose that high levels of CMT predict low rsFC between vmPFC and amygdala, which, in turn, predicts higher levels of anxiety and depression. Since vmPFC connections to the amygdala are involved in implicit or automatic emotion regulation (see Figure 8.6), these results support the view that CMT is associated with deficits in implicit emotion regulation, which may then lead to amygdala hyperactivity (above normal values) in response to potentially threatening stimuli.

Marusak, Martin, Etkin, and Thomason (2015) studied two groups of 12- to 13-year-old children, primarily composed of females. The control group did not have a history of CMT. Parent and child retrospective reports assessed the level of CMT in the group with such a history. On a neuropsychological task that measures implicit emotion regulation abilities, in comparison to controls, the CMT group exhibited deficits in implicit emotion regulation. While in a scanner during this emotion task, the CMT group exhibited greater amygdala activity and decreased negative functional connectivity between the vmPFC (pgACC; area 32) and the amygdala. Marusak et al. (2015) proposed that exposure to child abuse and/or neglect prevents the full development of negative connectivity between the vmPFC and the amygdala, which interferes with the ability of the vmPFC to downregulate amygdala activity to negative emotional stimuli, with a resultant deficit in implicit emotion regulation abilities.

Marusak et al. (2015) also reported that the heightened amygdala activity in the CMT group during tasks that required implicit emotion regulation was negatively associated with a behavioral measure of reward sensitivity. That is, heightened amygdala activity was associated with decreased positive affect in response to rewarding stimuli. This finding is reminiscent of the proposal by Fareri and Tottenham (2016) that the abnormal development of mPFC-to-amygdala functional connectivity in previously institutionalized children may interfere with the normal development of amygdala connectivity with the NA-VP circuit, and this proposal was supported, in part, by the findings of Goff et al. (2013; also see the previous discussion of the development of emotion regulating neural systems in children that have been raised in orphanages). Importantly, several studies have reported that adolescents of both sexes who reported a history of CMT, in comparison to controls, demonstrate decreased BOLD responses to reward-related stimuli across neural regions that encompass the NA-VP circuit, and in most cases this neural change was associated with decreases in positive affect and anhedonia (Corral-Frias et al., 2015; Dennison et al., 2016; Dillon et al., 2009; Hanson, Hariri, & Williamson, 2015). In most of these studies, rewards or reward-related stimuli were presented in a nonsocial context. However, the study by Dennison et al. used social stimuli. While in the scanner, the participants in the Dennison et al. study passively viewed positive and neutral social stimuli, such as happy versus neutral facial expressions. Those individuals who had a history of CMT and who also scored higher on measures of depression, exhibited decreased BOLD responses in the left pallidum, which presumably included the ventral pallidum, to positive social stimuli.

To the best of my knowledge, studies do not exist that have examined the neural reactivity across the NA-VP circuit to infant stimuli in postpartum women with a history of CMT. However, it is expected that certain postpartum women with a history of CMT would demonstrate decreased responsiveness within the NA-VP circuit to infant stimuli. Perhaps such a decrease would be associated with decreases in the joys and pleasures of maternal caretaking activities and with the development of a weak mother–infant bond. Such changes, if severe enough, could lead to a mother neglecting the needs of her infant and would reflect an intergenerational transmission of abnormal maternal behavior. There is some indirect evidence, conducted on postpartum women, that lends support to these views. Su, Leerkes, and Augustine (2018) examined the interactions between primiparous postpartum women and their 2-year-old infants, and maternal sensitivity during these interactions was scored. Each mother also provided self-reports of early adverse life experiences, which included instances of physical abuse, death of a parent, and parental divorce. Finally, salivary samples were obtained from each mother to detect variations in the DA D4 receptor gene (DRD4 gene). Note that DA receptors (DR) fall into two major classes, the D1 and D5 class and the D2, D3, and D4 class, and that the receptors in each class respond similarly to DA (Missale, Nash, Robinson, Jaber, & Caron, 1998). Recall from Chapter 5, that DA action on D2 receptors in NAs is involved in the formation of an enduring mother–infant bond. Perhaps DA action on D4 receptors would exert similar effects. As indicated by Su et al., the DRD4 gene is associated with two major variants or alleles, a long (DRD4L) and a short (DRD4S) allele, and there is evidence that the DRD4L allele results in the expression of fewer D4 receptors. The results of the study by Su et al. provided evidence for a G × E interaction. Mothers who had a history of early life adversity and who also carried one or two copies of DRD4L allele demonstrated lower levels of maternal sensitivity while interacting with their infants in comparison to mothers without early life adversity or mothers with a history of early life adversity who carried two copies of DRD4S. Although these results do not directly inform us about brain mechanisms and do not inform us about neural sites where DA action on D4 receptors may be occurring, they are consistent with the hypothesis that a history of CMT (neglect or abuse) may interfere with the development of the neural mechanisms that underpin reward processing. The observation of a significant G × E interaction also contributes to our understanding of the fact that not all mothers who experience CMT end up showing less than adequate maternal behavior toward their own children. Although highly speculative, the coupling of CMT with a less efficient DRD4 (due to lower levels of receptor expression) may have interfered with the formation of a strong mother–infant bond by affecting the strength of amygdala-to-VP connectivity (see Figure 5.15).

The Relationships Between Childhood Maltreatment and CRF and 5-HT Neural Systems

In Chapter 9, I proposed that exposure of the young infant to maternal neglect activates a hypersecretion of CRF within the brain. Abnormal CRF levels, in turn, were proposed to activate supernormal levels of 5-hydroxytryptamine (5-HT) release into the mPFC, leading to the development of deficits in emotion regulation and heightened anxiety due to the faulty development of mPFC-to-amygdala relationships. In the previous discussion, I presented evidence that suboptimal parenting is associated with emotional dysregulation in the affected offspring, and I related this dysregulation to alterations in mPFC-to-amygdala connectivity. In this section, I will explore those human studies that have examined the relationships between CMT and CRF/5-HT neural systems and their associated behavioral/psychological correlates.

One important aspect of the animal studies reviewed in Chapter 9 was related to epigenetic effects of mothering on GR expression in the hippocampus of the mother's offspring. Low levels of maternal care were related to increased methylation within the promoter region of the GR gene of the affected young, which was associated with decreased GR expression in the hippocampus and an increase in the physiological stress response due to the decreased ability of corticosterone to exert negative feedback effects at the level of the hippocampus. Is there evidence that CMT exerts similar effects in humans? (Also refer to the work of Conradt et al., 2018, which was previously presented in this chapter in the subsection on the evidence that normal variations in parental behavior are related to the development of the parental brain in offspring as measured in childhood.)

Several, but not all, studies have indicated that a history of CMT is associated with a heightened physiological stress response in the affected children (Fisher et al., 2016). A particularly instructive study was conducted by Kuhlman, Geiss, Vargas, and Lopez-Duran (2015). Children (9–16 years of age) with and without a prior history of maltreatment completed the Socially Evaluated Cold Pressor Task. During this task, each child submerged their hand into ice water for 3 minutes while being observed by a researcher. Cortisol levels from saliva samples were obtained before, during, and after exposure to this stressful event. The parent of each child also completed an Early Trauma Inventory to indicate the types of maltreatment that their child might have been exposed to. A childhood history of emotional abuse (being persistently ridiculed and insulted by a parent) was associated with a slower decline in cortisol levels following the acute stress, suggesting an impaired cortisol negative feedback regulation of HPA axis reactivity.

A history of CMT is also associated with increased DNA methylation within the promoter region of the GR gene, although not all studies have reported this effect (Barker, Walton, & Cecil, 2018; Turecki & Meaney, 2016). In most of these studies, DNA samples were obtained from peripheral tissues, although some studies have utilized brain tissue. The research of McGowan et al. (2009) provides findings that closely match the animal data presented in Chapter 9. In this study, postmortem analysis of hippocampal tissue was conducted on three groups of adult subjects. Group 1 was composed of individuals with a history of CMT and who committed suicide. Group 2 was composed of adults who committed suicide but were not exposed to CMT. Group 3 was comprised of control subjects who did not experience CMT and did not commit suicide, but died suddenly from other causes. In comparison to the other groups, the subjects in Group 1 had the lowest amount of hippocampal GR messenger ribonucleic acid (mRNA) expression and the highest levels of DNA cytosine methylation within the promoter region of the GR gene.

Parent et al. (2017) conducted an interesting study on a group of male and female preschool children (3–5 years old). Fifty percent of these children had substantiated cases, as indicated in child welfare records, of CMT within the 6-month period prior to the onset of the study. DNA samples from saliva were obtained at the onset of the study (baseline sample) and again 6 months later, which would be within 1 year of CMT. At the onset of the study, at baseline, DNA methylation within the promoter region of the GR gene was higher in the CMT group than in the comparison control, but this group difference did not exist at the 6-month follow-up time point.

How can we interpret these results? One possibility, as suggested by Parent et al. (2017), is that interventions by state welfare agencies may have prevented the continued occurrence of adverse experiences by the formerly maltreated children. Such an outcome may have, in turn, reversed the hypermethylation status of the children in the maltreated group. There is recent research (O'Donnell et al., 2018) that analyzed genome-wide DNA methylation and supports this hypothesis. Therefore, individuals who are exposed to CMT and who also exhibit GR gene hypermethylation in adulthood, such as those in the McGowan et al. (2009) study, may not have received any intervention-type support during childhood, perhaps because it was not reported to state child welfare agencies, with the result that such individuals may have had a history of prolonged periods of childhood abuse and/or neglect.

It is worth considering, however, that even brief periods of hypermethylation of the GR gene during childhood may still exert a negative impact on the emotional development of the maltreated child. If decreased expression of GRs within the hippocampus for brief periods of time during early childhood causes brief periods of HPA axis hyperactivity and if such enhanced CRF and cortisol release occurs during particularly sensitive periods of brain development, then detrimental neurobehavioral outcomes may occur (cf. Dunn et al., 2019). Recall the findings of Moriceau et al. (2006) from Chapter 9, where an abnormal increase in corticosterone levels within the amygdala of young rats resulted in a premature maturation of amygdala-regulated fear learning, which, in turn, may have interfered with the normal development of the mPFC circuits that regulate and restrain amygdala reactivity to emotional stimuli.

The studies reviewed so far show that CMT is associated with increased DNA methylation within the promoter region of the GR gene. However, there is also correlational evidence that the association between a history of CMT and the subsequent development of internalizing disorders (anxiety and depression) in maltreated children of both sexes is only partially mediated by increased methylation of the GR gene (Barker et al., 2018; Parade et al., 2016). These findings indicate that factors other than increased methylation of the GR gene may mediate the relationship between CMT and the subsequent development of psychopathology.

Overall, this research fits with the proposal that the enhanced cortisol and CRF release that may occur in individuals with a history of CMT, due to epigenetic processes that blunt cortisol negative feedback regulation of CRF release, may ultimately contribute to emotional dysregulation in the maltreated individuals. In Chapter 9, I related the heightened physiological stress response in offspring that had received low levels of maternal care to the concurrent development of increased behavioral stress reactivity and anxiety: Due to disruptions in the negative feedback effects of corticosterone on CRF release, I suggested that increased CRF release within the brain, originating from the PVN and/or CeA, could result in enhanced anxiety and stress reactivity. Importantly, adult women and men who self-reported a history of CMT, particularly in early childhood, have higher levels of CRF in cerebrospinal fluid than a comparison control group without such a history (Carpenter et al., 2004; Lee, Gollan, Kaschow, Geracioti, & Coccaro, 2006).

In Chapter 9 I also suggested that abusive and/or neglectful mothering, which would trigger the enhanced release of CRF in the affected offspring, would also result in CRF strongly activating dorsal raphe (DR) 5-HT neurons projecting to the mPFC. Evidence was presented that such supernormal 5-HT release into mPFC might disrupt the normal functional relationship between the mPFC and the amygdala, leading to the development of emotional dysregulation. Is there any evidence from human research that would conform to these proposals that were developed from animal research?

Recall that there are two important alleles of the 5-HTT gene, the short (s) allele and the long (l) allele and that the short allele is associated with a decreased expression of 5-HTT mRNA and protein. Therefore, individuals carrying the short allele would presumably have more long-lasting actions of 5-HT at neural targets where it is released because there would be fewer 5-HT transporters present to terminate serotonin's action. With this background, one could incorporate an understanding of G × E interactions to explain why some, but not all, individuals with a history of CMT go on to develop emotional disorders, such as anxiety and depression, which could then have a negative impact on their parental responsiveness toward their own children. Carriers of the short allele would presumably have a more long-lasting effect of 5-HT on mPFC development in response to stress-induced CRF release into DR caused by parental neglect and/or abuse.

The classic findings of Caspi et al. (2003) lend some support to the hypothetical sequence described in the previous paragraph. The incidence of depression in adult men and women who did or did not have a history of CMT was moderated by the individual's 5-HTT genotype. The occurrence of depression in homozygous l/l individuals was relatively low and did not vary with the presence or absence of a history of CMT. In contrast, while s/s individuals without a history of CMT had a low incidence of adult depression, similar to that of l/l individuals, exposure to CMT was associated with a dramatic increase in the occurrence of depression in s/s individuals. On the basis of these findings, the 5-HTT short allele has been referred to as a risk allele because carriers of the short allele would be more vulnerable to the deleterious effects of CMT and other life stressors, perhaps because of the enhanced action of 5-HT at critical neural sites during sensitive periods of brain development.

The findings of Caspi et al. (2003) have been confirmed in a number of subsequent studies (Nemeroff, 2016; Starr, Hammen, Conway, Raposa, & Brennan, 2014). A study that is relevant to maternal behavior was conducted by Mehta et al. (2012). During pregnancy, women completed a self-report questionnaire that assessed their exposure to negative events during their lifetime. These negative life events included not only a prior history of CMT, but also other types of stressors, such as death of a relative or friend. Women who carried the s allele (s/s, s/l) and had experienced at least one prior adverse life event had higher depression scores at 6 months postpartum than did l/l homozygous women. These two genetic groups did not differ in postpartum depression scores if they were not previously exposed to major negative life events.

Recent research suggests that the use of the term *risk* allele when referring to G × E interactions may not capture to full extent to which different alleles interact with the environment to result in distinct phenotypic outcomes. The differential susceptibility hypothesis proposes that for certain genes and their allelic variants, the term *susceptibility* allele may be more appropriate (Bakermans-Kranenburg & van Ijzendoorn, 2015; Belsky, Bakermans-Kranenburg, & van Ijzendoorn, 2007). According to the differential susceptibility hypothesis, the same genotype or allele that makes a person vulnerable to adversity may result in a disproportionately favorable phenotypic outcome when such an individual is exposed to positive environmental conditions. With respect to the serotonin-transporter-linked polymorphic region (5-HTTLPR), Figure 10.1 shows a model of how the differential susceptibility hypothesis might work. This figure shows the relative amounts of 5-HT that may be released at critical brain sites, such as the mPFC, early in life in response to different types of environmental conditions (early life stress vs. no stress or a positive early environment) in individuals with either and s/s or l/l genotype. Under early life stress, such as CMT, too much 5-HT may be released at critical brain sites (perhaps triggered by CRF) in s/s compared to l/l individuals, due to a less efficient 5-HT reuptake mechanism associated with the s/s genotype. However, too little 5-HT release into the brain during early life may also be associated with negative outcomes. Under nonstress or positive early life experiences, s/s individuals may have a more optimal 5-HT release into the brain than l/l individuals, leading to more favorable outcomes for individuals carrying the s/s genotype.

Is there any evidence to support this differential susceptibility hypothesis with respect to the 5-HTTLPR? Brett et al. (2015) compared the degree of externalizing behavior (impulsive aggression; a form of emotional dysregulation) observed in 4.5-year-old children who had been raised either in orphanages or

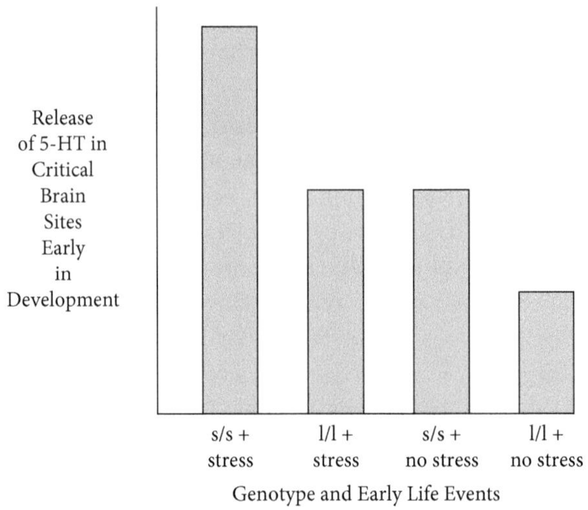

Figure 10.1. The differential susceptibility hypothesis as applied to the serotonin transporter-linked polymorphic region (5-HTTLPR) of the serotonin transporter gene (5-HTT gene). Research indicates that the short allele (s) of 5-HTTLPR results in the synthesis of lower levels of 5-HTT protein, which would enhance serotonin (5-HT) neural effects at its target sites. In contrast, the long (l) allele produces higher levels of 5-HTT protein, which decreases 5-HT neural effects. Too much or too little 5-HT release into the brain during early life may disrupt the normal development of critical brain regions, such as the medial prefrontal cortex. Moderate amounts of 5-HT release would be associated with normal brain development. Early life stressful experiences are proposed to stimulate 5-HT release. Gene by environment (G × E) interactions can modulate the release of 5-HT into the brain of a young organism. Under early life stress, such as being exposed to child maltreatment, individuals carrying the s/s genotype would have too much 5-HT release into the brain compared to individuals with the l/l genotype. In contrast, for individuals who are exposed to nonstress (normal) rearing conditions early in life, s/s individuals would have moderate amounts of 5-HT release into the brain, while l/l individuals would have too little 5-HT release into the brain. These proposed results support the differential susceptibility hypothesis in that the s/s genotype would result in abnormal brain development under adverse early life conditions, but would result in favorable brain development under normal rearing conditions. In contrast, l/l individuals would have normal brain development under early life adversity but would have less than adequate brain development under favorable early life conditions.

in good-quality foster care. For individuals raised in orphanages, externalizing behavior was much higher in s/s compared to l/l individuals. However, for those raised in good-quality foster care, externalizing behavior was higher in l/l individuals than in those carrying the s/s genotype.

Most important, the differential susceptibility model with respect to the 5-HTTLPR may also be applicable to mother–infant interactions. Bouvette-Turcot et al. (2015) examined the degree of negative emotionality and behavioral dysregulation in 36-month-old-infants as reported by their mothers. Each mother's experience of childhood adversity, determined by their responses to the CTQ, was also assessed. Children with the s/s genotype had higher levels of emotional dysregulation than those with the l/l genotype if their mothers had experienced high levels of childhood adversity. In contrast, children with the l/l genotype had higher levels of emotional dysregulation than those with the s/s genotype when their mothers reported an absence of childhood adversity. Although mother–infant interactions were not examined in this study, one interpretation of these results, consistent with the differential susceptibility model, is that mothers who experienced CMT were less nurturing toward their infants in comparison to mothers without such a history. The emotional development of infants with the s/s genotype may have been adversely affected by poor mothering, but may have been positively affected by supportive and sensitive maternal caretaking. Finally, Mileva-Seitz et al. (2011) explored relationship between the mother's 5-HTTLPR genotype and her current maternal behavior in the context of her early experiences with her own parents. Several questionnaire responses were obtained from each mother. First, each mother reported on her perceived attachment to her 6-month-old infant. Second, each mother provided retrospective self-reports on her experience of CMT and on the quality of parental care she received during her childhood. Mothers with the s/s genotype who reported that they experienced poor quality early care from their parents perceived a lower attachment to their infants than did l/l mothers with similar low levels of experienced early care from their parents. These relationships were reversed, however, for those mothers who experienced good parental care during their childhood: s/s mother felt more strongly attached to their infants than did l/l mothers. These results are interesting not only because they support the differential susceptibility model, but because they suggest that 5-HT neural systems may not only impact the development of emotionality but may also affect the development of mother-infant attachments.

The results reviewed in this section, although correlational in nature, when viewed in the context of the animal research, support the view that when the experience of CMT enhances 5-HT release in the brain to supernormal levels, such an effect interferes with the normal development of emotionality; the emotional dysregulation that ensues may then affect maternal responsiveness in the next generation.

The Relationships Between Childhood Maltreatment and OT Neural Systems

Given the role of OT in parental behavior and the involvement of OT neural systems in the intergenerational transmission of normal variations in maternal behavior, it makes sense that researchers have proposed that the way a human mother treats her child will likely influence the development of OT neural systems in her child, which will then impact the parental behavior of the child when it reaches adulthood and has its own children (Toepfer et al., 2017). The purpose of this section is to review and evaluate research that shows that CMT is associated with alterations in the development of OT neural systems, which could then impact the parental behavior of the affected child.

Before I begin this analysis, a few points will be re-emphasized. First, given OT's role in the maternal behavior of animals and humans, CMT effects that produce a deficient OT neural system in the affected child may directly disrupt parental neural circuits in the next generation. Second, given the known anxiolytic effects of OT, CMT effects that result in a deficient OT neural system in the affected child may indirectly disrupt parental behavior in the next generation by promoting heightened anxiety and deficits in emotion regulation. As we will see, most research has been focused on this second proposal. Third, the research reviewed in Chapter 6 provided evidence that OT action on CeA may restrain the release of CRF, in this way reducing anxiety. This finding may provide an important link between CMT effects on the development of OT neural systems and developmental effects of CMT on CRF and 5-HT neural systems. If CMT interferes with the ability of OT to suppresses CRF release, such an effect may co-act with the potentially direct effects of CMT on enhancing CRF release in the maltreated child. The resultant impact of these two effects on CRF release may enhance 5-HT release at the level of the mPFC, causing heightened anxiety and deficits in emotion regulation.

Heim et al. (2009) have provided evidence that deficiencies in the central release of OT may result from a history of CMT. These researchers examined CSF levels of OT in two groups of healthy adult women, one with a prior history of early childhood abuse and/or neglect and the other without such a history (based on retrospective reports from the subjects). The women who reported being maltreated by their parents had significantly lower levels of CSF OT than did the women who were not maltreated. Although emotional abuse, emotional neglect, and physical abuse were each associated with lower adult OT levels, the association between childhood emotional abuse and lower adult CSF OT levels was particularly prominent. Unfortunately, the maternal behavior of the participants, who ranged in age from 18 to 45 years (mean = 31 years) was not examined in this study, and it was not indicated whether any of these women were mothers. Nonetheless, these findings lend support to the view that decreases in central neural OT function may contribute to the intergenerational transmission of faulty maternal behavior in women.

Postpartum depression is typically associated with poor maternal behavior and the experience of CMT contributes to the etiology of postpartum depression. Steube et al. (2013) found that blood plasma OT levels in postpartum women were negatively correlated with their depression and anxiety symptoms (also see the summary to the section on the cortical neural regions and circuits relevant to the maternal behavior in women in Chapter 8). In the context of the potential role of OT in the intergenerational transmission of faulty maternal behavior, Pratt et al. (2015) studied depressed and nondepressed mothers and their children over the first 6 postpartum years. Depressed mothers and their children, in comparison to nondepressed mothers, exhibited lower baseline urinary OT levels and an attenuated OT response during mother–infant interactions. To the extent that urinary OT is representative of OT release within the brain, these correlational results suggest that maternal depression, and the poor maternal behavior that is typically associated with such depression, are correlated with the decreased release of OT in the mother's brain, which, in turn, is correlated with low levels of OT release within the brain of her child.

Fan et al. (2014) studied adult males with and without a prior history of CMT, as measured through retrospective self-reports. In the subjects with a history of CMT, particularly emotional abuse, an fMRI analysis indicated that the resting state functional connectivity between the mPFC (pgACC and sgACC) and amygdala was lower than that observed in the control subjects. Importantly, high levels of early life emotional abuse and low connectivity between the mPFC and amygdala were correlated with high levels of anxiety. Finally, intranasal administration of OT altered the functional connectivity between these neural regions in control subjects, but had no effect in the subjects in the CMT group. Although correlational in nature, these results conform with a proposal that exposure to CMT, particularly emotional abuse, decreases top–down inhibitory regulation of the amygdala by the mPFC, in part due to decreased responsiveness of this neural system to OT, resulting in deficits in emotion regulation and heightened anxiety. It would be important to show that these relationships also occur in women with a prior history of CMT. In comparing these results to those of Heim et al. (2009), the implication is that CMT may not only depress the central release of OT in the affected offspring, but may also decrease the responsiveness of critical brain regions to OT. Perhaps CMT influences the expression of OTRs within critical brain regions, such as CeA and mPFC.

Previously in this chapter, I reviewed the work of Riem et al. (2012, 2016), which suggested that nulliparous women with an insecure-anxious attachment style find infant cry sounds to be aversive, as indicated by increased amygdala activation and the use of excessive force on a handgrip, in comparison to women with a secure attachment style. Importantly, intranasal administration of OT decreased these presumed aversive neural and behavioral responses. These results should be compared to those of Bakermans-Kranenburg, van Ijzendoorn, Riem, Tops, and Alink (2012). In this study healthy adult nulliparous women, through retrospective reports, described their childhood experiences of harsh discipline or gentle discipline by their parents. While listening to infant cry sounds, intranasal application of OT decreased the force used on the handgrip by women in the gentle discipline group but had no effect on the excessive force used by the women in the harsh discipline group. One interpretation of these combined results is that low levels of parental care that fall within the normal range may decrease the central release of OT in the affected offspring, while more severe forms of harsh discipline might also decrease the expression or function of central OTRs in the affected offspring. Each of these effects, in turn, influences a women's responsiveness to infant stimuli.

How might CMT or harsh childhood discipline decrease the expression of OTRs in the affected child? The animal research conducted by Champagne and her colleagues (see section on normal variations in maternal licking/grooming of pups affect the development of the MPOA-to-VTA-to-NA circuit in rodent offspring in Chapter 9) indicated that epigenetic processes mediate the effects of normal variations in rodent maternal behavior on the phenotypic outcomes observed in the mother's offspring: Lower levels of maternal attentiveness were associated with increased DNA methylation within the estrogen receptor-alpha gene in the mother's offspring, which reduced the expression of this gene. Decreased estrogen receptor-alpha expression, in turn, depressed the expression of OTRs in the MPOA of the affected offspring. OTRs were also decreased in the CeA of offspring that received lower levels of maternal care. Since the expression of OTRs in CeA is not affected by estradiol, some other process must have resulted in lower levels of maternal care being associated with the decreased expression of OTRs within this region in the affected young. One possibility, as suggested in Chapter 9, is that maternal treatment effects, via epigenetic processes, result in increased DNA methylation directly within the OT receptor gene (OXTR) of the affected offspring, which would presumably decrease the expression of OTR mRNA and protein. This particular epigenetic process has been explored in a few human studies.

In research by Unternaehrer et al. (2015), adult women and men completed a questionnaire that measured their recollections of the maternal care they received as children. On the basis of their scores, these subjects were divided into low and high maternal care groups. The degree of DNA methylation within the OXTR gene was determined from samples taken from peripheral blood cells. The results found greater OXTR gene methylation in the low versus the high maternal care group, and this effect was stronger in women than in men. To the extent that this DNA methylation reduced the expression of the OTR, these results indicate an association between lower levels of experienced maternal care and decreased expression of the OTR in the offspring receiving such care. Certain issues need to be considered in evaluating this study. First, since the expression of OTR mRNA and protein was not measured, it is not possible to conclude that the observed increased methylation of the gene actually reduced its expression. Second, the methylation pattern observed in blood cells may not match that which might occur in the brain. Third, there are many cytosine sites within the OXTR gene, and it is certainly possible that methylation of certain cytosine islands would have different effects than methylation at other cytosine sites. It will be important for future research to relate DNA methylation patterns to the actual expression of OTRs (see Kusui et al., 2001).

In a related study, Ein-Door, Verbeke, Mokry, and Vrticka (2018) reported a higher level of DNA methylation within the promoter region of the OXTR gene (obtained from cells in saliva samples) in adult men and women with an insecure attachment style when compared to those that with a secure attachment. The same considerations mentioned with respect to the Unternaehrer et al. (2015) study also apply to the interpretation of these results. However, both studies are consistent with the view that lower levels of maternal care may result in decreased OTR expression in the mother's children. Unfortunately, neither of these studies examined the parenting behavior of their participants. Such an examination would be important for providing evidence for the involvement of OXTR gene methylation in the intergenerational transmission of parental behavior in humans.

Both of these studies involve the presumed effects of normal variations in maternal behavior on OXTR gene methylation in humans. I am aware of only study that has examined the effects of CMT on such epigenetic processes. Smearman et al. (2016) studied of group of adult women and men, some of whom (about 50%) had a prior history of moderate to severe childhood abuse (physical, emotional, and/or sexual abuse) as determined through retrospective self-reports. DNA analysis was conducted on blood cells. Psychiatric diagnoses for depression and anxiety were also made. One major finding of this study was that the degree of DNA methylation of the OXTR gene per se did not directly predict the occurrence of psychiatric disorders. However, the occurrence of childhood abuse did interact with methylation status at particular DNA cytosine sites within the OXTR gene to predict depression and anxiety. That is, methylation was associated with psychiatric symptoms if the individual also reported abuse.

Interestingly, while high methylation at certain sites within the OXTR gene was associated with increased depression and anxiety in subjects with a history of abuse, high methylation at other sites was associated with a decreased incidence of psychiatric symptoms in participants with a history of abuse. These results show why it is so important to not only obtain measures of DNA methylation, but to also obtain measures of gene expression (OTR mRNA and protein). As indicated by Smearman et al., it is possible that DNA methylation at certain sites within the OXTR gene depresses transcription and expression, as is typically expected, while methylation at other sites within the gene might actually enhance transcription and OTR production.

Puglia, Lillard, Morris, and Connelly (2015) have reported on the association between DNA methylation levels within the promoter region of the OXTR gene (obtained from peripheral blood cells) in healthy men and women and amygdala reactivity to angry/fearful faces. While in an fMRI scanner, subjects with higher OXTR gene methylation exhibited greater amygdala BOLD responses to negative emotional facial expressions than did subjects with lower levels of methylation. These results are possibly related to the known anxiolytic effects of OT. If increased DNA methylation decreased the expression of OTRs in the amygdala, or in mPFC regions which suppress amygdala hyperactivity to aversive stimuli, then one would predict such enhanced amygdala responsiveness. Relatedly, Lancaster, Goldbeck, Puglia, Morris, and Connelly (2018) found that increases in OXTR gene methylation were associated with increases in the gray matter volume in the CeA of adult women and men, which, in turn, were associated with increases in a measure of anxiety. Perhaps these neural processes contribute, in part, to the more serious psychiatric symptoms (depression, severe anxiety, deficits in emotion regulation) that result from CMT or institutional rearing.

In the previously reviewed studies, I have emphasized the effects of early life experiences on the degree of DNA methylation within the OXTR gene. It is also possible that early life experiences influence the degree of DNA methylation within the OT gene, which might then decrease the expression of the OT protein within the PVN and other brain regions. This would be another route through which early parental rearing experiences might affect the development of the parental brain in offspring, and there is some evidence to support this possibility (Haas et al., 2016; cf. Toepfer et al., 2019).

Finally, another way in which OT neural systems may contribute to the intergenerational transmission of faulty maternal behavior and the psychopathological conditions associated with poor mothering is through G × E interactions. Research has focused on the proposal that certain alleles of the OXTR gene may be risk or susceptibility alleles. Individuals with certain OXTR gene variants may be more susceptible to the negative impacts of CMT than individuals that carry alternative alleles. Mechanisms that could underlie this relationship are that different alleles of the OXTR gene might produce OTR proteins that differ in their sensitivity to endogenous OT or that different alleles of the OXTR gene may express different amounts of OTR mRNA and protein. One way in which different alleles of a gene can be formed is through a single nucleotide polymorphism (SNP). A SNP results from the presence of different nucleotides at a single site or locus within the nucleotide sequence of a gene. In many cases a correlation is detected between a particular SNP and a behavioral or psychological trait, but the functional effects of that SNP, at the molecular and cellular level, are not known. Therefore, one can only offer hypotheses about the underlying mechanism through which the SNP might affect behavior.

The OXTR gene in humans is located on chromosome 3 (Meyer-Lindenberg, Domes, Kirsh, & Heinrichs, 2011), and one particular SNP, referred to as rs53576, has received much attention with respect to its role in moderating the effects of CMT on the behavioral and psychological development of the affected child. At the rs53576 site, which is located in the third intron of the OXTR gene, either guanine (G) or adenine (A) can be present, and the resultant genotypes containing one of these two different nucleotides on each homologous chromosome 3 would be GG, AA, or AG (Tost et al., 2010). Note that introns within a gene are not transcribed into mRNA (Numan, 2015), and therefore the exact influence of this SNP on the function of the translated OTR protein is not known. It is certainly possible, however, that alterations in the nucleotide sequence within introns of a gene can influence those portions of the gene (exons) that are transcribed into mRNA (Jakubauskiene, Janaviciute, Peciuliene, Soderkvist, & Kanopka, 2012). One can conceive of a particular intronic SNP as either enhancing or depressing OTR mRNA synthesis, in this way altering the number of OTRs produced and, therefore, the sensitivity of particular brain regions to OT.

Initial research by Tost et al. (2010) provided evidence that the A allele at site rs53576 within the OXTR gene is a risk allele with respect to social behavior. When healthy adult men and women were administered a personality questionnaire that measured prosocial temperament, homozygous GG individuals exhibited the highest prosociality scores while AA individuals had the lowest scores. Li et al. (2015) also review evidence that G allele homozygotes demonstrate higher levels of general sociality than do A allele carriers. In contrast to these results, most studies that have examined the behavioral and psychological outcomes associated with a history of CMT have provided evidence that the G allele at rs53576 is the risk allele.

Dannlowski et al. (2016) administered the CTQ to healthy adult men and women. DNA samples were obtained from blood cells. A strong interaction between rs53576 genotype and a prior history of child maltreatment was observed with respect to certain psychological and neural outcomes. GG homozygous men and women, but not A allele carriers, exhibited large gray matter volume reductions (as measured with magnetic resonance imaging) in the ventral striatum with increasing CTQ scores. This decrease in the gray matter volume of the nucleus accumbens region was associated with a psychological measure that reflected low levels of social reward dependence (represented by cold, socially detached traits). The questionnaire that was used to determine social reward dependence was that same as that used by Tost et al. (2010). Since OTRs are located at each site along the MPOA-to-VTA-to-NA circuit, it is possible that alterations of OTR function along this neural pathway in GG individuals exposed to CMT resulted in deficits in the development of the NA and in the processing of social rewards.

In comparing their results with those of Tost et al. (2010), Dannlowski et al. (2016) propose that the G allele at rs53576 of the OXTR gene might be a susceptibility allele rather than a risk allele. More specifically, they propose that the G allele might be beneficial when individuals develop in stable and supportive social environments (see Bradley, David, Wingo, Mercer, & Ressler, 2013, for research that adds support for this view) but might increase an individual's vulnerability to childhood maltreatment.

In my previous discussion of infant-to-mother and adult attachment styles, I referred to secure, insecure-avoidant, and insecure-anxious attachments. There is a fourth type of attachment style that may be particularly relevant to those individuals that have been exposed to CMT: insecure-disorganized (Bradley et al., 2011). An insecure-disorganized attachment, whether in children or adults, represents a type of approach-avoidance conflict within social relationships that presumably results from a developmental history where a child, at times, receives support/comfort from its caregiver but also fears the caregiver due to the occurrence of episodes of maltreatment. Importantly, Bradley et al. (2011) reported that adult male and female GG carriers of the OXTR gene at site rs53576, in comparison to individuals with the AA or AG genotype, exhibited an enhanced insecure-disorganized adult attachment style and large deficits in emotion regulation (emotional dysregulation) if they had been exposed to severe childhood maltreatment. These authors conclude that A allele carriers are resilient against the effects of CMT on the development of emotional dysregulation and disorganized attachment. Alternatively, a double dose of the G allele places one at risk for these negative developmental outcomes in the face of CMT. Similar results with respect to the development of poor emotion regulation in children experiencing lower levels of maternal care have been reported by Augustine, Leerkes, Smolen, and Calkins (2018).

The previously described studies did not examine the parental behavior of individuals with different rs53576 genotypes who also had a history of CMT. It certainly would be predicted that individuals with emotional dysregulation and/

or deficits in social reward processing would show less than adequate parental behavior toward their own children. Recent studies relevant to this important issue indicate that an OXTR gene by CMT interaction may influence maternal behavior across generations in women. Ludmer et al. (2018) studied mother–infant dyads at 17 months postpartum. At this time, the infants were subjected to the Strange Situation Procedure, which determined the attachment style of the infant as either disorganized or not disorganized. Mothers who carried the GG genotype of rs53576 and who were also exposed to a prior history of maltreatment during their childhood were more likely to have infants that exhibited an insecure-disorganized attachment style in comparison to mothers with the AA genotype who were also exposed to CMT. Significantly, mothers with the GG genotype had infants with lower insecure disorganization scores than did AA mothers in the context of low maltreatment histories. Although these researchers did not directly examine maternal behavior, the occurrence of an insecure-disorganized infant-to-mother attachment is suggestive of lower levels of appropriate maternal caretaking. This suggestion receives additional support from related research by Fugiwara et al. (2019). Overall, these results favor a differential susceptibility model where the G allele has a negative effect under early life adversity but may have a positive effect in the context of an early nurturing environment. The suggestion that the G allele at rs53576 of OXTR gene may foster warm and sensitive maternal behavior in mothers without a history of child maltreatment is supported by other research (Klahr, Klump, & Burt, 2015).

Given the nature of these studies, which are correlational and which do not provide any evidence with respect to how the different rs53576 alleles might affect the quantity and quality of OTRs in the brain, I can only conjecture about the mechanisms that might underpin a differential susceptibility hypothesis with respect to G × E interactions involving this SNP of the OXTR gene. Let's propose that the GG genotype, in comparison to the AA genotype, results in the expression of more OTRs in the brain and/or in the production of OTRs in the brain that are individually more sensitive to OT. For GG individuals who are raised by loving parents, the associated enhanced responsiveness of target neurons to OT may promote better emotion regulation, prosocial behavior, and parental behavior. However, when GG individuals are exposed to abusive parenting, perhaps high levels of OT are released into the brain during development to reduce anxiety in the young child. Perhaps such long-term release of OT in a young child with more sensitive OTRs permanently desensitizes or downregulates OTRs in those parts of the brain (amygdala; mPFC) that are involved in depressing hyperactivity in amygdala fear circuits (Bales & Perkeybile, 2012; Freeman et al., 2018; Grotegut et al., 2016), with the result that in adulthood critical neural systems are actually less responsive to OT, leading to increased anxiety and deficits in emotion regulation, with ultimate impacts on parental behavior. But how can this hypothesis explain the lower levels of prosociality associated with the GG genotype in individuals with a history of CMT? To be wildly speculative, when a child is exposed to a parent who is loving and nurturing on some occasions and severely abusive on other occasion, perhaps there is a hypersecretion of OT throughout the child's MPOA-VTA-NA prosocial circuits during periods of positive parenting to compensate for the effects of abusive parenting, with such hypersecretion then desensitizing OTRs in these regions to a much greater extent in those individuals with the GG genotype that produce more sensitive OTRs.

Lastly, it is worth considering the possibility that the G allele of rs53576, while producing a more effective OTR under healthy rearing conditions, might render the OTR gene more susceptible to hypermethylation under harmful rearing conditions, which might then suppress its expression throughout the brain. There is some evidence in support of this proposal (Smearman et al., 2016).

Conclusions

The research reviewed in this chapter supports the view that the manner in which parents treat their children can affect the development of neural circuits that (a) regulate emotionality and (b) directly underpin parental motivation and parental love/empathy in their children. Such developmental effects would then influence the nature of parental behavior in the affected offspring once they reach adulthood and have their own children. The majority of the studies have focused on the first process. Much more research needs to be devoted to examining the direct effects of parenting on the development within their offspring of the neural circuits that are known to regulate parental motivation and empathy and parent–infant bond formation.

One important theme was that poor parenting is associated with emotional dysregulation and deficits in implicit emotion regulation in the affected child. In particular, deficits in the ability of the mPFC to downregulate the amygdala's reactivity to threatening stimuli, leading to heightened anxiety, has been observed in individuals with a prior history of poor parental care. Such heightened anxiety and poor emotion regulation would be expected to disrupt parental behavior in these individuals under challenging environmental conditions.

One way in which poor parental care could cause heightened emotionality in the affected offspring is by causing heightened cortisol and CRF release in an abused or neglected child. Cortisol action on the amygdala, coupled with CRF-induced release of high levels of 5-HT into the mPFC, may cause the premature and deficient development of mPFC inhibitory control over the amygdala.

Another important theme was that parental treatment effects could influence the development of the child's OT neural system, with poor parenting being related to decreases in endogenous OT release and/or decreases in OTR sensitivity in the affected child. OT action within the brain has two major effects. The first is its known anxiolytic effects. The association between poor parenting and a deficient development within those OT neural systems that promote anxiety reduction would likely be related to the emotional dysregulation that is observed in individuals with a prior history of poor parental care. OT neural system downregulation might increase anxiety by enhancing CRF release from the CeA, since OT is presumed to restrain such release, and/or by removing a stimulatory influence of OT on mPFC neurons that downregulate amygdala reactivity to threatening stimuli.

Another major effect of OT, of course, is its role in promoting parental motivation and the parent–infant bond. Clearly, a downregulation of OT action along parental neural circuits in an individual with a prior history of poor parental care should directly affect the parental behavior of that individual toward her or his own children. The best evidence for this effect comes from the research of Strathearn et al. (2009), previously described in the subsection on the evidence that normal variations in parental behavior are related to the development of the parental brain in offspring as measured in adulthood. Given the research of Champagne and her colleagues reviewed in Chapter 9 (see section on normal variations in maternal licking/grooming of pups affect the development of the MPOA-to-VTA-to-NA circuit in rodent offspring), more research needs to be devoted to this important issue. Progress along these lines awaits the development of radioligands for OTRs that are capable of crossing the blood brain barrier and that can be safely used in humans. Once developed, these ligands, used in conjunction with positron emission tomography scans would allow us to determine whether CMT is associated with decreases in OTR expression in parental circuit regions such as the MPOA, VTA, and NA.

This chapter also emphasized the involvement of epigenetic processes as a potential mediator of the effects of CMT on the behavioral and neural development of the affected offspring. I emphasized research on CMT being associated with increased DNA methylation within the GR gene and the OXTR gene. Such effects are typically conceived as downregulating the expression of GRs and OTRs. A downregulation of GRs would likely be associated with enhanced CRF release within the brain, leading to emotional dysregulation. A downregulation of OTRs would likely be associated with effects described in the previous two paragraphs.

Finally, this chapter explored the involvement of G × E interactions with regard to the effects of CMT on the development of the maltreated offspring. Since not all maltreated children develop poor parenting and/or psychological disorders, it is possible that certain genotypes make an individual vulnerable to the negative impacts of exposure to maltreatment, while other genotypes protect an individual from these negative impacts. In my analysis, I emphasized different alleles of the 5-HTT gene and the OXTR gene. Research, for the most part, supported the view that the short allele of the 5-HTTLPR and the G allele at rs53576 of the OXTR gene were each associated with an increased susceptibility of the carriers of these alleles to the negative impacts of poor parenting.

A major limitation of the majority of the studies that I have reviewed is that they primarily explored the influences of a prior history of being exposed to parental abuse and/or neglect on brain function and behavior in nonparents. Although many of the associated neural effects, such as a disruption of neural circuits involved in emotion regulation or reward processing, would be expected to affect parental behaviors, studies on parents are needed to obtain a more complete understanding of the neural modifications that might mediate the intergenerational continuity of faulty parental behavior. This focus on nonparents, as opposed to parents, is probably related to researchers' primary interest in the etiology of psychopathology in children and adolescents, as well as logistical and methodological issues. However, it is interesting to speculate that exposure to CMT may also decrease an individual's desire to become a parent and have children of their own.

11
Evolutionary Perspectives on the Parental Brain

Introduction

In this chapter I will focus on two major issues. First, since alloparental behavior, which is an adaptive characteristic (see Chapter 7 of this volume and the later discussion in this chapter on behavioral and psychological analysis), occurs in cooperatively breeding species, including humans, such behavior has been emancipated from strict control by the physiological events associated with late pregnancy, parturition, and lactation. I will re-examine (based on the data that I presented in Chapter 7 of this volume) the potential neural modifications that may have occurred throughout parental brain circuits, under the guidance of natural selection, that may have allowed for the relatively prompt parental responsiveness that occurs in species that show alloparental behavior.

Second, since the uniparental maternal care system is the primordial aid-giving system that is present in all mammals, I will examine the proposal that this system provided a neural foundation upon which natural selection acted to promote the evolution of other types of strong prosocial bonds whenever such bond formation had adaptive significance for a particular species.

Brain Modifications That May Underpin Alloparental Behavior

In Chapter 7, I examined allomaternal behavior in virgin female laboratory mice, where experimental genetic selection (inbreeding and selective breeding) has produced virgin females that show spontaneous maternal behavior in home cage tests. In that chapter I also examined what we know about the control of naturally occurring alloparental behavior in cooperatively breeding species such as prairie voles and marmosets. In all of these cases, experimental genetic selection or natural selection has created alternative routes through which infant stimuli can gain access to parental brain circuits without requiring pregnancy, parturition, and lactation.

The Parental Brain. Michael Numan, Oxford University Press (2020). © Oxford University Press.
DOI: 10.1093/oso/9780190848675.001.0001

As outlined in Chapter 7, for alloparental behavior to occur, the defensive neural system must be downregulated so that infant stimuli are not aversive, while the maternal motivational neural system must be upregulated so that infant stimuli can gain easy access to those neural systems that make infant stimuli attractive and rewarding. In this section, I want to focus on the latter process.

Olazabal (2014, 2018) has been a strong proponent of the view that oxytocin (OT) action on OT receptors (OTRs) in the nucleus accumbens (NA) may be the most important factor that drives alloparental motivation. He notes that when alloparental mammalian species are compared with mammalian species that do not show alloparental behavior, OTR expression is higher in the NA of alloparental species. The implication is that evolutionary forces may have resulted in an upregulation of OTRs in NA, which then promotes alloparental motivation. However, this research is correlational in nature and the only experimental evidence that supports this view is the research that shows that the experimental manipulation of OTR expression in the NA can alter alloparenting in virgin female prairie voles that are tested in a novel environment (see Chapter 7 of this volume). Indeed, recent research (Horie et al., 2019) has suggested that OTRs may not be necessary for alloparental behavior in male prairie voles. Therefore, there are likely to be alternate and/or multiple routes where modifications occur along the medial preoptic area (MPOA)-to-ventral tegmental area (VTA)-dopamine (DA)-to-NA circuit and the MPOA-to-paraventricular nucleus (PVN)-OT circuit to allow for the occurrence of alloparental motivation in virgin females and males of particular mammalian species.

I would like to present additional ideas with respect to the proposal presented in Chapter 7 that modifications of MPOA reactivity to infant stimuli are an important mechanism that contributes to alloparental behavior. Such modifications may allow infant stimuli to gain direct access to "parental" MPOA neurons so that alloparental behavior can occur in the absence of pregnancy and parturition. Such an "open" MPOA parental system likely co-evolved with a corresponding downregulation of those avoidance/rejection neural systems that depress parental responses to infant stimuli.

To present the evidence supporting the proposal that experimental genetic selection or natural selection may have altered the way the MPOA responds to infant stimuli in sexually naïve mammals, I will rely on research from virgin laboratory mice because the role of the MPOA in the naturally occurring alloparental behavior of cooperatively breeding species has not been explored.

In my analysis of maternal behavior in rats, it was noted that estradiol action on the MPOA is a major factor in the stimulation of the onset of maternal behavior. In virgin female laboratory mice, however, it has been found that ovariectomized mice and mice with a null mutation of the aromatase gene show spontaneous maternal behavior in home cage tests (Stolzenberg & Rissman, 2011). These results indicate that estradiol is not essential for spontaneous allomaternal behavior in virgin laboratory mice when tested under nonstressful home-cage procedures. However, these results do not rule out the possibility that pup stimuli may induce ligand-independent activation of estrogen receptors (ERs) within MPOA neurons and that such activation, by itself, may contribute to spontaneous maternal behavior in virgin laboratory mice. Ligand-independent activation of ERs means that some factor other than estradiol (which is the natural [cognate] ligand for the ER) may activate ERs in MPOA neurons to promote maternal responsiveness.

To the best of my knowledge, Broad, Curley, and Keverne (2006) were the first to propose that ligand-independent activation of hormone receptors in the brain may contribute to the occurrence of alloparental behavior (also see Numan, 2015). They based their proposal on the findings of Mani and her colleagues (see Mani & Portillo, 2010, for a review), who showed that a DA-D1 receptor agonist (SKF 38393) was able to substitute for progesterone action in the brain to enhance sexual behavior (lordosis) in female laboratory mice. The research of Mani's group found that the action of DA on D1 receptors in the brain, presumably at the level of the ventromedial hypothalamic nucleus, activates the intracellular cyclic adenosine monophosphate (cAMP)-protein kinase A (PKA) signaling pathway that ultimately phosphorylates coactivators of the intracellular progesterone receptor, in this way stimulating the activity of progesterone receptors in a ligand-independent fashion.

Based on this background, is it possible that ligand-independent activation of ERs in MPOA contributes to spontaneous maternal behavior in virgin female laboratory mice? To start this analysis, note that there is a body of evidence that supports a role for ER-alpha in mouse maternal behavior: (a) Ogawa et al. (1998) reported that virgin female laboratory mice with a null mutation of the ER-alpha gene demonstrated severe deficits in maternal behavior toward pups, with many of the females exhibiting infanticide; (b) Ribeiro et al. (2012) reported that suppression of ER-alpha messenger ribonucleic acid within MPOA neurons depressed maternal behavior in mice; and (c) optogenetic inhibition of ER-alpha containing MPOA neurons depresses maternal responsiveness in virgin female laboratory mice (Fang et al., 2018; Wei et al., 2018).

Given that ER-alpha in MPOA appears important for spontaneous maternal behavior in virgin laboratory mice, is there any evidence that ligand-independent activation of ER-alpha in MPOA promotes maternal behavior in such mice? The evidence that does exist is supportive, but not conclusive. First, there is evidence that DA-D1 receptor agonists, such as SKF 38393, can activate the ER through a cAMP-PKA phosphorylation cascade (Gangolli, Conneely, & O'Malley, 1997; Olesen & Auger, 2008). Therefore, it is possible that DA release into MPOA may facilitate the onset of maternal behavior by substituting for estradiol to activate ER-alpha intracellular receptors in MPOA. Relevant to this possibility, in Chapter 5 I described research that showed that microinjections of SKF 38393 into the MPOA of female rats can substitute for estradiol to stimulate the onset of maternal behavior in females that were suboptimally primed with hormones (15 day pregnant hysterectomized and ovariectomized rats; Stolzenberg et al., 2007). Although rats need some hormone priming for a D1 agonist in MPOA to facilitate the onset of maternal behavior (Stolzenberg et al., 2007), perhaps in virgin female laboratory mice, DA action on D1 receptors in MPOA is sufficient to stimulate the initiation of maternal responsiveness and perhaps this stimulatory effect is the result of ER-alpha activation.

If DA action on D1 receptors in MPOA stimulates maternal behavior in mice, what DA neural input to MPOA might be involved? In the discussion of neural inputs to MPOA relevant to maternal behavior in Chapter 5, I described research that indicated that DA neurons in the anteroventral periventricular nucleus (AVPV) of the hypothalamus project to MPOA and that olfactory inputs can reach AVPV via the medial amygdala. Therefore, it is possible that when virgin female laboratory mice are exposed to pups for the first time, pup-related olfactory inputs to AVPV activate DA release into MPOA. Finally, there are the findings by Scott et al. (2015) that 6-hydroxydopamine (a neurotoxin that destroys DA neurons) injections into AVPV disrupted the spontaneous onset of maternal responsiveness in virgin female laboratory mice, but did not have this effect in day 4 postpartum lactating mice (see the discussion of neural inputs to MPOA relevant to maternal behavior in Chapter 5 of this volume).

Significantly, there is a sex difference in the number of tyrosine hydroxylase-containing neurons (which are presumably DA neurons) on the AVPV of laboratory mice, with virgin females containing significantly more of these neurons than sexually inexperienced males (Scott et al., 2015). Perhaps this difference, along with the other factors described in Chapter7, relates to the fact that while virgin female mice show allomaternal behavior, virgin male mice attack or avoid pups upon their initial exposure to them.

Putting all of the pieces of evidence together, I would like to propose a mechanism through which ligand-independent activation of ER-alpha in MPOA may promote the onset of spontaneous maternal behavior in virgin female laboratory mice when they are tested under nonstressful home cage conditions: Exposure to pup stimuli activates AVPV-DA neurons that project to MPOA. DA action on D1 receptors in MPOA then stimulates a cAMP-PKA phosphorylation cascade, which then activates ER-alpha to stimulate the onset of maternal behavior. This proposal is schematically shown in Figure 11.1. I also propose that ligand-independent activation of ER-alpha in MPOA through a DA-D1 receptor mechanism is primarily involved in the initiation of maternal responsiveness, but is no longer required for maternal motivation once the behavior has been established.

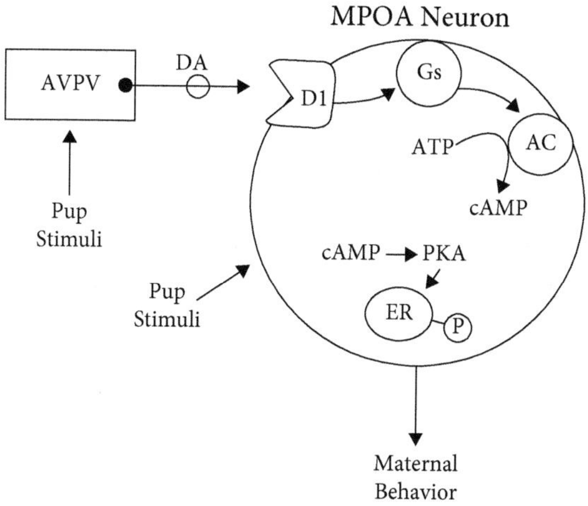

Figure 11.1. A neural model explaining how ligand-independent activation of estrogen receptor (ER)-alpha might stimulate the rapid onset of allomaternal behavior in virgin female laboratory mice. Pup stimuli are shown as activating presumed dopamine (DA) neurons in the anteroventral periventricular nucleus of the hypothalamus (AVPV). These neurons are shown as projecting to the medial preoptic area (MPOA) where DA acts on its D1 receptor to stimulate the intracellular cAMP-PKA signaling pathway within MPOA neurons: The stimulated D1 receptor activates a stimulatory G protein (Gs), which, in turn, activates the enzyme adenylate cyclase (AC). AC catalyzes the conversion of adenosine triphosphate (ATP) to cyclic adenosine monophosphate (cAMP). cAMP then activates protein kinase A (PKA), which is proposed to phosphorylate ER. The phosphorylated ER is then active in the absence of estradiol and through genomic and/or nongenomic actions it alters the structure and/or function of MPOA neurons, allowing them to respond to pup stimuli so that maternal behavior occurs. P = phosphate group.

The importance of this proposal is that it describes a mechanism through which evolutionary forces could modify MPOA reactivity to infant stimuli to promote alloparental behavior in those species where alloparenting has adaptive significance. Further experimentation, of course, is needed to validate my proposal. For example, would application of a DA-D1 receptor antagonist to MPOA block spontaneous maternal behavior in laboratory mice and, in the absence of pup stimuli, would SKF 38393 phosphorylate coactivators that stimulate ER-alpha activity in MPOA? Most important, the involvement of ligand-independent activation of hormone receptors in MPOA in the alloparental behavior of species such as prairie voles and marmosets that show this behavior naturally needs to be explored.

If ligand-independent activation of ERs in MPOA stimulates the onset of spontaneous maternal behavior in laboratory mice, what effects of the activated ERs might be involved in altering the functional activity of MPOA neurons in response to pup stimuli? In Chapter 5, I described research that showed that one of the genomic effects of ER-alpha is to stimulate the expression of OTRs in MPOA. However, for laboratory mice, I described the findings that show that when tested under nonstressful home cage conditions, the OT-OTR complex is not required for spontaneous maternal behavior (Yoshihara et al., 2018). Perhaps ER-alpha activation of Fos gene expression in MPOA is involved, but this begs the question of what genomic effects Fos proteins have to modify MPOA functional reactivity to infant stimuli. Finally, given the short-latency onset of maternal behavior in virgin mice, it is also possible that fast-acting nongenomic effects of ER-alpha activation are involved (Rainville, Pollard, & Vasudevan, 2015; Stolzenberg et al., 2009). Much more research is needed to shed light on these important issues.

In Chapter 4, I reviewed the evidence that the OT-OTR system was important when virgin laboratory mice were tested under challenging environmental conditions. These results suggest that under more natural ecological conditions, OT neural systems and the induction of OTRs in MPOA may indeed be involved in the alloparental behavior shown by a variety of mammalian species. Further, in both laboratory mice and virgin marmosets, brief maternal experiences with infants further boost allomaternal behavior (Pryce, 1993; Stolzenberg & Rissman, 2011). Research by Okabe et al. (2017) on virgin female laboratory mice has shown that OT action on OTRs in MPOA during periods of maternal experience are necessary for experience-based improvements in the allomaternal behavior of virgin female mice.

In Chapter 3, I described the research of Stolzenberg and Rissman (2011) that indicated that while virgin female mice promptly retrieve pups in their home cages on the first day they were exposed to pups, they would not initially retrieve pups in a novel T-maze. However, after 4 days of maternal experience with pups under home cage conditions, such females will retrieve pups in a novel T-maze. Stolzenberg and Rissman concluded that 4 days of maternal experience modified the parental brain circuitry in such a way as to boost maternal motivation, which allowed the virgins to care for pups under challenging environmental conditions. Subsequent research from Stolzenberg's group (Stolzenberg & Mayer, 2019; Stolzenberg, Stevens, & Rissman, 2012) presented evidence that this experience-induced improvement in allomaternal behavior in laboratory virgin mice was due to experience-induced epigenetic processes that were associated with increased OT messenger ribonucleic acid levels in the MPOA (their MPOA region probably included the anterior commissural nucleus of the anterior PVN).

Based on these cumulative findings, I offer the following hypotheses concerning the neural mechanisms that may underpin allomaternal behavior in laboratory mice and perhaps in other species that demonstrate alloparenting. A baseline level of high maternal responsiveness is shown in virgin females under nonstress conditions as a result of infant stimuli inducing the ligand-independent activation of ER-alpha in MPOA. Although such activated ERs may induce OTRs in MPOA neurons, these receptors are not necessary for baseline maternal responsiveness. Allomaternal experience subsequently boosts allomaternal motivation, which allows for competent allomaternal responsiveness under challenging environmental conditions and the OT-OTR system is involved in this effect. Perhaps under nonstress baseline conditions, MPOA activation of the mesolimbic DA system is sufficient for allomaternal behavior. Under stressful conditions, MPOA activation of the PVN-OT system allows OT to act on OTRs at certain sites throughout the MPOA-VTA-NA circuit to further boost allomaternal motivation, and the particular sites where OT acts may be dependent on the species being examined and its species-typical distribution of OTRs in its brain. Allomaternal experience may further boost maternal motivation by increasing the synthesis of OT in the anterior PVN via epigenetic processes, at least in mice.

It is interesting to note that OTRs are expressed at very low levels in the MPOA of prairie voles and common marmosets, but both of these species demonstrate alloparental behavior and express OTRs in NA (Ahern & Young, 2009; Olazabal, 2014; Schorscher-Petcu et al., 2009). However, both of these species also express high levels of the V1a vasopressin receptor in MPOA/ventral part of the bed nucleus of the stria terminalis (vBST) region (Schorscher-Petcu et al., Wang et al., 1997; Wang, Young, Liu, & Insel, 1997). As suggested previously, perhaps high endogenous levels of OT activate V1a receptors in the MPOA/vBST region of marmosets and prairie voles and this activation, along with OT action on OTRs in NA, contribute to alloparental behavior in these species under challenging environmental conditions (such as being tested in a novel cage or being examined under natural ecological conditions). More specifically, infant stimuli may activate MPOA projections to PVN in alloparental voles and marmosets. PVN-OT projections to MPOA may stimulate V1a receptors, while PVN-OT projections to NA act on OTRs located in that region. These actions may supplement the effects of MPOA stimulation of the mesolimbic DA system to promote alloparental responsiveness to infant stimuli.

It is certainly possible that ligand-independent activation of other MPOA hormone receptors is also involved in the process of alloparental motivation, such as ligand-independent activation of MPOA prolactin receptors (Brown et al., 2017; Horseman et al., 1997; Lucas et al., 1998). The reader is referred to Chapter 3 for a critical analysis of this possibility. It is also possible that DA action on MPOA neurons, in the absence of ER-alpha, is involved. These possibilities highlight a broader point. Depending on the species, there may be alternate and multiple routes through which MPOA neurons may be modified so that they respond to infant stimuli without being primed by the physiological events of late pregnancy and parturition.

The research I have described in laboratory mice suggests that experimental genetic selection may have resulted in the MPOA becoming an open system in the sense that it can be activated by infant stimuli in virgin females. I have emphasized the potential neuromolecular changes in MPOA because of its dominant role in the maternal behavior of mammals. Much more research needs to be done on the role of the MPOA in the alloparental behavior that naturally occurs in cooperatively breeding species such as prairie voles, marmosets, and humans to determine whether natural selection has resulted in similar functional changes in the MPOA of these species (see Glasper et al., 2018). It should also be considered that for certain species, estradiol action on ERs in MPOA, in the absence of the other hormones of pregnancy, may be a trigger for alloparental behavior. There is some evidence for this conjecture with respect to the alloparental behavior that occurs in naked mole rats (Watarai et al., 2018).

In contrast to mice, virgin female rats do not show spontaneous maternal behavior, but can be sensitized to show maternal behavior after about 7 days of constant exposure to young pups. In a recent report, Gallagher, Nephew, Poirier, King, and Bridges (2019) found that virgin rats with a null mutation of the ER-alpha gene showed normal sensitization latencies. These results suggest that while ligand-independent activation of ER-alpha may be involved in short-latency alloparental behavior, it is not involved, at least in rats, in the pup-stimulated induction of long-latency maternal behavior (sensitized maternal behavior). Therefore, longer-term experience with infants may be able to induce parental behavior in the absence of ER-alpha signaling.

Finally, there is one study that has provided suggestive, but preliminary evidence that ER-alpha may be involved in human maternal behavior. Lahey et al. (2012) have reported that a particular single nucleotide polymorphism (SNP) within the ER-alpha gene is associated with a harsh parenting style in women. It is interesting to speculate that ligand-independent activation of ER-alpha may be involved in human alloparenting and that postpartum women with this SNP may not bond effectively with their child, which might then lead to a harsh parenting style. Whether maternal experience with their infant ameliorated some of the negative effects of this SNP remains to be determined. This issue could be examined through the use of a longitudinal study.

MPOA Interactions With the Mesolimbic DA System Regulate the Appetitive Aspects of Maternal Behavior, Male Sexual Behavior, and Female Sexual Behavior

In this section, I want to review the evidence that MPOA interaction with the mesolimbic DA system is an evolutionarily conserved neural system that is involved in the appetitive aspects of all types of reproductive behaviors: parental, male sexual, and female sexual behaviors. More specifically, in a variety of reproductive contexts, the MPOA appears to process socially relevant cues from conspecifics and the interactions of such MPOA neurons with the mesolimbic DA system, primarily through projections to the VTA, allow for the occurrence of the prosocial appetitive aspects of parental behavior and male and female sexual behaviors. Since the MPOA-mesolimbic DA connection is importantly involved in the social attraction between conspecifics in reproductive contexts, this core circuit may have provided a neural foundation upon which natural selection could operate to allow for the evolution of broader types of prosocial appetitive interactions between conspecifics that extend beyond the boundaries of reproduction.

In a prescient paper, Stolzenberg and Numan (2011) proposed that MPOA interactions with the mesolimbic DA system might not only be involved in maternal behavior, but may also be involved in the appetitive aspects of male and female sexual behaviors (also see Numan, 2015). They stipulated four pieces of evidence that would be needed to support their proposal: (a) Neuroanatomical data should show that the MPOA projects to VTA; (b) behavioral evidence should indicate that the MPOA is involved in the appetitive aspects of the behavior in question; (c) behavioral evidence should indicate that the mesolimbic DA system is involved in the appetitive aspects of the behavior in question; and (d) neurobehavioral evidence should show that the MPOA is functionally linked with the mesolimbic DA system to regulate the appetitive aspects of the behavior in question. For the appetitive aspects of parental behaviors, all of these requirements had been met, as reviewed in Chapter 5. At the time that Stolzenberg and Numan published their paper (2011), the first three requirements were satisfied for the appetitive aspects of male and female sexual behaviors, but neurobehavioral evidence that the MPOA is functionally linked with the mesolimbic DA system to regulate that appetitive aspects of sexual behaviors had not been met. Importantly, recent evidence has now shown that MPOA interactions with the mesolimbic DA system do regulate the appetitive aspects of male and female sexual behaviors (Lyilikci, Balthazart, & Ball, 2017; McHenry et al., 2017; Wei et al., 2018).

Given that MPOA projections to the mesolimbic DA system influence the appetitive aspects of all three reproductive behaviors, an important question is whether the same or different MPOA neurons are involved in each behavior. One view, which I and my colleagues (Olazabal et al., 2013) have called the common population view, proposes that the same population of MPOA neurons projects to the VTA to influence the appetitive aspects of each type of reproductive behavior. If that were true, what kind of process would regulate motivational specificity? As one example, while a lactating female rat who is in estrus on day 1 postpartum is attracted to pup stimuli and to a sexually active male, a lactating female on day 5 postpartum, after the termination of her postpartum estrus, shows strong maternal motivation, but is no longer interested in mating with males. The common population view might argue that the particular hormonal environment that MPOA neurons are exposed to regulates the particular stimuli that are capable of activating the common MPOA neural population that projects to VTA. For females on day 1 postpartum, when estradiol has recently primed MPOA neurons, both pup stimuli and male stimuli would be able to activate the common MPOA neural population, while on day 5 postpartum, when the effects of estradiol have waned, only pup stimuli would be able to effectively activate this common MPOA neural population.

In contrast to the common population view, an alternative has been referred to as the labeled-line point of view, which proposes that distinct MPOA neurons, each of which projects to the VTA, are activated by either infant stimuli, male stimuli, or female stimuli to regulate the appetitive aspects of parental behavior, female sexual behavior, or male sexual behavior under particular physiological conditions (Stolzenberg & Numan, 2011). Recent evidence supports the labeled-line point of view rather than the common population view, and it appears that different populations of MPOA neurons regulate the appetitive aspects of parental, male sexual, and female sexual behaviors (Moffitt et al., 2018; Wei et al., 2018; Wu et al., 2014).

Although different intermingled MPOA neuron populations, each of which activates the mesolimbic DA system, are involved in the appetitive phases of each type of reproductive behavior in mammals and other vertebrate species (see Lyilikci et al., 2017), the most important point is that MPOA input to the mesolimbic DA system is crucially involved in regulating social attraction between conspecifics during all aspects of reproduction, from mating to caring for offspring. This core circuit may have been appropriated and modified by natural selection to allow for the evolution of other types of strong social attractions and prosocial behaviors between conspecifics, outside the boundaries of reproduction, if such prosocial interactions had adaptive significance.

Modifications to the Core MPOA-to-VTA-DA Circuit and the Formation of an Enduring Mother–Infant Bond in Mammals

Distinct neural populations within the MPOA, each of which responds to unique hormonal and/or sensory stimuli, interact with the mesolimbic DA system to influence the appetitive aspects of mating and parental behaviors. For most mammalian species, however, there is a major difference between the regulation of mating behaviors and parental behaviors. As I indicated in Chapter 2, most mammalian species exhibit either a polygynous or promiscuous mating system and the typical parental care system is a hormone-dependent uniparental maternal care system. In these species, the male and female separate after mating, and the impregnated female raises her offspring on her own. Therefore, for most mammals, although opposite-sex conspecifics are attracted to one another during mating, which is under hormonal control, they do not form an enduring social bond to one another (Numan, 2015). In sharp contrast, although the onset of maternal behavior that occurs in such species is triggered by the physiological events associated with late pregnancy and parturition, the mother ultimately forms an enduring prosocial, aid-giving bond with her infants after the physiological events that triggered the behavior have waned, and this bond persists at least until the young are weaned. In Chapter 5, I outlined the underlying mechanisms that appear to underpin this enduring mother–infant bond. Briefly, I proposed that MPOA stimulation of both VTA-DA neurons and PVN-OT neurons activates their projections to the NA. The co-action of DA and OT on NA neurons was conceived as depressing NA GABAergic inhibitory input to the ventral pallidum (VP), which then allowed for the strengthening of synapses between the amygdala and the VP. Subsequently, during the maintenance phase of maternal behavior, and in the absence of continued hormonal stimulation of "maternal" MPOA neurons, MPOA activation of DA input to NA becomes sufficient to stimulate mother–infant attraction and the persistence of the appetitive aspects of maternal behavior throughout the postpartum period because this MPOA-to-VTA-DA projection, coupled with the strengthened amygdala-to-VP synapses, allows the VP to be stimulated by infant-related sensory inputs from the amygdala.

In addition to the previously described mechanism and in relation to the research described in the previous discussion in this chapter on brain modifications that may underpin alloparental behavior, it is likely that other mechanisms also underpin the enduring mother–infant bond that occurs in mammalian species with a uniparental maternal care system that requires hormonal stimulation for its onset. Perhaps mother–infant interactions during the early postpartum period modify the organization and function of certain MPOA neurons so that they become an open system in the sense that "maternal" MPOA neurons can ultimately be activated by infant stimuli during the postpartum period in the absence of continued hormonal stimulation.

The Neural Mechanisms of the Mother–Infant Bond: A Potential Neural Foundation for the Enduring Pair Bond That Forms Between Mates in Socially Monogamous Mammalian Species

Although a core MPOA-to-VTA-DA circuit underlies the appetitive aspects of mating behaviors, in mammalian species with a polygynous or promiscuous mating system, the mechanisms described for creating an enduring mother–infant bond are not operative between male and female mates, and the sexes do not form a persistent attraction to one another after mating has been consummated.

In contrast to most mammalian species, approximately 10% of mammalian species exhibit social monogamy, but biparental care of offspring occurs in only about 50% of these monogamous species. Therefore, social monogamy with biparental care occurs in about 5% of mammalian species. I made the important point in my discussion of an evolutionary perspective in Chapter 7 that the social monogamy that occurs in the context of biparental care of offspring probably includes a stronger affiliative bond between the male and female partners than does the relatively exclusive mating that occurs in socially monogamous species where only the mother cares for the young. This difference may represent a distinction between pair bonding and pair living, respectively (Tecot, Singletary, & Eadie, 2016), and the former is my main concern in this section. The most essential relationship between the mating partners in a monogamous mating system with biparental care is the formation of a strong and enduring affiliative pair bond between the male and female mates that persists after the termination of sexual behavior and the hormonal events that triggered their sexual interactions. This pair bond includes the mated pair living in close proximity to one another and exhibiting high levels of affiliative behaviors and a strong partner preference for one another (French, Cavanaugh, Mustoe, Carp, & Womack, 2018; Numan, 2015; Numan & Young, 2016). Note some of the similarities between the monogamous pair bond and the mother–infant bond. Initial interactions between the individuals involved are typically triggered by hormonal events, which then result in the formation of a strong prosocial bond between the individuals that persists long after the hormonal events that initially triggered their initial social interactions have waned.

Since the strong attraction between a mother and her infant(s) is common to all mammals, while the strong affiliative pair bond that occurs in monogamous biparental mammals is rare, it has been proposed that the neural circuitry and mechanisms that underlie the long-term mother–infant bond may have provided the initial neural foundation upon which the strong affiliative pair bond was built (Numan, 2012b, 2015; Numan & Young, 2016). More specifically, the neural mechanisms present in all mammals that promote the mother–infant bond may have been utilized by natural selection to establish the ability to form a selective affiliative bond between mating partners during the evolution of certain monogamous mating systems. If this is the case, although complete commonality should not necessarily be expected, there should be significant similarities between the neural mechanisms that promote the formation of mother–infant bonds and pair bonds in mammals. In this section, I want to review the evidence that supports this proposal.

Most of the detailed experimental analyses of the neural mechanisms that underpin the pair bond that forms between mates in socially monogamous species has been conducted on the prairie vole (*Microtus ochrogaster*), which is one of many rodent species that make up the Microtus genus (Young, Gobrogge, Liu, & Wang, 2011; Young & Wang, 2004; Walum & Young, 2018). As indicated in Chapter 7, prairie voles are a socially monogamous species that also exhibit biparental care of offspring. Mating in prairie voles typically involves 15 to 30 bouts of copulation during a 24-hour period and, in the laboratory, the formation of a pair bond between the mates is measured by using a partner preference test (Numan & Young, 2016). In this test, after a period of cohabitation between mates, one of these mates is tethered in one chamber of a three-chambered testing arena and an unfamiliar "stranger" of the same sex as the tethered mate is tethered in an opposite chamber. The experimental subject, the other mating partner, is placed in a center chamber and is allowed to associate with either stimulus prairie vole (the original mate or the stranger) over a 3-hour period. When prairie voles are tested in this manner, both males and females choose to spend most of the 3-hours test period in contact with the individual with whom they mated, and this preference is taken as a measure of pair bond formation. Importantly, once a partner preference has been established, it persists in the absence of further partner presence: The pair that formed a bond to one another can be separated for up to 2 weeks and still display a partner preference for one another once they are reunited (Insel & Hulihan, 1995; Sun, Smith, Lei, Liu, & Wang, 2014). This enduring bond between mates is similar to the effects of mother–infant bond formation, which allows for the continued occurrence of maternal responsiveness after the mother has been separated from her young for varying time intervals and then reunited with them, as described in the section on the maintenance of maternal behavior and the onset-maintenance dichotomy in rats, rabbits, and sheep in Chapter 3.

While prairie voles exhibit social monogamy with biparental care, some other microtine vole species, such as montane voles, exhibit the typical nonmonogamous uniparental maternal care system, and pair bonds do not form between mates (Young & Wang, 2004). Under natural conditions, prairie voles live in different habitats than do montane voles, with the former occupying habitats with a low density of resources, while the latter live in areas where the density of resources, such as food, is much higher. It has been proposed that a monogamous mating system with biparental care and delayed dispersal of young after they are weaned may be adapted to a low density of suitable territories and food (prairie voles), while plentiful resources allow for a nonmonogamous mating system, uniparental maternal care, and dispersal of young after weaning (montane voles; Carter, DeVries, & Getz, 1995). This proposal implies that natural selection has affected the social behavior of closely related vole species by affecting the organization of brain mechanisms that control sociality and pair bonding. According to the perspective being developed in this section, it is likely that the neural mechanisms present in all mammals that promote the mother–infant bond have been operated upon by natural selection to establish the ability to form enduring pair bonds between mating partners when the evolution of monogamous mating systems has adaptive significance.

Figure 11.2 shows receptor autoradiograms of the distribution of OTRs and V1a vasopressin receptors in the brains of prairie voles and montane voles (Young & Wang, 2004).

This figure shows that OTR density and V1aR density is high in the NA and VP, respectively, of prairie voles, while these neuropeptide receptors are virtually absent within the NA-VP circuit of montane voles. Given the importance of the NA-VP circuit for maternal motivation and the role of OT action on OTRs in NA for the formation of an enduring mother–infant bond in rats (see Chapter 5 of this volume), these anatomical data provide the first hint of neural commonality between the mother–infant bond and the pair bond in mammals. These data also show how differences in the neurochemical architecture of the NA-VP circuit might influence differences in the social organization of vole species.

Figure 11.3 depicts the neural circuitry and mechanisms that may underlie the formation of pair bonds in prairie voles and that also closely approximates the neural circuitry and mechanisms that underpin the formation of an enduring mother–infant bond, as described in Chapter 5 (see Figure 5.15). In this model (Numan, 2015; Numan & Young, 2016), during appetitive and consummatory mating interactions between male and female prairie voles, the MPOA is

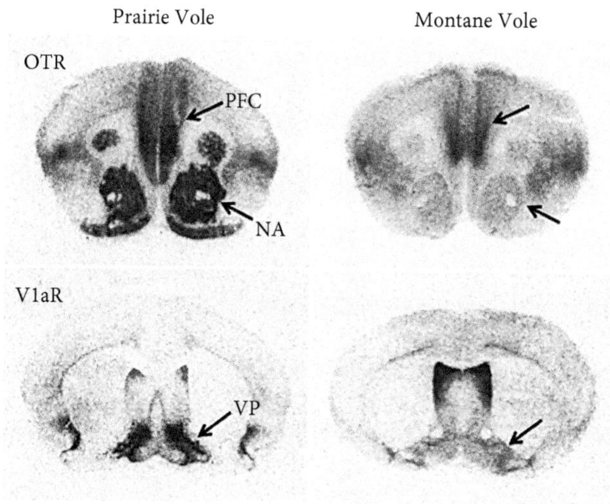

Figure 11.2. Receptor autoradiograms showing species differences in oxytocin receptor (OTR; top row) and vasopressin V1a receptor (V1aR; bottom row) densities in the nucleus accumbens (NA) and ventral pallidum (VP) of monogamous prairie voles (left panels) and nonmonogamous montane voles (right panels). These species differences in distribution of receptors in NA-VP circuit are thought to mediate species differences in the ability to form pair bonds in voles. There are no sex differences in receptor density in NA and VP. PFC = prefrontal cortex.

Source: Reprinted from Figure 4 in Numan and Young (2016), with permission from Elsevier and based on data presented in Young and Wang (2004).

active, since it controls the appetitive aspects of male and female sexual behavior in response stimuli from the mating partner. The MPOA is shown as activating VTA-DA neurons that project to the NA. The MPOA is also shown as activating PVN-OT neurons that project to both the NA and amygdala. According to this model, OT action at the level of the amygdala is involved in a selective recognition process: It acts to strengthen the synapses between the specific incoming sensory stimuli from the mating partner and neurons in medial amygdala (MeA) and basolateral amygdala (BLA)/basomedial amygdala (BMA) that project to the NA-VP circuit. Such OT action may also excite these specific amygdala projections to the NA-VP circuit. Further, OT and DA actions at the level of the NA result in a potent inhibition of NA GABAergic projections to VP. In Chapter 5, I described research that suggests that DA-D2 receptors and OTRs can form heteromers on NA medium spiny neurons (MSNs) and that the combined action of OT and DA on these heteromers strongly inhibits NA GABAergic MSNs. The resultant disinhibition of VP allows the VP to be superexcited by

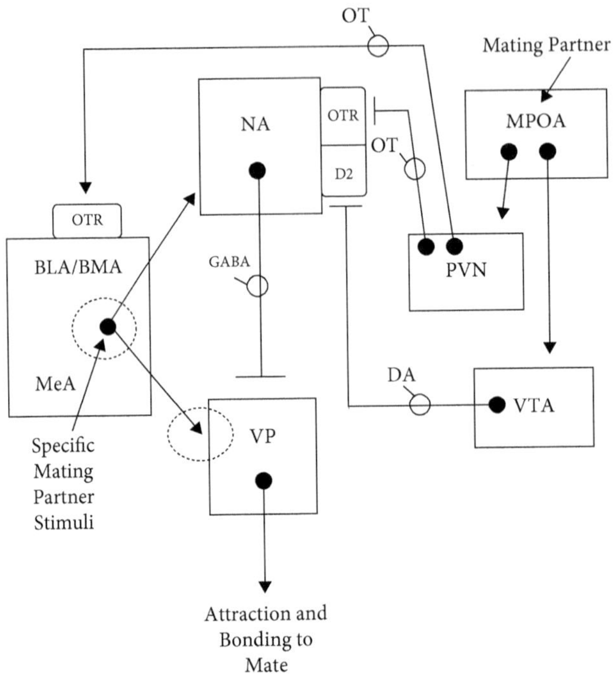

Figure 11.3. A proposed neural model depicting the neural circuitry that may underlie the formation of pair bonds in prairie voles. This neural circuitry closely approximates the neural circuits and mechanisms that underlie the formation of an enduring mother-infant bond, as described in Chapter 5 (see Figure 5.15). During mating between a male and a female, the medial preoptic area activates both ventral tegmental area (VTA) dopamine (DA) neurons and paraventricular nucleus of the hypothalamus (PVN) oxytocin (OT) neurons. OT action at the level of the medial amygdala (MeA) and basolateral/basomedial amygdala (BLA/BMA) is proposed to regulate the selective recognition process, in that it promotes synaptic strengthening (a dashed circle) between the specific sensory characteristics of the mating partner and neurons in MeA and BLA/BMA. These amygdala neurons, in turn, project to the nucleus accumbens (NA)-ventral pallidum (VP) circuit. OT and DA action at the level of NA result in a potent inhibition of NA GABAergic projections to VP. The resultant disinhibition of VP allows the VP to be superexcited by its inputs from the amygdala, which carry the specific sensory signature of the mating partner. Such superexcitation strengthens the synapses between the amygdala and VP (a dashed circle). Once mating is complete, a selective partner preference and pair bond is formed as a result of the fact that the specific sensory stimuli from the mating partner are capable of effectively activating the VP, resulting in a strong affiliative attraction between the mates. D2 = D2 DA receptor; OTR = oxytocin receptor. Neural effects ending in a bar are inhibitory and those ending in an arrow are excitatory.

Source: Modified from Figure 6.2 in Numan (2015), with permission from Elsevier.

incoming excitatory axon terminals from the amygdala, which carry the specific sensory signature of the mating partner. This superexcitation strengthens these particular synapses between the amygdala and the VP. Once mating has occurred, a selective partner preference and pair bond is formed because the specific sensory stimuli from the partner are capable of strongly activating the VP, leading to the strong affiliative attraction that occurs between mates. A strange prairie vole would not have this effect, because its specific sensory signature would not be capable of strongly activating amygdala projections to the VP. The neural model described in Figure 11.3 involves a dual process: The development of a selective recognition process that occurs at the level of the amygdala and the development of an enduring attraction to the recognized stimulus at the level of the VP. When analyzing this model in the context of the research on maternal behavior, one should be able to see that the model is a synthesis of the mechanisms that underlie the enduring mother–infant bond that occurs in sheep and rats (Numan, 2015; Numan & Young, 2016). Rats develop an enduring attraction to a generic infant stimulus but sheep, due to synaptic plasticity that occurs in the olfactory bulb (see Chapter 4 of this volume) and amygdala (Keller, Perrin, Meurisse, Ferreira, & Levy, 2004), develop an enduring bond to a specific lamb stimulus. The important point is that the type of social stimuli that results in an enduring activation of the VP determines the social stimulus toward which an organism forms a long-term attraction.

Most of the evidence in support of this model has focused on the importance of the NA-VP circuit for pair bond formation in prairie voles. Initial research indicated that OT action on OTRs in NA is necessary for pair bond formation in female prairie voles, while vasopressin action on V1aRs in the VP is necessary for pair bond formation in male prairie voles (Young & Wang, 2004). More recent research has indicated, however, that OT action on OTRs in NA is also important for pair bond formation in male prairie voles (Duclot, Wang, Youssef, Wang, & Kabbai, 2016; Johnson et al., 2016; Johnson, Walum, Xiao, Riefkohl, & Young, 2017; King, Walum, Inoue, Eyrich, & Young, 2016; Modi et al., 2015). While not denying the importance of vasopressin action on VP for pair bond formation in male prairie voles, I want to concentrate on the importance of OT action on OTRs in NA for pair bond formation in female and male prairie voles because this mechanism maps onto the role of OT action on the NA for the formation of an enduring mother–infant bond (see Chapter 5 of this volume).

Research has shown that DA action on DA-D2 receptors in NA is also required for pair bond formation in both male and female prairie voles (Aragona, Liu, Curtis, Stephan, & Wang, 2003; Gingrich, Liu, Cascico, Wang, & Insel, 2000). Importantly, pair bond formation requires the activation of both DA-D2 receptors and OTRs in NA (Liu & Wang, 2003). These findings fit with the model presented in Figure 11.3, where it is proposed that D2R-OTR heteromers, when activated by DA and OT, result in a potent inhibition of NA GABAergic projections to VP, and that the disinhibited VP allows for the development of a strong selective attachment to one's mating partner. In this regard, recent neurophysiological research has provided evidence that the neural excitability of NA MSNs is decreased in male prairie voles that form a pair bond (Willett et al., 2018). In Chapter 5, evidence showed that DA and OT action on NA is also required for the development of an enduring mother–infant bond, although this research indicated that DA action on both D2 and D1 DA receptors is involved.

In Chapter 5, I described the research on rats that indicated that it is the output of VP that is of primary importance for the mother–infant bond. In prairie voles, the main experimental evidence for a role of the VP in pair bonding is derived from research that shows that vasopressin acts on V1a receptors in VP to promote pair bond formation and maintenance in male prairie voles (Donaldson, Spiegel, & Young, 2010; Young & Wang, 2004). The question of whether vasopressin acts on V1aRs in VP to influence pair bonding in female prairie voles has not been investigated. It would be important to determine whether vasopressin action is stimulating the output of VP neurons to other brain regions to regulate the formation and maintenance of the pair bond in male (and perhaps female) prairie voles. As in maternal rats, would neural inactivation of VP neurons disrupt the formation and maintenance of pair bonds in prairie voles, and would this occur in both sexes? These are important questions to investigate to validate aspects of the neural model depicted in Figure 11.3. Additionally, although there is some evidence that vasopressin is involved in certain aspects of maternal behavior in rats (Bosch & Neumann, 2008), whether vasopressin acts on VP to influence maternal behavior remains to be determined. Answers to these questions will help us determine the degree of commonality between the neural mechanisms that regulate the mother–infant bond and the pair bond.

In the model shown in Figure 11.3, it is proposed that OT action on OTRs in the amygdala is involved in the development of a selective recognition process in prairie voles. For voles, there is some indirect evidence to support this role of OT in the amygdala in the formation of a selective recognition of one's mating partner: Sexually naïve male and female prairie voles exhibit high OT binding sites in the BLA region of the amygdala, while such binding is virtually absent in sexually naïve nonmonogamous montane voles (Insel & Shapiro, 1992). Therefore, in addition to the differences in OTR and V1a receptor distributions in NA and VP in these two species, shown in Figure 11.2, there is also a prominent difference in OTR expression in BLA. These differences support the view, presented in Figure 11.3, that OT action on OTRs in both the NA and the amygdala region is involved in pair bond formation in prairie voles. To continue the comparison with the mother–infant bond, in rats it has been shown that BLA/

BMA functional connections with VP are necessary for an enduring maternal attraction to a generic pup stimulus (see Chapter 5 of this volume). It would be important to demonstrate that this mechanism, likely also to include MeA neurons, is also operative in the formation and maintenance of selective pair bonds in prairie voles.

Similar to the formation of an enduring mother–infant bond, Figure 11.3 proposes that the MPOA and its projections to the PVN and VTA are also involved in the development of pair bonds in prairie voles. Correlational evidence has shown that gene and protein expression in the MPOA is modified during mating behavior and pair bond formation in monogamous voles (Lim & Young, 2004; Seelke et al., 2018; Wang, Li, Wu, Zhang, & Tai, 2015). Surprisingly, however, the direct involvement of MPOA in pair bonding, an important aspect of the proposed model that would show clear parallelism with mother–infant bond formation, has not been experimentally examined. One possible reason for this lack of investigation is probably related to the fact that MPOA inactivation would interfere with sexual behavior. However, recall that the MPOA is essential for the continued maintenance of maternal behavior after a mother–infant bond has become established. I would predict that MPOA inactivation would disrupt an already established pair bond in both male and female prairie voles.

Another important question remains to be answered. Since rats, and other nonmongamous mammalian females, can form a mother–infant bond, why don't they form pair bonds? One possible explanation is related to the fact that although OTRs are present in the NA of virgin female rats, their density is lower than that which exists in virgin prairie voles (Olazabal & Young, 2006b; 2008). Perhaps OTRs become elevated in NA during mother–infant interactions in the early postpartum period, with this increase then contributing to OT's role in the formation of the long-lasting mother–infant bond. It is also possible that epigenetic processes are involved in such a potential increase (Wang, Duclot, Liu, Wang, & Kabbas, 2013). To the best of my knowledge, no one has carefully explored whether changes occur in the density of OTRs in the NA of pregnant and postpartum rats. OT release into NA may also be higher at parturition than at mating in rats. Another interesting, and related, possibility is derived from the findings of Insel and Shapiro (1992). Although virgin female montane voles do not express OTRs in BLA, maternal parturient females express high levels of OTRs in this region, levels that match those that occur in virgin prairie voles. Therefore, OT action in BLA may not be able to activate the proposed connection between BLA/BMA and VP during mating in montane voles, but does activate this connection in parturient females. In this way, OT action on BLA/BMA might facilitate mother–infant bond formation but not the formation of the pair bond.

Is there evidence that the mechanisms underlying pair bond formation in species other than monogamous voles share some similarities with the mechanisms that regulate the mother–infant bond? In answering this question, a basic principle should be taken into consideration. Although a basic MPOA-to-mesolimbic DA circuit may undergo long-term changes in synaptic strength to regulate pair bond formation in a variety of species, the exact mechanisms through which such changes are realized may vary between species.

Most of the experimental research dealing with pair bond formation and maintenance in species other than prairie voles has examined the importance of OT-related peptides. Since OT is important for the mother–infant bond, evidence that OT is also involved in pair bond formation in a variety of species provides some evidence for commonality between these two types of social bonds.

In contrast to mammals, social monogamy is the dominant mating system in birds, but research into the mechanisms that contribute to the pair bond in birds is scant. In a preliminary study that examined pair bond formation in zebra finches, chronic antagonism of central OT-like receptors (mesotocin receptors), via intracerebroventricular injections of an OTR antagonist, disrupted pair bond formation in both males and females (Klatt & Goodson, 2013). Unfortunately, as I indicated in Chapter 7, I am not aware of any experimental evidence that central mesotocin receptors are also involved in parental behavior in birds, including zebra finches, although some correlational evidence is supportive.

In Chapter 4, I reviewed the evidence that OT appears to be involved in certain aspects of maternal responsiveness in common marmosets. There is also evidence that OT neural systems are involved in pair bond formation/maintenance in the socially monogamous biparental common marmoset. Smith, Agmo, Birnie, and French (2010) administered one of the following three treatments to male and female marmosets who had been paired together for 3 weeks: intranasal (IN) administration of OT (INOT); oral administration of a nonpeptide OTR antagonist, capable of crossing the blood–brain barrier; or control (placebo) treatments. Compared to the control conditions, affiliative interactions between mating partners were reduced by the OTR antagonist and enhanced by INOT. A similar effect of an OTR antagonist has recently been reported by Cavanaugh, Mustoe, and French (2018). Recall that a high density of OT-binding sites exists in the NA of common marmosets (Schorscher-Petcu et al., 2009).

The New World titi monkey also exhibits a monogamous and biparental social system and the sexes form long-term pair bonds (Carp et al., 2016). In a correlational positron emission tomography study, Maninger et al. (2017) measured resting-state glucose uptake in various brain regions as an indication of changes in neural activity in paired compared to unpaired male titi monkeys. Pair-bonded males showed a significant increase in radiolabeled glucose uptake in several brain regions, which included MPOA, MeA, and the NA-VP circuit, after

1 week of being pair-bonded. These results certainly fit with the neural model shown in Figure 11.3. Interestingly, the OTR is not widely distributed in the titi monkey brain and is absent from the NA-VP circuit, while the vasopressin V1a receptor is highly expressed in the NA (Freeman et al., 2014b). These results suggest that vasopressin, rather than OT, may act on NA to promote pair bonding in titi monkeys. It is also possible, however, that high levels of OT are released within the brain during mating and pair bonding and that these high levels are capable of effectively binding to V1a receptors in NA. In support of a role for vasopressin, Jarcho, Mendoza, Mason, Yang, and Bales (2011) have found that IN administration of vasopressin to male titi monkeys increased affiliative behavior directed toward their partner. However, the potential effects of INOT were not examined in this study. To resolve this important question, further studies might measure the actual neuropeptide, OT or AVP, that is released into NA during the formation of a pair bond in these monkeys.

Social monogamy with biparental care of offspring is common, but not universal, across all human societies (Fernandez-Duque et al., 2009; French et al., 2018). However, social monogamy with some degree of biparental care is universal in most of the world's highly developed countries (Heinrich, Boyd, & Richerson, 2012). Many evolutionary theorists have argued that social monogamy without biparental care initially evolved in humans as a male mate guarding strategy and then subsequently social monogamy with paternal behavior (and therefore biparental care) developed as a secondary adaptation (Coxworth, Kim, McQueen, & Hawkes, 2015; Schacht & Bell, 2016). Since all of the nonhuman great apes exhibit a polygynous (or promiscuous) mating system with uniparental maternal care, it can be suggested that pair living without paternal behavior initially evolved in humans from a polygynous/promiscuous ancestor. Under conditions where the occurrence of paternal behavior enhanced the reproductive success of the pair-living males, then it is assumed that long-term pair bonds with biparental behavior developed. Given this analysis, I want to explore whether the neural systems that have been shown to contribute to pair bonding in other biparental mammals are also involved in human pair bonding. Recall from Chapter 8 that OT and MPOA interactions with the mesolimbic DA system appear to play an important role in mother–infant bonding in humans and that there is evidence that OTRs are located in the MPOA, ventral midbrain, basolateral amygdala, and the NA-VP circuit in the human brain.

All of the studies I am about to discuss have been performed using participants from modern industrialized societies. In an functional magnetic resonance imaging (fMRI) study, Acevedo, Aron, Fisher, and Brown (2012) showed men and women who were married for many years photographs of their spouse or photos of an acquaintance. Images of the spouse, in comparison to those of an acquaintance, resulted in significantly greater blood-oxygen–level dependent (BOLD) responses in the hypothalamus, VTA, amygdala, and NA-VP circuit. These are the same brain regions that one would expect to be involved in pair bond formation and maintenance in prairie voles (see Figure 11.3) and in the formation and maintenance of the mother–infant bond. In recent work, Acevedo, Poulin, and Brown (2019) have emphasized the particular importance of BOLD activations in VP when a spouse views facial images of their partner expressing romantic love in comparison to an acquaintance with a neutral facial expression. Schneiderman, Zagoory-Sharon, Leckman, and Feldman (2012) measured plasma OT levels, as a proxy for the central release of OT in the brain, in heterosexual men and women who were in a romantic relationship for 3 months or who were unattached single adults. They reported significantly higher OT levels in both men and women who were in a relationship compared to the unattached singles. Interestingly, attached couples who had higher levels of OT were more likely to remain romantically attached after the 3-month period than those couples that had lower OT levels. These two correlational studies support the view that OT may alter neural activity in the human brain across the circuits depicted in Figure 11.3 to positively influence pair bond formation and maintenance in men and women.

Several studies have examined the effects of INOT on human pair bonds. When OT is administered to either men or women who are in a strong heterosexual romantic relationship, OT treatment, in comparison to placebo, caused the participants to rate a photo of their partner, but not a photo of an unfamiliar opposite sex adult, as more attractive. This effect was paralleled by increased BOLD responses in the VTA and NA when viewing images of the partner (Scheele, Plota, Stoffel-Wagner, Maier, & Hurlemann, 2016; Scheele et al., 2013). These results support the view that OT contributes to pair bonding in both men and women and that activation of the mesolimbic DA system may be involved in this effect. Finally, Kreuder et al. (2017) administered INOT to either men or women. Subsequently, while the subjects were in an fMRI scanner, they were gently touched by an individual who they assumed was either their opposite sex romantic partner or an unfamiliar person of the opposite sex. The subjects could not see the actual individual touching them, and in fact, all subjects were touched by the same experimenter. OT, in comparison to placebo, only increased the subjective pleasantness of the touch of the assumed partner, and this OT effect was associated with increased BOLD responses in NA. All of these studies, taken together, suggest that OT may not only contribute to the formation of human pair bonds, but may also be involved in the maintenance of such bonds (also see Scheele et al., 2012), presumably by enhancing the reward value of partner-related stimuli.

The research described in this section has offered good support for the proposal that the neural circuitry and mechanisms that underlie the long-term mother–infant bond may have provided the neural foundation upon which the strong affiliative pair bond has been built, although much more research needs to be performed. When viewed in total, the role of OT seems to be particularly important. It should be noted, however, that aspects of this general proposal, and the evidence used to support it, has not been free from criticism. Focusing on the known importance for vasopressin binding to V1a receptors in the VP for pair bond formation and maintenance in male prairie voles, Turner et al. (2010) compared several species of deer mice (*Peromyscus* genus), some of which are monogamous and form pair bonds, while others are promiscuous and do not form pair bonds. They reported that all of these species expressed high levels of V1a receptors in VP, and there were no differences between species. These researchers suggested that monogamy and pair bonding in mammals has likely evolved through multiple mechanisms. In evaluating this study, it is important to differentiate between necessary and sufficient mechanisms. Vasopressin action on V1a receptors in VP may be necessary but not sufficient for pair bond formation and maintenance. Many neural processes contribute to pair bond formation (see Figure 11.3), and a particular type of anatomical organization and function within neural sites outside the VP may also be required for pair bond formation even when V1a receptors are expressed at high levels in VP. Also, since more recent research has suggested the importance of OT for pair bond formation in male prairies voles, it would be interesting to determine the nature of OTR binding at various neural sites, such as the NA, VP, and BLA, in the deer mice studied by Turner et al. to determine whether differences in deer mice mating systems are related to differences in OT binding within the brain. It is very possible that high levels of OTR and/or V1aR expression within the NA-VP circuit are necessary, but not sufficient, for social monogamy and pair bonding in mammals, including humans. In my review of the literature (Numan, 2015), I was not able to find evidence for the existence of a monogamous pair bonding mammalian species that displays an absence of both OT and vasopressin binding within this crucial neural circuit. If such a mammalian species could be found, then the view expressed by Turner et al. could be modified to say that mechanisms other than OT and/or vasopressin action within the NA-VP circuit may have evolved to promote the formation of pair bonds, although such a finding would not rule out the possibility that synaptic plasticity needs to occur within this circuit, through alternative mechanisms, for strong affiliative pair bonds to occur.

With respect to the proposal that is central to this chapter, that the neural circuitry and mechanisms that regulate mother–infant bonding may have provided a rudimentary neural foundation for the formation and maintenance of pair bonds in socially monogamous species, Nowicki, Pratchett, Walker, Coker, and O'Connell (2017) have made the important point that certain teleost fishes, such as butterflyfishes (*Chaetodon* genus) exhibit a socially monogamous mating system but that parental care of eggs and hatchlings does not occur. They also emphasize that parental care, when it exists, did not precede the evolution of social monogamy in the butterflyfish genus. It is therefore possible that social monogamy in certain nonmammalian vertebrates is not based on parental neural circuits. However, it is also possible that social monogamy in butterflyfish is an example of pair living rather than pair bonding, with only the latter relying on a strong affiliative attachment between mating partners. For example, Frick (1986) found that in a monogamous butterflyfish (*Chaetodon chrysurus*) that does not display any type of brood care, when either the male or the female was experimentally removed, all mates successfully acquired new mating partners within a short period of time, which was as short as 2 hours. Such quick repairing with a new mate (short partner fidelity) is not likely to occur in pair bonded prairie voles (Sun, Smith, Lei, Liu, & Wang, 2014; Thomas & Wolff, 2004). Fricke suggested that pair living and social monogamy probably evolved in butterflyfish as a mutual mate-guarding strategy coupled with cooperative defense of the territory within which the pair resides. More research is needed on this important question, and it is certainly possible that in certain nonmammalian vertebrate species that do not exhibit parental behavior, social monogamy with a strong affiliative pair bond between mates can still form (see Bull, 2000).

The Potential Contribution of Maternal Neural Circuits to the Neural Basis of Human Hyper-Cooperation and Hyper-Prosociality

A Behavioral and Psychological Analysis

Compared to the nonhuman great apes (orangutans, gorillas, chimpanzees, and bonobos), the nearest living relatives to the human lineage, human societies are unique in their display of high levels of collaboration, cooperation, and aid-giving responses that occur between unrelated (nonkin) individuals that live together within a particular social group. Such prosociality includes the development of mutual trust between individuals, affection for close friends, and the occurrence of empathy for individuals in need of aid (Jaeggi, Burkart & van Schaik, 2010; Lee, 2018; Marean, 2015; Mitani, 2009; Rosati, DiNicola, & Buckholtz, 2018; Tomasello & Vaish, 2013). The mother–infant bonding system may have not only provided the basic neural scaffold for pair bonding, as previously described, but may have also provided the foundation for the strong social bonds that occur between unrelated individuals in human societies, with such friendships fostering high levels of cooperative and aid-giving responses between individuals living within large social networks. This general idea has been around for a long time (for reviews, see Brown, Brown, & Penner, 2012; Decety, Norman, Bernston, & Cacioppo, 2012; Eisenberger, 2013; Marsh, 2019; Numan, 2012b, 2015; Numan & Young, 2016; Pedersen, 2004; Preston, 2013; Ross & Young, 2009), but my intent is to build on these previous reports to present a detailed neural circuitry analysis that supports this fundamental idea.

A series of experiments on chimpanzees and human children has asked a simple question that examines the nature of prosocial behavior (aid-giving behavior to another individual) in these two species: Will a particular individual provide a reward (e.g., a desired food item) to a fellow group member if the donor of the reward does not receive anything in return (Burkart et al., 2014; Jaeggi et al., 2010; Jensen, Hare, Call, & Tomasello, 2006; Silk & House, 2011). These experiments presumably measure the degree to which an individual shows concern for the welfare of other group members. An important conclusion from this research is that while human children are highly prosocial in this context, our closest primate relative, the chimpanzee, is not. Some significant socioecological change must have occurred during human evolution to result in this crucial difference in the quality of prosociality between chimpanzees and humans.

With respect to human prosociality, an interesting experiment on 17-month-old human infants was conducted by Jin and Baillargeon (2017). The infants viewed two unfamiliar women who identified themselves as belonging to the same social group. The infants observed these women engaging in two types of social interactions. In one case, one woman needed help in reaching a particular goal and the other woman helped her. In the other case, help was not offered to the woman who needed help. Infants looked significantly longer at the interaction where no help was offered in comparison to the scenario where helping behavior occurred. Interestingly, if the two women were identified themselves as belonging to different groups, infants showed no difference in the amount of time they viewed the helping and non-helping interactions. The authors concluded that infants expect individuals within the same group to help one another, and they are surprised when such helping does not occur. In contrast, they have no particular expectation of the occurrence of helping behavior in individuals that belong to different groups. The authors proposed that at a very early age infants have an expectation of in-group favoritism during human interactions and that individuals within a particular group are expected to help one another.

Burkart, Hrdy, and van Schaik (2009) and Hrdy (2009) have presented evidence that aligns with the idea that the evolution of hyper-prosocial behavior in humans, in comparison to other great apes, may be related to the degree to which the human maternal care neural system has become an open system, which would allow certain social stimuli to gain access to this aid-giving system in the absence of priming by the physiological events of late pregnancy and parturition.

More specifically, in a comparison of a variety of primate species, including humans, evidence was presented that when high levels of prosociality occur in primates, they are associated with the occurrence of alloparental behavior. Based on these data, these authors have proposed the cooperative breeding hypothesis, which predicts that high levels of prosociality are linked to the amount of alloparental behavior that is displayed by a particular primate species.

Cooperative breeding coupled with high levels of alloparental behavior occur in callitrichids (marmosets and tamarins), and high levels of alloparental behavior occur in diverse human societies, including extant hunter-gather human societies, which presumably represent the type of conditions that existed during early human evolution (Burkart et al., 2009; Hrdy, 2009; Kramer, 2011; Kramer & Otarola-Castillo, 2015; Sear & Mace, 2008; Weisner & Gillmore, 1977). Because of the occurrence of alloparenting, humans are considered to be cooperative breeders in the sense that individuals other than the child's parents provide significant help in caring for the child (Burkart et al., 2009; Hrdy, 2009). In contrast, nonhuman great apes exhibit a uniparental maternal care system, and alloparental behavior does not occur (Burkart et al., 2009; Kramer & Otarola-Castillo, 2015).

Burkart et al. (2014) have designed tests to measure what they refer to as unsolicited proactive prosociality. In these tests, individuals assist others—for example, by providing food—without direct gains for themselves and without being solicited to provide such aid. Burkart and her colleagues measured such prosociality in a number of primate species, including humans, using the group service paradigm. In this procedure, individuals are tested in their social group and can provide fellow group members with food without obtaining food for themselves. The primate species examined included lemurs, Old World monkeys (*Macaca* species), callitrichids, apes, and humans. The findings indicated that the extent of alloparental behavior that occurred in a particular species was the best predictor of proactive prosociality: Proactive prosociality occurred at high levels in those primate species that also displayed alloparental behavior, but was low or absent in those species without alloparental behavior. Marmosets, tamarins, and human children exhibited the highest levels of prosocial behavior, while such prosociality was virtually absent in nonhuman great apes. Since we know that the MPOA is involved in alloparental behavior (see Chapter 7 of this volume), it would be incredibly interesting to determine whether experimental manipulations of MPOA function would influence unsolicited proactive prosociality in marmosets and tamarins, or whether changes in MPOA neural activity are correlated with the expression of such prosociality.

Why does alloparenting occur in humans, but not in other great apes? Humans distinguish themselves from the other great apes by the fact that they wean their offspring before they are fully independent, while weaned nonhuman great ape juveniles can forage for food on their own without help from others. Human mothers, therefore, have to care for multiple offspring at the same time, that is, their current infants and their older but still dependent juveniles. This may have been a factor leading to the development of human alloparenting during early human evolution, where older *independent* siblings, as well as the mother's sisters, brothers, and parents (the aunts, uncles and grandparents of dependent offspring) helped care for dependent young (Hrdy, 2009; Kramer, 2011; Kramer & Otarola-Castillo, 2015). It is also possible that nonkin group members, whom the mother trusted, also participated in alloparenting.

Given that alloparental care exists in both callitrichids and humans, but not in great apes, that such alloparenting is linked to the occurrence of proactive prosociality, and that callitrichids and humans are distantly related, it appears that convergent evolutionary processes contributed to the emergence of high levels of prosociality in small-brained and cognitively unsophisticated callitrichid monkey species and in the cognitively sophisticated human. Callitrichids give birth to multiple offspring (usually twins) and therefore require help from others in raising their young, while humans, in comparison to other great apes, have relatively short interbirth intervals and must care for both their current offspring and weaned but dependent juveniles, which also requires help from others. This need for alloparental support in both callitrichids and humans appears to have fostered the emergence of proactive prosociality in both of these primate groups.

The Cooperative Breeding Hypothesis for the evolution of hyper-prosociality in primates is primarily a behavioral/psychological hypothesis, not a neural hypothesis. I am adding a neural dimension to this hypothesis by proposing that by opening the social bond formation neural mechanism beyond the mother–infant bond and the pair bond, so that it includes forming enduring social bonds between alloparents and infants, a neural preadaptation may have been set that would ultimately allow evolutionary processes in humans to foster the occurrence of strong social bonds between adult group members even if they were not kin or mates (also see Marsh, 2019; Numan, 2018). That is, with alloparenting, the neural behavioral bonding and aid-giving system that forms the basis of maternal behavior in most mammals was made more flexible and less dependent upon pregnancy and parturition. Under the proper socioecological conditions, natural selection may have acted on this basic alloparenting neural foundation to allow for the evolution of the complex forms of cooperation, collaboration, and aid-giving behaviors that occur in humans. In a further elaboration of their view, Burkart et al. (2009) have suggested that in humans, because of the evolution of alloparenting, a highly prosocial motivation was added to a cognitively sophisticated brain inherited from our great ape relatives and that the combination of such hyper-prosocial behavior with enhanced cognitive abilities led to the emergence of a human form of hyper-cooperation and prosociality that rose to levels far above that which is observed in other primates, including callitrichids.

The combination of prosocial motivation with enhanced cognition in humans has significant implications. It is important to distinguish prosocial behavior from prosocial cognitions and feeling states, such as cognitive and emotional empathy, and sentiments of affection and care for the well-being of others. Given that humans appear to have developed a more open subcortical network that regulates unsolicited proactive prosocial behavior against a background of enhanced cognitive abilities, this combination may have led to the emergence of a coupling of mentalizing and cognitive/emotional empathy with a more open prosocial behavioral network. Hrdy (2009) has proposed that a further enhancement of mentalizing and empathy, which are regulated by cortical brain regions (see Chapter 8 of this volume), above the rudimentary levels of such states that might exist in nonhuman great apes, may have emerged in humans in the context of alloparenting. In particular, she argues that dependent juveniles who were cared for by nonmothers needed to understand what others were feeling and thinking to assess the intentions of potential caregivers so that they would accept aid from those that would help them, but avoid those individuals who might harm them. I would like to add that such enhanced mentalizing and empathy in humans should have also evolved in the offspring's biological mother since the mother had to develop the ability to relinquish her biological offspring to individuals that she perceived to be reliable and trustworthy—individuals who would properly care for, rather than harm, her young and would also return her young to her when she desired reunion with them. A mother's biological offspring are a much more valued resource than a desired food item. Nonhuman great ape mothers are highly possessive of their young and rarely share their young with others. The ability to develop trust, in a cognitive (cortically based) sense, toward a broad range of other individuals, rather than being a simple behavioral proclivity to share one's offspring with others, with the latter being based on subcortical mechanisms that were probably modified by kin selection in such cooperative breeding extended family groups such as callitrichid species, may be a uniquely human characteristic.

Once enhanced cortically based mentalizing and empathy evolved along with a more open prosocial subcortical network in the context of human alloparenting, then the emergence of these characteristics may have set the stage for the further evolution of high levels of cooperation and collaboration in humans. As human ecological conditions changed so that group hunting of large prey became adaptive, then hyper-cooperation between adults, even if nonkin, might have proven to be adaptive in early human hunter-gather groups. Indeed, Tomasello, Melis, Tennie, Wyman, and Hermann (2012) have proposed that cooperation between nonkin may have evolved in the context of group hunting, which resulted in the development of joint intentionality, so that individuals within a group would work together to achieve the mutually beneficial goal of acquiring large amounts of meat that would then be fairly shared with all group members. My proposal is that the evolution of enhanced cortical mentalizing and empathy neural circuits that were coupled to a more open subcortical prosocial behavioral network, which emerged in a human alloparenting context, may have set the stage for the general ability to judge the trustworthiness of fellow group members and to interact with them accordingly. This proposed process could then lead to hyper-cooperation, mutual aid-giving, and the formation of strong emotional, cognitive, and behavioral social bonds between adults, even when they were unrelated to one another (in a genetic sense).

Wild chimpanzees show limited degrees of cooperation/collaboration, and this usually occurs in the context of competition and self-interest (Mitani, 2009). For example, male chimpanzees may form alliances with another male in order to rise to a higher dominance level within their social group. It is interesting to propose that self-interested cooperation in chimpanzees, as opposed to joint intentionality and sharing in humans, is due to the fact that chimpanzees have not evolved an open subcortical prosocial behavioral network with the result that their cognitive skills are used mainly to serve their own self-interest.

With respect to the human parental brain, Figures 8.3 and 8.4 show how cortical mentalizing and cognitive and emotional empathy systems might interact with subcortical parental behavior neural circuits, with the interconnections between these systems being mediated by the medial regions of the prefrontal cortex (mPFC), to translate maternal cognitive and emotional empathy states into maternal caretaking responses directed toward the mother's infant. In Figure 11.4, I want to build upon these figures, in an abbreviated form, to show how human parental brain neural circuits might have been utilized by natural selection to promote the evolution of hyper-prosociality and hyper-cooperation within human social groups under the particular socioecological conditions that I have previously described and that have been proposed to have occurred during early human evolution. This figure shows that in humans, social stimuli from other members of an individual's social group are capable of reaching and influencing cognitive and emotional empathy systems. These systems, in turn, are capable of activating subcortical prosocial behavioral networks so that one's regard for the welfare of other group members results in appropriate prosocial aid-giving responses directed toward members of one's social group.

As indicated in Chapter 8, Ashar et al. (2017) have made a distinction between emotional empathy, cognitive empathy, and empathic care. Empathic care refers to an individual's desire to act prosocially toward another person. Empathic care results in affiliative emotional states associated with caring for the welfare of others. Such states can then promote prosocial aid-giving responses to others.

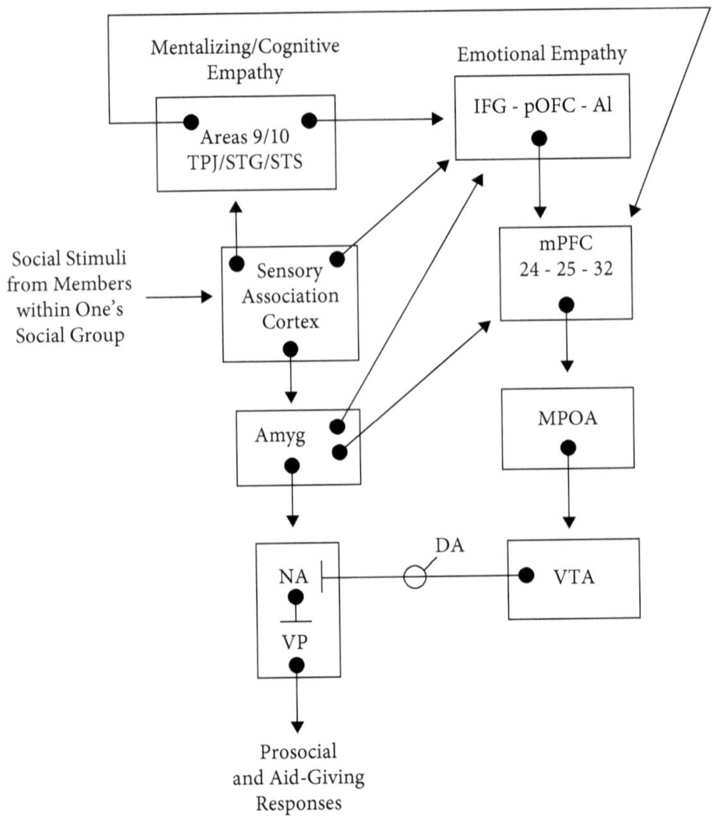

Figure 11.4. A hypothetical neural model that shows how human parental brain circuits (see Figures 8.3 and 8.4) might have been appropriated and modified by natural selection to promote the evolution of hyper-prosociality and hyper-cooperation within human social groups. Social stimuli from other group members in one's social group activate cortical mentalizing and cognitive/emotional empathy neural systems. These systems, in turn, activate subcortical prosocial behavioral networks so that one's regard for the welfare of other group members results in prosocial aid-giving responses toward them, as well as cooperation with other members of one's group. Although not shown in this figure, social stimuli- and medial preoptic area-induced activation of paraventricular nucleus oxytocin release would be able to influence activity at several nodes in the proposed neural network. 9/10 = cortical areas within the dorsomedial prefrontal cortex; AI = anterior insula; Amyg = amygdala; DA = dopamine; IFG = inferior frontal gyrus; mPFC = medial prefrontal cortex containing areas 24, 25, and 32; MPOA = medial preoptic area; NA = nucleus accumbens; pOFC = posterior orbitofrontal cortex; TPJ = temporoparietal junction; STG = superior temporal gyrus; STS = superior temporal sulcus; VP = ventral pallidum; VTA = ventral tegmental area. Neural effects ending in an arrow signify excitation while those ending in a bar represent inhibition. See text for details and supporting evidence.

Source: Modified from Figure 7.2 in Numan (2015) with permission from Elsevier.

This view can be expanded to include the activation of prosocial feeling states between two or more individuals once they appreciate their common goals based on the occurrence of shared intentions that have resulted from the prior development of mutual trust. It can be proposed that once an appropriate understanding of another individual's emotional state and intentions are appreciated, so that one understands the needs of another and their desire to be cooperative, then the connections between mentalizing and empathy neural systems become capable of effectively stimulating empathic care neural networks within mPFC. The mPFC would then activate prosocial subcortical behavioral networks to which it projects. As I indicated in Chapter 8, certain parts of the mPFC appear to be a crucial neural link that translates mentalizing and empathy into prosocial behavior. Therefore, particular neurons within the mPFC may comprise an important integrative neural site that represents feelings of empathic care. In the sections that follow, I will present the evidence that supports these proposals, which are derived from the neural circuit model shown in Figure 11.4.

Overlap Between Parental Brain Neural Circuits and the Neural Circuits That Underpin Human Prosociality

Oxytocin

Given the role of OT in promoting prosocial cognitive, emotional, and motivational states within the mother–infant context (see Chapter 8 of this volume), if the proposal that the mother–infant bonding system provided the foundation for the strong social bonds that occur between unrelated individuals in human societies is accurate, then one would predict that OT neural systems play a broader role, outside the mother–infant context, to influence prosocial human interactions more generally. There is a large body of evidence that supports this perspective and many of these studies involve examining the effects of INOT administration on human prosocial cognitions, emotions, and behavior.

In the context of Figure 11.4, recall that OTRs are likely to be expressed in the following regions of the human brain: mentalizing/cognitive empathy regions (dorsomedial prefrontal cortex [dmPFC], cortical area 9; superior temporal gyrus); cortical regions involved in translating cognitive and emotional empathy into prosocial aid-giving responses (mPFC [cortical areas 24, 25, 32]); subcortical brain regions involved in the expression of prosocial behavior (MPOA, VTA, NA-VP circuit, BLA). Therefore, there are multiple sites within the human brain where OT could act to promote the cognitive, emotional, and motivational processes involved in human prosociality. More specifically, OT release in the human brain during social interactions between individuals who are not kin and who are not mating/romantic partners may play a role in the development of mutual trust and the resultant formation of social bonds that underlie strong friendships, which, in turn, influences the occurrence of shared intentionality, mutual cooperation, and aid-giving responses.

In a seminal study, Kosfeld, Heinrichs, Zak, Fischbacher, and Fehr (2005) investigated the effects of INOT or IN placebo administration on the behavioral responses of men who acted as investors in an economic trust game. Each subject (the investor) was given a standard amount of money that he could invest with an unfamiliar trustee. The trustee might share the profits made from the investment with the investor. Alternatively, the trustee could act selfishly and keep the entire investment and profits for himself. The dependent variable in this study was the proportion of the original monetary allotment that each investor gave to the trustee during an initial investment and without any prior knowledge of the trustworthiness of the trustee. Participants in the placebo condition only invested a small amount of money with the trustee, while OT-treated subjects invested a larger proportion of their endowment. Kosfeld et al. proposed that INOT administration increased OT levels within the brain and that the neural effects of OT increased the trusting behavior of the investor (see Nave, Camerer, & McCullough, 2015, for a critical evaluation of this study).

These findings can be interpreted within the context of a reciprocal altruism explanation. Under natural conditions, if a donor person were to aid a recipient for the first time, a small or moderate amount of aid might be offered (e.g., to a new neighbor who just moved next door). If the recipient reciprocated and helped the donor in the future (the recipient was trustworthy), then upon subsequent interactions, the original donor might provide greater aid to the original recipient due to the development of increased trust. This mechanism probably contributes to how friendships between individuals normally develop: Social bonds between individuals are enhanced by reciprocal altruism (or mutual cooperation), which promotes trust between the cooperating individuals, with the understanding that they can depend upon one another. It can be proposed that as social bonds and trust between individuals become stronger, social stimuli from one individual activate endogenous OT neural systems in the friend, with OT then favoring prosocial cooperative behaviors. OT may be acting in the brain during social interactions to both promote prosocial motivation (aid-giving responses) and to foster the development of an enduring social bond between the interacting individuals. It is possible that the subjects in the Kosfeld et al. (2005) study who received INOT may have invested a large part of their money with the trustee in the absence of a naturally formed friendship because the exogenous OT replicated the internal state that would have normally occurred only after a true friendship had developed, where endogenous OT levels would be high.

Based on the research I presented in this chapter, it should be realized that OT presumably has these effects because it is operating within the context of an open and flexible prosocial behavioral network neural network.

Several other studies conform with the interpretation of the Kosfeld et al. (2005) study. Pornpattananangkul, Zhang, Chen, Kok, and Yu (2017) asked human participants to characterize various social relationships between themselves and others. These relationships varied along a continuum from close relationships to distant relationships to strangers. In a monetary task, the participants could either keep a monetary endowment for themselves or share part of the money with others. Subjects in the IN-placebo group behaved as would be expected. They shared money with close friends but were not likely to share money with strangers or with individuals with whom they had distant social relationships. In contrast, subjects that were administered INOT were more likely to share some of their money with distantly related individuals, even if they were complete strangers. In a somewhat related study, Cohen and Shamay-Tsoory (2018) presented three images on a computer screen. One image represented the subject in the study, one image represented a friend, and a third image was that of a stranger. Each subject was asked to press the space bar on the computer keyboard that moved the image of themselves toward the friend or the stranger. Under either the placebo or INOT condition, subjects moved themselves very close to their friend. With respect to the image of the stranger, subjects who were administered INOT moved the image of themselves closer to the stranger in comparison to subjects that were administered the placebo. Although this task is not a monetary task, it is related to social approach motivation and the results indicate that when endogenous OT is presumably low, exogenous treatment with OT enhances social approach motivation. In each of these studies, it would have been interesting to have obtained some measure of endogenous OT levels, perhaps by measuring plasma OT. I would predict that endogenous plasma OT levels would probably be higher in an individual who was interacting with a close friend in comparison to interactions with a stranger. However, I will qualify this possibility later in this chapter when I discuss the fact that OT not only influences prosocial motivation, but also has effects on defensiveness. By measuring OT in plasma, one is not getting an accurate measure of OT release within particular brain regions. Therefore, it is possible that plasma OT levels could be high in both prosocial (reactions to a friend) and defensive behavioral contexts (reactions to a stranger who might appear threatening).

There are studies, however, that do provide evidence that endogenous OT levels are correlated with human prosocial feeling states toward others and social bond formation under conditions where defensive reactions are presumably not involved. Barraza and Zak (2009) had participants view video images of emotional and nonemotional social interactions between people. Self-reported measures of empathy were higher when subjects viewed the emotional scenes, and such empathic states were correlated with higher plasma OT levels. In a study by Wolfe, Deruelle, and Chaminade (2018), participants viewed facial images of siblings, best friends, or unknown individuals. fMRI analysis indicated increased BOLD responses in a hypothalamic region that contained the PVN when the subjects viewed sibling or friend faces in comparison to unknown faces, and the degree of familiarity (time spent together) was positively associated with the BOLD response. The authors suggest that their results support the view that PVN-OT neurons are involved in social attachment, although no specific measures of OT were taken (also see Sugiura et al., 2001).

In an interesting study, Bernaerts et al. (2017) administered INOT or placebo to men who measured high on an attachment avoidance scale. Long-term treatment with INOT over a 2-week period improved attachment security and decreased attachment avoidance. It is likely that these subjects with an insecure-avoidant attachment style had low endogenous OT release within brain regions that regulate prosocial motivation and behavior (see Chapter 10 of this volume; also see Haas et al., 2016; Kohlhoff et al., 2017; Samuel et al., 2015; Strathearn et al., 2009). It is also possible that such individuals express OTRs that less effectively bind OT and that they need higher levels of OT to effectively activate these receptors. In a related study, Bartz et al. (2010) asked men to complete a social competency questionnaire. Some men were rated as prosocial while others were rated as relatively asocial, perhaps representing secure and insecure-avoidant attachment styles, respectively. Each subject then viewed videos of target individuals discussing emotional events in their lives, and each subject used keyboard presses on a computer to rate how positive or negative they thought each target felt at each moment during the target's narrative. This latter test was used to measure the empathic accuracy of each subject (presumably a measure of both cognitive and emotional empathy). When subjects received an IN-placebo treatment, those who scored high on social competence also scored high on empathic accuracy, while those who scored low on social competence scored low on empathic accuracy. When the participants received INOT, then the empathic accuracy scores of those subjects that had previously scored low on social competency improved to levels above that which occurred under the placebo condition. These results suggest that endogenous OT probably promotes empathic accuracy in human subjects who are socially competent. In contrast, socially incompetent individuals probably have low endogenous brain OT levels and/or OTRs that weakly bind endogenous OT, with the result that they have deficits in empathic accuracy, with empathy being improved by the exogenous administration of OT. These results are relevant to our understanding of prosociality because a critical level of OT-related cognitive and emotional empathy may be needed to result in empathic care, which then leads to the translation of empathy into prosocial motivation and behavior. These results also show that personality variables (individual differences in sociality), which may be affected by individual differences in endogenous activity across specific OT neural pathways, can influence the effects of exogenous OT administration on prosociality.

With respect to the idea that differences in the ability of the OTR to effectively bind OT may be related to individual differences in prosociality, Li et al. (2015) have reported that differences in SNPs within the OXTR gene at the rs53576 site are related to individual differences in measures of human prosociality. Similar findings have been reported by others (Gong et al., 2017; Kogan et al., 2011; Krueger et al., 2012; Rodrigues, Saslow, Garcia, John, & Keltner, 2009; Smith, Porges, Norman, Connelly, & Decety, 2014; Woods, Bedard, McQuaid, Matheson, & Anisman, 2018).

Riem, Bakermans-Kranenburg, Huffmeijer, and van Ijzendoorn (2013) have provided direct evidence that INOT can enhance prosocial helping behavior. This study examined the effects of OT on prosocial behavior during a computerized virtual ball-tossing game referred to as cyberball. They examined the effects of OT on helping behavior toward a socially excluded person who was known to the participant. All the subjects were women, and under placebo conditions the participants compensated for a player's ostracism by tossing a ball more often to the excluded player. Importantly, INOT further increased the number of ball throws to the excluded person. The authors suggest that one way in which OT may enhance prosocial behavior is by increasing cognitive and emotional empathy for the excluded individual. This process could give rise to empathic care and prosocial behavior. Interestingly, and related to maternal treatment effects on the development of OT neural systems (see Chapters 9 and 10 of this volume), INOT did not improve prosocial behavior in subjects who received poor maternal care during their childhood.

In an important study, De Dreu et al. (2010; also see De Dreu & Kret, 2016; Ne'eman, Perach-Barzilay, Fischer-Shofty, Atias, & Shamay-Tsoory, 2016) examined the effects of INOT on monetary sharing between individual human subjects under conditions where individuals were assigned, on the basis of a trivial criterion, to one of two 3-person groups, with one group being an in-group (the group to which the subject was assigned) and the other group being an out-group (the group to which the subject was not assigned). Subjects that received INOT, when compared to the placebo condition, increased the amount of their monetary endowment that they shared with in-group members, but it did not increase their prosocial behavior toward out-group members. This finding appears to be related to the work of Jin and Baillargeon (2017) that I described earlier in this chapter, where it was found that young infants have a general expectation of in-group favoritism. Significantly, De Dreu et al. also found that if monetary payoff conditions were manipulated so that out-group members could take money away from the other group (exploit the other group), then INOT not only enhanced in-group favoritism, but also promoted out-group aggression, where an in-group participant, under the influence of INOT, selected choices that would not only provide money to their in-group, but would also subtract money from the out-group. Such "aggression" did not occur if out-group members were not able to exploit the in-group.

How should we interpret these results? De Dreu et al. (2010) propose that the behavioral context to which an individual is exposed can influence the effects of exogenous OT administration on prosocial behavior. INOT promotes in-group favoritism but does not enhance prosociality toward an out-group. However, if the out-group becomes a potential threat, then OT can act to enhance antisocial responses toward the out-group. These findings have given rise to the view that OT does not have ubiquitously prosocial effects (Bethlehem, Baron-Cohen, van Honk, Auyeung, & Bos, 2014). This view is not surprising from a neural perspective, since OT can act on different brain regions to have different behavioral effects, and the context to which an individual is exposed influences the neural circuits that become engaged and active. Because of the human bias toward in-group favoritism, OT presumably acts on the circuits shown in Figure 11.4 to selectively promote in-group prosociality because out-group social stimuli are assumed to not have easy access to these circuits (see Numan, 2015). However, when the out-group becomes a threat, then other neural systems that regulate defensive aggression are presumably engaged, and OT acts on these circuits (not shown in Figure 11.4) to promote defensive aggression. As I indicated in Chapter 6, OT acts on separate neural circuits in the brain, different from those that regulate prosocial maternal motivation, to promote maternal aggression (also see Bakermans-Kranenburg & van Ijzendoorn, 2018). Similarly, in a nonmaternal context, OT might act on circuits in the brain analogous to those that underpin maternal aggression to promote antisocial responses toward out-group members that are potentially threatening to one's in-group. My perspective is that when a subject in administered INOT, OT acts on those circuits that are primed and active by a particular situation or context (see DeWall et al., 2014). These findings do not detract from the clearly prosocial effects that OT has within certain contexts, but informs us that, by acting on brain regions that regulate defensiveness and protective responses, OT can also have antisocial effects in other situations. Furthermore, personality variables might also affect the degree to which prosocial or antisocial neural pathways are engaged and susceptible to OT effects in a given situation or context (Declerck, Boone, & Kiyonari, 2014).

It should be obvious that the influence of INOT on prosocial cognitions, feeling states, and behavior are complex and can be influenced by social context, personality variables, maternal treatment developmental effects, and genetic factors. Sex differences also appear to have impacts on the effects of INOT on cooperative behaviors in humans, since women tend to have higher baseline endogenous OT levels than do men (Borland, Rilling, Frantz, & Albers, 2019).

I do have one important caveat to the views I have just expressed. There is a large literature that indicates that the central effects of vasopressin at certain neural sites promote aggressive and antisocial behaviors (Numan, 2015). Since OT has some affinity for vasopressin receptors, it is possible that INOT promotes antisocial responses because it is actually acting on vasopressin receptors and its associated aggression-related neural systems under threatening social contexts. Therefore, in studies that examine the potential prosocial and antisocial effects of INOT, three groups should be utilized: IN administration of placebo, OT, or vasopressin. If OT is actually acting on vasopressin receptors to promote antisocial responses, then I would predict that at low doses, IN vasopressin but not INOT, would promote such responses because vasopressin has a higher affinity for vasopressin receptors than does OT (cf. Nishina, Takagishi, Takahashi, Sakagami, & Inoue-Murayama, 2019).

Because the goal of this chapter is to examine the overlap between the parental brain and the neural circuits that underpin human hyper-prosociality, the remainder of this chapter will focus on the brain circuits that regulate prosociality and cooperation in humans to compare these circuits with parental brain neural circuits.

Neural Circuits

In this selective review of the literature, I will present the human data that support the involvement of the circuits shown in Figure 11.4 in human prosociality.

In relation to emotional empathy, the first general point I want to make is that although the anterior insular (AI) (inferior frontal gyrus/posterior orbitofrontal cortex/AI)-anterior cingulate cortex circuit (ACC) has usually been associated with vicariously sharing others' pain and suffering, which could lead to aid-giving responses to those in need, research indicates that this critical brain region is also involved in sharing others' positive emotional states, which would be an example of vicarious positive emotional empathy (Jabbi, Swart, & Keysers, 2007; Lockwood, 2016; Morelli, Sacchet, & Zaki, 2015). For example, in the meta-analytic review performed by Morelli et al. (2015), studies were included only if subjects directly observed or imagined another person receiving a reward outcome, such as receiving praise, a pleasant touch (hug), or being involved in a positive emotional event (getting married). During such vicarious positive affective states, brain activation was observed across several brain regions, which included the amygdala and the AI-ACC circuit. I am emphasizing this point because human prosociality not only involves giving aid to those in need, but also includes cooperative behaviors, which are presumably associated with reciprocal and mutual pleasurable feeling states whenever working together results in the achievement of a common goal. Perhaps related to this idea is the finding of Lewis, Kanai, Rees, and Bates (2014), who found that the gray matter volume of the insular cortex was positively related to the degree to which individuals perceived themselves as having positive social relationships with others.

The second general point is that although separate neural circuits have been shown to be involved in cognitive empathy and emotional empathy (Tusche, Bocker, Kanske, Trautwein, & Singer, 2016), anatomical studies indicate that there are several anatomical connections that allow these two functional regions to interact (see Chapter 8 of this volume and Figure 11.4). For example, Figure 11.4 shows that mentalizing and cognitive empathy regions project to the AI. Given this anatomical relationship, one would predict that AI dysfunction might disrupt both emotional and cognitive empathy. In support, research has shown that patients with damage to the insular cortex, in comparison to control subjects, have deficits in both emotional and cognitive empathy (Boucher et al., 2015; Chen et al., 2016).

Although empathy for others can influence prosocial behavior, it can be proposed that a critical level of empathy must be reached to activate empathic care, which, in turn, activates the subcortical brain mechanisms that promote prosocial behavior (Ashar et al., 2017; Bethlehem, Allison et al., 2017). Although a baseline level of empathy for others may exist in most humans, several factors have been shown to influence one's degree of empathy and the prosocial behaviors that such empathy may induce. Such factors include whether the person being observed by the empathizer is a member of one's group (in-group favoritism) or is a close friend.

In an fMRI study, Meyer et al. (2013) measured the emotional and neural responses of subjects who observed either their best friend or a stranger being included or excluded from a computerized ball-tossing game. Subjects empathized more with the presumed negative emotional experiences of their best friend being excluded in comparison to their empathy for the stranger, and this greater degree of empathy was associated with a larger BOLD response in the AI-ACC circuit. That such greater empathy and AI BOLD responses toward an individual exposed to social exclusion is related to subsequent acts of prosocial behavior toward the excluded individual is supported by the findings of Masten, Morelli, and Eisenberger (2011). Prosocial behavior toward the excluded person, as measured by sending that person a sympathetic email, was associated with greater BOLD responses not only in the AI, but also in the dmPFC (cortical area 10), suggesting that an interaction between cognitive and emotional empathy promoted prosocial behavior.

With respect to in-group favoritism, Hein, Silani, Preuschoff, Batson, and Singer (2010) have reported results that converge with the findings I have just reviewed. While in an fMRI scanner, subjects observed members of their in-group (fans of the participants' favorite sports team) or an out-group (fans of an opposing team) receive electric shocks. Subjects reported higher empathy levels for their in-group in comparison to the out-group, and this greater empathy was positively correlated with a greater AI BOLD response when observing in-group members receiving shock. Subsequently, while outside the scanner, subjects were given the opportunity to perform an aid-giving prosocial response toward members of each group: They could receive shock to decrease the shock received by the other individual. Subjects were more likely to act prosocially toward in-group members, and the degree of their prosocial behavior was positively correlated with the AI BOLD response that they exhibited during the first part of the study. These results are consistent with a proposal that when AI empathy-related neural activity reaches a critical threshold, then AI inputs to the mPFC result in the activation of subcortical neural circuits that underpin prosocial behaviors (see Figure 11.4). It is certainly possible that a critical level of both cognitive and emotional empathy is needed to fully activate mPFC (areas 24, 25, 32) projections to the subcortical neural circuits that regulate prosocial behaviors.

Low levels of empathy toward members of a perceived out-group have resulted in serious social conflicts throughout human history, and such conflicts clearly persist today. In a very interesting study, Hein, Engelmann, Vollberg, and Tobler (2016) examined whether help offered by an out-group member to a person of another group could modify the empathy of this person toward the out-group member. This study consisted of three phases: preintervention, intervention, and postintervention. Swiss participants were paired with subjects of Swiss descent (in-group; identified by their names) or subjects of Balkan descent (out-group; identified by their names; such individuals represent a significant minority population in Switzerland). To measure empathic brain responses, these researchers assessed each subject's brain responses while they were in a scanner and observed pain in an in-group or out-group member. In the first phase, subjects had more positive impressions of in-group members and their AI BOLD response was greater when they observed an in-group member, compared to an out-group member, in pain. In the intervention phase, each subject was expecting to receive a shock, but another individual, identified by their name, could give up money to save the participant from pain. In the third phase, if an in-group member had received aid from an out-group member, their positive impressions of the out-group increased, and their AI BOLD response also increased, when compared to the preintervention phase, when the in-group member observed an out-group member receiving pain. This increased AI BOLD response was positively correlated with increases in empathic care for the out-group member that was receiving a painful stimulus.

Although only AI responses were measured in this study, it should be clear that cognitive processes had to be involved in the observed effects and that the interaction between cognitive empathy and mentalizing neural systems with emotional empathy systems contributed to the enhanced empathic responses of the in-group toward the out-group. The Hein et al. (2016) study is important because it shows that negative biases toward the "other" can be positively modified by social interactions and the formation of social bonds.

In examining Figure 11.4, my model suggests that when a certain amount of cognitive and emotional empathy for others reaches a critical threshold, then the mPFC (areas 24, 25, 32) is activated, resulting in the experience of empathic care, which then motivates prosocial behavior through mPFC projections to subcortical brain regions that underpin prosocial behavior. What is the evidence that supports a role for the mPFC in prosociality? First, there is a large body of evidence that shows that patients with damage to the mPFC, which includes damage to areas 25 and 32, demonstrate antisocial behavioral traits (Anderson, Bechara, Damasio, Tranel, & Damasio, 1999; Boes et al., 2011; Numan, 2015). Such individuals lack empathy and frequently behave in ways that enhance their own self-interest. Such individuals may also harm and/or manipulate others to reach the particular goals that they desire.

With respect to healthy human subjects, fMRI research also demonstrates an important role of the mPFC in prosocial sentiments and behavior. Rigney, Koski, and Beer (2018) found that the BOLD response increased in the mPFC (areas 24, 25, 32) when participants identified positive traits associated with in-group members. In the study by Ashar et al. (2017), while in a scanner, participants listened to biographical narratives describing true stories of suffering individuals. Outside the scanner, they listened to these same narratives again and indicated their moment-to-moment level of empathic care. Empathic care was defined as responding to the distress and suffering of others with feelings of tenderness, warmth, care, compassion, and sympathy. Subsequent analysis indicated that the degree of empathic care was positively correlated with the BOLD response in the ventromedial prefrontal cortex that occurred while subjects listened to the narratives. This ventromedial prefrontal cortex region appeared to include parts of area 32 and the medial OFC that lies ventral to area 32. Similar results have been reported by FeldmanHall, Dalgleish, Evans, and Mobbs (2015), where activity in areas 25 and 32 was associated with empathic care.

Moll et al. (2006) performed a study where subjects, while in an fMRI scanner, could either receive a monetary reward or donate money to a charitable organization of their choice. The mPFC was not active when subjects received a monetary reward, but it was highly active, based on the BOLD response, when the participants made charitable donations. The active area of the mPFC appeared to be at the border between areas 25 and 32. These results are consistent with the view that activity in mPFC can drive prosocial behavior.

An important issue is the involvement of the MPOA in human prosociality. Figure 11.4 proposes that mPFC input to MPOA drives MPOA activation of the mesolimbic DA system, in this way promoting prosocial behavioral responses in humans. In an excellent review, Eisenberger (2013) presented the animal research that demonstrates that MPOA interactions with the mesolimbic DA system are critically involved in maternal motivation. With respect to the potential involvement of the MPOA in human prosociality, she indicates that the small size of the MPOA makes it difficult to reliably isolate this region with neuroimaging methods, and therefore activity in this region is not usually analyzed in studies on human sociality (but recall that Atzil et al., 2017, did relate MPOA activity to human maternal behavior). The volume of the entire human hypothalamus is about 1 cubic centimeter, and that portion of the hypothalamus that comprises the preoptic region has a volume of about 0.2 cubic centimeter (Makaris et al., 2013; Schindler et al., 2013). Therefore, in neuroimaging studies, because of the inherent spatial resolution capacity of typical fMRI methods, it is usually difficult to spatially separate potential MPOA BOLD responses from the dorsally adjoining septal area and the laterally adjoining VP. However, recent magnetic resonance imaging methods have clearly been able to image the MPOA and measure its volume (Makaris et al., Schindler et al.). Given these advances, it would certainly be interesting to correlate the volume of the MPOA with the degree of empathic care and prosocial behavior that individuals display, since there are important individual differences in these characteristics. Improvements in fMRI methods, adapted from the previously described structural magnetic resonance imaging methods, may also eventually allow the selective analysis of MPOA BOLD responses in a social context.

Despite the current limitations of fMRI methods, there is functional imaging research that supports a role for the MPOA in human prosociality. In a study by Moll et al. (2012), adult men and women, most of whom did not have children of their own, read short social scenarios while in an fMRI scanner. The scenarios were categorized into dimensions that included the degree to which they evoked a positive emotional state (positive valence) and the degree to which they evoked strong affiliative or attachment emotions represented by feelings of tenderness and care. The participants were asked to vividly imagine themselves performing the actions being portrayed. In Part A of Table 11.1, examples of affiliative and nonaffiliative positive scenarios and an example of a neutral social scenario are shown. Part B of Table 11.1 shows the BOLD responses observed in the preoptic area and adjoining septal area that were associated with reading each of these scenario types. Significantly, the preoptic area-septal area was selectively activated only when the subjects read and imagined themselves performing the social affiliation-related scenarios. Hopefully, future research along these lines, using fMRI methods with enhanced spatial resolution, will be able to zero in on the BOLD response that occurs selectively in MPOA. These data do suggest, however, that the MPOA is a region that is active in situations where strong social

Table 11.1 The Effects of Reading Various Social Scenarios on the Blood–Oxygen Level–Dependent Signal in the Septal-Preoptic Area of Human Participants

A. *Examples of social scenario types*

1. Affiliative—positive scenario

   You put together your old photos from childhood and made an album for your mother

2. Nonaffiliative—positive scenario

   You delivered a beautiful speech and the audience stood up to applaud you

3. Neutral scenario

   You ordered items from a drug store and someone delivered them to you

B. *BOLD response (fMRI data)*

Scenario	*Septal-POA BOLD response*
Affiliative-positive	High
Nonaffiliative-positive	Low
Neutral	Low

Notes: BOLD = blood–oxygen level–dependent. fMRI = functional magnetic response imaging. POA = preoptic area.

Source: The information in this table is derived from research by Moll et al (2012).

bonds are imagined to exist. I do not think that the MPOA is a site that represents prosocial feeling states because I conceive it as being part of subcortical social behavior network. Accordingly, I think that the increased activity in the MPOA region was associated with imagining the performance of prosocial acts toward others with whom strong social bonds have been formed.

In an interesting study performed by Rameson, Morelli, and Lieberman (2012), healthy men and women, while in a scanner, read a sentence describing a sad situation that was then followed by photos showing different individuals in that particular situation. The participants were asked to take the perspective of the individuals being depicted and to imagine how those individuals felt in their particular situations. This procedure was meant to induce empathic care in the subjects, and after the scanning procedure, each subject rated their empathic care for the targets they observed. Prior to the scanning procedure, the participants also completed a daily diary that assessed their daily helping behavior directed toward a friend or a stranger. Based on my analysis of their brain scans, it appears that significant BOLD responses across several brain regions occurred when participants were experiencing high levels of empathic care, and these responses were positively correlated with daily helping toward a friend (but not a stranger).

These brain regions included dmPFC (area 9), temporoparietal junction, AI, mPFC (including area 32), septal-preoptic area, and NA. This analysis certainly fits with the model shown in Figure 11.4, where the interaction of mentalizing and cognitive and emotional empathy can induce empathic care-related neural activity in the mPFC, which, in turn, activates MPOA interactions with the mesolimbic DA system to promote prosocial aid-giving responses. Please note that these authors did not emphasize and consider the BOLD responses that appear to be clearly evident in the anterior part of the hypothalamus (preoptic area) and adjoining septal region. However, in more recent work, Morelli, Rameson, and Lieberman (2014) have reported that septal area activity (probably including the preoptic area and the vBST) is indeed predictive of daily prosocial helping behavior, and related findings have been reported by Zahn et al. (2009).

In addition to understanding the neural systems involved in empathic care and prosocial behavior toward an individual in need of aid, my perspective on human prosociality also involves an understanding of mutualism/cooperation between individuals and the development of trust between members of a social group so that they can work together to achieve a common goal. In this regard, an important study was performed by Krueger et al. (2007). These investigators examined the BOLD responses that occurred in individuals who participated in a multiround trust game, where one individual could invest or not invest money with a trustee, and the trustee could either reciprocate an investment or not reciprocate (keeping all of the investment for himself/herself). The most important finding of this study, based on the neural model presented in Figure 11.4, was that pairs of individuals who showed the highest trust–reciprocate history, based on the frequency of their trust and reciprocation responses, also showed the highest activation in the septal-preoptic region. Another important finding was that decisions of an investor to trust (give part of their endowment to a trustee), in comparison to decisions not to trust, were associated with greater BOLD responses in medial prefrontal areas 9 and 32 and in the septal-preoptic region. These results suggest that as trust develops between partners, mentalizing and cognitive empathy brain regions (cortical area 9 of the dmPFC) may interact with the mPFC (area 32), which, in turn, activates the septal-preoptic region, which then regulates the occurrence of prosocial trusting behavioral responses.

Figure 11.4 indicates that the MPOA interacts with the mesolimbic DA to promote prosocial behavior. In the context of a multiround trust game, Phan, Sripada, Angstadt, and McGabe (2010) examined BOLD responses in the ventral striatal region when investors played with a fair trustee (one who reciprocated often) or an unfair trustee (one who tended not to reciprocate an investment). By the fifth round of this game, investors trusted the fair partner (invested part of their monetary endowment with the trustee) much more than they trusted the unfair partner, and the occurrence of trusting behavior toward the fair partner was associated with an enhanced BOLD response in the ventral striatum.

With respect to the finding of Kosfeld et al. (2005) that INOT was found to increase trust in an individual playing a trust game, where might OT act to produce such effects? Based on the location of OTRs in the human brain, OT could be acting at several sites within the neural network shown in Figure 11.4 (Paloyelis et al., 2016), including mentalizing/cognitive empathy regions (dmPFC area 9 and superior temporal gyrus; see Mackinnon et al., 2014; Rochetti et al., 2014), mPFC, MPOA, and/or NA-VP circuit to influence the development of trust between individuals. With respect to the maternal behavior research reviewed in Chapter 5, it is interesting to speculate that OT may contribute to the development of mutual trust and friendships between two interacting individuals by producing at least three major effects: (a) the promotion of prosocial motivation and behavior through actions on the MPOA; (b) the consolidation of an enduring social bond between two individuals through actions at the level of the NA; (c) and the translation of mentalizing (perspective-taking) and empathy into overt prosocial behaviors through activation of mPFC projections to MPOA.

The research reviewed in this section certainly supports the model shown in Figure 11.4, although it is obvious that much more research needs to be done. Because of the extensive research on the importance of the MPOA for maternal behavior, more research needs to be done on the role of the MPOA in human prosociality, and neural responses that occur in MPOA must be clearly distinguished from those that occur in the septal area. This last point is important because although the septal area has not been shown to play a crucial role in the maternal motivation of animals (Numan, 1985; cf. Cruz & Beyer, 1972), it is involved in other aspects of animal social behavior (Sheehan & Numan, 2000).

Conclusions

The goals of this chapter have been twofold: (a) to explore the neural changes that may have occurred within the parental neural circuitry to allow for the evolution of alloparental behavior and (b) to provide evidence for the proposal that the parental brain neural circuitry that regulates mother–infant bond formation provided the basic neural foundation for the evolution of other types of strong social bonds between individuals under those socioecological conditions where such bond formation has adaptive significance.

With respect to the neuromolecular changes that may have evolved to allow for the occurrence of alloparental behavior in mammals, which has adaptive significance for such species as prairie voles, callitrichids (marmosets and tamarins), and humans, several proposals, along with supporting evidence, were offered.

Most important, I provided evidence to support the view that for alloparental behavior to occur, the MPOA may have been modified to become a relatively open system with respect to the ability of infant stimuli to gain access to MPOA neural circuits in the absence of exposure to the physiological events associated with late pregnancy and parturition. I suggested that ligand-independent activation of ERs within the MPOA may have been involved in this process. In addition, evolutionary modifications within OT neural systems may have also contributed to the emergence of alloparenting, with such modifications likely including the enhanced expression of OTRs in NA and MPOA (or V1a vasopressin receptors in MPOA that can be affected by high levels of endogenous OT) of sexually naïve mammals. For competent alloparental behavior to occur under natural conditions, it is possible that ligand-independent activation of ERs in MPOA stimulates the expression of OTRs in MPOA and that OT action on MPOA then allows the MPOA to activate OT and DA release into NA through its projections to the PVN and VTA, respectively. The upregulation of OTRs in MPOA and NA, where OT acts to boost parental motivation, is conceived as being particularly important for the occurrence of effective alloparental behavior under challenging environmental conditions, and such conditions are likely to occur within natural ecological contexts.

With respect to the parental neural circuitry providing the rudimentary neural foundation for the evolution of enduring social bonds other than the mother–infant bond, I presented the following analysis:

(a) MPOA interactions with the mesolimbic DA system were shown to be an evolutionarily conserved neural system involved in the appetitive aspects of male and female sexual behaviors and parental behavior in mammals. This finding is important because it demonstrates the prominent role of the MPOA in social motivational processes in mammals, although I noted that it is likely that different subpopulations of MPOA neurons, each of which projects to the mesolimbic DA system, are selectively involved in each type of reproductive behavior.

(b) Although MPOA interactions with the mesolimbic DA system contribute to the appetitive aspects of sexual and maternal behaviors, for most mammals enduring social bonds are not formed between mating partners, but such enduring bonds do occur between a mother and her infant(s). As reviewed in Chapter 5, enduring mother–infant bonds are the result of the neural plasticity changes that occur across the MPOA-to-VTA-DA-to-NA-VP circuit and the MPOA-to-PVN-OT-to-NA-VP circuit. Such plasticity does not occur between mating partners of most mammalian species that demonstrate a polygynous/promiscuous mating system with uniparental maternal care. However, pair bonding and strong and enduring bonds between mating partners do occur in about 5% of mammalian species, and the neural circuitry and neural plasticity that underpins such pair bonding between mating partners shares many commonalities with the neural mechanisms that underlie the formation of the mother–infant bond (see Figure 11.3).

(c) The hyper-prosociality and hyper-cooperation that occurs in humans, including the formation of strong social bonds between individuals other than kin and mating partners, is likely to be rooted in the evolution of human alloparenting and in the neural linking of an open subcortical system that regulates prosocial behavior (consisting of MPOA projections to the mesolimbic DA system and to the PVN-OT system) with highly developed cortical mentalizing and empathy neural systems, inherited in part from our great ape relatives and further enhanced in the context of human alloparenting. Importantly, the neural systems that underpin human hyper-prosociality (see Figure 11.4) show many commonalities with the human parental brain neural circuitry described in Chapter 8 (see Figures 8.3 and 8.4). Given the role of OT in mother–infant bonding, it is also important to emphasize that OT appears to act at several nodes in the circuits depicted in Figure 11.4 to promote trust, various aspects of empathy, and the formation of strong social bonds between cooperating partners. OT release into these brain sites may be stimulated, in part, by MPOA activation of PVN-OT neurons.

References

Abraham, E., Hendler, T., Shapira-Lichter, I., Kanat-Maymon, Y., Zagoory-Sharon, O., & Feldman, R. (2014). Father's brain is sensitive to childcare experiences. *Proceedings of the National Academy of Sciences USA, 111*(27), 9792–9797.

Abraham, E., Hendler, T., Zagoory-Sharon, O., & Feldman, R. (2016). Network integrity of the parental brain in infancy supports the development of children's social competencies. *Social, Cognitive, and Affective Neuroscience, 11*(11), 1707–1718.

Abraham, E., Raz, G., Zagoory-Sharon, O., & Feldman, R. (2018). Empathy networks in the parental brain and their long-term effects on children's stress reactivity and behavior adaptation. *Neuropsychologia, 116*(Pt A), 75–85.

Acevedo, B. P., Aron, A., Fisher, H. E., & Brown, L. L. (2012). Neural correlated of long-term intense love. *Social, Cognitive, and Affective Neuroscience, 7*, 145–159.

Acevedo, B. P., Poulin, M. J., & Brown, L. L. (2019). Beyond romance: Neural and genetic correlates of altruism in pair-bonds. *Behavioral Neuroscience, 133*(1), 18–31.

Adamec, R. E., & McKay, D. (1993). The effects of CRF and alpha-helical CRF on anxiety in normal and hypophysectomized rats. *Journal of Psychopharmacology, 7*(4), 346–354.

Afonso, V. M., Grella, S. L., Chatterjee, D., & Fleming, A. S. (2008). Previous maternal experience affects accumbal dopaminergic responses to pup-stimuli. *Brain Research, 1198*, 115–123.

Afonso, V. M., King, S., Chatterjee, D., & Fleming, A. S. (2009). Hormones that increase maternal responsiveness affect accumbal dopaminergic responses to pup- and food-stimuli in the female rat. *Hormones and Behavior, 56*(1), 11–23.

Afonso, V. M., King, S. J., Novakov, M., Burton, C. L., & Fleming, A. S. (2011). Accumbal dopamine function in postpartum rats that were raised without their mothers. *Hormones and Behavior, 60*(5), 632–643.

Afonso, V. M., Shams, W. M., Jin, D., & Fleming, A. S. (2013). Distal pup cues evoke dopamine responses in hormonally primed rats in the absence of pup experience or ongoing maternal behavior. *Journal of Neuroscience, 33*(6), 2305–2312.

Agrati, D., & Lonstein, J. S. (2016). Affective changes during the postpartum period: Influences of genetic and experiential factors. *Hormones and Behavior, 77*, 141–152.

Aguirre, J., Meza, E., & Caba, M. (2017). Dopaminergic activation anticipates daily nursing in the rabbit. *European Journal of Neuroscience, 45*(11),1396–1409.

Ahern, T. H., Hammock, E. A., & Young, L. J. (2011). Parental division of labor, coordination, and the effects of family structure on parenting in monogamous prairie voles (Microtus ochrogaster). *Developmental Psychobiology, 53*(2), 118–131.

Ahern, T. H., & Young, L. J. (2009). The impact of early life family structure on adult social attachment, alloparental behavior, and the neuropeptide systems regulating affiliative behaviors in the monogamous prairie vole (Microtus ochrogaster). *Frontiers in Behavioral Neuroscience, 3*, 17.

Ainsworth, M. D. S., Blehar, M. C., Waters, E., & Wall, S. (1978). *Patterns of attachment: A psychological study of the strange situation*. Oxford, England: Erlbaum.

Akther, S., Fakhrul, A. A., & Higashida, H. (2014). Effects of electrical lesions of the medial preoptic area and the ventral pallidum on mate-dependent paternal behavior in mice. *Neuroscience Letters, 570*, 21–25.

Akther, S., Huang, Z., Liang, M., Zhong, J., Fakhrul, A. A., Yuhi, T., . . . Higashida, H. (2015). Paternal retrieval behavior regulated by brain estrogen synthetase (aromatase) in mouse sires that engage in communicative interactions with pairmates. *Frontiers in Neuroscence, 9*, 450.

Akther, S., Korshnova, N., Zhong, J., Liang, M., Cherepanov, S. M., Lopatina, O., . . . Higashida, H. (2013). CD38 in the nucleus accumbens and oxytocin are related to paternal behavior in mice. *Molecular Brain, 6*, 41.

Alhassen, L., Phan, A., Alhassen, W., Nyuyen, P., Lo, A., Shaharuddin, H., . . . Alachkar M. (2019). The role of olfaction in MCH-regulated spontaneous maternal responses. *Brain Research, 1719*, 71–76.

Amico, J. A., Challinor, S. M., & Cameron, J. L. (1990). Pattern of oxytocin concentrations in the plasma and cerebrospinal fluid of lactating rhesus monkeys (*Macaca mulatta*): Evidence for functionally independent oxytocinergic pathways in primates. *Journal of Clinical Endocrinology and Metabolism, 71*(6), 1531–1535.

Amico, J. A., Mantella, R. C., Vollmer, R. R., & Li, X. (2004). Anxiety and stress responses in female oxytocin deficient mice. *Journal of Neuroendocrinology, 16*(4), 319–324.

Amico, J. A., Thomas, A., & Hollingshead, D. J. (1997). The duration of estradiol and progesterone exposure prior to progesterone withdrawal regulates oxytocin mRNA levels in the paraventricular nucleus of the rat. *Endocrine Research, 23*(3), 141–156.

Anderson, S. W., Bechara, A., Damasio, H., Tranel, D., & Damasio, A. R. (1999). Impairment of social and moral behavior related to early damage in the human prefrontal cortex. *Nature Neuroscience, 2*, 1032–1037.

Anniverno, R., Bramante, A., Mencacci, C., & Durbano, F. (2013) . Anxiety disorders in pregnancy and the postpartum period. In F. Durbano (Ed.), *New insights into anxiety disorders*. (pp. 259–285). Rijeka, Croatia: IntechOpen.

Anthony, T. E., Dee, N., Bernard, A., Lerchner, W., Heintz, N., & Anderson, D. J. (2014). Control of stress-induced persistent anxiety by an extra-amygdala septohypothalamic circuit. *Cell, 156*(3), 522–536.

Aragona, B. J., Liu, Y., Curtis, J. T., Stephan, F. K., & Wang, Z. (2003). A critical role for nucleus accumbens dopamine in partner preference formation in male prairie voles. *Journal of Neuroscience, 23*, 3483–3490.

Arrati, P. G., Carmona, C., Dominguez, G., Beyer, C., & Rosenblatt, J. S. (2006). GABA receptor agonists in the medial preoptic area and maternal behavior in lactating rats. *Physiology & Behavior, 87*(1), 51–65.

Ashar, Y. K., Andrews-Hanna, J. R., Dimidjian, S., & Wager, T. D. (2017). Empathic care and distress: Predictive brain markers and dissociable brain systems. *Neuron, 94*(6), 1263–1273.

Asok, A., Draper, A., Hoffman, A. F., Schulkin, J., Lupica, C. R., & Rosen, J. B. (2018). Optogenetic silencing of a corticotropin-releasing factor pathway from the central amygdala to the bed nucleus of the stria terminalis disrupts sustained fear. *Molecular Psychiatry, 23*(4), 914–922.

Assink, M., Spruit, A., Schuts, M., Lindauer, R., van der Put, C. E., & Stams, G. J. M. (2018). The intergenerational transmission of child maltreatment: A three-level meta-analysis. *Child Abuse & Neglect, 84*, 131–145.

Attwell, D., & Iadecola, C. (2002). The neural basis of functional brain imaging signals. *Trends in Neurosciences, 25*(12), 621–625.

Atzil, S., Hendler, T., & Feldman, R. (2011). Specifying the neurobiological basis of human attachment: Brain, hormones, and behavior in synchronous and intrusive mothers. *Neuropsychopharmacology, 36*(13), 2603–2615.

Atzil, S., Touroutoglou, A., Rudy, T., Salcedo, S., Feldman, R., Hooker, J. M., . . . Barrett, L. F. (2017). Dopamine in the medial amygdala network mediates human bonding. *Proceedings of the National Academy of Sciences USA, 114*(9), 2361–2366.

Augustine, M. E., Leerkes, E. M., Smolen, A., & Calkins, S. D. (2018). Relations between early maternal sensitivity and toddler self-regulation: Exploring variation by oxytocin and dopamine D2 receptor genes. *Developmental Psychobiology, 60*(7), 789–804.

Augustine, R. A., Ladyman, S. R., Bouwer, G. T., Alyousif, Y., Sapsford, T. J., Scott, V., . . . Brown, C. H. (2017). Prolactin regulation of oxytocin neurone activity in pregnancy and lactation. *Journal of Physiology, 595*(11), 3591–3605.

Augustine, R. A., Seymour, A. J., Campbell, R. E., Grattan, D. R., & Brown, C. H. (2018). Integrative neuro-humoral regulation of oxytocin neuron activity in pregnancy and lactation. *Journal of Neuroendocrinology, 30*(8), e12569.

Babcock Fenerci, R. L., & Allen, B. (2018). From mother to child: Maternal betrayal trauma and risk for maltreatment and psychopathology in the next generation. *Child Abuse & Neglect, 82*, 1–11.

Bahr, N. I., Martin, R. D., & Pryce, C. R. (2001). Peripartum sex steroid profiles and endocrine correlates of postpartum maternal behavior in captive gorillas (*Gorilla gorilla gorilla*). *Hormones and Behavior, 40*(4), 533–541.

Baker, M., Lindell, S. G., Driscoll, C. A. Zhou, Z., Yuan, Q. Schwandt, M. L., . . . Barr, C. S. (2017). Early rearing history influences oxytocin receptor epigenetic regulation in rhesus macaques. *Proceedings of the National Academy of Sciences USA, 114*(44), 11769–11774.

Bakermans-Kranenburg, M. J., & van Ijzendoorn, M. H. (2015). The hidden efficacy of intervention: Gene x environment experiments from a differential susceptibility perspective. *Annual Review of Psychology, 66*, 381–409.

Bakermans-Kranenburg, M. J., & van Ijzendoorn, M. H. (2017). Protective parenting: Neurobiological and behavioral dimensions. *Current Opinion in Psychology, 15*, 45–49.

Bakermans-Kranenburg, M. J., & van Ijzendoorn, M. H. (2018). Oxytocin and human sensitive and protective parenting. *Current Topics in Behavioral Neurosciences, 35*, 421–448.

Bakermans-Kranenburg, M. J., van Ijzendoorn, M. H., Riem, M. M., Tops, M., & Alink, L. R. (2012). Oxytocin decreases handgrip force in reaction to infant crying in females without harsh parenting experiences. *Social, Cognitive, and Affective Neuroscience, 7*(8), 951–957.

Bakowska, J. C., & Morrell, J. I. (1995). Quantitative autoradiographic analysis of D1 and D2 dopamine receptors in rat brain in early and late pregnancy. *Brain Research, 703*(1-2), 191–200.

Bakowska, J. C., & Morrell, J. I. (1997). Atlas of the neurons that express mRNA for the long form of the prolactin receptor in the forebrain of the female rat. *Journal of Comparative Neurology, 386*(2), 161–177.

Bakowska, J. C., & Morrell, J. I. (2003). The distribution of mRNA for the short form of the prolactin receptor in the forebrain of the female rat. *Molecular Brain Research, 116*(1-2), 50–58.

Bale, T. L., Davis, A. M., Auger, A. P., Dorsa, D. M., & McCarthy, M. M. (2001). CNS region-specific oxytocin receptor expression: Importance in regulation of anxiety and sex behavior. *Journal of Neuroscience, 21*(7), 2546–2552.

Bales, K., Dietz, J., Baker, A., Miller, K., & Tardif, S. D. (2000). Effects of allocare-givers on fitness of infants and parents in callitrichid primates. *Folia Primatologia (Basel), 71*(1-2), 27–38.

Bales, K. L., Kim, A. J., Lewis-Reese, A. D., & Sue Carter, C. (2004). Both oxytocin and vasopressin may influence alloparental behavior in male prairie voles. *Hormones and Behavior, 45*(5), 354–361.

Bales, K. L., & Perkeybile, A. M. (2012). Developmental experiences and the oxytocin receptor system. *Hormones & Behavior, 61*(3), 313–319.

Balfour, M. E., Brown, J. L., Yu, L., & Coolen, L. M. (2006). Potential contributions of efferents from medial prefrontal cortex to neural activation following sexual behavior in the male rat. *Neuroscience, 137*(4), 1259–1276.

Balshine, S. (2012). Patterns of parental care in vertebrates. In N. J. Royle, P. T. Smiseth, & M. Kölliker (Eds.), *The Evolution of Parental Care* (pp. 62–80). Oxford, England: Oxford University Press.

Banerjee, S. B., & Liu, R. C. (2013). Storing maternal memories: Hypothesizing an interaction of experience and estrogen on sensory cortical plasticity to learn infant cues. *Frontiers in Neuroendocrinology, 34*(4), 300–314.

Barbas, H., Ghashghaei, H., Dombrowski, S. M., & Rempel-Clower, N. L. (1999). Medial prefrontal cortices are unified by common connections with superior temporal cortices and distinguished by input from memory-related areas in the rhesus monkey. *Journal of Comparative Neurology, 410*(3), 343–367.

Barbas, H., Zikopoulos, B., & Timbie, C. (2011). Sensory pathways and emotional context for action in primate prefrontal cortex. *Biological Psychiatry, 69*(12), 1133–1139.

Barbosa, M. N., & da Silva Mota, M. T. (2013). Alloparental responsiveness to newborns by nonreproductive, adult male, common marmosets (*Callithrix jacchus*). *American Journal of Primatology, 75*(2), 145–152.

Bardi, M., Petto, A. J., & Lee-Parritz, D. E. (2001). Parental failure in captive cotton-top tamarins (*Saguinus Oedipus*). *American Journal of Primatology, 54*(3), 159–169.

Bardi, M., Shimizu, K., Barrett, G. M., Borgognini-Tarli, S. M., & Huffman, M. A. (2003). Peripartum sex steroid changes and maternal style in rhesus and Japanese macaques. *General and Comparative Endocrinology, 133*(3), 323–331.

Barker, D. J., Miranda-Barrientos, J., Zhang, S., Root, D. H., Wang, H.-L., Liu, B., . . . Morales, M. (2017). Lateral preoptic control of the lateral habenula through convergent glutamate and GABA transmission. *Cell Reports, 21*(7), 1757–1769.

Barker, E. D., Walton, E., & Cecil, C. A. M. (2018). Annual research review: DNA methylation as a mediator in the association between risk exposure and child and adolescent psychopathology. *Journal of Child psychology and Psychiatry, 59*(4), 303–322.

Barraza, J. A., & Zak, P. J. (2009). Empathy toward strangers triggers oxytocin release and subsequent generosity. *Annals of the New York Academy of Sciences, 1167*, 182–189.

Barrett, J. Wonch, K. E., Gonzalez, A., Ali, N., Steiner, M., Hall, G. B., & Fleming., A. S. (2012). Maternal affect and quality of parenting experiences are related to amygdala response to infant faces. *Social Neuroscience, 7*(3), 252–268.

Bartlett, J. D., Kotake, C., Fauth, R., & Easterbrooks, M. A. (2017). Intergenerational transmission of child abuse and neglect: Do maltreatment type, perpetrator, and substantiation status matter? *Child Abuse & Neglect, 63*, 84–94.

Bartz, J. A., Zaki, J., Bolger, N., Hollander, E., Ludwig, N. N., Kolevzon, A., & Ochsner, K. N. (2010). Oxytocin selectively improves empathic accuracy. *Psychological Science, 21*(10), 1426–1428.

Basurto, E., Hoffman, K., Lemus, A. C., & Gonzalez-Mariscal, G. (2018). Electrolytic lesions to the anterior hypothalamus-preoptic area disrupt maternal nest-building in intact and ovariectomized, steroid-treated rabbits. *Hormones and Behavior, 102*, 48–54.

Bauer, J. H. (1983). Effects of maternal state on the responsiveness to nest odors of hooded rats. *Physiology & Behavior, 30*(2), 229–232.

Bayerl, D. S., & Bosch, O. J. (2019). Brain vasopressin signaling modulates aspects of maternal behavior in lactating rats. *Genes, Brain, and Behavior, 18*(1), e12517.

Beach, F. A., & Jaynes, J. (1956). Studies of maternal retrieving in rats III: Sensory cues involved in the lactating female's response to her young. *Behavior, 10*, 104–125.

Bean, N. J., & Wysocki, C. J. (1989). Vomeronasal organ removal and female mouse aggression: The role of experience. *Physiology & Behavior, 45*(5), 875–882.

Beier, K. T., Steinberg, E. E., DeLoach, K. E., Xie, S., Miyamichi, K., Schwarz, L., . . . Luo, L. (2015). Circuit architecture of VTA dopamine neurons revealed by systematic input-output mapping. *Cell, 162*(3), 622–634.

Belsky, J., Bakermans-Kranenburgh, M. J., & van Ijzendoorn, M. H. (2007). For better or for worse: Differential susceptibility to environmental influences. *Current Directions in Psychological Science, 16*, 300–304.

Belsky, J., & Pasco Fearon, R. M. (2002). Infant-mother attachment security, contextual risk, and early development: A moderation analysis. *Development and Psychopathology, 14*, 293–310.

Belugin, S., Diogenes, A. R., Patil, M. J., Ginsburg, E., Henry, M. A., & Akopian, A. N. (2013). Mechanisms of transient signaling via short and long prolactin receptor isoforms in female and male sensory neurons. *Journal of Biological Chemistry, 288*(48), 34943–34955.

Benekareddy, M., Goodfellow, N. M., Lambe, E. K., & Vaidya, V. A. (2010). Enhanced function of prefrontal serotonin 5-HT2 receptors in a rat model of psychiatric vulnerability. *Journal of Neuroscience, 30*(36), 12138–12150.

Benekareddy, M., Vadodaria, K. C., Nair, A. R., & Vaidya, V. A. (2011). Postnatal serotonin type 2 receptor blockade prevents the emergence of anxiety behavior, dysregulated stress-induced immediate early gene responses, and specific transcriptional changes that arise following early life stress. *Biological Psychiatry, 70*(11), 1024–1032.

Benedetto, L., Pereira, M., Ferreira, A., & Torterolo, P. (2014). Melanin-concentrating hormone in the medial preoptic area reduces active components of maternal behavior in rats. *Peptides, 58*, 20–25.

Bennett, A. J., Lesch, K. P., Heils, A., Long, J. C., Lorenz, J. G., Shoaf, S. E., . . . Higley, J. D. (2002). Early experience and serotonin transporter gene variation interact to influence primate CNS function. *Molecular Psychiatry, 7*(1), 118–122.

Benoit, D. (2004). Infant–parent attachment: Definition, types, antecedents, measurement and outcome. *Pediatric Child Health, 9*(8), 541–545.

Bergquist, F., & Ludwig, M. (2008). Dendritic transmitter release: A comparison of two model systems. *Journal of Neuroendocrinology, 20*(6), 677–686.

Berlin, L. J., Appleyard, K., & Dodge, K. A. (2011). Intergenerational continuity in child maltreatment: Mediating mechanisms and implications for prevention. *Child Development, 82*(1), 162–176.

Bernaerts, S., Prinsen, J., Berra, E. Bosmans, G., Steyaert, J., & Alaerts, K. (2017). Long-term oxytocin administration enhances the experience of attachment. *Psychoneuroendocrinology, 78*, 1–9.

Bernard, K., Nissim, G., Vaccaro, S., Harris, J. L., & Lindhiem, O. (2018). Association between maternal depression and maternal sensitivity from birth to 12 months: A meta-analysis. *Attachment & Human Development, 20*(6), 578–599.

Bernhardt, B. C., & Singer, T. (2012). The neural basis of empathy. *Annual Review of Neuroscience, 35*, 1–23.

Berridge, K. C. (2004). Motivation concepts in behavioral neuroscience. *Physiology & Behavior, 81*(2), 179–209.

Berridge, K. C. (2007). The debate over dopamine's role in reward: The case for incentive salience. *Psychopharmacology (Berl), 191*(3), 391–431.

Bester-Meredith, J. K., Burns, J. N., Conley, M. F., Mammarella, G. E., & Ng, N. D. (2017). Peromyscus as a model system for understanding the regulation of maternal behavior. *Seminars in Cell & Developmental Biology, 61*, 99–106.

Bethlehem, R. A., Baron-Cohen, S., van Honk, J., Auyeung, B., & Bos, P. A. (2014). The oxytocin paradox. *Frontiers in Behavioral Neuroscience, 8*, 48.

Bethlehem, R. A. I., Allison, C., van Andel, E. M., Coles, A. I., Neil, K., & Baron-Cohen, S. (2017). Does empathy predict altruism in the wild? *Social Neuroscience, 12*(6), 743–750.

Bethlehem, R. A. I., Lombardo, M. V., Lai, M. C., Auyeung, B., Crockford, S. K., Deakin, J., . . . Baron-Cohen, S. (2017). Intranasal oxytocin enhances intrinsic corticostriatal functional connectivity in women. *Translational Psychiatry, 7*(4), e1099.

Bian, X. (2013). Physiological and morphological characterization of GABAergic neurons in the medial amygdala. *Brain Research, 1509*, 8–19.

Bian, X., Yanagawa, Y., Chen, W. R., & Luo, M. (2008). Cortical-like functional organization of the pheromone-processing circuits in the medial amygdala. *Journal of Neurophysiology, 99*(1), 77–86.

Bitran, D., Hilvers, R. J., & Kellogg, C. K. (1991). Ovarian endocrine status modulates the anxiolytic potency of diazepam and the efficacy of gamma-aminobutyric acid-benzodiazepine receptor-mediated chloride ion transport. *Behavioral Neuroscience, 105*(5), 653–662.

Blank, T., Nijholt, I., Grammatopoulos, D. K., Randeva, H. S., Hillhouse, E. W., & Spiess, J. (2003). Corticotropin-releasing factor receptors couple to multiple G-proteins to activate diverse intracellular signaling pathways in mouse hippocampus: Role in neuronal excitability and associative learning. *Journal of Neuroscience, 23*(2), 700–707.

Blankenship, S. L., Chad-Friedman, E., Riggins, T., & Dougherty, L. R. (2019). Early parenting predicts hippocampal subregion volume via stress reactivity in childhood. *Developmental Psychobiology, 61*, 125–140.

Boccia, M. L., & Pedersen, C. A. (2001). Brief vs. long maternal separations in infancy: Contrasting relationships with adult maternal behavior and lactation levels of aggression and anxiety. *Psychoneuroendocrinology, 26*(7), 657–672.

Boccia, M. L., Petrusz, P., Suzuki, K., Marson, L., & Pedersen, C. A. (2013). Immunohistochemical localization of oxytocin receptors in human brain. *Neuroscience, 253*, 155–164.

Boes, A. D., Grafft, A. H., Joshi, C., Chuang, N. A., Nopoulos, P., & Anderson, S. W. (2011). Behavioral effects of congenital ventromedial prefrontal cortex malformation. *BMC Neurology, 11*, 151.

Bole-Feysot, C., Goffin, V., Edery, M., Binart, N., & Kelly, P. A. (1998). Prolactin (PRL) and its receptor: Actions, signal transduction pathways and phenotypes observed in PRL receptor knockout mice. *Endocrine Reviews, 19*(3), 225–268.

Borland, J. M., Rilling, J. K., Frantz, K. J., & Albers, H. E. (2019). Sex-dependent regulation of social reward by oxytocin: An inverted U hypothesis. *Neuropsychopharmacology, 44*(1), 97–110.

Born, J., Lange, T., Kern, W., McGregor, G. P., Bickel, U., & Fehm, H. L. (2002). Sniffing neuropeptides: A transnasal approach to the human brain. *Nature Neuroscience, 5*(6), 514–516.

Bos, P. A., Spencer, H., & Montoya, E. R. (2018). Oxytocin reduces neural activation in response to infant faces in nulliparous young women. *Social Cognitive and Affective Neuroscience, 13*(10), 1099–1109.

Bosch, O. J., Meddle, S. L., Beiderbeck, D. I., Douglas, A. J., & Neumann, I. D. (2005). Brain oxytocin correlates with maternal aggression: Link to anxiety. *Journal of Neuroscience, 25*(29), 6807–6815.

Bosch, O. J., & Neumann, I. D. (2008). Brain vasopressin is an important regulator of maternal behavior independent of dams' trait anxiety. *Proceedings of the National Academy of Sciences USA, 105*(44), 17139–17144.

Boucher, O., Rouleau, I., Lassonde, M., Lepore, F., Bouthillier, A., & Nguyen, D. K. (2015). Social information processing following resection of the insular cortex. *Neuropsychologia, 71*, 1–10.

Bouvette-Turcot, A. A., Fleming, A. S., Unternaehrer, E., Gonzalez, A., Atkinson, L., Gaudreau, H., . . . Meaney, M. J. (2019). Maternal symptoms of depression and sensitivity mediate the relation between maternal history of early adversity and her child temperament: The inheritance of circumstance. *Development and Psychopathology*. doi:10.1017/S0954579419000488. [Epub ahead of print]

Bouvette-Turcot, A. A., Fleming, A. S., Wazana, A., Sokolowski, M. B., Gaudrean, H., Gonzalez, A., . . . Meaney, M. J. (2015). Maternal childhood adversity and child temperament: An association moderated by child 5-HTTLPR genotype. *Genes, Brain, and Behavior, 14*(3), 229–237.

Bowlby, J. (1973). *Attachment and loss: Vol. 2, Separation: Anxiety and anger*. New York, NY: Basic Books.

Bowlby, J. (1980). *Attachment and loss: Vol. 3, Loss: Sadness and depression*. New York, NY: Basic Books.

Bowlby, J. (1983). *Attachment and Loss: Vol. 1, Attachment* (2nd ed). New York, NY: Basic Books.

Bradley, B., Davis, T. E., Wingo, A. P., Mercer, K. B., & Ressler, K. J. (2013). Family environment and adult resilience: Contributions of positive parenting and the oxytocin receptor gene. *European Journal of Psychotraumatology, 4*, 21659.

Bradley, B., Westen, D., Mercer, K. B., Binder, E. B., Jovanovic, T., Crain, D., . . . Heim, C. (2011). Association between childhood maltreatment and adult emotional dysregulation in a low-income, urban, African American sample: Moderation by oxytocin receptor gene. *Development and Psychopathology, 23*(2), 439–452.

Breton, J. M., Charbit, A. R., Snyder, B. J., Fong, P. T. K., Dias, E. V., Himmels, P., . . . Margolis, E. B. (2019). Relative contributions and mapping of ventral tegmental area dopamine and GABA neurons by projection target in the rat. *Journal of Comparative Neurology, 527*(5), 916–941.

Brett, Z. H., Humphreys, K. L., Smyke, A. T., Gleason, M. M., Nelson, C. A., Zeanah, C. H., ... Drury, S. S. (2015). Serotonin transporter linked polymorphic region (5-HTTLPR) genotype moderates the longitudinal impact of early caregiving on externalizing behavior. *Development and Psychopathology, 27*(1), 7–18.

Bridges, R., Rigero, B., Byrnes, E., Yang, L., & Walker, A. (2001). Central infusions of the recombinant human prolactin receptor antagonist, S179D-PRL, delay the onset of maternal behavior in steroid-primed, nulliparous female rats. *Endocrinology, 142*(2), 730–739.

Bridges, R., Zarrow, M. X., Gandelman, R., & Denenberg, V. H. (1972). Differences in maternal responsiveness between lactating and sensitized rats. *Developmental Psychobiology, 5*(2), 123–127.

Bridges, R. S. (1975). Long-term effects of pregnancy and parturition upon maternal responsiveness in the rat. *Physiology & Behavior, 14*(3), 245–249.

Bridges, R. S. (1977). Parturition: Its role in the long term retention of maternal behavior in the rat. *Physiology & Behavior, 18*(3), 487–490.

Bridges, R. S. (1978). Retention of rapid onset of maternal behavior during pregnancy in primiparous rats. *Behavioral Biology, 24*(1), 113–117.

Bridges, R. S. (1984). A quantitative analysis of the roles of dosage, sequence, and duration of estradiol and progesterone exposure in the regulation of maternal behavior in the rat. *Endocrinology, 114*(3), 930–940.

Bridges, R. S., & Grattan, D. R. (2019). 30 years after: CNS actions of prolactin: Sources, mechanisms and physiological significance. *Journal of Neuroendocrinology, 31*(3), e12669.

Bridges, R. S., Mann, P. E., & Coppeta, J. S. (1999). Hypothalamic involvement in the regulation of maternal behaviour in the rat: Inhibitory roles for the ventromedial hypothalamus and the dorsal/anterior hypothalamic areas. *Journal of Neuroendocrinology, 11*(4), 259–266.

Bridges, R. S., Numan, M., Ronsheim, P. M., Mann, P. E., & Lupini, C. E. (1990). Central prolactin infusions stimulate maternal behavior in steroid-treated, nulliparous female rats. *Proceedings National Academy of Sciences USA, 87*(20), 8003–8007.

Bridges, R. S., Robertson, M. C., Shiu, R. P., Friesen, H. G., Stuer, A. M., & Mann, P. E. (1996). Endocrine communication between conceptus and mother: Placental lactogen stimulation of maternal behavior. *Neuroendocrinology, 64*(1), 57–64.

Bridges, R. S., Robertson, M. C., Shiu, R. P., Sturgis, J. D., Henriquez, B. M., & Mann, P. E. (1997). Central lactogenic regulation of maternal behavior in rats: Steroid dependence, hormone specificity, and behavioral potencies of rat prolactin and rat placental lactogen I. *Endocrinology, 138*(2), 756–763.

Bridges, R. S., & Ronsheim, P. M. (1990). Prolactin (PRL) regulation of maternal behavior in rats: Bromocriptine treatment delays and PRL promotes the rapid onset of behavior. *Endocrinology, 126*(2), 837–848.

Bridges, R. S., & Russell, D. W. (1981). Steroidal interactions in the regulation of maternal behaviour in virgin female rats: Effects of testosterone, dihydrotestosterone, oestradiol, progesterone and the aromatase inhibitor, 1,4,6-androstatriene-3,17-dione. *Journal of Endocrinology, 90*(1), 31–40.

Bridgett, D. J., Burt, N. M., Edwards, E. S., & Deater-Deckard, K. (2015). Intergenerational transmission of self-regulation: A multidisciplinary review and integrative conceptual framework. *Psychological Bulletin, 141*(3), 602–654.

Brinschwitz, K., Dittgen, A., Madai, V. I., Lommel, R., Geisler, S., & Veh, R. W. (2010). Glutamatergic axons from the lateral habenula mainly terminate on GABAergic neurons of the ventral midbrain. *Neuroscience, 168*(2), 463–476.

Broad, K. D., Curley, J. P., & Keverne, E. B. (2006). Mother–infant bonding and the evolution of mammalian social relationships. *Philosophical Transactions of the Royal Society of London B Biological Sciences, 361*(1476), 2199–2214.

Broad, K. D., Levy, F., Evans, G., Kimura, T., Keverne, E. B., & Kendrick, K. M. (1999). Previous maternal experience potentiates the effect of parturition on oxytocin receptor mRNA expression in the paraventricular nucleus. *European Journal of Neuroscience, 11*(10), 3725–3737.

Brown, J. L., Morales, V., & Summers, K. (2010). A key ecological trait drove the evolution of biparental care and monogamy in an amphibian. *American Naturalist, 175*(4), 436–446.

Brown, P. L., Palacorolla, H., Brady, D., Riegger, K., Elmer, G. I., & Shepard, P. D. (2017). Habenula-induced inhibition of midbrain dopamine neurons is diminished by lesions of the rostromedial tegmental nucleus. *Journal of Neuroscience, 37*(1), 217–225.

Brown, P. L., & Shepard, P. D. (2016). Functional evidence for a direct excitatory projection from the lateral habenula to the ventral tegmental area in the rat. *Journal of Neurophysiology, 116*(3), 1161–1174.

Brown, R. S., Herbison, A. E., & Grattan, D. R. (2015). Effects of prolactin and lactation on A15 dopamine neurones in the rostral preoptic area of female mice. *Journal of Neuroendocrinology, 27*(9), 708–717.

Brown, R. S., Kokay, I. C., Herbison, A. E., & Grattan, D. R. (2010). Distribution of prolactin-responsive neurons in the mouse forebrain. *Journal of Comparative Neurology, 518*(1), 92–102.

Brown, R. S., Piet, R., Herbison, A. E., & Grattan, D. R. (2012). Differential actions of prolactin on electrical activity and intracellular signal transduction in hypothalamic neurons. *Endocrinology, 153*(5), 2375–2384.

Brown, R. S. E., Aoki, M., Ladyman, S. R., Phillipps, H. R., Wyatt, A., Boehm, U., & Grattan, D. R. (2017). Prolactin action in the medial preoptic area is necessary for postpartum maternal nursing behavior. *Proceedings of the National Academy of Sciences USA, 114*(40), 10779–10784.

Brown, S. L., Brown, R. M., & Penner, L. A. (2012). *Moving Beyond self-interest: Perspectives from evolutionary biology, neuroscience, and the social sciences*. Oxford, England: Oxford University Press.

Brummelte, S., & Galea, L. A. (2016). Postpartum depression: Etiology, treatment and consequences for maternal care. *Hormones and Behavior, 77*, 153–166.

Brummelte, S., McGlanaghy, E., Bonnin, A., & Oberlander, T. F. (2017). Developmental changes in serotonin signaling: Implications for early brain function, behavior and adaptation. *Neuroscience, 342*, 212–231.

Bull, C. M. (2000). Monogamy in lizards. *Behavioural Processes, 51*(1-3), 7–20.

Buntin, J. D., Becker, G. M., & Ruzycki, E. (1991). Facilitation of parental behavior in ring doves by systemic or intracranial injections of prolactin. *Hormones and Behavior, 25*, 424–444.

Buntin, L., Berghman, L. R., & Buntin, J. D. (2006). Patterns of fos-like immunoreactivity in the brains of parent ring doves (*Streptopelia risoria*) given tactile and nontactile exposure to their young. *Behavioral Neuroscience, 120*(3), 651–664.

Buonfiglio, D. C., Ramos-Lobo, A. M., Silveira, M. A., Furigo, I. C., Hennighausen, L., Frazao, R., & Donato, J., Jr. (2015). Neuronal STAT5 signaling is required for maintaining lactation but not for postpartum maternal behaviors in mice. *Hormones and Behavior, 71*, 60–68.

Burkart, J. M., Allon, O., Amici, F., Fichtel, C., Finenwirth, C. Heschl, A., . . . van Schaik, C. P. (2014). The evolutionary origin of human hyper-cooperation. *Nature Communications, 5*, 4747.

Burkart, J. M., Hrdy, S. B., & van Schaik, C. P. (2009). Cooperative breeding and human cognitive evolution. *Evolutionary Anthropology, 18*, 175–186.

Byrnes, J. J., Gleason, E. D., Schoen, M. K., Lovelock, D. F., Carini, L. M., Byrnes, E. M., & Bridges, R. S. (2011). Accelerated maternal responding following intra-VTA pertussis toxin treatment. *Behavioural Brain Research, 223*(2), 322–328.

Caba, M., Melo, A. I., Fleming, A., & Meza, E. (2019). Maternal care activates the ventral tegmental area but not dopaminergic cells in the rat. *Journal of Neuroendocrinology, 39*(9), e12713.

Caldji, C., Francis, D., Sharma, S., Plotsky, P. M., & Meaney, M. J. (2000). The effects of early rearing environment on the development of GABAA and central benzodiazepine receptor levels and novelty-induced fearfulness in the rat. *Neuropsychopharmacology, 22*(3), 219–229.

Caldwell, H. K., Lee, H. J., Macbeth, A. H., & Young, W. S., III. (2008). Vasopressin: Behavioral roles of an "original" neuropeptide. *Progress in Neurobiology, 84*(1), 1–24.

Calhoon, G. G., & Tye, K. M. (2015). Resolving the neural circuits of anxiety. *Nature Neuroscience, 18*(10), 1394–1404.

Calhoun, J. B. (1962). *The ecology and sociology of the Norway rat.* Bethesda, MD: U.S. Department of Health, Education, and Welfare.

Callaghan, B. L., & Richardson, R. (2011). Maternal separation results in early emergence of adult-like fear and extinction learning in infant rats. *Behavioral Neuroscience, 125*(1), 20–28.

Callaghan, B. L., & Richardson, R. (2013). Early experiences and the development of emotional learning systems in rats. *Biology of Mood & Anxiety Disorders, 3*, 8.

Callaghan, B. L., Sullivan, R. M., Howell, B., & Tottenham, N. (2014). The international society for developmental psychobiology Sackler symposium: Early adversity and the maturation of emotion circuits—a cross-species analysis. *Developmental Psychobiology, 56*(8), 1635–1650.

Cameron, N. M., Champagne, F. A., Parent, C., Fish, E. W., Ozaki-Kuroda, K., & Meaney, M. J. (2005). The programming of individual differences in defensive responses and reproductive strategies in the rat through variations in maternal care. *Neuroscience and Biobehavioral Reviews, 29*(4–5), 843–865.

Cant, M. A. (2012). Cooperative Breeding Systems. In N. J. Royle, P. T. Smiseth, & M. Kölliker (Eds.), *The evolution of parental care* (pp. 206–225). Oxford, England: Oxford University Press.

Canteras, N. S. (2002). The medial hypothalamic defensive system: Hodological organization and functional implications. *Pharmacology, Biochemistry, and Behavior, 71*(3), 481–491.

Canteras, N. S., Simerly, R. B., & Swanson, L. W. (1992). Projections of the ventral premammillary nucleus. *Journal of Comparative Neurology, 324*(2), 195–212.

Canteras, N. S., Simerly, R. B., & Swanson, L. W. (1994). Organization of projections from the ventromedial nucleus of the hypothalamus: A Phaseolus vulgaris-leucoagglutinin study in the rat. *Journal of Comparative Neurology, 348*(1), 41–79.

Canteras, N. S., Simerly, R. B., & Swanson, L. W. (1995). Organization of projections from the medial nucleus of the amygdala: A PHAL study in the rat. *Journal of Comparative Neurology, 360*(2), 213–245.

Carini, L. M., & Nephew, B. C. (2013). Effects of early life social stress on endocrinology, maternal behavior, and lactation in rats. *Hormones and Behavior, 64*(4), 634–641.

Carp, S. B., Rothwell, E. S., Bourdon, A., Freeman, S. M., Ferrer, E., & Bales, K. L. (2016). Development of a partner preference test that differentiates between established pair bonds and other relationships in socially monogamous titi monkeys (*Callicebus cupreus*). *American Journal of Primatology, 78*, 326–339.

Carpenter, L. L., Tyrka, A. R., McDougle, C. J., Malison, R. T., Owens, M. J., Nemeroff, C. B., & Price, L. H. (2004). Cerebrospinal fluid corticotropin-releasing factor and perceived early-life stress in depressed patients and healthy control subjects. *Neuropsychopharmacology, 29*(4), 777–784.

Carter, C. S., De Vries, A., & Getz, L. (1995), Physiological substrates of mammalian monogamy: The prairie vole model. *Neuroscience and Biobehavioral Reviews, 19*, 303–314.

Caspi, A., McClay, J., Moffitt, T. E., Mill, J., Martin, J., Craig, I. W., . . . Poulton, R. (2002). Role of genotype in the cycle of violence in maltreated children. *Science, 297*(5582), 851–854.

Caspi, A., Sugden, K., Moffitt, T. E., Taylor, A., Craig, I. W., Harrington, H., . . . Poulton, R. (2003). Influence of life stress on depression: Moderation by a polymorphism in the 5-HTT gene. *Science, 301*(5631), 386–389.

Cavanaugh, J., Mustoe, A., & French, J. A. (2018). Oxytocin regulates reunion affiliation with a pairmate following social separation in marmosets. *American Journal of Primatology, 80*(10), e22750.

Chabert, T., Colin, A., Aubin, T., Shacks, V., Bourquin, S. L., Elsey, R. M., . . . Mathevon, N. (2015). Size does matter: Crocodile mothers react more to the voice of smaller offspring. *Science Reports, 5*, 15547.

Chalfin, L., Dayan, M., Levy, D. R., Austad, S. N., Miller, R. A., Iraqi, F. A., . . . Kimchi, T. (2014). Mapping ecologically relevant social behaviours by gene knockout in wild mice. *Nature Communications, 5*, 4569.

Champagne, F. A. (2008). Epigenetic mechanisms and the transgenerational effects of maternal care. *Frontiers in Neuroendocrinology, 29*(3), 386–397.

Champagne, F., Diorio, J., Sharma, S., & Meaney, M. J. (2001). Naturally occurring variations in maternal behavior in the rat are associated with differences in estrogen-inducible central oxytocin receptors. *Proceedings of the National Academy of Sciences USA, 98*(22), 12736–12741.

Champagne, F. A., Chretien, P., Stevenson, C. W., Zhang, T. Y., Gratton, A., & Meaney, M. J. (2004). Variations in nucleus accumbens dopamine associated with individual differences in maternal behavior in the rat. *Journal of Neuroscience, 24*(17), 4113–4123.

Champagne, F. A., Francis, D. D., Mar, A., & Meaney, M. J. (2003). Variations in maternal care in the rat as a mediating influence for the effects of environment on development. *Physiology & Behavior, 79*(3), 359–371.

Champagne, F. A., & Meaney, M. J. (2006). Stress during gestation alters postpartum maternal care and the development of the offspring in a rodent model. *Biological Psychiatry, 59*(12), 1227–1235.

Champagne, F. A., & Meaney, M. J. (2007). Transgenerational effects of social environment on variations in maternal care and behavioral response to novelty. *Behavioral Neuroscience, 121*(6), 1353–1363.

Champagne, F. A., Weaver, I. C., Diorio, J., Dymov, M., Szyf, M., & Meaney, M. J. (2006). Maternal care associated with methylation of the estrogen receptor-α1b promoter and estrogen receptor-α expression in the medial preoptic area of female offspring. *Endocrinology, 147*, 2909–2915.

Champagne, F. A., Weaver, I. C., Diorio, J., Sharma, S., & Meaney, M. J. (2003). Natural variations in maternal care are associated with estrogen receptor alpha expression and estrogen sensitivity in the medial preoptic area. *Endocrinology, 144*(11), 4720–4724.

Chang, S. E., Smedley, E. B., Stansfield, K. J., Stott, J. J., & Smith, K. S. (2017). Optogenetic inhibition of ventral pallidum neurons impairs context-driven salt seeking. *Journal of Neuroscience, 37*(23), 5670–5680.

Chang, S. W., Barter, J. W., Ebitz, R. B., Watson, K. K., & Platt, M. L. (2012). Inhaled oxytocin amplifies both vicarious reinforcement and self reinforcement in rhesus macaques (*Macaca mulatta*). *Proceedings of the National Academy of Sciences USA, 109*(3), 959–964.

Charara, A., & Grace, A. A. (2003). Dopamine receptor subtypes selectively modulate excitatory afferents from the hippocampus and amygdala to rat nucleus accumbens neurons. *Neuropsychopharmacology, 28*(8), 1412–1421.

Chen, J. Evans, A. N., Liu, Y., Honda, M., Saavedra, J. M., & Aguilera, G. (2012). Maternal deprivation in rats is associated with corticotrophin-releasing hormone promoter hypomethylation and enhances in CRH transcriptional responses to stress in adulthood. *Journal of Neuroendocrinology, 24*, 1055–1064.

Chen, P. Wang, G., Ma, R., Jing, F., Zhang, Y. Wang, Y., . . . Zhang, X. (2016). Multidimensional assessment of empathic abilities in patients with insular glioma. *Cognitive, Affective, & Behavioral Neuroscience, 16*, 962–975.

Chen, P. B., Hu, R. K., Wu, Y. E., Pan, L., Huang, S., Micevych, P. E., & Hong, W. (2019). Sexually dimorphic control of parenting behavior by the medial amygdala. *Cell, 176*(5), 1206–1221.

Chirino, R., Beyer, C., & Gonzalez-Mariscal, G. (2007). Lesion of the main olfactory epithelium facilitates maternal behavior in virgin rabbits. *Behavioural Brain Research, 180*(2), 127–132.

Chirino, R., & Gonzalez-Mariscal, G. (2015). Changes in responsiveness to kit odors across pregnancy: Relevance for the onset of maternal behavior. *World Rabbit Science, 23*, 103–109.

Chokchaloemwong, D., Prakobsaneg, N., Sartsoongnoen, N., Kosonsiriluk, S., El Halawai, M., & Chaiseha, V. (2013). Mesotocin and maternal care in Thai hens. *Hormone and Behavior, 64*, 53–69.

Cittern, D., & Edalat, A. (2017). A neural model of empathic states in attachment-based psychotherapy. *Computational Psychiatry, 1*, 132–167.

Cohen, D., & Shamay-Tsoory, S. G. (2018). Oxytocin regulated social approach. *Social Neuroscience, 13*(6), 680–687.

Condes-Lara, M., Veinante, P., Rabai, M., & Freund-Mercier, M. J. (1994). Correlation between oxytocin neuronal sensitivity and oxytocin-binding sites in the amygdala of the rat: Electrophysiological and histoautoradiographic study. *Brain Research, 637*(1–2), 277–286.

Conradt, E., Hawes, K., Guerin, D., Armstrong, D. A., Marsit, C. J., Tronick, E., & Lester, B. M. (2016). The contributions of maternal sensitivity and maternal depressive symptoms to epigenetic processes and neuroendocrine functioning. *Child Development, 87*(1), 73–85.

Consiglio, A. R., & Lucion, A. B. (1996). Lesion of hypothalamic paraventricular nucleus and maternal aggressive behavior in female rats. *Physiology & Behavior, 59*(4-5), 591–596.

Constantin, S., & Wray, S. (2016). Galanin activates G protein gated inwardly rectifying potassium channels and suppresses kisspeptin-10 activation of GnRH neurons. *Endocrinology, 157*(8), 3197–3212.

Coplan, J. D., Abdallah, C. G., Kaufman, J., Gelernter, J., Smith, E. L., Perera, T. D., . . . Nemeroff, C. B. (2011). Early-life stress, corticotropin-releasing factor, and serotonin transporter gene: A pilot study. *Psychoneuroendocrinology, 36*(2), 289–293.

Coplan, J. D., Andrews, M. W., Rosenblum, L. A., Owens, M. J., Friedman, S., Gorman, J. M., & Nemeroff, C. B. (1996). Persistent elevations of cerebrospinal fluid concentrations of corticotropin-releasing factor in adult nonhuman primates exposed to early-life stressors: Implications for the pathophysiology of mood and anxiety disorders. *Proceedings of the National Academy of Sciences USA, 93*(4), 1619–1623.

Corcoran, C. A., Pierre, P. J., Haddad, T., Bice, C., Suomi, S. J., Grant, K. A., . . . Bennett, A. J. (2012). Long-term effects of differential early rearing in rhesus macaques: Behavioral reactivity in adulthood. *Developmental Psychobiology, 54*(5), 546–555.

Cornil, C. A., Ball, G. F., & Balthazart, J. (2015). The dual action of estrogen hypothesis. *Trends in Neurosciences, 38*(7), 408–416.

Corodimas, K. P., Rosenblatt, J. S., Canfield, M. E., & Morrell, J. I. (1993). Neurons in the lateral subdivision of the habenular complex mediate the hormonal onset of maternal behavior in rats. *Behavioral Neuroscience, 107*(5), 827–843.

Corodimas, K. P., Rosenblatt, J. S., & Morrell, J. I. (1992). The habenular complex mediates hormonal stimulation of maternal behavior in rats. *Behavioral Neuroscience, 106*(5), 853–865.

Corona, R., & Levy, F. (2015). Chemical olfactory signals and parenthood in mammals. *Hormones and Behavior, 68*, 77–90.

Corral-Frias, N. S., Nikolova, Y. S., Michalski, L. J., Baranger, D. A., Hariri, A. R., & Bogdan, R. (2015). Stress-related anhedonia is associated with ventral striatum reactivity to reward and transdiagnostic psychiatric symptomatology. *Psychological Medicine, 45*(12), 2605–2617.

Costa, H. C., Da-Silva, J. M., Diniz, G. B., Motta-Teixeira, L. C., Da-Silva, R. J., de-Moraes Machado. C., . . . Bittencourt, J. C. (2019). Characterization and origins of melanin-concentrating hormone immunoreactive fibres of the posterior lobe of the pituitary and median eminence during lactation in the Long-Evans rat. *Journal of Neuroendocrinology, 31*(9), e12723.

Coutellier, L. Logemann, A., Kuo, J., Rusnak, M., & Usdin, T. B. (2011). TIP 39 modulates effects of novelty-induced arousal on memory. *Genes, Brain, and Behavior, 10*(1), 90–99.

Coutellier, L., Logemann, A., Rusnak, M., & Usdin, T. B. (2011). Maternal absence of the parathyroid hormone 2 receptor affects postnatal pup development. *Journal of Neuroendocrinology, 23*(7), 612–619.

Coxworth, J. E., Kim, P. S., McQueen, J. S., & Hawkes, K. (2015). Grandmothering life histories and human pair bonding. *Proceedings of the National Academy of Sciences USA, 112*(38), 11806–11811.

Craig, A. D. (2009). How do you feel—now? The anterior insula and human awareness. *Nature Reviews Neuroscience, 10*(1), 59–70.

Croy, I., Mohr, T., Weidner, K., Hummel, T., & Junge-Hoffmeister, J. (2019). Mother–child bonding is associated with the maternal perception of the child's body odor. *Physiology & Behavior, 198*, 151–157.

Cruz, M. L., & Beyer, C. (1972). The effects of septal lesions on maternal behavior and lactation in the rabbit. *Physiology & Behavior, 9*, 361–365.

Cservenak, M., Bodnar, I., Usdin, T. B., Palkovits, M., Nagy, G. M., & Dobolyi, A. (2010). Tuberoinfundibular peptide of 39 residues is activated during lactation and participates in the suckling-induced prolactin release in rat. *Endocrinology, 151*(12), 5830–5840.

Cservenak, M., Keller, D., Kis, V., Fazekas, E. A., Ollos, H., Leko, A. H., . . . Dobolyi, A. (2017). A thalamo-hypothalamic pathway that activates oxytocin neurons in social contexts in female rats. *Endocrinology, 158*(2), 335–348.

Cservenak, M., Kis, V., Keller, D., Dimen, D., Menyhart, L., Olah, S., . . . Dobolyi, A. (2017). Maternally involved galanin neurons in the preoptic area of the rat. *Brain Structure and Function, 222*(2), 781–798.

Cservenak, M., Szabo, E. R., Bodnar, I., Leko, A., Palkovits, M., Nagy, G. M., . . . Dobolyi, A. (2013). Thalamic neuropeptide mediating the effects of nursing on lactation and maternal motivation. *Psychoneuroendocrinology, 38*(12), 3070–3084.

Culot, L., Lledo-Ferrer, Y., Hoelscher, O., Munoz Lazo, F. J., Huynen, M. C., & Heymann, E. W. (2011). Reproductive failure, possible maternal infanticide, and cannibalism in wild moustached tamarins. *Saguinus mystax. Primates, 52*(2), 179–186.

Cyr, C., Euser, E. M., Bakermans-Kranenburg, M. J., & van Ijzendoorn, M. H. (2010). Attachment security and disorganization in maltreating and high-risk families: A series of meta-analyses. *Development and Psychopathology, 22*(1), 87–108.

Da Costa, A. P., Broad, K. D., & Kendrick, K. M. (1997). Olfactory memory and maternal behaviour-induced changes in c-fos and zif/268 mRNA expression in the sheep brain. *Molecular Brain Research, 46*(1–2), 63–76.

Da Costa, A. P., Guevara-Guzman, R. G., Ohkura, S., Goode, J. A., & Kendrick, K. M. (1996). The role of oxytocin release in the paraventricular nucleus in the control of maternal behaviour in the sheep. *Journal of Neuroendocrinology, 8*(3), 163–177.

da Silva Mota, M. T., Franci, C. R., & de Sousa, M. B. (2006). Hormonal changes related to paternal and alloparental care in common marmosets (*Callithrix jacchus*). *Hormones and Behavior, 49*(3), 293–302.

Dabrowska, J., Hazra, R., Ahern, T. H., Guo, J. D., McDonald, A. J., Mascagni, F., . . . Rainnie, D. G. (2011). Neuroanatomical evidence for reciprocal regulation of the corticotrophin-releasing factor and oxytocin systems in the hypothalamus and the bed nucleus of the stria terminalis of the rat: Implications for balancing stress and affect. *Psychoneuroendocrinology, 36*(9), 1312–1326.

Daniel, S. E., & Rainnie, D. G. (2016). Stress modulation of opposing circuits in the bed nucleus of the stria terminalis. *Neuropsychopharmacology, 41*(1), 103–125.

Daniels, W. M., Fairbairn, L. R., van Tilburg, G., McEvoy, C. R., Zigmond, M. J., Russell, V. A., & Stein, D. J. (2009). Maternal separation alters nerve growth factor and corticosterone levels but not the DNA methylation status of the exon 1(7) glucocorticoid receptor promoter region. *Metabolic Brain Diseases, 24*(4), 615–627.

D'Anna, K. L., & Gammie, S. C. (2009). Activation of corticotropin-releasing factor receptor 2 in lateral septum negatively regulates maternal defense. *Behavioral Neuroscience, 123*(2), 356–368.

Dannlowski, U., Kugel, H., Grotegerd, D., Redlich, R., Opel, N., Dohm, K., . . . Baune, B. T. (2016). Disadvantage of social sensitivity: Interaction of oxytocin receptor genotype and child maltreatment on brain structure. *Biological Psychiatry, 80*(5), 398–405.

Dannlowski, U., Stuhrmann, A., Beutelmann, V., Zwanzger, P., Lenzen, T., Grotegerd, D., . . . Kugel, H. (2012). Limbic scars: Long-term consequences of childhood maltreatment revealed by functional and structural magnetic resonance imaging. *Biological Psychiatry, 71*(4), 286–293.

Davis, M., Walker, D. L., Miles, L., & Grillon, C. (2010). Phasic vs sustained fear in rats and humans: Role of the extended amygdala in fear vs anxiety. *Neuropsychopharmacology, 35*(1), 105–135.

D'Cunha, T. M., King, S. J., Fleming, A. S., & Levy, F. (2011). Oxytocin receptors in the nucleus accumbens shell are involved in the consolidation of maternal memory in postpartum rats. *Hormones and Behavior, 59*(1), 14–21.

de Jong, T. R., Chauke, M., Harris, B. N., & Saltzman, W. (2009). From here to paternity: Neural correlates of the onset of paternal behavior in California mice (*Peromyscus californicus*). *Hormones and Behavior, 56*(2), 220–231.

de Jong, T. R., Korosi, A., Harris, B. N., Perea-Rodriguez, J. P., & Saltzman, W. (2012). Individual variation in paternal responses of virgin male California mice (*Peromyscus californicus*): Behavioral and physiological correlates. *Physiological and Biochemical Zoology, 85*(6), 740–751.

DeAngelis, R., Dodd, L., Snyder, A., & Rhodes, J. S. (2018). Dynamic regulation of brain aromatase and isotocin receptor gene expression depends on parenting status. *Hormones and Behavior, 103*, 62–70.

DeAngelis, R., Gogola, J., Dodd, L., & Rhodes, J. S. (2017). Opposite effects of nonapeptide antagonists on paternal behavior in the teleost fish *Amphiprion ocellaris*. *Hormones and Behavior, 90*, 113–119.

Declerck, C. H., Boone, C., & Kiyonari, T. (2014). The effect of oxytocin on cooperation in a prisoner's dilemma depends on the social context and a person's social value orientation. *Social, Cognitive, and Affective Neuroscience, 9*(6), 802–809.

Decety, J., Norman, G. J., Berntson, G. G., & Cacioppo, J. T. (2012). A neurobehavioral evolutionary perspective on the mechanisms underlying empathy. *Progress in Neurobiology, 98*(1), 38–48.

De Dreu, C. K., Greer, L. L., Handgraaf, M. J., Shalvi, S., van Kleef, G. A., Baas, M., . . . Feith, S. W. (2010). The neuropeptide oxytocin regulates parochial altruism in intergroup conflict among humans. *Science, 328*(5984), 1408–1411.

De Dreu, C. K., & Kret, M. E. (2016). Oxytocin conditions intergroup relations through upregulated in-group empathy, cooperation, conformity, and defense. *Biological Psychiatry, 79*(3), 165–173.

Delgado, M. R., Nearing, K. I., Ledoux, J. E., & Phelps, E. A. (2008). Neural circuitry underlying the regulation of conditioned fear and its relation to extinction. *Neuron, 59*(5), 829–838.

DeLuycker, A. M. (2014). Observations of a daytime birthing event in wild titi monkeys (*Callicebus oenanthe*): Implications of the male parental role. *Primates, 55*, 59–67.

Dennison, M. J., Sheridan, M. A., Busso, D. S., Jenness, J. L., Peverill, M., Rosen, M. L., & McLaughlin, K. A. (2016). Neurobehavioral markers of resilience to depression amongst adolescents exposed to child abuse. *Journal of Abnormal Psychology, 125*(8), 1201–1212.

Devi, Y. S., & Halperin, J. (2014). Reproductive actions of prolactin mediated through short and long receptor isoforms. *Molecular and Cell Endocrinology, 382*(1), 400–410.

DeWall, C. D., Gillath, O., Pressman, S. D., Black, L. L., Bartz, J. A., Moskovitz, J., & Stetler. D. A. (2014). When the love hormone leads to violence: Oxytocin increases intimate partner violence inclinations among high trait aggressive people. *Social Psychological and Personality Science, 5*(6), 691–697.

Diaz-Munoz, S. L. (2016). Complex cooperative breeders: Using infant care costs to explain variability in callitrichine social and reproductive behavior. *American Journal of Primatology, 78*(3), 372–387.

Digby, L. (1995). Infant care, infanticide, and female reproductive strategies in polygynous groups of common marmosets (*Callithrix jacchus*). *Behavioral Ecology and Sociobiology, 37*(1), 51–61.

Dillon, D. G., Holmes, A. J., Birk, J. L., Brooks, N., Lyons-Ruth, K., & Pizzagalli, D. A. (2009). Childhood adversity is associated with left basal ganglia dysfunction during reward anticipation in adulthood. *Biological Psychiatry, 66*(3), 206–213.

Dobi, A., Margolis, E. B., Wang, H. L., Harvey, B. K., & Morales, M. (2010). Glutamatergic and nonglutamatergic neurons of the ventral tegmental area establish local synaptic contacts with dopaminergic and nondopaminergic neurons. *Journal of Neuroscience, 30*(1), 218–229.

Dobner, P. R., Fadel, J., Deitemeyer, N., Carraway, R. E., & Deutch, A. Y. (2001). Neurotensin-deficient mice show altered responses to antipsychotic drugs. *Proceedings of the National Academy of Sciences USA, 98*(14), 8048–8053.

Dobolyi, A. (2011). Novel potential regulators of maternal adaptations during lactation: Tuberoinfundibular peptide 39 and amylin. *Journal of Neuroendocrinology, 23*(11), 1002–1018.

Dobolyi, A., Cservenak, M., & Young, L. J. (2018). Thalamic integration of social stimuli regulating parental behavior and the oxytocin system. *Frontiers in Neuroendocrinology, 51*, 102–115.

Dodhia, S., Hosanagar, A., Fitzgerald, D. A., Labuschagne, I., Wood, A. G., Nathan, P. J., & Phan, K. L. (2014). Modulation of resting-state amygdala-frontal functional connectivity by oxytocin in generalized social anxiety disorder. *Neuropsychopharmacology, 39*(9), 2061–2069.

Dolen, G. (2015). Oxytocin: Parallel processing in the social brain? *Journal of Neuroendocrinology, 27*(6), 516–535.

Dolen, G., Darvishzadeh, A., Huang, K. W., & Malenka, R. C. (2013). Social reward requires coordinated activity of nucleus accumbens oxytocin and serotonin. *Nature, 501*(7466), 179–184.

Dominguez, J. M., & Hull, E. M. (2001). Stimulation of the medial amygdala enhances medial preoptic dopamine release: Implications for male rat sexual behavior. *Brain Research, 917*(2), 225–229.

Dominguez, M., Aguilar-Roblero, R., & Gonzalez-Mariscal, G. (2017). Bilateral lesions of the paraventricular hypothalamic nucleus disrupt nursing behavior in rabbits. *European Journal of Neuroscience, 46*(5), 2133–2140.

Donaldson, Z. R., Spiegel, L., & Young, L. J. (2010). Central vasopressin V1a receptor activation is independently necessary for both partner preference formation and expression in socially monogamous male prairie voles. *Behavioral Neuroscience, 124*(1), 159–163.

Dong, H. W., Petrovich, G. D., & Swanson, L. W. (2001). Topography of projections from amygdala to bed nuclei of the stria terminalis. *Brain Res Reviews, 38*(1–2), 192–246.

Dong, H. W., Petrovich, G. D., Watts, A. G., & Swanson, L. W. (2001). Basic organization of projections from the oval and fusiform nuclei of the bed nuclei of the stria terminalis in adult rat brain. *Journal of Comparative Neurology, 436*(4), 430–455.

Dong, H. W., & Swanson, L. W. (2006). Projections from bed nuclei of the stria terminalis, dorsomedial nucleus: Implications for cerebral hemisphere integration of neuroendocrine, autonomic, and drinking responses. *Journal of Comparative Neurology, 494*(1), 75–107.

Donner, N., Bredewold, R., Maloumby, R., & Neumann, I. D. (2007). Chronic intracerebral prolactin attenuates neuronal stress circuitries in virgin rats. *European Journal of Neuroscience, 25*(6), 1804–1814.

Douglas, A. J., Leng, G., & Russell, J. A. (2002). The importance of oxytocin mechanisms in the control of mouse parturition. *Reproduction, 123*(4), 543–552.

Douglas, A. J., & Meddle, S. L. (2008). Fast delivery: A central role for oxytocin. In R. S. Bridges (Ed.), *Neurobiology of the parental brain* (pp. 225–234). Burlington, MA: Academic Press.

Duclot, F., Wang, H., Youssef, C., Wang, Z., & Kabbai, M. (2016). Trichostatin A (TSA) facilitates formation of partner preference in male prairie voles (*Microtus orchrogaster*). *Hormones and Behavior, 81*, 68–73.

Dulac, C., O'Connell, L. A., & Wu, Z. (2014). Neural control of maternal and paternal behaviors. *Science, 345*(6198), 765–770.

Dulac, C., & Wagner, S. (2006). Genetic analysis of brain circuits underlying pheromone signaling. *Annual Review of Genetics, 40*, 449–467.

Dumartin, B., Doudnikoff, E., Gonon, F., & Bloch, B. (2007). Differences in ultrastructural localization of dopaminergic D1 receptors between dorsal striatum and nucleus accumbens in the rat. *Neuroscience Letters, 419*(3), 273–277.

Dunn, A. J., & Berridge, C. W. (1990). Physiological and behavioral responses to corticotropin-releasing factor administration: Is CRF a mediator of anxiety or stress responses? *Brain Research Reviews, 15*(2), 71–100.

Dunn, E. C., Soare, T. W., Zhu, Y., Simpkin, A. J., Suderman, M. J., Klengel, T., . . . Relton, C. L. (2019). Sensitive periods for the effect of childhood adversity on DNA methylation: Results from a prospective, longitudinal study. *Biological Psychiatry, 85*, 838–849.

Dwyer, C. M. (2008). Individual variation in the expression of maternal behaviour: A review of the neuroendocrine mechanisms in the sheep. *Journal of Neuroendocrinology, 20*(4), 526–534.

Eapen, V., Dadds, M., Barnett, B., Kohlhoff, J., Khan, F., Radom, N., & Silove, D. M. (2014). Separation anxiety, attachment, and inter-personal representations: Disentangling the role of oxytocin in the perinatal period. *PloS One, 9*(9), e107745.

Edwards, N. J., Tajeda, H. A., Pignatelli, M. Zhang, S. McDevitt, R. A., Wu, J., . . . Bonci, A. (2017). Circuit specificity in the inhibitory architecture of the VTA regulates cocaine-induced behavior. *Nature Neuroscience, 20*(3), 438–448.

Ein-Dor, T., Verbeke, W., Mokry, M., & Vrticka, P. (2018). Epigenetic modification of the oxytocin and glucocorticoid receptor genes is linked to attachment avoidance in young adults. *Attachment &Human Development, 20*(4), 439–454.

Eisenberger, N. I. (2013). An empirical review of the neural underpinnings of receiving and giving social support: Implications for health. *Psychosomatic Medicine, 75*(6), 545–556.

Elmadih, A., Wan, M. W., Downey, D., Elliott, R., Swain, J. E., & Abel, K. M. (2016). Natural variation in maternal sensitivity is reflected in maternal brain responses to infant stimuli. *Behavioral Neuroscience, 130*(5), 500–510.

Elyada, Y. M., & Mizrahi, A. (2015). Becoming a mother-circuit plasticity underlying maternal behavior. *Current Opinion in Neurobiology, 35*, 49–56.

Emlen, S. T. (1995). An evolutionary theory of the family. *Proceedings of the National Academy of Sciences USA, 92*(18), 8092–8099.

Erskine, M. S., Barfield, R. J., & Goldman, B. D. (1978). Intraspecific fighting during late pregnancy and lactation in rats and effects of litter removal. *Behavioral Biology, 23*(2), 206–218.

Erskine, M. S., Barfield, R. J., & Goldman, B. D. (1980a). Postpartum aggression in rats: I: Effects of hypophysectomy. *Journal of Comparative and Physiological Psychology, 94*(3), 484–494.

Erskine, M. S., Barfield, R. J., & Goldman, B. D. (1980b). Postpartum aggression in rats: II: Dependence on maternal sensitivity to young and effects of experience with pregnancy and parturition. *Journal of Comparative and Physiological Psychology, 94*(3), 495–505.

Etkin, A., Prater, K. E., Hoeft, F., Menon, V., & Schatzberg, A. F. (2010). Failure of anterior cingulate activation and connectivity with the amygdala during implicit regulation of emotional processing in generalized anxiety disorder. *American Journal of Psychiatry, 167*(5), 545–554.

Factor, E. M., Mayer, A. D., & Rosenblatt, J. S. (1993). Peripeduncular nucleus lesions in the rat: I. Effects on maternal aggression, lactation, and maternal behavior during pre- and postpartum periods. *Behavioral Neuroscience, 107*(1), 166–185.

Fadok, J. P., Krabbe, S., Markovic, M., Courtin, J., Xu, C., Massi, L., . . . Luthi, A. (2017). A competitive inhibitory circuit for selection of active and passive fear responses. *Nature, 542*(7639), 96–100.

Faget, L., Zell, V., Souter, E., McPherson, A., Ressler, R., Gutierrez-Reed, N., . . . Hnasko, T. S. (2018). Opponent control of behavioral reinforcement by inhibitory and excitatory projections from the ventral pallidum. *Nature Communications, 9*(1), 849.

Fahrbach, S. E., Morrell, J. I., & Pfaff, D. W. (1984). Oxytocin induction of short-latency maternal behavior in nulliparous, estrogen-primed female rats. *Hormones and Behavior, 18*(3), 267–286.

Fahrbach, S. E., Morrell, J. I., & Pfaff, D. W. (1985). Possible role for endogenous oxytocin in estrogen-facilitated maternal behavior in rats. *Neuroendocrinology, 40*(6), 526–532.

Fahrbach, S. E., Morrell, J. I., & Pfaff, D. W. (1986). Identification of medial preoptic neurons that concentrate estradiol and project to the midbrain in the rat. *Journal of Comparative Neurology, 247*(3), 364–382.

Fahrbach, S. E., & Pfaff, D. W. (1986). Effect of preoptic region implants of dilute estradiol on the maternal behavior of ovariectomized, nulliparous rats. *Hormones and Behavior, 20*(3), 354–363.

Fakhoury, M. (2018). The tail of the ventral tegmental area in behavioral processes and in the effect of psychostimulants and drugs of abuse. *Progress in Neuropsychopharmacology and Biological Psychiatry, 84*(Pt A), 30–38.

Fan, Y., Herrera-Melendez, A. L., Pestke, K., Feeser, M., Aust, S., Otte, C., . . . Grimm. S. (2014). Early life stress modulates amygdala-prefrontal functional connectivity: Implications for oxytocin effects. *Human Brain Mapping, 35*, 5328–5339.

Fang, Y. Y., Yamaguchi, T., Song, S. C., Tritsch, N. X., & Lin, D. (2018). A hypothalamic midbrain pathway essential for driving maternal behaviors. *Neuron, 98*(1), 192–207.

Fareri, D. S., & Tottenham, N. (2016). Effects of early life stress on amygdala and striatal development. *Developmental Cognitive Neuroscience, 19*, 233–247.

Febo, M., Felix-Ortiz, A. C., & Johnson, T. R. (2010). Inactivation or inhibition of neuronal activity in the medial prefrontal cortex largely reduces pup retrieval and grouping in maternal rats. *Brain Research, 1325*, 77–88.

Feierstein, C. E. (2012). Linking adult olfactory neurogenesis to social behavior. *Frontiers in Neuroscience, 6*, 173.

Feierstein, C. E., Lazarini, F., Wagner, S., Gabellec, M. M., de Chaumont, F., Olivo-Marin, J. C., . . . Gheusi, G. (2010). Disruption of adult neurogenesis in the olfactory bulb affects social interaction but not maternal behavior. *Frontiers in Behavioral Neuroscience, 4*, 176.

Feldman, R., Bamberger, E., & Kanat-Maymon, Y. (2013). Parent-specific reciprocity from infancy to adolescence shapes children's social competence and dialogical skills. *Attachment & Human Development, 15*(4), 407–423.

Feldman, R., Braun, K., & Champagne, F. A. (2019). The neural mechanisms and consequences of paternal caregiving. *Nature Reviews Neuroscience, 20*, 205–224.

Feldman, R., Gordon, I., Influs, M., Gutbir, T., & Ebstein, R. P. (2013). Parental oxytocin and early caregiving jointly shape children's oxytocin response and social reciprocity. *Neuropsychopharmacology, 38*(7), 1154–1162.

Feldman, R., Gordon, I., Schneiderman, I., Weisman, O., & Zagoory-Sharon, O. (2010). Natural variations in maternal and paternal care are associated with systematic changes in oxytocin following parent–infant contact. *Psychoneuroendocrinology, 35*(8), 1133–1141.

Feldman, R., Zagoory-Sharon, O., Weisman, O., Schneiderman, I., Gordon, I., Maoz, R., . . . Ebstein, R. P. (2012). Sensitive parenting is associated with plasma oxytocin and polymorphisms in the OXTR and CD38 genes. *Biological Psychiatry, 72*(3), 175–181.

FeldmanHall, O., Dalgleish, T., Evans, D., & Mobbs, D. (2015). Empathic concern drives costly altruism. *Neuroimage, 105*, 347–356.

Feng, X., Wang, L., Yang, S., Qin, D., Wang, J., Li, C., . . . Hu, X. (2011). Maternal separation produces lasting changes in cortisol and behavior in rhesus monkeys. *Proceedings of the National Academy of Sciences USA, 108*(34), 14312–14317.

Fernandez-Duque, E., Valeggia, C. R., & Mendoza, S. P. (2009). The biology of paternal care in human and nonhuman primates. *Annual Review of Anthropology, 38*, 115–130.

Ferreira, A., Dahlof, L. G., & Hansen, S. (1987). Olfactory mechanisms in the control of maternal aggression, appetite, and fearfulness: Effects of lesions to olfactory receptors, mediodorsal thalamic nucleus, and insular prefrontal cortex. *Behavioral Neuroscience, 101*(5), 709–717, 746.

Ferreira, A., & Hansen, S. (1986). Sensory control of maternal aggression in Rattus norvegicus. *Journal of Comparative Psychology, 100*(2), 173–177.

Ferreira, A., Pereira, M., Agrati, D., Uriarte, N., & Fernandez-Guasti, A. (2002). Role of maternal behavior on aggression, fear and anxiety. *Physiology & Behavior, 77*(2-3), 197–204.

Ferri, S. L., Kreibich, A. S., Torre, M., Piccoli, C. T., Dow, H., Pallathra, A. A., . . . Brodkin, E. S. (2016). Activation of basolateral amygdala in juvenile C57BL/6J mice during social approach behavior. *Neuroscience, 335*, 184–194.

Figueira, R. J., Peabody, M. F., & Lonstein, J. S. (2008). Oxytocin receptor activity in the ventrocaudal periaqueductal gray modulates anxiety-related behavior in postpartum rats. *Behavioral Neuroscience, 122*(3), 618–628.

Finkenwirth, C., Martins, E., Deschner, T., & Burkart, J. M. (2016). Oxytocin is associated with infant-care behavior and motivation in cooperatively breeding marmoset monkeys. *Hormones and Behavior, 80*, 10–18.

Fisher, P. A., Beauchamp, K. G., Roos, L. E., Noll, L. K., Flannery, J., & Delker, B. C. (2016). The neurobiology of intervention and prevention in early adversity. *Annual Review of Clinical Psychology, 12*, 331–357.

Fleming, A., Korsmit, M., & Deller, M. (1994). Rat pups are potent reinforcers to the maternal animal: Effects of experience, parity, hormones, and dopamine function. *Psychobiology, 22*, 44–53.

Fleming, A., Vaccarino, F., Tambosso, L., & Chee, P. (1979). Vomeronasal and olfactory system modulation of maternal behavior in the rat. *Science, 203*(4378), 372–374.

Fleming, A. S., Cheung, U., Myhal, N., & Kessler, Z. (1989). Effects of maternal hormones on 'timidity' and attraction to pup-related odors in female rats. *Physiology & Behavior, 46*(3), 449–453.

Fleming, A. S., Gavarth, K., & Sarker, J. (1992). Effects of transections to the vomeronasal nerves or to the main olfactory bulbs on the initiation and long-term retention of maternal behavior in primiparous rats. *Behavioral and Neural Biology, 57*(3), 177–188.

Fleming, A. S., & Luebke, C. (1981). Timidity prevents the virgin female rat from being a good mother: Emotionality differences between nulliparous and parturient females. *Physiology & Behavior, 27*(5), 863–868.

Fleming, A. S., & Rosenblatt, J. S. (1974a). Maternal behavior in the virgin and lactating rat. *Journal of Comparative and Physiological Psychology, 86*(5), 957–972.

Fleming, A. S., & Rosenblatt, J. S. (1974b). Olfactory regulation of maternal behavior in rats: I. Effects of olfactory bulb removal in experienced and inexperienced lactating and cycling females. *Journal of Comparative and Physiological Psychology, 86*(2), 221–232.

Fleming, A. S., & Rosenblatt, J. S. (1974c). Olfactory regulation of maternal behavior in rats: II. Effects of peripherally induced anosmia and lesions of the lateral olfactory tract in pup-induced virgins. *Journal of Comparative and Physiological Psychology, 86*(2), 233–246.

Fleming, A. S., Ruble, D., Krieger, H., & Wong, P. Y. (1997). Hormonal and experiential correlates of maternal responsiveness during pregnancy and the puerperium in human mothers. *Hormones and Behavior, 31*(2), 145–158.

Fleming, A. S., Vaccarino, F., & Luebke, C. (1980). Amygdaloid inhibition of maternal behavior in the nulliparous female rat. *Physiology & Behavior, 25*(5), 731–743.

Forster, G. L., & Blaha, C. D. (2000). Laterodorsal tegmental stimulation elicits dopamine efflux in the rat nucleus accumbens by activation of acetylcholine and glutamate receptors in the ventral tegmental area. *European Journal of Neuroscience, 12*(10), 3596–3604.

Forster, G. L., Pringle, R. B., Mouw, N. J., Vuong, S. M., Watt, M. J., Burke, A. R., . . . Renner, K. J. (2008). Corticotropin-releasing factor in the dorsal raphe nucleus increases medial prefrontal cortical serotonin via type 2 receptors and median raphe nucleus activity. *European Journal of Neuroscience, 28*(2), 299–310.

Francis, D., Diorio, J., Liu, D., & Meaney, M. J. (1999). Nongenomic transmission across generations of maternal behavior and stress responses in the rat. *Science, 286*(5442), 1155–1158.

Francis, D. D., Champagne, F. C., & Meaney, M. J. (2000). Variations in maternal behaviour are associated with differences in oxytocin receptor levels in the rat. *Journal of Neuroendocrinology, 12*(12), 1145–1148.

Francis, D. D., & Meaney, M. J. (1999). Maternal care and the development of stress responses. *Current Opinion in Neurobiology, 9*(1), 128–134.

Fraser, E. J., & Shah, N. M. (2014). Complex chemosensory control of female reproductive behaviors. *PLoS One, 9*(2), e90368.

Freeman, S. M., Inoue, K., Smith, A. L., Goodman, M. M., & Young, L. J. (2014). The neuroanatomical distribution of oxytocin receptor binding and mRNA in the male rhesus macaque (*Macaca mulatta*). *Psychoneuroendocrinology, 45*, 128–141.

Freeman, S. M., Ngo, J., Singh, B., Masnaghetti, M., Bales, K. L., & Blevins, J. E. (2018). Effects of chronic oxytocin administration and diet composition on oxytocin and vasopressin 1a receptor binding in the rat brain. *Neuroscience, 392*, 241–251.

Freeman, S. M., Walum, H., Inoue, K., Smith, A. L., Goodman, M. M., Bales, K. L., & Young, L. J. (2014). Neuroanatomical distribution of oxytocin and vasopressin 1a receptors in the socially monogamous coppery titi monkey (Callicebus cupreus). *Neuroscience, 273*, 12–23.

Freeman, S. M., & Young, L. J. (2016). Comparative perspectives on oxytocin and vasopressin receptor research in rodents and primates: Translational implications. *Journal of Neuroendocrinology, 28*(4).

Freemark, M., Driscoll, P., Andrews, J., Kelly, P. A., & Royster, M. (1996). Ontogenesis of prolactin receptor gene expression in the rat olfactory system: Potential roles for lactogenic hormones in olfactory development. *Endocrinology, 137*(3), 934–942.

French, J. A., Cavanaugh, J., Mustoe, A. C., Carp, S. B., & Womack, S. L. (2018). Social monogamy in nonhuman primates: Phylogeny, phenotype, and physiology. *Journal of Sex Research, 55*(4–5), 410–434.

French, J. A., Koban, T., Rukstalis, M., Ramirez, S. M., Bardi, M., & Brent, L. (2004). Excretion of urinary steroids in pre- and postpartum female baboons. *General and Comparative Endocrinology, 137*(1), 69–77.

French, J. A., Taylor, J. H., Mustoe, A. C., & Cavanaugh, J. (2016). Neuropeptide diversity and the regulation of social behavior in New World primates. *Frontiers in Neuroendocrinology, 42*, 18–39.

Freund-Mercier, M. J., Stoeckel, M. E., & Klein, M. J. (1994). Oxytocin receptors on oxytocin neurones: Histoautoradiographic detection in the lactating rat. *Journal of Physiology, 480*(Pt 1), 155–161.

Frick, H. W. (1986). Pair swimming and mutual partner guarding in monogamous butterfly fish (Pisces, Chaetodontidae): A joint advertisement for territory. *Ethology, 73*, 307–333.

Fugiwara, T., Weisman, O., Ochi, M., Shirai, K., Matsumoto, K, Noguchi, E., & Feldman, R. (2019). Genetic and peripheral markers of the oxytocin system and parental care jointly support cross-generational transmission of bonding across three generations. *Psychoneuroendocrinology, 102*, 172–181.

Fujimoto, A., Hori, Y., Nagai, Y., Kikuchi, E., Oyama, K., Suhara, T., & Minamimoto, T. (2019). Signaling incentive and drive in the primate ventral pallidum for motivational control of goal-directed action. *Journal of Neuroscience, 39*(10), 1793–1804.

Fuster, J. (2008). *The Prefrontal Cortex* (4th ed.). San Diego, CA: Academic Press.

Gallagher, J. M., Nephew, B. C. Poirer, G., King, J. A., & Bridges, R. S. (2019). Estrogen receptor-alpha knockouts and maternal memory in nulliparous rats. *Hormones and Behavior, 110,* 40–45.

Gallo, E. A. G., Munhoz, T. N., Loret de Mola, C., & Murray, J. (2018). Gender differences in the effects of childhood maltreatment on adult depression and anxiety: A systematic review and meta-analysis. *Child Abuse & Neglect, 79,* 107–114.

Gamer, M., Zurowski, B., & Buchel, C. (2010). Different amygdala subregions mediate valence-related and attentional effects of oxytocin in humans. *Proceedings of the National Academy of Sciences USA, 107*(20), 9400–9405.

Gammie, S. C., D'Anna, K. L., Gerstein, H., & Stevenson, S. A. (2009). Neurotensin inversely modulates maternal aggression. *Neuroscience, 158*(4), 1215–1223.

Gammie, S. C., & Lonstein, J. S. (2006). Maternal aggression. In R. J. Nelson (Ed.), *The biology of aggression.* (pp. 250–274). New York, NY: Oxford University Press.

Gammie, S. C., Negron, A., Newman, S. M., & Rhodes, J. S. (2004). Corticotropin-releasing factor inhibits maternal aggression in mice. *Behavioral Neuroscience, 118*(4), 805–814.

Gandelman, R. (1973). Maternal behavior in the mouse: Effect of estrogen and progesterone. *Physiology & Behavior, 10,* 153–155.

Gandelman, R., & vom Saal, F. S. (1975). Pup-killing in mice: The effects of gonadectomy and testosterone administration. *Physiology & Behavior, 15*(6), 647–651.

Gandelman, R., Zarrow, M. X., Denenberg, V. H., & Myers, M. (1971). Olfactory bulb removal eliminates maternal behavior in the mouse. *Science, 171*(3967), 210–211.

Gangolli, E. A., Conneely, O. M., & O'Malley, B. W. (1997). Neurotransmitters activate the human estrogen receptor in a neuroblastoma cell line. *Journal of Steroid Biochemistry and Molecular Biology, 61*(1–2), 1–9.

Gao, P., Groenewegen, H. J., Vanderschuren, L., & Voorn, P. (2018). Heterogeneous neuronal activity in the lateral habenula after short- and long-term cocaine self-administration in rats. *European Journal of Neuroscience, 47*(1), 83–94.

Gee, D. G., Gabard-Durnam, L. J., Flannery, J., Goff, B., Humphreys, K. L., Telzer, E. H., . . . Tottenham, N. (2013). Early developmental emergence of human amygdala-prefrontal connectivity after maternal deprivation. *Proceedings of the National Academy of Sciences USA, 110*(39), 15638–15643.

Gee, D. G., Gabard-Durnam, L., Telzer, E. H., Humphreys, K. L., Goff, B., Shapiro, M., . . . Tottenham, N. (2014). Maternal buffering of human amygdala-prefrontal circuitry during childhood but not during adolescence. *Psychological Science, 25*(11), 2067–2078.

Gee, D. G., Humphreys, K. L., Flannery, J., Goff, B., Telzer, E. H., Shapiro, M., . . . Tottenham, N. (2013). A developmental shift from positive to negative connectivity in human amygdala-prefrontal circuitry. *Journal of Neuroscience, 33*(10), 4584–4593.

Geisler, S., Derst, C., Veh, R. W., & Zahm, D. S. (2007). Glutamatergic afferents of the ventral tegmental area in the rat. *Journal of Neuroscience, 27*(21), 5730–5743.

Geisler, S., & Zahm, D. S. (2006). Neurotensin afferents of the ventral tegmental area in the rat: [1] Re-examination of their origins and [2] responses to acute psychostimulant and antipsychotic drug administration. *European Journal of Neuroscience, 24*(1), 116–134.

Geisler, S., & Zahm, D. S. (2006). On the retention of neurotensin in the ventral tegmental area (VTA) despite destruction of the main neurotensinergic afferents of the

VTA—Implications for the organization of forebrain projections to the VTA. *Brain Research, 1087*(1), 87–104.

Getz, L. L., McGuire, B., Pizzuto, T., Hofmann, J. E., & Frase, B. (1993). Social organization of the prairie vole (*Microtus ochrogaster*). *Journal of Mammalogy, 74*(1), 44–58.

Gimpl, G., & Fahrenholz, F. (2001). The oxytocin receptor system: Structure, function, and regulation. *Physiological Reviews, 81*(2), 629–683.

Gingrich, B., Liu, Y., Cascio, Wang, Z., & Insel, T. R. (2000). Dopamine D2 receptors in the nucleus accumbens are important for social attachment in female prairie voles (*Microtus ochrogaster*). *Behavioral Neuroscience, 114*(1), 173–183.

Giovenardi, M., Padoin, M. J., Cadore, L. P., & Lucion, A. B. (1998). Hypothalamic paraventricular nucleus modulates maternal aggression in rats: Effects of ibotenic acid lesion and oxytocin antisense. *Physiology & Behavior, 63*(3), 351–359.

Giuliani, N. R., Beauchamp, K. G., Noll, L. K., & Fisher, P. A. (2019). A preliminary study investigating maternal neurocognitive mechanisms underlying a child-supportive parenting intervention. *Frontiers in Behavioral Neuroscience, 13*, 16.

Glasper, E. R., Kenkel, W. M., Bick, J., & Rilling, J. K. (2019). More than just mothers: The neurobiological and neuroendocrine underpinnings of allomaternal caregiving. *Frontiers in Neuroendocrinology, 53*, 100741.

Glocker, M. L., Langleben, D. D., Ruparel, K., Loughead, J. W., Valdez, J. N., Griffin, M. D., . . . Gur, R. C. (2009). Baby schema modulates the brain reward system in nulliparous women. *Proceedings of the National Academy of Sciences USA, 106*(22), 9115–9119.

Glynn, L. M., Davis, E. P., & Sandman, C. A. (2013). New insights into the role of perinatal HPA-axis dysregulation in postpartum depression. *Neuropeptides, 47*(6), 363–370.

Glynn, L. M., Davis, E. P., Sandman, C. A., & Goldberg, W. A. (2016). Gestational hormone profiles predict human maternal behavior at 1-year postpartum. *Hormones and Behavior, 85*, 19–25.

Glynn, L. M., & Sandman, C. A. (2014). Evaluation of the association between placental corticotrophin-releasing hormone and postpartum depressive symptoms. *Psychosomatic Medicine, 76*, 355–362.

Goff, B., Gee, D. G., Telzer, E. H., Humphreys, K. L., Gabard-Durnam, L., Flannery, J., & Tottenham, N. (2013). Reduced nucleus accumbens reactivity and adolescent depression following early-life stress. *Neuroscience, 249*, 129–138.

Gonçalves, L., Sego, C., & Metzger, M. (2012). Differential projections from the lateral habenula to the rostromedial tegmental nucleus and ventral tegmental area in the rat. *Journal of Comparative Neurology, 520*(6), 1278–1300.

Gong, P., Fan, H., Liu, J., Yang, X., Zhang, K., & Zhou, X. (2017). Revisiting the impact of OXTR rs53576 on empathy: A population-based study and meta-analysis. *Psychoneuroendocrinology, 80*, 131–136.

Gonzalez, A., Lovic, V., Ward, G. R., Wainwright, P. E., & Fleming, A. S. (2001). Intergenerational effects of complete maternal deprivation and replacement stimulation on maternal behavior and emotionality in female rats. *Developmental Psychobiology, 38*(1), 11–32.

Gonzalez, A., & Smeets, W. J. (1992). Comparative analysis of the vasotocinergic and mesotocinergic cells and fibers in the brain of two amphibians, the anuran Rana ridibunda and the urodele Pleurodeles waltlii. *Journal of Comparative Neurology, 315*(1), 53–73.

Gonzalez, A., & Smeets, W. J. (1997). Distribution of vasotocin- and mesotocin-like immunoreactivities in the brain of *Typhlonectes compressicauda* (Amphibia, gymnophiona): Further assessment of primitive and derived traits of amphibian neuropeptidergic systems. *Cell and Tissue Research, 287*(2), 305–314.

González-Mariscal, G. (2001). Neuroendocrinology of maternal behavior in the rabbit. *Hormones and Behavior, 40*(2), 125–132.

González-Mariscal, G., Caba, M., Hoffman, K., & Melo, A. (2017). Parental behavior. In D. W. Pfaff & M. Joëls (Eds.), *Hormones, brain and behavior: Vol. 1, Mammalian hormone-behavior systems* (3rd ed., pp. 83–116). Amsterdam, The Netherland: Academic Press.

González-Mariscal, G., Caba, M., Martinez-Gomez, M., Bautista, A., & Hudson, R. (2016). Mothers and offspring: The rabbit as a model system in the study of mammalian maternal behavior and sibling interactions. *Hormones and Behavior, 77*, 30–41.

González-Mariscal, G., Chirino, R., Beyer, C., & Rosenblatt, J. S. (2004). Removal of the accessory olfactory bulbs promotes maternal behavior in virgin rabbits. *Behavioural Brain Research, 152*(1), 89–95.

González-Mariscal, G., Chirino, R., Flores-Alonso, J. C., Rosenblatt, J. S., & Beyer, C. (2004). Intracerebroventricular injections of prolactin counteract the antagonistic effect of bromocriptine on rabbit maternal behaviour. *Journal of Neuroendocrinology, 16*(12), 949–955.

Gonzalez-Mariscal, G., Chirino, R., Rosenblatt, J. S., & Beyer, C. (2005). Forebrain implants of estradiol stimulate maternal nest-building in ovariectomized rabbits. *Hormones and Behavior, 47*(3), 272–279.

González-Mariscal G., & Gallegos, J. A., G. (2007). New Zealand white rabbits show non-selective nursing in various types of nests. *World Rabbit Science, 15*, 167–172.

Gonzalez-Mariscal, G., Jimenez, A., Chirino, R., & Beyer, C. (2009). Motherhood and nursing stimulate c-FOS expression in the rabbit forebrain. *Behavioral Neuroscience, 123*(4), 731–739.

González-Mariscal, G., Lemus, A. C., & Aguilar-Roblero, R. (2015). Contribution of suckling stimulation to the daily periodic display of nursing behavior in non-lactating virgin rabbits. *Journal of Neurology and Neurophysiology, 6*(6), 327.

Gonzalez-Mariscal, G., Melo, A. I., Chirino, R., Jimenez, P., Beyer, C., & Rosenblatt, J. S. (1998). Importance of mother/young contact at parturition and across lactation for the expression of maternal behavior in rabbits. *Developmental Psychobiology, 32*(2), 101–111.

González-Mariscal, G., Melo, A. I., Jimenez, P., Beyer, C., & Rosenblatt, J. S. (1996). Estradiol, progesterone, and prolactin regulate maternal nest-building in rabbits. *Journal of Neuroendocrinology, 8*(12), 901–907.

Gonzalez-Mariscal, G., Melo, A. I., Parlow, A. F., Beyer, C., & Rosenblatt, J. S. (2000). Pharmacological evidence that prolactin acts from late gestation to promote maternal behavior in rabbits. *Journal of Neuroendocrinology, 12*, 983–992, 2000.

Goodson, J. L., Evans, A. K., & Bass, A. H. (2003). Putative isotocin distributions in sonic fish: Relation to vasotocin and vocal–acoustic circuitry. *Journal of Comparative Neurology, 462*(1), 1–14.

Goodwin, N. B., Balshine-Earn, S., & Reynolds, J. D. (1998). Evolutionary transitions in parental care in cichlid fish. *Proceedings of the Royal Society Biological Sciences, 265*(1412), 2265.

Gordon, I., Zagoory-Sharon, O., Leckman, J. F., & Feldman, R. (2010). Oxytocin, cortisol, and triadic family interactions. *Physiology & Behavior, 101*(5), 679–684.

Gore, F., Schwartz, E. C., Brangers, B. C., Aladi, S., Stujenske, J. M., Likhtik, E., . . . Axel, R. (2015). Neural representations of unconditioned stimuli in basolateral amygdala mediate innate and learned responses. *Cell, 162*(1), 134–145.

Grasso, D. J., Moser, J. S., Dozier, M., & Simons, R. (2009). ERP correlates of attention allocation in mothers processing faces of their children. *Biological Psychology, 81*(2), 95–102.

Grattan, D. R. (2001). The actions of prolactin in the brain during pregnancy and lactation. *Progress in Brain Research, 133*, 153–171.

Grattan, D. R. (2002). Behavioural significance of prolactin signalling in the central nervous system during pregnancy and lactation. *Reproduction, 123*(4), 497–506.

Grattan, D. R., Pi, X. J., Andrews, Z. B., Augustine, R. A., Kokay, I. C., Summerfield, M. R., . . . Bunn, S. J. (2001). Prolactin receptors in the brain during pregnancy and lactation: Implications for behavior. *Hormones and Behavior, 40*(2), 115–124.

Grebe, N., E. Sarafin, R., Strenth, C., & Zilioli, S. (2019). Pair-bonding, fatherhood, and the role of testosterone: A meta-analytic review. *Neuroscience and Biobehavioral Reviews, 98*, 221–233.

Grieb, Z. A., Tierney, S. M., & Lonstein, J. S. (2017). Postpartum inhibition of ovarian steroid action increases aspects of maternal caregiving and reduces medial preoptic area progesterone receptor expression in female rats. *Hormones and Behavior, 96*, 31–41.

Grienberger, C., & Konnerth, A. (2012). Imaging calcium in neurons. *Neuron, 73*(5), 862–885.

Grinevich, V., Knobloch-Bollmann, H. S., Eliava, M., Busnelli, M., & Chini, B. (2016). Assembling the puzzle: Pathways of oxytocin signaling in the brain. *Biological Psychiatry, 79*(3), 155–164.

Grone, B. P., Carpenter, R. E., Lee, M., Maruska, K. P., & Fernald, R. D. (2012). Food deprivation explains effects of mouthbrooding on ovaries and steroid hormones, but not brain neuropeptide and receptor mRNAs, in an African cichlid fish. *Hormones and Behavior, 62*(1), 18–26.

Gross, C. T., & Canteras, N. S. (2012). The many paths to fear. *Nature Reviews Neuroscience, 13*(9), 651–658.

Grotegut, C. A., Mao, L., Pierce, S. L., Swamy, G. K., Heine, R. P., & Murtha, A. P. (2016). Enhanced uterine contractability and stillbirth in mice lacking G protein-coupled receptor kinase 6 (GRK6): Implications for oxytocin receptor desensitization. *Molecular Endocrinology, 30*(4), 455–468.

Gruber, C. J., Gruber, D. M., Gruber, I. M., Wieser, F., & Huber, J. C. (2004). Anatomy of the estrogen response element. *Trends in Endocrinology and Metabolism, 15*(2), 73–78.

Gu, X., Hof, P. R., Friston, K. J., & Fan, J. (2013). Anterior insular cortex and emotional awareness. *Journal of Comparative Neurology, 521*(15), 3371–3388.

Gubernick, D. J. (1981). Parent and infant attachment in mammals. In D. J. Gubernick & P. H. Klopfer (Eds.), *Parental care in mammals* (pp. 243–305). New York, NY: Plenum Press.

Gubernick, D. J. (1990). A maternal chemosignal maintains paternal behaviour in the biparental California mouse, *Peromyscus californicus. Animal Behaviour, 39* (5), 936–942.

Gubernick, D. J., & Alberts, J. R. (1987). The biparental care system of the California mouse, Peromyscus californicus. *Journal of Comparative Psychology, 101*(2), 169–177.

Gubernick, D. J., & Alberts, J. R. (1989). Postpartum maintenance of paternal behaviour in the biparental California mouse, *Peromyscus californicus. Animal Behaviour, 37*(4), 656–664.

Gubernick, D. J., & Nelson, R. (1989). Prolactin and paternal behavior in the biparental California mouse, *Peromyscus californicus. Hormones and Behavior, 23*(2), 203–210.

Gubernick, D. J., Schneider, K. A., & Jeannotte, L. A. (1994). Individual differences in the mechanisms underlying the onset and maintenance of paternal behavior and the inhibition of infanticide in the monogamous biparental California mouse, *Peromyscus californicus*. *Behavioral Ecology and Sociobiology 34*(3), 225–231.

Guo, C., Moses-Kolko, E., Phillips, M., Swain, J. E., & Hipwell, A. E. (2018). Severity of anxiety moderates the association between neural circuits and maternal behaviors in the postpartum period. *Cognitive, Affective, & Behavioral Neuroscience, 18*(3), 426–436.

Gyurak, A., Gross, J. J., & Etkin, A. (2011). Explicit and implicit emotion regulation: A dual-process framework. *Cognition & Emotion, 25*(3), 400–412.

Haas, B. W., Filkowski, M. M., Cochran, R. N., Denison, L., Ishak, A., Nishitani, S., & Smith, A. K. (2016). Epigeneitc modification of OXT and human sociability. *Proceedings of the National Academy of Sciences USA, 113*(27), E3816–E3823.

Hagen, E. H. (1999). The functions of postpartum depression. *Evolution & Human Behavior, 20*, 325–359.

Hahn, J. D., & Swanson, L. W. (2015). Connections of the juxtaventromedial region of the lateral hypothalamic area in the male rat. *Frontiers in Systems Neuroscience, 9*, 66.

Hansen, S. (1989). Medial hypothalamic involvement in maternal aggression of rats. *Behavioral Neuroscience, 103*(5), 1035–1046.

Hansen, S., Bergvall, A. H., & Nyiredi, S. (1993). Interaction with pups enhances dopamine release in the ventral striatum of maternal rats: A microdialysis study. *Pharmacology, Biochemistry, and Behavior, 45*(3), 673–676.

Hansen, S., & Ferreira, A. (1986). Food intake, aggression, and fear behavior in the mother rat: Control by neural systems concerned with milk ejection and maternal behavior. *Behavioral Neuroscience, 100*(1), 64–70.

Hansen, S., Harthon, C., Wallin, E., Lofberg, L., & Svensson, K. (1991). Mesotelencephalic dopamine system and reproductive behavior in the female rat: Effects of ventral tegmental 6-hydroxydopamine lesions on maternal and sexual responsiveness. *Behavioral Neuroscience, 105*(4), 588–598.

Hanson, J. L., Hariri, A. R., & Williamson, D. E. (2015). Blunted ventral striatum development in adolescence reflects emotional neglect and predicts depressive symptoms. *Biological Psychiatry, 78*(9), 598–605.

Hashikawa, Y., Hashikawa, K., Falkner, A. L., & Lin, D. (2017). Ventromedial hypothalamus and the generation of aggression. *Frontiers in Systems Neuroscience, 11*, 94.

Hauser, H., & Gandelman, R. (1985). Lever pressing for pups: Evidence for hormonal influence upon maternal behavior of mice. *Hormones and Behavior, 19*(4), 454–468.

Hausfater, G., & Hrdy, S. B. (1984). Infanticide. New York, NY: Aldine.

Hayes, U. L., & De Vries, G. J. (2007). Role of pregnancy and parturition in induction of maternal behavior in prairie voles (*Microtus ochrogaster*). *Hormones and Behavior, 51*(2), 265–272.

He, Z., Young, L., Ma, X-M., Guo, Q., Wang, L., Yang, Y., . . . Tai, F. (2019). Increased anxiety and decreased sociability induced by paternal deprivation involve the PVN-PrL OTergic pathway. *eLife, 8*, e44026.

Heatherton, T. F., & Wagner, D. D. (2011). Cognitive neuroscience of self-regulation failure. *Trends in Cognitive Sciences, 15*(3), 132–139.

Heim, C., Young, L. J., Newport, D. J., Mletzko, T., Miller, A. H., & Nemeroff, C. B. (2009). Lower CSF oxytocin concentrations in women with a history of childhood abuse. *Molecular Psychiatry, 14*(10), 954–958.

Hein, G., Engelmann, J. B., Volberg, M. C., & Tobler, P. N. (2016). How learning shapes the empathic brain. *Proceedings of the National Academy of Sciences USA, 113*(1), 80–85.

Hein, G., Silani, G., Preuschoff, K., Batson, C. D., & Singer, T. (2010). Neural responses to ingroup and outgroup members' suffering predict individual differences in costly helping. *Neuron, 68*(1), 149–160.

Hein, T. C., & Monk, C. S. (2017). Research review: Neural response to threat in children, adolescents, and adults after child maltreatment: A quantitative meta-analysis. *Journal of Child Psychology and Psychiatry, 58*(3), 222–230.

Heinrich, J., Boyd, R., & Richerson, P. J. (2012). The puzzle of monogamous marriage. *Philosophical Transactions of the Royal Society of London B Biological Sciences, 367*(1589), 657–669.

Hernandez-Gonzalez, M., Prieto-Beracoechea, C., Navarro-Meza, M., Ramos-Guevara, J. P., Reyes-Cortes, R., & Guevara, M. A. (2005). Prefrontal and tegmental electrical activity during olfactory stimulation in virgin and lactating rats. *Physiology & Behavior, 83*(5), 749–758.

Herrenkohl, L. R., & Rosenberg, P. A. (1972). Exteroceptive stimulation of maternal behavior in the naive rat. *Physiology & Behavior, 8*(4), 595–598.

Herrenkohl, L. R., & Rosenberg, P. A. (1974). Effects of hypothalamic deafferentation late in gestation on lactation and nursing behavior in the rat. *Hormones and Behavior, 5*(1), 33–41.

Herpertz, S. C., & Bertsch, K. (2015). A new perspective on the pathophysiology of borderline personality disorder: A model of the role of oxytocin. *American Journal of Psychiatry, 172*(9), 840–851.

Herringa, R. J., Birn, R. M., Ruttle, P. L., Burghy, C. A., Stodola, D. E., Davidson, R. J., & Essex, M. J. (2013). Childhood maltreatment is associated with altered fear circuitry and increased internalizing symptoms by late adolescence. *Proceedings of the National Academy of Sciences USA, 110*(47), 19119–19124.

Hidema, S., Fukuda, T., Mizukami, H., Hayashi, R., Otsuka, A., Suzuki, S., . . . Nishimori, K. (2016). Generation of Oxtr cDNA(HA)-ires-cre mice for gene expression in an oxytocin specific manner. *Journal of Cellular Biochemistry, 117*(5), 1099–1111.

Hinde, R. A. (1970). *Animal behaviour*. New York, NY: McGraw-Hill.

Ho, S. S., & Swain, J. E. (2017). Depression alters maternal extended amygdala response and functional connectivity during distress signals in attachment relationship. *Behavioural Brain Research, 325*(Pt B), 290–296.

Hoistad, M., & Barbas, H. (2008). Sequence of information processing for emotions through pathways linking temporal and insular cortices with the amygdala. *Neuroimage, 40*(3), 1016–1033.

Homberg, J. R., & Lesch, K. P. (2011). Looking on the bright side of serotonin transporter gene variation. *Biological Psychiatry, 69*(6), 513–519.

Hong, W., Kim, D.-W., & Anderson, D. J. (2014). Antagonistic control of social versus repetitive self-grooming behaviors by separable amygdala neuronal subsets. *Cell, 158*(6), 1348–1361.

Horie, K., Inoue, K., Suzuki, S. Adachi, S., Yada, S., Hirayama, T., . . . Nishimori, K. (2019). Oxytocin receptor knockout prairie voles by CRISPR/Cas9 editing show reduced preference for social novelty and exaggerated repetitive behaviors. *Hormones and Behavior, 111*, 60–69.

Horrell, N. D., Hickmott, P. W., & Saltzman, W. (2019). Neural regulation of paternal behavior in mammals: Sensory, neuroendocrine, and experiential influences on the paternal brain. *Current Topics in Behavioral Neurosciences*. doi:10.1007/7854_2018_55 [Epub ahead of print]

Horrell, N. D., Perea-Rodriguez, J. P., Harris, B. N., & Saltzman, W. (2017). Effects of repeated pup exposure on behavioral, neural, and adrenocortical responses to pups in male California mice (*Peromyscus californicus*). *Hormones and Behavior, 90*, 56–63.

Horrell, N. D., Saltzman, W., & Hickmott, P. W. (2019). Plasticity of paternity: Effects of fatherhood on synaptic, intrinsic and morphological characteristics of neurons in the medial preoptic area of male California mice. *Behavioural Brain Research, 365*, 89–102.

Horseman, N. D., Zhao, W., Montecino-Rodriguez, E., Tanaka, M., Nakashima, K., Engle, S. J., . . . Dorshkind, K. (1997). Defective mammopoiesis, but normal hematopoiesis, in mice with a targeted disruption of the prolactin gene. *EMBO Journal, 16*(23), 6926–6935.

Hrdy, S. B. (1979). Infanticide among animals: A review, classification, and examination of the implications for the reproductive strategies of females. *Ethology and Sociobiology, 1*(1), 13–40.

Hrdy, S. B. (2009). *Mothers and others: The evolutionary origins of mutual understanding*. Cambridge, MA: Belknap Press of Harvard University Press.

Hrdy, S. B. (2016). Variable postpartum responsiveness among humans and other primates with "cooperative breeding": A comparative and evolutionary perspective. *Hormones and Behavior, 77*, 272–283.

Huber, D., Veinante, P., & Stoop, R. (2005). Vasopressin and oxytocin excite distinct neuronal populations in the central amygdala. *Science, 308*(5719), 245–248.

Hume, J. M., & Wynne-Edwards, K. E. (2005). Castration reduces male testosterone, estradiol, and territorial aggression, but not paternal behavior in biparental dwarf hamsters (*Phodopus campbelli*). *Hormones and Behavior, 48*(3), 303–310.

Hume, J. M., & Wynne-Edwards, K. E. (2006). Paternal responsiveness in biparental dwarf hamsters (*Phodopus campbelli*) does not require estradiol. *Hormones and Behavior, 49*(4), 538–544.

Humphries, M. D., & Prescott, T. J. (2010). The ventral basal ganglia, a selection mechanism at the crossroads of space, strategy, and reward. *Progress in Neurobiology, 90*(4), 385–417.

Hung, L. W., Neuner, S., Polepalli, J. S., Beier, K. T., Wright, M., Walsh, J. J., . . . Malenka, R. C. (2017). Gating of social reward by oxytocin in the ventral tegmental area. *Science, 357*(6358), 1406.

Hyder, S. M., Chiappetta, C., & Stancel, G. M. (1998). The 3-flanking region of the mouse c-fos gene contains a cluster of GGTCA hormone response-like elements. *Molecular Biology Reports, 25*(3), 189–191.

Imayoshi, I., Sakamoto, M., Ohtsuka, T., Takao, K., Miyakawa, T., Yamaguchi, M., . . . Kageyama, R. (2008). Roles of continuous neurogenesis in the structural and functional integrity of the adult forebrain. *Nature Neuroscience, 11*(10), 1153–1161.

Insel, T. R. (2010). The challenge of translation in social neuroscience: A review of oxytocin, vasopressin, and affiliative behavior. *Neuron, 65*(6), 768–779.

Insel, T. R., & Harbaugh, C. R. (1989). Lesions of the hypothalamic paraventricular nucleus disrupt the initiation of maternal behavior. *Physiology & Behavior, 45*(5), 1033–1041.

Insel, T. R., & Hulihan, T. J. (1995). A gender-specific mechanism for pair bonding: Oxytocin and partner preference formation in monogamous voles. *Behavioral Neuroscience, 109*(4), 782–789.

Insel, T. R., & Shapiro, L. E. (1992). Oxytocin receptor distribution reflects social organization in monogamous and polygamous voles. *Proceedings of the National Academy of Sciences USA, 89*(13), 5981–5985.

Ishikawa, J., Nishimura, R., & Ishikawa, A. (2015). Early-life stress induces anxiety-like behaviors and activity imbalances in the medial prefrontal cortex and amygdala in adult rats. *European Journal of Neuroscience, 41*(4), 442–453.

Isogai, Y., Wu, Z., Love, M. I., Ahn, M. H., Bambah-Mukku, D., Hua, V., . . . Dulac, C. (2018). Multisensory logic of infant-directed aggression by males. *Cell, 175*(7), 1827–1841.e1817.

Izquierdo, M. A., Collado, P., Segovia, S., Guillamon, A., & del Cerro, M. C. (1992). Maternal behavior induced in male rats by bilateral lesions of the bed nucleus of the accessory olfactory tract. *Physiology & Behavior, 52*(4), 707–712.

Jabbi, M., & Keysers, C. (2008). Inferior frontal gyrus activity triggers anterior insula response to emotional facial expressions. *Emotion, 8*(6), 775–780.

Jabbi, M., Swart, M., & Keysers, C. (2007). Empathy for positive and negative emotions in the gustatory cortex. *Neuroimage, 34*(4), 1744–1753.

Jacobson, C. D., Terkel, J., Gorski, R. A., & Sawyer, C. H. (1980). Effects of small medial preoptic area lesions on maternal behavior: Retrieving and nest building in the rat. *Brain Research, 194*(2), 471–478.

Jaeggi, A. V., Burkart, J. M., & van Schaik, C. P. (2010). On the psychology of cooperation in humans and other primates: Combining the natural history and experimental evidence of prosociality. *Philosophical Transactions of the Royal Society of London B Biological Sciences, 365*(1553), 2723–2735.

Jakubauskiene, E., Janaviciute, V., Peciuliene, I., Soderkvist, P., & Kanopka, A. (2012). G/A polymorphism in intronic sequence affects the processing of MAO-B gene in patients with Parkinson disease. *FEBS Letters, 586*(20), 3698–3704.

Jakubowski, M., & Terkel, J. (1982). Infanticide and caretaking in non-lactating Mus musculus: Influence of genotype, family group and sex. *Animal Behaviour, 30*(4), 1029–1035.

Jakubowski, M., & Terkel, J. (1986). Establishment and maintenance of maternal responsiveness in postpartum Wistar rats. *Animal Behaviour, 34*, 256–262.

Janak, P. H., & Tye, K. M. (2015). From circuits to behaviour in the amygdala. *Nature, 517*(7534), 284–292.

Jarcho, M. R., Mendoza, S. P., Mason, W. A., Yang, X., & Bales, K. L. (2011). Intranasal vasopressin affects pair bonding and peripheral gene expression in male *Callicebus cupreus*. *Genes, Brain, and Behavior, 10*(3), 375–383.

Jasnow, A. M., Davis, M., & Huhman, K. L. (2004). Involvement of central amygdalar and bed nucleus of the stria terminalis corticotropin-releasing factor in behavioral responses to social defeat. *Behavioral Neuroscience, 118*(5), 1052–1061.

Jensen, K. Hare, B., Call, J., & Tomasello, M. (2006). What's in it for me? Self-regard precludes altruism and spite in chimpanzees. *Proceedings of the Royal Society B, 273*, 1013–1021.

Jhou, T. C., Geisler, S., Marinelli, M., Degarmo, B. A., & Zahm, D. S. (2009). The mesopontine rostromedial tegmental nucleus: A structure targeted by the lateral habenula that projects to the ventral tegmental area of Tsai and substantia nigra compacta. *Journal of Comparative Neurology, 513*(6), 566–596.

Jimenez, A., Young, L. J., Triana-Del Rio, R., LaPrairie, J. L., & Gonzalez-Mariscal, G. (2015). Neuroanatomical distribution of oxytocin receptor binding in the female rabbit forebrain: Variations across the reproductive cycle. *Brain Research, 1629*, 329–339.

Jin, K. S., & Baillargeon, R. (2017). Infants possess an abstract expectation of ingroup support. *Proceedings of the National Academy of Sciences USA, 114*(31), 8199–8202.

Jirik-Babb, P., Manaker, S., Tucker, A. M., & Hofer, M. A. (1984). The role of the accessory and main olfactory systems in maternal behavior of the primiparous rat. *Behavioral and Neural Biology, 40*(2), 170–178.

Johnson, Z. V., Walum, H., Jamal, Y. A., Xiao, Y., Keebaugh, A. C., Inoue, K., & Young, L. J. (2016). Central oxytocin receptors mediate mating-induced partner preferences and enhance correlated activation across forebrain nuclei in male prairie voles. *Hormones and Behavior, 79*, 8–17.

Johnson, Z. V., Walum, H., Xiao, Y., Riefkohl, P. C., & Young, L. J. (2017). Oxytocin receptors modulate social salience neural network in male prairie voles. *Hormones and Behavior, 87*, 16–24.

Jones, J. D., Cassidy, J., & Shaver, P. R. (2015). Parents' self-reported attachment styles: A review of links with parenting behaviors, emotions, and cognitions. *Personality and Social Psychology Review, 19*(1), 44–76.

Jurek, B., & Neumann, I. D. (2018). The oxytocin receptor: From intracellular signaling to behavior. *Physiological Reviews, 98*(3), 1805–1908.

Kalamatianos, T., Faulkes, C. G., Oosthuizen, M. K., Poorun, R., Bennett, N. C., & Coen, C. W. (2010). Telencephalic binding sites for oxytocin and social organization: A comparative study of eusocial naked mole-rats and solitary cape mole-rats. *Journal of Comparative Neurology, 518*(10), 1792–1813.

Kalinichev, M., Rosenblatt, J. S., & Morrell, J. I. (2000). The medial preoptic area, necessary for adult maternal behavior in rats, is only partially established as a component of the neural circuit that supports maternal behavior in juvenile rats. *Behavioral Neuroscience, 114*(1), 196–210.

Kan, J. M., Callaghan, B. L., & Richardson, R. (2016). A mother's past can predict her offspring's future: Previous maternal separation leads to the early emergence of adult-like fear behavior in subsequent male infant rat offspring. *Behavioral Neuroscience, 130*(5), 511–520.

Kaneko, T., Kaneda, K., Ohno, A., Takahasi, D. Hara, T., Amano, T., . . . Minami, M. (2016). Activation of adenylate cyclase-cyclic AMP-protein kinase A signaling by corticotropin-releasing factor within the dorsolateral bed nucleus of the stria terminalis is involved in pain-induced aversion. *European Journal of Neuroscience, 44*(11), 2914–2924.

Kanes, S., Colquhoun, H., Gunduz-Bruce, H., Raines, S., Arnold, R., Schacterle, A., . . . Meltzer-Brody, S. (2017). Brexanolone (SAGE-547 injection) in post-partum depression: A randomised controlled trial. *Lancet, 390*(10093), 480–489.

Kanske, P., Bockler, A., Trautwein, F. M., Parianen Lesemann, F. H., & Singer, T. (2016). Are strong empathizers better mentalizers? Evidence for independence and interaction between the routes of social cognition. *Social, Cognitive, and Affective Neuroscience, 11*(9), 1383–1392.

Kaufling, J., Veinante, P., Pawlowski, S. A., Freund-Mercier, M. J., & Barrot, M. (2009). Afferents to the GABAergic tail of the ventral tegmental area in the rat. *Journal of Comparative Neurology, 513*(6), 597–621.

Kawamata, M., Mitsui-Saito, M., Kimura, T., Takayanagi, Y., Yanagisawa, T., & Nishimori, K. (2003). Vasopressin-induced contraction of uterus is mediated solely by the oxytocin receptor in mice, but not in humans. *European Journal of Pharmacology, 472*, 229–234.

Keebaugh, A. C., Barrett, C. E., Laprairie, J. L., Jenkins, J. J., & Young, L. J. (2015). RNAi knockdown of oxytocin receptor in the nucleus accumbens inhibits social attachment and parental care in monogamous female prairie voles. *Social Neuroscience, 10*(5), 561–570.

Keebaugh, A. C., & Young, L. J. (2011). Increasing oxytocin receptor expression in the nucleus accumbens of pre-pubertal female prairie voles enhances alloparental responsiveness and partner preference formation as adults. *Hormones and Behavior, 60*(5), 498–504.

Keer, S. E., & Stern, J. M. (1999). Dopamine receptor blockade in the nucleus accumbens inhibits maternal retrieval and licking, but enhances nursing behavior in lactating rats. *Physiology & Behavior, 67*(5), 659–669.

Keller, M., Perrin, G., Meurisse, M., Ferreira, G., & Levy, F. (2004). Cortical and medial amygdala are both involved in the formation of olfactory offspring memory in sheep. *European Journal of Neuroscience, 20*(12), 3433–3441.

Kempadoo, K. A., Tourino, C., Cho, S. L., Magnani, F., Leinninger, G. M., Stuber, G. D., . . . Bonci, A. (2013). Hypothalamic neurotensin projections promote reward by enhancing glutamate transmission in the VTA. *Journal of Neuroscience, 33*(18), 7618–7626.

Kendrick, K. M. (2000). Oxytocin, motherhood and bonding. *Experimental Physiology, 85*(Suppl.), 111S–124S.

Kendrick, K. M., Da Costa, A. P., Broad, K. D., Ohkura, S., Guevara, R., Levy, F., & Keverne, E. B. (1997). Neural control of maternal behaviour and olfactory recognition of offspring. *Brain Research Bulletin, 44*(4), 383–395.

Kendrick, K. M., & Keverne, E. B. (1991). Importance of progesterone and estrogen priming for the induction of maternal behavior by vaginocervical stimulation in sheep: Effects of maternal experience. *Physiology & Behavior, 49*(4), 745–750.

Kendrick, K. M., Keverne, E. B., & Baldwin, B. A. (1987). Intracerebroventricular oxytocin stimulates maternal behaviour in the sheep. *Neuroendocrinology, 46*(1), 56–61.

Kendrick, K. M., Keverne, E. B., Hinton, M. R., & Goode, J. A. (1992). Oxytocin, amino acid and monoamine release in the region of the medial preoptic area and bed nucleus of the stria terminalis of the sheep during parturition and suckling. *Brain Research, 569*(2), 199–209.

Kendrick, K. M., Levy, F., & Keverne, E. B. (1991). Importance of vaginocervical stimulation for the formation of maternal bonding in primiparous and multiparous parturient ewes. *Physiology & Behavior, 50*(3), 595–600.

Kenkel, W. M., Paredes, J., Yee, J. R., Pournajafi-Nazarloo, H., Bales, K. L., & Carter, C. S. (2012). Neuroendocrine and behavioural responses to exposure to an infant in male prairie voles. *Journal of Neuroendocrinology, 24*(6), 874–886.

Kenkel, W. M., Perkeybile, A. M., & Carter, C. S. (2017). The neurobiological causes and effects of alloparenting. *Developmental Neurobiology, 77*(2), 214–232.

Keshavarzi, S., Power, J. M., Albers, E. H., Sullivan, R. K., & Sah, P. (2015). Dendritic organization of olfactory inputs to medial amygdala neurons. *Journal of Neuroscience, 35*(38), 13020–13028.

Keshavarzi, S., Sullivan, R. K., Ianno, D. J., & Sah, P. (2014). Functional properties and projections of neurons in the medial amygdala. *Journal of Neuroscience, 34*(26), 8699–8715.

Keverne, E. B., Levy, F., Poindron, P., & Lindsay, D. R. (1983). Vaginal stimulation: An important determinant of maternal bonding in sheep. *Science, 219*(4580), 81–83.

Keyser-Marcus, L., Stafisso-Sandoz, G., Gerecke, K., Jasnow, A., Nightingale, L., Lambert, K. G., . . . Kinsley, C. H. (2001). Alterations of medial preoptic area neurons following pregnancy and pregnancy-like steroidal treatment in the rat. *Brain Research Bulletin, 55*(6), 737–745.

Kim, J., Lee, S., Fang, Y., Shin, A., Park, S., Hashikawa, K., . . . Suh, G. S. B. (2019). Rapid, biphasic CRF neuronal responses encode positive and negative valence. *Nature Neuroscience, 22*, 576–585.

Kim, P., Capistrano, C., & Congleton, C. (2016). Socioeconomic disadvantages and neural sensitivity to infant cry: Role of maternal distress. *Social, Cognitive, and Affective Neuroscience, 11*(10), 1597–1607.

Kim, P., Capistrano, C. G., Erhart, A., Gray-Schiff, R., & Xu, N. (2017). Socioeconomic disadvantage, neural responses to infant emotions, and emotional availability among first-time new mothers. *Behavioural Brain Research, 325*(Pt B), 188–196.

Kim, P., Feldman, R., Mayes, L., Eicher, V., Thompson, N., Leckman, J., & Swain, J. (2011). Breastfeeding, brain activation to own infant cry, and maternal sensitivity. *Journal of Child Psychology and Psychiatry, 52*(8), 907–15.

Kim, P., Leckman, J., Mayes, L., Feldman, R., Wang, X., & Swain, J. (2010). The plasticity of human maternal brain: Longitudinal changes in brain anatomy during the early postpartum period. *Behavioral Neuroscience, 124*(5), 695–700.

Kim, P., Leckman, J., Mayes, L., Newman, M. A., Feldman, R., & Swain, J. (2010). Perceived quality of maternal care in childhood and structure and function of mothers' brain. *Developmental Science, 13* (4), 662–673.

Kim, P., Rigo, P., Mayes, L., Feldman, R., Leckman, J., & Swain, J. (2014). Neural plasticity in fathers of human infants. *Social Neuroscience, 9*(5), 522–35.

Kim, S., Adhikari, A., Lee, S. Y., Marshel, J. H., Kim, C. K., Mallory, C. S., . . . Deisseroth, K. (2013). Diverging neural pathways assemble a behavioural state from separable features of anxiety. *Nature, 496*, 219–223.

Kim, S., Fonagy, P., Allen, J., & Strathearn, L. (2014). Mothers' unresolved trauma blunts amygdala response to infant distress. *Social Neuroscience, 9*(4), 352–363.

Kim, S., Soeken, T. A., Cromer, S. J., Martinez, S. R., Hardy, L. R., & Strathearn, L. (2014). Oxytocin and postpartum depression: Delivering on what's known and what's not. *Brain Research, 1580*, 219–232.

Kim, S., & Strathearn, L. (2017). Trauma, mothering, and intergenerational transmission: A synthesis of behavioral and oxytocin research. *Psychoanalytic Study of the Child, 70*(1), 200–223.

King, L. B., Walum, H., Inoue, K., Eyrich, N. W., & Young, L. J. (2016). Variation in the oxytocin receptor gene predicts brain region-specific expression and social attachment. *Biological Psychiatry, 80*(2), 160–169.

Kinnally, E. L., Ceniceros, L., & Martinez, S. J. (2018). Genetic and environmental factors in the intergenerational transmission of maternal care in rhesus macaques: Preliminary findings. *American Journal of Primatology, 80*(12), e22939.

Kinsley, C. H., & Bridges, R. S. (1990). Morphine treatment and reproductive condition alter olfactory preferences for pup and adult male odors in female rats. *Developmental Psychobiology, 23*(4), 331–347.

Kirkpatrick, B., Kim, J. W., & Insel, T. R. (1994). Limbic system fos expression associated with paternal behavior. *Brain Research, 658*(1–2), 112–118.

Kirsch, P., Esslinger, C., Chen, Q., Mier, D., Lis, S., Siddhanti, S., . . . Meyer-Lindenberg, A. (2005). Oxytocin modulates neural circuitry for social cognition and fear in humans. *Journal of Neuroscience, 25*(49), 11489–11493.

Klahr, A. M., Klump, K., & Burt, S. A. (2015). A constructive replication of the association between the oxytocin receptor genotype and parenting. *Journal of Family Psychology, 29*(1), 91–99.

Klampfl, S. M., & Bosch, O. J. (2019). Mom doesn't care: When increased brain CRF system activity leads to maternal neglect in rodents. *Frontiers in Neuroendocrinology, 53*, 100735.

Klampfl, S. M., Brunton, P. J., Bayerl, D. S., & Bosch, O. J. (2014). Hypoactivation of CRF receptors, predominantly type 2, in the medial-posterior BNST is vital for adequate maternal behavior in lactating rats. *Journal of Neuroscience, 34*(29), 9665–9676.

Klampfl, S. M., Neumann, I. D., & Bosch, O. J. (2013). Reduced brain corticotropin-releasing factor receptor activation is required for adequate maternal care and maternal aggression in lactating rats. *European Journal of Neuroscience, 38*(5), 2742–2750.

Klampfl, S. M., Schramm, M. M., Gassner, B. M., Hubner, K., Seasholtz, A. F., Brunton, P. J., . . . Bosch, O. J. (2018). Maternal stress and the MPOA: Activation of CRF receptor 1 impairs maternal behavior and triggers local oxytocin release in lactating rats. *Neuropharmacology, 133*, 440–450.

Klatt, J. D., & Goodson, J. L. (2013). Oxytocin-like receptors mediate pair bonding in a socially monogamous songbird. *Proceedings in Biological Sciences, 280*(1750), 20122396.

Kleiman, D. G. (1977). Monogamy in mammals. *Quarterly Review of Biology, 52*(1), 39–69.

Knobloch, H. S., Charlet, A., Hoffmann, L. C., Eliava, M., Khrulev, S., Cetin, A. H., . . . Grinevich, V. (2012). Evoked axonal oxytocin release in the central amygdala attenuates fear response. *Neuron, 73*(3), 553–566.

Knobloch, H. S., & Grinevich, V. (2014). Evolution of oxytocin pathways in the brain of vertebrates. *Frontiers in Behavioral Neuroscience, 8*, 31.

Kogan, A., Saslow, L. R., Impett, E. A., Oveis, C., Keltner, D., Rodrigues Saturn, S. (2011). Thin-slicing study of the oxytocin receptor (OXTR) gene and the evaluation and expression of the prosocial disposition. *Proceedings of the National Academy of Sciences USA, 108*(48), 189–192.

Kohl, J., Babayan, B. M., Rubinstein, N. D., Autry, A. E., Marin-Rodriguez, B., Kapoor, V., . . . Dulac, C. (2018). Functional circuit architecture underlying parental behaviour. *Nature, 556*(7701), 326–331.

Kohlhoff, J., Eapen, V., Dadds, M., Khan, F., Silove, D., & Barnett, B. (2017). Oxytocin in the postnatal period: Associations with attachment and maternal caregiving. *Comprehensive Psychiatry, 76*, 56–68.

Kolber, B. J., Boyle, M. P., Wieczorek, L., Kelley, C. L., Onwuzurike, C. C., Nettles, S. A., . . . Muglia, L. J. (2010). Transient early-life forebrain corticotropin-releasing hormone elevation causes long-lasting anxiogenic and despair-like changes in mice. *Journal of Neuroscience, 30*(7), 2571–2581.

Kolber, B. J., Roberts, M. S., Howell, M. P., Wozniak, D. F., Sands, M. S., & Muglia, L. J. (2008). Central amygdala glucocorticoid receptor action promotes fear-associated CRH activation and conditioning. *Proceedings of the National Academy of Sciences USA, 105*(33), 12004–12009.

Komisaruk, B. R. (1967). Effects of local brain implants of progesterone on reproductive behavior in ring doves. *Journal of Comparative and Physiological Psychology, 64*(2), 219–224.

Kopel, H., Schechtman, E., Groysman, M., & Mizrahi, A. (2012). Enhanced synaptic integration of adult-born neurons in the olfactory bulb of lactating mothers. *Journal of Neuroscience, 32*(22), 7519–7527.

Korte, S. M. (2001). Corticosteroids in relation to fear, anxiety and psychopathology. *Neuroscience and Biobehavioral Reviews, 25*(2), 117–142.

Kosfeld, M., Heinrichs, M., Zak, P. J., Fischbacher, U., & Fehr, E. (2005). Oxytocin increases trust in humans. *Nature, 435*(7042), 673–676.

Kramer, K. L. (2011). The evolution of human parental care and recruitment of juvenile help. *Trends in Ecology & Evolution, 26*(10), 533–540.

Kramer, K. L., & Otarola-Castillo, E. (2015). When mothers need others: The impact of hominin life history evolution on cooperative breeding. *Journal of Human Evolution, 84*, 16–24.

Kramer, K. L., & Russell, A. F. (2015). Was monogamy a key step on the hominin road? Reevaluating the monogamy hypothesis in the evolution of cooperative breeding. *Evolutionary Anthropology, 24*(2), 73–83.

Kramer, K. L., & Veile, A. (2018). Infant allocare in traditional societies. *Physiology & Behavior, 193*(Pt A), 117–126.

Krehbiel, D., Poindron, P., Levy, F., & Prud'Homme, M. J. (1987). Peridural anesthesia disturbs maternal behavior in primiparous and multiparous parturient ewes. *Physiology & Behavior, 40*(4), 463–472.

Kremarik, P., Freund-Mercier, M. J., & Stoeckel, M. E. (1995). Oxytocin and vasopressin binding sites in the hypothalamus of the rat: Histoautoradiographic detection. *Brain Research Bulletin, 36*(2), 195–203.

Kreuder, A. K., Scheele, D., Wassermann, L., Wollseifer, M., Stoffell-Wagner, B., Lee, M. R., . . . Hurlemann, R. (2017). How the brain codes intimacy: The neurobiological substrates of romantic touch. *Human Brain Mapping, 38*(9), 4525–4534.

Krishnan, K., Lau, B. Y. B., Ewall, G., Huang, J., & Shea, S. D. (2017). MECP2 regulates cortical plasticity underlying a learned behaviour in adult female mice. *Nature Communications, 8*, 14077.

Krueger, F., McCabe, K., Moll, J., Kriegeskorte, N., Zahn, R., Strenziok, M., . . . Graffman, J. (2007). Neural correlated of trust. *Proceedings of the National Academy of Sciences USA, 104*(50), 20084–20088.

Krueger, F., Parasuraman, R., Iyengar, V., Thornburg, M., Weel, J., Lin, M., . . . Lipsky, R. H. (2012). Oxytocin receptor genetic variation promotes trust behavior. *Frontiers in Human Neuroscience, 6*, 4.

Kuhlman, K. R., Geiss, E. G., Vargas, I., & Lopez-Duran, N. L. (2015). Differential associations between childhood trauma subtypes and adolescent HPA-axis functioning. *Psychoneuroendocrinology, 54*, 103–114.

Kunwar, P. S., Zelikowsky, M., Remedios, R., Cai, H., Yilmaz, M., Meister, M., & Anderson, D. J. (2015). Ventromedial hypothalamic neurons control a defensive emotion state. *Elife, 4*, e06633

Kuroda, K. O., Tachikawa, K., Yoshida, S., Tsuneoka, Y., & Numan, M. (2011). Neuromolecular basis of parental behavior in laboratory mice and rats: With special emphasis on technical issues of using mouse genetics. *Progress in Neuropsychopharmacology & Biological Psychiatry, 35*(5), 1205–1231.

Kusui, C., Kimura, T., Ogita, K., Nakamura, H., Matsumura, Y., Koyama, M., . . . Murata, Y. (2001). DNA methylation of the human oxytocin receptor gene promoter regulates tissue-specific gene suppression. *Biochemical and Biophysical Research Communications, 289*(3), 681–686.

LaBuda, C. J., Dobolyi, A., & Usdin, T. B. (2004). Tuberoinfundibular peptide of 39 residues produces anxiolytic and antidepressant actions. *Neuroreport, 15*(5), 881–885.

Labuschagne, I., Phan, K. L., Wood, A., Angstadt, M., Chua, P., Heinrichs, M., . . . Nathan, P. J. (2010). Oxytocin attenuates amygdala reactivity to fear in generalized social anxiety disorder. *Neuropsychopharmacology, 35*(12), 2403–2413.

Lahey, B. B., Michalska, K. J., Liu, C., Chen, Q., Hipwell, A. E., Chronis-Tuscano, A., & Decety, J. (2012). Preliminary genetic imaging study of the association between estrogen receptor-α gene polymorphisms and harsh human maternal parenting. *Neuroscience Letters 525*(1), 17–22.

Lancaster, K., Goldbeck, C., Puglia, M. H., Morris, J. P., & Connelly, J. S. (2018). DNA methylation of OXTR is associated with parasympathetic nervous system activity and amygdala morphology. *Social, Cognitive, and Affective Neuroscience, 13*(11), 1155–1162.

Landgraf, R., & Neumann, I. D. (2004). Vasopressin and oxytocin release within the brain: A dynamic concept of multiple and variable modes of neuropeptide communication. *Frontiers in Neuroendocrinology, 25*(3–4), 150–176.

Landgraf, R., Neumann, I., & Pittman, Q. J. (1991). Septal and hippocampal release of vasopressin and oxytocin during late pregnancy and parturition in the rat. *Neuroendocrinology, 54*(4), 378–383.

Larsen, C. M., & Grattan, D. R. (2010). Prolactin-induced mitogenesis in the subventricular zone of the maternal brain during early pregnancy is essential for normal postpartum behavioral responses in the mother. *Endocrinology, 151*(8), 3805–3814.

Larsen, C. M., & Grattan, D. R. (2012). Prolactin, neurogenesis, and maternal behaviors. *Brain Behavior, & Immunity, 26*(2), 201–209.

Larsen, C. M., Kokay, I. C., & Grattan, D. R. (2008). Male pheromones initiate prolactin-induced neurogenesis and advance maternal behavior in female mice. *Hormones and Behavior, 53*(4), 509–517.

Laurent, H. K., & Ablow, J. C. (2012a). A cry in the dark: Depressed mothers show reduced neural activation to their own infant's cry. *Social Cognitive, and Affective Neuroscience, 7*(2), 125–134.

Laurent, H. K., & Ablow, J. C. (2012b). The missing link: Mothers' neural response to infant cry related to infant attachment behaviors. *Infant Behavior and Development, 35*(4), 761–772.

Laurent, H. K., & Ablow, J. C. (2013). A face a mother could love: Depression-related maternal neural responses to infant emotion faces. *Social Neuroscience, 8*(3), 228–239.

Laurent, H. K., Harold, G. T., Leve, L., Shelton, K. H., & Van Goozen, S. H. (2016). Understanding the unfolding of stress regulation in infants. *Development and Psychopathology, 28*(4 Pt 2), 1431–1440.

Laurent, H. K., Stevens, A., & Ablow, J. C. (2011). Neural correlates of hypothalamic-pituitary-adrenal regulation of mothers with their infants. *Biological Psychiatry, 70*(9), 826–832.

Le Neindre, P., Poindron, P., & Delouis, C. (1979). Hormonal induction of maternal behavior in non-pregnant ewes. *Physiology & Behavior, 22*(4), 731–734.

Lee, A., Clancy, S., & Fleming, A. S. (2000). Mother rats bar-press for pups: Effects of lesions of the MPOA and limbic sites on maternal behavior and operant responding for pup-reinforcement. *Behavioural Brain Research, 108*(2), 215–231.

Lee, A. W., & Brown, R. E. (2002). Medial preoptic lesions disrupt parental behavior in both male and female California mice (*Peromyscus californicus*). *Behavioral Neuroscience, 116*(6), 968–975.

Lee, A. W., & Brown, R. E. (2007). Comparison of medial preoptic, amygdala, and nucleus accumbens lesions on parental behavior in California mice (Peromyscus californicus). *Physiology & Behavior, 92*(4), 617–628.

Lee, A., & Hankin, B. L. (2009). Insecure attachment, dysfunctional attitudes, and low self-esteem predicting prospective symptoms of depression and anxiety during adolescence. *Journal of Clinical Child and Adolescent Psychology, 38*(2), 219–231.

Lee, G., & Gammie, S. C. (2009). GABA(A) receptor signaling in the lateral septum regulates maternal aggression in mice. *Behavioral Neuroscience, 123*(6), 1169–1177.

Lee, H. J., Caldwell, H. K., Macbeth, A. H., Tolu, S. G., & Young, W. S., 3rd. (2008). A conditional knockout mouse line of the oxytocin receptor. *Endocrinology, 149*(7), 3256–3263.

Lee. M. R., Scheidweiler, Diao, X. X., Akhlaghi, F., Cummins, A., Huestis, M. A., Leggio, L., & Averbeck, B. B. (2018). Oxytocin by intranasal and intravenous routes reached the cerebrospinal fluid in rhesus macaques: Determination using a novel oxytocin assay. *Molecular Psychiatry, 23*, 115–122.

Lee, R. B. (2018). Hunter-gatherers and human evolution: New light on old debates. *Annual Review of Anthropology, 47*, 513–531.

Lee, R. J., Gollan, J., Kasckow, J., Geracioti, T., & Coccaro, E. F. (2006). CSF corticotropin-releasing factor in personality disorder: Relationship with self-reported parental care. *Neuropsychopharmacology, 31*(10), 2289–2295.

Leerkes, E. M., & Siepak, K. J. (2006). Attachment linked predictors of women's emotional and cognitive responses to infant distress. *Attachment & Human Development, 8*(1), 11–32.

Lei, K., Liu, Y., Smith, A. S., Lonstein, J. S., & Wang, Z. (2017). Effects of pair bonding on parental behavior and dopamine activity in the nucleus accumbens in male prairie voles. *European Journal of Neuroscience, 46*(7), 2276–2284.

Leng, G., & Ludwig, M. (2016). Intranasal oxytocin: Myths and delusions. *Biological Psychiatry, 79*(3), 243–250.

Lenzi, D., Trentini, C., Macaluso, E., Graziano, S., Speranza, A. M., Pantano, P., & Ammaniti, M. (2016). Mothers with depressive symptoms display differential brain activations when empathizing with infant faces. *Psychiatry Research: Neuroimaging, 249*, 1–11.

Lenzi, D., Trentini, C., Pantano, P., Macaluso, E., Lenzi, G. L., & Ammaniti, M. (2013). Attachment models affect brain responses in areas related to emotions and empathy in nulliparous women. *Human Brain Mapping, 34*(6), 1399–1414.

Leonetti, F. (2014). What promotes secure attachment? The protective roles of infants' temperament and adoptive parents' attachment. *Attachment & Human Development, 16*(6), 573–589.

Lepri, J. J., Wysocki, C. J., & Vandenbergh, J. G. (1985). Mouse vomeronasal organ: Effects on chemosignal production and maternal behavior. *Physiology & Behavior, 35*(5), 809–814.

Leuner, B., & Sabihi, S. (2016). The birth of new neurons in the maternal brain: Hormonal regulation and functional implications. *Frontiers in Neuroendocrinology, 41*, 99–113.

Levy, F. (2008). Neural substrates involved in the onset of maternal responsiveness and selectivity in sheep. In R. S. Bridges (Ed.), *Neurobiology of the parental brain.* (pp. 23–37). Boston, MA: Academic Press.

Levy, F. (2016). Neuroendocrine control of maternal behavior in non-human and human mammals. *Annales d'Endocrinologie (Paris), 77*(2), 114–125.

Levy, F., Gervais, R., Kindermann, U., Orgeur, P., & Piketty, V. (1990). Importance of beta-noradrenergic receptors in the olfactory bulb of sheep for recognition of lambs. *Behavioral Neuroscience, 104*(3), 464–469.

Levy, F., Kendrick, K. M., Keverne, E. B., Piketty, V., & Poindron, P. (1992). Intracerebral oxytocin is important for the onset of maternal behavior in inexperienced ewes delivered under peridural anesthesia. *Behavioral Neuroscience, 106*(2), 427–432.

Levy, F., Locatelli, A., Piketty, V., Tillet, Y., & Poindron, P. (1995). Involvement of the main but not the accessory olfactory system in maternal behavior of primiparous and multiparous ewes. *Physiology & Behavior, 57*(1), 97–104.

Levy, F., Melo, A. I., Galef, B. G., Jr., Madden, M., & Fleming, A. S. (2003). Complete maternal deprivation affects social, but not spatial, learning in adult rats. *Developmental Psychobiology, 43*(3), 177–191.

Levy, F., Porter, R. H., Kendrick, K. M., Keverne, E. B., & Romeyer, A. (1996). Physiological, sensory, and experiential factors of parental care in sheep. *Advances in the Study of Behavior, 25*, 385–422.

Lewis, G. J., Kanai, R., Rees, G., & Bates, T. C. (2014). Neural correlates of the "good life": Eudaimonic well-being is associated with insular cortex volume. *Social, Cognitive, and Affective Neuroscience, 9*(5), 615–618.

Lewis, S. E., & Pusey, A. E. (1997). Factors influencing the occurrence of communal care in plural breeding mammals. In N. G. Solomon & J. A. French (Eds.), *Cooperative breeding in mammals* (pp. 335–363). Cambridge, England: Cambridge University Press.

Li, J., Zhao, Y., Li, R., Broster, L. S. Zhou, C., & Yang, S. (2015). Association of oxytocin receptor gene (OXTR) re53576 polymorphism with sociality: A meta-analysis. *PloS One, 10*(6), e0131820.

Li, M., & Fleming, A. S. (2003). The nucleus accumbens shell is critical for normal expression of pup-retrieval in postpartum female rats. *Behavioural Brain Research, 145*(1–2), 99–111.

Li, S., Kim, J. H., & Richardson, R. (2012). Differential involvement of the medial prefrontal cortex in the expression of learned fear across development. *Behavioral Neuroscience, 126*(2), 217–225.

Li, T., Chen, X., Mascaro, J., Haroon, E., & Rilling, J. K. (2017). Intranasal oxytocin, but not vasopressin, augments neural responses to toddlers in human fathers. *Hormones and Behavior, 93*, 193–202.

Li, T., Horta, M., Mascoaro, J. S. Bijanki, K., Arnal, L. H. Adams, M., Barr, R. G., & Rilling, J. K. (2018). Explaining individual variation in paternal brain responses to infant cries. *Physiology & Behavior, 193*, 43–54.

Li, Y., Long, Z., Cao, D., & Cao, F. (2017). Maternal history of child maltreatment and maternal depression risk in the perinatal period: A longitudinal study. *Child Abuse & Neglect, 63*, 192–201.

Lim, M. M., & Young, L. J. (2004). Vasopressin-dependent neural circuits underlying pair bond formation in the monogamous prairie vole. *Neuroscience, 125*(1), 35–45.

Liu, D., Diorio, J., Tannenbaum, B., Caldji, C., Francis, D., Freedman, A., . . . Meaney, M. J. (1997). Maternal care, hippocampal glucocorticoid receptors, and hypothalamic-pituitary-adrenal responses to stress. *Science, 277*(5332), 1659–1662.

Liu, H., Lopatina, O., Higashida, C., Fugimoto, H., Akther, S., Inzhutova, A., . . . Higashida, H. (2013). Displays of paternal mouse pup retrieval following communicative interaction with maternal mates. *Nature Communications, 4*, 1346.

Liu, Y., & Wang, Z. X. (2003). Nucleus accumbens oxytocin and dopamine interact to regulate pair bond formation in female prairie voles. *Neuroscience, 121*(3), 537–544.

Lockwood, P. L., (2016). The anatomy of empathy: Vicarious experience and disorders of social cognition. *Behavioural Brain Research, 311*, 255–266.

Lodge, D. J., & Grace, A. A. (2006). The laterodorsal tegmentum is essential for burst firing of ventral tegmental area dopamine neurons. *Proceedings of the National Academy of Sciences USA, 103*(13), 5167–5172.

Lomanowska, A. M., Boivin, M., Hertzman, C., & Fleming, A. S. (2017). Parenting begets parenting: A neurobiological perspective on early adversity and the transmission of parenting styles across generations. *Neuroscience, 342*, 120–139.

Lonstein, J. S. (2002). Effects of dopamine receptor antagonism with haloperidol on nurturing behavior in the biparental prairie vole. *Pharmacology, Biochemistry, and Behavior, 74*(1), 11–19.

Lonstein, J. S. (2005). Reduced anxiety in postpartum rats requires recent physical interactions with pups, but is independent of suckling and peripheral sources of hormones. *Hormones and Behavior, 47*(3), 241–255.

Lonstein, J. S. (2007). Regulation of anxiety during the postpartum period. *Frontiers in Neuroendocrinology, 28*(2–3), 115–141.

Lonstein, J. S., & De Vries, G. J. (1999). Sex differences in the parental behaviour of adult virgin prairie voles: Independence from gonadal hormones and vasopressin. *Journal of Neuroendocrinology, 11*(6), 441–449.

Lonstein, J. S., & De Vries, G. J. (2000). Maternal behaviour in lactating rats stimulates c-fos in glutamate decarboxylase-synthesizing neurons of the medial preoptic area, ventral bed nucleus of the stria terminalis, and ventrocaudal periaqueductal gray. *Neuroscience, 100*(3), 557–568.

Lonstein, J. S., & De Vries, G. J. (2001). Social influences on parental and nonparental responses toward pups in virgin female prairie voles (*Microtus ochrogaster*). *Journal of Comparative Psychology, 115*(1), 53–61.

Lonstein, J. S., & Gammie, S. C. (2002). Sensory, hormonal, and neural control of maternal aggression in laboratory rodents. *Neuroscience and Biobehavioral Reviews, 26*(8), 869–888.

Lonstein, J. S., Greco, B., De Vries, G. J., Stern, J. M., & Blaustein, J. D. (2000). Maternal behavior stimulates c-fos activity within estrogen receptor alpha-containing neurons in lactating rats. *Neuroendocrinology, 72*(2), 91–101.

Lonstein, J. S., Levy, F., & Fleming, A. S. (2015). Common and divergent psychobiological mechanisms underlying maternal behaviors in non-human and human mammals. *Hormones and Behavior, 73*, 156–185.

Lonstein, J. S., Pereira, M., Morrell, J. I., & Marler, C. A. (2015). Parenting behavior. In T. M. Plant & A. J. Zeleznik (Eds.), *Knobil and Neill's physiology of reproduction* (4th ed., pp. 2371–2437). San Diego, CA: Academic Press.

Lonstein, J. S., Simmons, D. A., & Stern, J. M. (1998). Functions of the caudal periaqueductal gray in lactating rats: Kyphosis, lordosis, maternal aggression, and fearfulness. *Behavioral Neuroscience, 112*(6), 1502–1518.

Lonstein, J. S., Simmons, D. A., Swann, J. M., & Stern, J. M. (1998). Forebrain expression of c-Fos due to active maternal behaviour in lactating rats. *Neuroscience, 82*(1), 267–281.

Lonstein, J. S., & Stern, J. M. (1997a). Role of the midbrain periaqueductal gray in maternal nurturance and aggression: c-Fos and electrolytic lesion studies in lactating rats. *Journal of Neuroscience, 17*(9), 3364–3378.

Lonstein, J. S., & Stern, J. M. (1997b). Somatosensory contributions to c-fos activation within the caudal periaqueductal gray of lactating rats: Effects of perioral, rooting, and suckling stimuli from pups. *Hormones and Behavior, 32*(3), 155–166.

Lonstein, J. S., Wagner, C. K., & De Vries, G. J. (1999). Comparison of the "nursing" and other parental behaviors of nulliparous and lactating female rats. *Hormones and Behavior, 36*(3), 242–251.

Loup, F., Tribollet, E., Dubois-Dauphin, M., & Dreifuss, J. J. (1991). Localization of high-affinity binding sites for oxytocin and vasopressin in the human brain. An autoradiographic study. *Brain Research, 555*(2), 220–232.

Loup, F., Tribollet, E., Dubois-Dauphin, M., Pizzolato, G., & Dreifuss, J. J. (1989). Localization of oxytocin binding sites in the human brainstem and upper spinal cord: An autoradiographic study. *Brain Res, 500*(1–2), 223–230.

Lovejoy, M. C., Graczyk, P. A., O'Hare, E., & Neuman, G. (2000). Maternal depression and parenting behavior: A meta-analytic review. *Clinical Psychology Review, 20*(5), 561–592.

Lovic, V., & Fleming, A. S. (2004). Artificially-reared female rats show reduced repulse inhibition and deficits in the attentional set shifting task: Reversal of effects with maternal-like licking stimulation. *Behavioural Brain Research, 148*(1–2), 209–219.

Lovic, V., Gonzalez, A., & Fleming, A. S. (2001). Maternally separated rats show deficits in maternal care in adulthood. *Developmental Psychobiology, 39*(1), 19–33.

Lubin, M., Leon, M., Moltz, H., & Numan, M. (1972). Hormones and maternal behavior in the male rat. *Hormones and Behavior, 3*(4), 369–374.

Lucas, B. K., Ormandy, C. J., Binart, N., Bridges, R. S., & Kelly, P. A. (1998). Null mutation of the prolactin receptor gene produces a defect in maternal behavior. *Endocrinology, 139*(10), 4102–4107.

Ludmer, J. A., Gonzalez, A., Kennedy, J., Masellis, M., Meinz, P., & Atkinson, L. (2018). Association between maternal childhood maltreatment and mother–infant attachment disorganization: Moderation by maternal oxytocin receptor gene and cortisol secretion. *Hormones and Behavior, 102*, 23–33.

Lukas, D., & Clutton-Brock, T. H. (2013). The evolution of social monogamy in mammals. *Science, 341*(6145), 526–530.

Lukkes, J. L., Forster, G. L., Renner, K. J., & Summers, C. H. (2008). Corticotropin-releasing factor 1 and 2 receptors in the dorsal raphe differentially affect serotonin release in the nucleus accumbens. *European Journal of Pharmacology, 578*(2-3), 185–193.

Lutz, P. E., Gross, J. A., Dhir, S. K., Maussion, G., Yang, J., Bramoulle, A., . . . Turecki, G. (2018). Epigenetic regulation of the kappa opioid receptor by child abuse. *Biological Psychiatry, 84*(10), 751–761.

Lyilikci, O., Balthazart, J., & Ball, G. F. (2017). Medial preoptic regulation of the ventral tegmental area related to the control of sociosexual behaviors. eNeuro, *3*(6).

Lynch, K. S., O'Connell, L. A., Louder, M. I. M., Balakrishnan, C. N., & Fischer, E. K. (2019). Understanding the loss of maternal care in avian brood parasites using preoptic area transcriptome comparisons in brood parasitic and non-parasitic blackbirds. *G3 Genes, Genomes, Genetics, 9*, 1075–1084.

Lynn, S. E. (2016). Endocrine and neuroendocrine regulation of fathering behavior in birds. *Hormones and Behavior, 77*, 237–248.

Macbeth, A. H., Stepp, J. E., Lee, H. J., Young, W. S., III, & Caldwell, H. K. (2010). Normal maternal behavior, but increased pup mortality, in conditional oxytocin receptor knockout females. *Behavioral Neuroscience, 124*(5), 677–685.

MacKinnon, A. L., Carter, C. S., Feeley, N., Gold, I., Hayton, B., Santhakumaran, S., & Zelkowitz, P. (2018). Theory of mind as a link between oxytocin and maternal behavior. *Psychoneuroendocrinology, 92*, 87–94.

MacKinnon, A. L., Gold, I., Feeley, N., Hayton, B., Carter, C. S., & Zelkowitz, P. (2014). The role of oxytocin in mothers' theory of mind and interactive behavior during the perinatal period. *Psychoneuroendocrinology, 48*, 52–63.

Maestripieri, D. (1994). Social structure, infant handling, and mothering styles in group living old world monkeys. *International Journal of Primatology, 15*, 531–553.

Maestripieri, D. (2005). Early experience affects the intergenerational transmission of infant abuse in rhesus monkeys. *Proceedings of the National Academy of Sciences USA, 102*(27), 9726–9729.

Maestripieri, D. (2011). Emotions, stress, and maternal motivation in primates. *American Journal of Primatology, 73*(6), 516–529.

Maestripieri, D., Hoffman, C. L., Anderson, G. M., Carter, C. S., & Higley, J. D. (2009). Mother-infant interactions in free-ranging rhesus macaques: Relationships between physiological and behavioral variables. *Physiology & Behavior, 96*(4-5), 613–619.

Maestripieri, D., Lindell, S. G., & Higley, J. D. (2007). Intergenerational transmission of maternal behavior in rhesus macaques and its underlying mechanisms. *Developmental Psychobiology, 49*(2), 165–171.

Mah, B. L., Bakermans-Kranenburg, M. J., van Ijzendoorn, M. H., & Smith, R. (2015). Oxytocin promotes protective behavior in depressed mothers: A pilot study with the enthusiastic stranger paradigm. *Depression and Anxiety, 32*(2), 76–81.

Mah, B. L., van Ijzendoorn, M. H., Smith, R., & Bakermans-Kranenburg, M. J. (2013). Oxytocin in postnatally depressed mothers: Its influence on mood and expressed emotion. *Progress in Neuropsychopharmacology & Biological Psychiatry, 40*, 267–272.

Makaris, N., Swaab, D. F., van der Kouwe, A., Abbs, B., Boriel, D., Handa, R. J., . . . Goldstein, J. M. (2013). Volumetric parcellation methodology of the human hypothalamus in neuroimaging: Normative data and sex differences. *Neuroimage, 69*, 1–10.

Malter Cohen, M., Jing, D. Yang, R. R., Tottenham, N., Lee, F. S., & Casey, B. J. (2013). Early-life stress has persistent effects on amygdala function and development in mice and humans. *Proceedings of the National Academy of Sciences USA, 110*(45), 18274–18278.

Mani, S., & Portillo, W. (2010). Activation of progestin receptors in female reproductive behavior: Interactions with neurotransmitters. *Frontiers in Neuroendocrinology, 31*(2), 157–171.

Maninger, N., Hinde, K., Mendoza, S. P., Mason, W. A., Larke, R. H., Ragen, B. J., . . . Bales, K. L. (2017). Pair bond formation leads to a sustained increase in global cerebral glucose metabolism in monogamous male titi monkeys (*Callicebus cupreus*). *Neuroscience, 348*, 302–312.

Mantella, R. C., Vollmer, R. R., Li, X., & Amico, J. A. (2003). Female oxytocin-deficient mice display enhanced anxiety-related behavior. *Endocrinology, 144*(6), 2291–2296.

Marean, C. W. (2015). An evolutionary anthropological perspective on modern human origins. *Annual Review of Anthropology, 44*, 533–556.

Marlin, B. J., Mitre, M., D'Amour J, A., Chao, M. V., & Froemke, R. C. (2015). Oxytocin enables maternal behaviour by balancing cortical inhibition. *Nature, 520*(7548), 499–504.

Marsh, A. A. (2019). The caring continuum: Evolved hormonal and proximal mechanisms explain prosocial and antisocial extremes. *Annual Review of Psychology, 70*, 347–371.

Martinon, D., & Dabrowska, J. (2018). Corticotropin-releasing factor receptors modulate oxytocin release in the dorsolateral bed nucleus of the stria terminalis (BNST) in male rats. *Frontiers in Neuroscience, 12*, 183.

Martin-Sanchez, A., Valera-Marin, G., Hernandez-Martinez, A., Lanuza, E., Martinez-Garcia, F., & Agustin-Pavon, C. (2015). Wired for motherhood: Induction of maternal care but not maternal aggression in virgin female CD1 mice. *Frontiers in Behavioral Neuroscience, 9*, 197.

Marusak, H. A., Martin, K. R., Etkin, A., & Thomason, M. E. (2015). Childhood trauma exposure disrupts the automatic regulation of emotional processing. *Neuropsychopharmacology, 40*(5), 1250–1258.

Mascaro, J. S., Hackett, P. D., Gouzoules, H., Lori, A., & Rilling, J. K. (2014). Behavioral and genetic correlates of the neural response to infant crying among human fathers. *Social, Cognitive, and Affective Neuroscience, 9*(11), 1704–1712.

Mascaro, J. S., Hackett, P. D., & Rilling, J. K. (2013). Testicular volume is inversely correlated with nurturing-related brain activity in human fathers. *Proceedings of the National Academy of Sciences USA, 110*(39), 15746–15751.

Masten, C. L., Morelli, S. A., & Eisenberger, N. I. (2011). An fMRI investigation of empathy for "social pain" and subsequent prosocial behavior. *Neuroimage, 55*(1), 381–388.

Matthews Felton, T., Linton, L. N., Rosenblatt, J. S., & Morrell, J. I. (1999a). Estrogen implants in the lateral habenular nucleus do not stimulate the onset of maternal behavior in female rats. *Hormones and Behavior, 35*(1), 71–80.

Matthews Felton, T., Linton, L. N., Rosenblatt, J. S., & Morrell, J. I. (1999b). First and second order maternal behavior related afferents to the lateral habenula. *NeuroReport, 10*, 883–887.

Matthews-Felton, T., Corodimas, K. P., Rosenblatt, J. S., & Morrell, J. I. (1995). Lateral habenula neurons are necessary for the hormonal onset of maternal behavior and for the display of postpartum estrus in naturally parturient female rats. *Behavioral Neuroscience, 109*(6), 1172–1188.

Mattson, B. J., & Morrell, J. I. (2005). Preference for cocaine- versus pup-associated cues differentially activates neurons expressing either Fos or cocaine- and amphetamine-regulated transcript in lactating, maternal rodents. *Neuroscience, 135*(2), 315–328.

Mayer, A. D., Carter, L., Jorge, W. A., Mota, M. J., Tannu, S., & Rosenblatt, J. S. (1987). Mammary stimulation and maternal aggression in rodents: Thelectomy fails to reduce pre- or postpartum aggression in rats. *Hormones and Behavior, 21*(4), 501–510.

Mayer, A. D., Freeman, N. C., & Rosenblatt, J. S. (1979). Ontogeny of maternal behavior in the laboratory rat: Factors underlying changes in responsiveness from 30 to 90 days. *Developmental Psychobiology, 12*(5), 425–439.

Mayer, A. D., & Rosenblatt, J. S. (1984). Prepartum changes in maternal responsiveness and nest defense in Rattus norvegicus. *Journal of Comparative Psychology, 98*(2), 177–188.

Mayer, H. S., Crepeau, M., Duque-Wilckens, N., Torres, L. Y., Trainor, B. C., & Stolzenberg, D. S. (2019). Histone deacetylase inhibitor treatment promotes spontaneous caregiving behavior in non-aggressive virgin male mice. *Journal of Neuroendocrinology*, e12734.

Mayer, H. S., Helton, J., Torres, L. Y., Cortina, I., Brown, W. M., & Stolzenberg, D. S. (2019). Histone deacetylase inhibitor treatment induces postpartum-like maternal behavior and immediate early gene expression in the maternal neural pathway in virgin mice. *Hormones and Behavior, 108*, 94–104.

McCarthy, M. M. (1990). Oxytocin inhibits infanticide in female house mice (*Mus domesticus*). *Hormones and Behavior, 24*(3), 365–375.

McCarthy, M. M., & vom Saal, F. S. (1985). The influence of reproductive state on infanticide by wild female house mice (*Mus musculus*). *Physiology & Behavior, 35*(6), 843–849.

McCormack, K., Newman, T. K., Higley, J. D., Maestripieri, D., & Sanchez, M. M. (2009). Serotonin transporter gene variation, infant abuse, and responsiveness to stress in rhesus macaque mothers and infants. *Hormones and Behavior, 55*(4), 538–547.

McCoy, C. R., Glover, M. E., Flynn, L. T., Simmons, R. K., Cohen, J. L., Ptacek, T., . . . Clinton, S. M. (2019). Altered DNA methylation in the developing brains of rats genetically prone to high versus low anxiety. *Journal of Neuroscience, 39*(16), 3144–3158.

McCrory, E. J., Gerin, M. I., & Viding, E. (2017). Annual research review: Childhood maltreatment, latent vulnerability and the shift to preventative psychiatry—the contribution of functional brain imaging. *Journal of Child Psychology and Psychiatry, 58*(4), 338–357.

McDonald, A. J. (1998). Cortical pathways to the mammalian amygdala. *Progress in Neurobiology, 55*(3), 257–332.

McGowan, P. O., Sasaki, A., D'Alessio, A. C., Dymov, S., Labonte, B., Szyf, M., . . . Meaney, M. J. (2009). Epigenetic regulation of the glucocorticoid receptor in human brain associates with childhood abuse. *Nature Neuroscience, 12*(3), 342–348.

McHenry, J. A., Otis, J. M., Rossi, M. A., Robinson, J. E., Kosyk, O., Miller, N. W., . . . Stuber, G. D. (2017). Hormonal gain control of a medial preoptic area social reward circuit. *Nature Neuroscience, 20*(3), 449–458.

McHenry, J. A., Rubinow, D. R., & Stuber, G. D. (2015). Maternally responsive neurons in the bed nucleus of the stria terminalis and medial preoptic area: Putative circuits for regulating anxiety and reward. *Frontiers in Neuroendocrinology, 38*, 65–72.

McLaughlin, K. A., Peverill, M., Gold, A. L., Alves, S., & Sheridan, M. A. (2015). Child maltreatment and neural systems underlying emotion regulation. *Journal of the American Academy of Child and Adolescent Psychiatry, 54*(9), 753–762.

Meaney, M. J., Aitken, D. H., Bodnoff, S. R., Iny, L. J., Tatarewicz, J. E., & Sapolsky, R. M. (1985). Early postnatal handling alters glucocorticoid receptor concentrations in selected brain regions. *Behavioral Neuroscience, 99*(4), 765–770.

Meddle, S. L., Bishop, V. R., Gkoumassi, E., van Leeuwen, F. W., & Douglas, A. J. (2007). Dynamic changes in oxytocin receptor expression and activation at parturition in the rat brain. *Endocrinology, 148*(10), 5095–5104.

Mehta, D., Quast, C., Fasching, P. A., Seifert, A., Voigt, F. Beckmann, M. W., . . . Goecke, T. W. (2012). The 5-HTTLPR polymorphism modulated the influence of environmental stressors on peripartum depression symptoms. *Journal of Affective Disorders, 136*(3), 1192–1197.

Melon, L. C., Hooper, A., Yang, X., Moss, S. J., & Maguire, J. (2018). Inability to suppress the stress-induced activation of the HPA axis during the peripartum period engenders deficits in postpartum behaviors in mice. *Psychoneuroendocrinology, 90*, 182–193.

Meltzer-Brody, S., Stuebe, A., Dole, N., Savitz, D., Rubinow, D., & Thorp, J. (2011). Elevated corticotropin releasing hormone (CRH) during pregnancy and risk of postpartum depression (PPD). *Journal of Clinical Endocrinology & Metabolism, 96*(1), E40–E47.

Mennella, J. A., & Moltz, H. (1988a). Infanticide in the male rat: The role of the vomeronasal organ. *Physiology & Behavior, 42*(3), 303–306.

Mennella, J. A., & Moltz, H. (1988b). Infanticide in rats: Male strategy and female counter-strategy. *Physiology & Behavior, 42*(1), 19–28.

Mennigen, J. A., Volkoff, H., Chang, J. P., & Trudeau, V. L. (2017). The nonapeptide isotocin in goldfish: Evidence for serotonergic regulation and functional roles in the control of food intake and pituitary hormone release. *General and Comparative Endocrinology, 254*, 38–49.

Menon, R., Grund, T., Zoicas, I., Althammer, F., Fiedler, D., Biermeier, V., . . . Neumann, I. D. (2018). Oxytocin signaling in the lateral septum prevents social fear during lactation. *Current Biology, 28*(7), 1066–1078.

Meyer, M. L., Masten, C. L., Ma, Y., Wang, C., Shi, Z., Eisenberger, N. I., & Han, S. (2013). Empathy for social suffering of friends and strangers recruits distinct patterns of brain activation. *Social, Cognitive, and Affective Neuroscience, 8*(4), 446–454.

Meyer-Lindenberg, A., Domes, G., Kirsch, P., & Heinrichs, M. (2011). Oxytocin and vasopressin in the human brain: Social neuropeptides for translational medicine. *Nature Reviews Neuroscience, 12*(9), 524–538.

Mielke, E. L., Neukel, C., Bertsch, K., Reck, C., Mohler, E., & Herpertz, S. C. (2016). Maternal sensitivity and the empathic brain: Influences of early life maltreatment. *Journal of Psychiatric Research, 77*, 59–66.

Mikhailova, M. A., Bass, C. E., Grinevich, V. P., Chappell, A. M., Deal, A. L., Bonin, K. D., . . . Budygin, E. A. (2016). Optogenetically-induced tonic dopamine release from VTA-nucleus accumbens projections inhibits reward consummatory behaviors. *Neuroscience, 333*, 54–64.

Milad, M. R., & Quirk, G. J. (2012). Fear extinction as a model for translational neuroscience: Ten years of progress. *Annual Review of Psychology, 63*, 129–151.

Mileva-Seitz, V., Kennedy, J., Atkinson, L., Steiner, M., Levitan, R., Matthews, S. G., . . . Fleming, A. S. (2011). Serotonin transporter allelic variation in mothers predicts maternal sensitivity, behavior and attitudes toward 6-month-old infants. *Genes, Brain, and Behavior, 10*(3), 325–333.

Miller, S. M., & Lonstein, J. S. (2005). Dopamine d1 and d2 receptor antagonism in the preoptic area produces different effects on maternal behavior in lactating rats. *Behavioral Neuroscience, 119*(4), 1072–1083.

Miller, S. M., Piasecki, C. C., Peabody, M. F., & Lonstein, J. S. (2010). GABA(A) receptor antagonism in the ventrocaudal periaqueductal gray increases anxiety in the anxiety-resistant postpartum rat. *Pharmacology, Biochemistry, & Behavior, 95*(4), 457–465.

Missale, C., Nash, S. R., Robinson, S. W., Jaber, M., & Caron, M. G. (1998). Dopamine receptors: From structure to function. *Physiological Reviews, 78*(1), 189–225.

Mitani, J. C. (2009). Cooperation and competition in chimpanzees: Current understanding and future challenges. *Evolutionary Anthropology, 18*, 215–227.

Mitre, M., Marlin, B. J., Schiavo, J. K., Morina, E., Norden, S. E., Hackett, T. A., . . . Froemke, R. C. (2016). A distributed network for social cognition enriched for oxytocin receptors. *Journal of Neuroscience, 36*(8), 2517–2535.

Modi, M. E., Inoue, K., Barrett, C. E., Kittelberger, K. A., Smithy, D. G., Landgraf, R., & Young, L. J. (2015). Melanocortin receptor agonists facilitate oxytocin-dependent partner preference formation in the prairie vole. *Neuropsychopharmacology, 40*(8), 1856–1865.

Moffitt, J. R., Bambah-Mukku, D., Eichhorn, S. W., Vaughn, E, Shekhar, K., Perez, J. D., . . . Zhuang, X. (2018). Molecular, spatial, and functional single-cell profiling of the hypothalamic preoptic region. *Science, 362*(6416), eaau5324.

Molenberghs, P., Johnson, H., Henry, J. D., & Mattingley, J. B. (2016). Understanding the minds of others: A neuroimaging meta-analysis. *Neuroscience and Biobehavioral Reviews, 65*, 276–291.

Moll, J., Bado, P., de Oliveira-Souza, R., Bramati, I. E., Lima, D. O., Paiva, F. F., . . . Zahn, R. (2012). A neural signature of affiliative emotion in the human septohypothalamic area. *Journal of Neuroscience, 32*(36), 12499–12505.

Moll, J., Krueger, F., Zahn, R., Pardini, M., de Oliveira-Souza, R., & Grafman, J. (2006). Human fronto-mesolimbic networks guide decisions about charitable donation. *Proceedings of the National Academy of Sciences USA, 103*(42), 15623–15628.

Moltz, H., Levin, R., & Leon, M. (1969). Differential effects of progesterone on the maternal behavior of primiparous and multiparous rats. *Journal of Comparative and Physiological Psychology, 67*(1), 36–40.

Moltz, H., Lubin, M., Leon, M., & Numan, M. (1970). Hormonal induction of maternal behavior in the ovariectomized nulliparous rat. *Physiology & Behavior, 5*(12), 1373–1377.

Moltz, H., & Weiner, E. (1966). Effects of ovariectomy on maternal behavior of primiparous and multiparous rats. *Journal of Comparative and Physiological Psychology, 62*, 382–387.

Mooney, S. J., Douglas, N. R., & Holmes, M. M. (2014). Peripheral administration of oxytocin increases social affiliation in the naked mole-rat (*Heterocephalus glaber*). *Hormones and Behavior, 65*(4), 380–385.

Mooney, S. J., & Holmes, M. M. (2013). Social condition and oxytocin neuron number in the hypothalamus of naked mole-rats (*Heterocephalus glaber*). *Neuroscience, 230*, 56–61.

Morelli, S. A., Rameson, L. T., & Lieberman, M. D. (2014). The neural components of empathy: Predicting daily prosocial behavior. *Social, Cognitive, and Affective Neuroscience, 9*(1), 39–47.

Morelli, S. A., Sacchet, M. D., & Zaki, J. (2015). Common and distinct neural correlates of personal and vicarious reward: A quantitative meta-analysis. *Neuroimage, 112*, 244–253.

Morgan, H. D., Fleming, A. S., & Stern, J. M. (1992). Somatosensory control of the onset and retention of maternal responsiveness in primiparous Sprague–Dawley rats. *Physiology & Behavior, 51*(3), 549–555.

Morgan, J. I., & Curran, T. (1991). Stimulus-transcription coupling in the nervous system: Involvement of the inducible proto-oncogenes fos and jun. *Annual Review of Neuroscience, 14*, 421–451.

Moriceau, S., Shionoya, K., Jakubs, K., & Sullivan, R. M. (2009). Early-life stress disrupts attachment learning: The role of amygdala corticosterone, locus ceruleus corticotropin releasing hormone, and olfactory bulb norepinephrine. *Journal of Neuroscience, 29*(50), 15745–15755.

Moriceau, S., Wilson, D. A., Levine, S., & Sullivan, R. M. (2006). Dual circuitry for odor-shock conditioning during infancy: Corticosterone switches between fear and attraction via amygdala. *Journal of Neuroscience, 26*(25), 6737–6748.

Mos, J., H. C. M. Lammers, J., M. van der Poel, A., Bermond, B., Meelis, W., & Kruk, M. R. (1983). Effects of midbrain central gray lesions on spontaneous and electrically induced aggression in the rat. *Aggressive Behavior, 9*(2), 133–155.

Motta, S. C., Guimaraes, C. C., Furigo, I. C., Sukikara, M. H., Baldo, M. V., Lonstein, J. S., & Canteras, N. S. (2013). Ventral premammillary nucleus as a critical sensory relay to the maternal aggression network. *Proceedings of the National Academy of Sciences USA, 110*(35), 14438–14443.

Muroi, Y., & Ishii, T. (2019). Glutamatergic signals in the dorsal raphe nucleus regulate maternal aggression and care in an opposing manner in mice. *Neuroscience, 400*, 33–47.

Musser, E., Laurent, H., & Ablow, J. (2012). The neural correlates of maternal sensitivity: An fMRI study. *Developmental Cognitive Neuroscience, 2*(4), 428–436.

Muzik, M., Bocknek, E. L., Broderick, A., Richardson, P., Rosenblum, K. L., Thelen, K., & Seng, J. S. (2013). Mother–infant bonding impairment across the first 6 months postpartum: The primacy of psychopathology in women with childhood abuse and neglect histories. *Archives of Women's Mental Health, 16*(1), 29–38.

Myers, B., McKlveen, J. M., & Herman, J. P. (2014). Glucocorticoid actions on synapses, circuits, and behavior: Implications for the energetics of stress. *Frontiers in Neuroendocrinology, 35*(2), 180–196.

Mykowycz, R., & Dudzinski, M. L. (1972). Aggressive and protective behaviour of adult rabbits Oryctolagus cuniculus towards juveniles. *Behaviour, 43*(1), 97–120.

Naber, F., van Ijzendoorn, M. H., Deschamps, P., van Engeland, H., & Bakermans-Kranenburg, M. J. (2010). Intranasal oxytocin increases fathers' observed responsiveness during play with their children: A double-blind within-subject experiment. *Psychoneuroendocrinology, 35*(10), 1583–1586.

Nagano, Y. Kaneda, K., Maruyama, C., Ide, S., Kato, F., & Minami, M. (2015). Corticotropin-releasing factor enhances inhibitory synaptic transmission to type III neurons in the bed nucleus of the stria terminalis. *Neuroscience Letters, 600*, 56–61.

Nave, G., Camerer, C., & McCullough, M. (2015). Does oxytocin increase trust in humans? A critical review of research. *Perspectives in Psychological Science, 10*(6), 772–789.

Ne'eman, R., Perach-Barzilay, N., Fischer-Shofty, Atias, A., & Shamay-Tsoory, S. G. (2016). Intranasal administration of oxytocin increases human aggressive behavior. *Hormones and Behavior, 80*, 125–131.

Neill, J. D., & Nagy, G. M. (1994). Prolactin secretion and its control. In E. Knobil & J. Neill (Eds.), *Physiology of reproduction* (Vol. 1, pp. 1833–1860). New York, NY: Raven Press.

Nelson-Coffey, S. K., Borelli, J. L., & River, L. M. (2017). Attachment avoidance, but not anxiety, minimizes the joys of caregiving. *Attachment & Human Development, 19*(5), 504–531.

Nemeroff, C. B. (2016). Paradise lost: The neurobiological and clinical consequences of child abuse and neglect. *Neuron, 89*(5), 892–909.

Nephew, B. C., & Bridges, R. S. (2011). Effects of chronic social stress during lactation on maternal behavior and growth in rats. *Stress, 14*(6), 677–684.

Nephew, B. C., Febo, M., Huang, W., Colon-Perez, L. M., Payne, L., Poirier, G. L., . . . King, J. A. (2018). Early life social stress and resting state functional connectivity in postpartum rat anterior cingulate circuits. *Journal of Affective Disorders, 229*, 213–223.

Neumann, I., Douglas, A. J., Pittman, Q. J., Russell, J. A., & Landgraf, R. (1996). Oxytocin released within the supraoptic nucleus of the rat brain by positive feedback action is involved in parturition-related events. *Journal of Neuroendocrinology, 8*(3), 227–233.

Neumann, I., Koehler, E., Landgraf, R., & Summy-Long, J. (1994). An oxytocin receptor antagonist infused into the supraoptic nucleus attenuates intranuclear and peripheral release of oxytocin during suckling in conscious rats. *Endocrinology, 134*(1), 141–148.

Neumann, I., & Landgraf, R. (1989). Septal and hippocampal release of oxytocin, but not vasopressin, in the conscious lactating rat during suckling. *Journal of Neuroendocrinology, 1*(4), 305–308.

Neumann, I. D. (2008). Brain oxytocin: A key regulator of emotional and social behaviours in both females and males. *Journal of Neuroendocrinology, 20*(6), 858–865.

Neumann, I. D., Kromer, S. A., & Bosch, O. J. (2005). Effects of psycho-social stress during pregnancy on neuroendocrine and behavioural parameters in lactation depend on the genetically determined stress vulnerability. *Psychoneuroendocrinology, 30*(8), 791–806.

Neumann, I. D., & Landgraf, R. (2012). Balance of brain oxytocin and vasopressin: Implications for anxiety, depression, and social behaviors. *Trends in Neurosciences, 35*(11), 649–659.

Neumann, I. D., Maloumby, R., Beiderbeck, D. I., Lukas, M., & Landgraf, R. (2013). Increased brain and plasma oxytocin after nasal and peripheral administration in rats and mice. *Psychoneuroendocrinology, 38*(10), 1985–1993.

Neumann, I. D., & Slattery, D. A. (2016). Oxytocin in general anxiety and social fear: A translational approach. *Biological Psychiatry, 79*(3), 213–221.

Neumann, I. D., Torner, L., & Wigger, A. (2000). Brain oxytocin: Differential inhibition of neuroendocrine stress responses and anxiety-related behaviour in virgin, pregnant and lactating rats. *Neuroscience, 95*(2), 567–575.

Nicholson, A. A., Rabellino, D., Densmore, M., Frewen, P. A., Paret, C., Kluetsch, R., . . . Lanius, R. A. (2017). The neurobiology of emotion regulation in posttraumatic stress disorder: Amygdala downregulation via real-time fMRI neurofeedback. *Human Brain Mapping, 38*(1), 541–560.

Nieuwenhuys, R. (2012). The insular cortex: A review. *Progress in Brain Research, 195*, 123–163.

Nishimori, K., Young, L. J., Guo, Q., Wang, Z., Insel, T. R., & Matzuk, M. M. (1996). Oxytocin is required for nursing but is not essential for parturition or reproductive behavior. *Proceedings of the National Academy of Sciences USA, 93*(21), 11699–11704.

Nishina, K., Takagashi, H. Takahashi, H., Sakagami, M., & Inoue-Murayama M. (2019). Association of polymorphism of arginine-vasopressin receptor 1A (AVPR1a) gene with trust and reciprocity. *Frontiers in Human Neuroscience, 13*, 230.

Nitschke, J. B., Nelson, E. E., Rusch, B. D., Fox, A. S., Oakes, T. R., & Davidson, R. J. (2004). Orbitofrontal cortex tracks positive mood in mothers viewing pictures of their newborn infants. *NeuroImage, 21*(2), 583–592.

Noonan, M., & Kristal, M. B. (1979). Effects of medial preoptic lesions on placentophagia and on the onset of maternal behavior in the rat. *Physiology & Behavior, 22*(6), 1197–1202.

Novakov, M., & Fleming, A. S. (2005). The effects of early rearing environment on the hormonal induction of maternal behavior in virgin rats. *Hormones and Behavior, 48*(5), 528–536.

Nowak, R., Keller, M., & Levy, F. (2011). Mother–young relationships in sheep: A model for a multidisciplinary approach of the study of attachment in mammals. *Journal of Neuroendocrinology, 23*(11), 1042–1053.

Nowak, R., Keller, M., Val-Laillet, D., & Levy, F. (2007). Perinatal visceral events and brain mechanisms involved in the development of mother–young bonding in sheep. *Hormones and Behavior, 52*(1), 92–98.

Nowicki, J. P., Pratchett, M. S., Walker, S. P. W., Coker, D. J., & O'Connell, L. A. (2017, November 6). Neurobiology of pair bonding in fishes; convergence of neural mechanisms across distant vertebrate lineages (2017). *bioRxiv.* Retrieved from https://www.biorxiv.org/content/10.1101/214759v1

Numan, M. (1974). Medial preoptic area and maternal behavior in the female rat. *Journal of Comparative and Physiological Psychology, 87*(4), 746–759.

Numan, M. (1978). Progesterone inhibition of maternal behavior in the rat. *Hormones and Behavior, 11*(2), 209–231.

Numan, M. (1985). Brain mechanisms and parental behavior. In N. Adler, D. Pfaff, & R. W. Goy (Eds.), *Handbook of behavioral neurobiology: Vol. 7, Reproduction* (pp. 537–605). New York, NY: Plenum Press.

Numan, M. (1990). Long-term effects of preoptic area knife cuts on the maternal behavior of postpartum rats. *Behavioral and Neural Biology, 53*(2), 284–290.

Numan, M. (1994). Maternal behavior. In E. Knobil & J. Neill (Eds.), *Physiology of Reproduction* (Vol. 2, pp. 221–302). New York, NY: Raven Press.

Numan, M. (2006). Hypothalamic neural circuits regulating maternal responsiveness toward infants. *Behavioral and Cognitive Neuroscience Reviews, 5*(4), 163–190.

Numan, M. (2010). Parental behavior. In G. Koob, M. Le Moal, & R. F. Thompson (Eds.), *Encyclopedia of behavioral neuroscience* (Vol. 3, pp. 14–23). Oxford, England: Elsevier.

Numan, M. (2012a). Maternal behavior: Neural circuits, stimulus valence, and motivational processes. *Parenting, 12*(2), 105–114.

Numan, M. (2012b). Neural circuits regulating maternal behavior: Implications for understanding the neural basis of social cooperation and competition. In S. L. Brown, R. M. Brown, & L. A. Penner (Eds.), *Moving beyond self-interest: Perspectives from evolutionary biology, neuroscience, and the social sciences* (pp. 89–108). New York, NY: Oxford University Press.

Numan, M. (2015). *Neurobiology of social behavior: Toward an understanding of the prosocial and antisocial brain.* Amsterdam, The Netherlands: Elsevier.

Numan, M. (2017). Parental behavior. In *Reference module in neuroscience and biobehavioral psychology* (pp. 1–15). Amsterdam, The Netherlands: Elsevier. ISBN 9780128093245.

Numan, M. (2018). Maternal bonding. In T. K. Schackelford & V. A. Weekes-Schackelford (Eds.), *Encyclopedia of evolutionary psychological science* (pp. 1–5). New York, NY: Springer.

Numan, M., Bress, J. A., Ranker, L. R., Gary, A. J., Denicola, A. L., Bettis, J. K., & Knapp, S. E. (2010). The importance of the basolateral/basomedial amygdala for goal-directed maternal responses in postpartum rats. *Behavioural Brain Research, 214*(2), 368–376.

Numan, M., & Callahan, E. C. (1980). The connections of the medial preoptic region and maternal behavior in the rat. *Physiology & Behavior, 25*(5), 653–665.

Numan, M., & Corodimas, K. P. (1985). The effects of paraventricular hypothalamic lesions on maternal behavior in rats. *Physiology & Behavior, 35*(3), 417–425.

Numan, M., Corodimas, K. P., Numan, M. J., Factor, E. M., & Piers, W. D. (1988). Axon-sparing lesions of the preoptic region and substantia innominata disrupt maternal behavior in rats. *Behavioral Neuroscience, 102*(3), 381–396.

Numan, M., Fleming, A. S., & Levy, F. (2006). Maternal Behavior. In J. D. Neill (Ed.), *Knobil and Neill's physiology of reproduction* (3rd ed., Vol. 2, pp. 1921–1993). St. Louis, MO: Academic Press.

Numan, M., & Insel, T. R. (2003). *The neurobiology of parental behavior.* New York, NY: Springer.

Numan, M., Leon, M., & Moltz, H. (1972). Interference with prolactin release and the maternal behavior of female rats. *Hormones and Behavior, 3*(1), 29–38.

Numan, M., McSparren, J., & Numan, M. J. (1990). Dorsolateral connections of the medial preoptic area and maternal behavior in rats. *Behavioral Neuroscience, 104*(6), 964–979.

Numan, M., & Numan, M. (1996). A lesion and neuroanatomical tract-tracing analysis of the role of the bed nucleus of the stria terminalis in retrieval behavior and other aspects of maternal responsiveness in rats. *Developmental Psychobiology, 29*(1), 23–51.

Numan, M., & Numan, M. J. (1991). Preoptic-brainstem connections and maternal behavior in rats. *Behavioral Neuroscience, 105*(6), 1013–1029.

Numan, M., & Numan, M. J. (1994). Expression of Fos-like immunoreactivity in the preoptic area of maternally behaving virgin and postpartum rats. *Behavioral Neuroscience, 108*(2), 379–394.

Numan, M., & Numan, M. J. (1995). Importance of pup-related sensory inputs and maternal performance for the expression of Fos-like immunoreactivity in the preoptic area and ventral bed nucleus of the stria terminalis of postpartum rats. *Behavioral Neuroscience, 109*, 135–149.

Numan, M., & Numan, M. J. (1997). Projection sites of medial preoptic area and ventral bed nucleus of the stria terminalis neurons that express Fos during maternal behavior in female rats. *Journal of Neuroendocrinology, 9*, 369–384.

Numan, M., Numan, M. J., & English, J. B. (1993). Excitotoxic amino acid injections into the medial amygdala facilitate maternal behavior in virgin female rats. *Hormones and Behavior, 27*(1), 56–81.

Numan, M., Numan, M. J., Pliakou, N., Stolzenberg, D. S., Mullins, O. J., Murphy, J. M., & Smith, C. D. (2005). The effects of D1 or D2 dopamine receptor antagonism in the medial preoptic area, ventral pallidum, or nucleus accumbens on the maternal retrieval response and other aspects of maternal behavior in rats. *Behavioral Neuroscience, 119*(6), 1588–1604.

Numan, M., Numan, M. J., Schwarz, J. M., Neuner, C. M., Flood, T. F., & Smith, C. D. (2005). Medial preoptic area interactions with the nucleus accumbens-ventral pallidum circuit and maternal behavior in rats. *Behavioural Brain Research, 158*(1), 53–68.

Numan, M., Roach, J. K., del Cerro, M. C., Guillamon, A., Segovia, S., Sheehan, T. P., & Numan, M. J. (1999). Expression of intracellular progesterone receptors in rat brain during different reproductive states, and involvement in maternal behavior. *Brain Research, 830*(2), 358–371.

Numan, M., Rosenblatt, J. S., & Komisaruk, B. (1977). The medial preoptic area and the onset of maternal behavior in the rat. *Journal of Comparative and Physiological Psychology, 91*(1), 146–64.

Numan, M., & Smith, H. G. (1984). Maternal behavior in rats: Evidence for the involvement of preoptic projections to the ventral tegmental area. *Behavioral Neuroscience, 98*(4), 712–727.

Numan, M., & Stolzenberg, D. S. (2009). Medial preoptic area interactions with dopamine neural systems in the control of the onset and maintenance of maternal behavior in rats. *Frontiers in Neuroendocrinology, 30*(1), 46–64.

Numan, M., Stolzenberg, D. S., Dellevigne, A. A., Correnti, C. M., & Numan, M. J. (2009). Temporary inactivation of ventral tegmental area neurons with either muscimol or baclofen reversibly disrupts maternal behavior in rats through different underlying mechanisms. *Behavioral Neuroscience, 123*(4), 740–751.

Numan, M., & Woodside, B. (2010). Maternity: Neural mechanisms, motivational processes, and physiological adaptations. *Behavioral Neuroscience, 124*(6), 715–741.

Numan, M., & Young, L. J. (2016). Neural mechanisms of mother–infant bonding and pair bonding: Similarities, differences, and broader implications. *Hormones and Behavior, 77*, 98–112.

Nusbaum, M. P., Blitz, D. M., & Marder, E. (2017). Functional consequences of neuropeptide and small-molecule co-transmission. *Nature Reviews Neuroscience, 18*(7), 389–403.

O'Connell, L. A., Matthews, B. J., & Hofmann, H. A. (2012). Isotocin regulates paternal care in a monogamous cichlid fish. *Hormones and Behavior, 61*(5), 725–733.

O'Donnell, K. J., Chen, L., MacIsaac, J. L., McEwen, L. M., Nguyen, T., Beckmann, K., . . . Meaney, M. J. (2018). DNA methylome variation in a perinatal nurse-visitation program that reduces child maltreatment: A 27-year follow-up. *Translational Psychiatry, 8*(1), 15.

Oettl, L. L., Ravi, N., Schneider, M., Scheller, M. F., Schneider, P., Mitre, M., . . . Kelsch, W. (2016). Oxytocin enhances social recognition by modulating cortical control of early olfactory processing. *Neuron, 90*(3), 609–621.

Ogawa, S. Eng, V., Taylor, J., Lubahn, D. B., Korach, K. S., & Pfaff, D. W. (1998). Roles of estrogen receptor-alpha gene expression in reproduction-related behaviors in female mice. *Endocrinology, 139*(12), 5070–5081.

O'Higgins, M., Roberts, I. S., Glover, V., & Taylor, A. (2013). Mother-child bonding at 1 year: Associations with symptoms of postnatal depression and bonding in the first few weeks. *Archives of Women's Mental Health, 16*(5), 381–389.

Okabe, S., Nagasawa, M., Kihara, T., Kato, M., Harada, T., Koshida, N., . . . Kikusui, T. (2013). Pup odor and ultrasonic vocalizations synergistically stimulate maternal attention in mice. *Behavavioral Neuroscience, 127*(3), 432–438.

Okabe, S., Tsuneoka, Y., Takahashi, A., Ooyama, R., Watarai, A., Maeda, S., . . . Kikusui, T. (2017). Pup exposure facilitates retrieving behavior via the oxytocin neural system in female mice. *Psychoneuroendocrinology, 79*, 20–30.

Okon, E. E. (1972). Factors affecting ultrasound production in rodents. *Journal of Zoology, 168*, 139–148.

Olazabal, D. E. (2014). Comparative analysis of oxytocin receptor density in the nucleus accumbens: An adaptation for female and male alloparental care? *Journal of Physiology (Paris), 108*(2–3), 213–220.

Olazabal, D. E. (2018). Role of oxytocin in parental behaviour. *Journal of Neuroendocrinology, 30*(7), e12594.

Olazabal, D. E., & Alsina-Llanes, M. (2016). Are age and sex differences in brain oxytocin receptors related to maternal and infanticidal behavior in naive mice? *Hormones and Behavior, 77*, 132–140.

Olazabal, D. E., Pereira, M., Agrati, D., Ferreira, A., Fleming, A. S., Gonzalez-Mariscal, G., . . . Uriarte, N. (2013). New theoretical and experimental approaches on maternal motivation in mammals. *Neuroscience and Biobehavioral Reviews, 37*(8), 1860–1874.

Olazabal, D. E., & Young, L. J. (2005). Variability in "spontaneous" maternal behavior is associated with anxiety-like behavior and affiliation in naive juvenile and adult female prairie voles (*Microtus ochrogaster*). *Developmental Psychobiology, 47*(2), 166–178.

Olazabal, D. E., & Young, L. J. (2006a). Oxytocin receptors in the nucleus accumbens facilitate "spontaneous" maternal behavior in adult female prairie voles. *Neuroscience, 141*(2), 559–568.

Olazabal, D. E., & Young, L. J. (2006b). Species and individual differences in juvenile female alloparental care are associated with oxytocin receptor density in the striatum and the lateral septum. *Hormones and Behavior, 49*(5), 681–687.

Olazabal, D. E., & Young, L. J. (2008). Oxytocin and individual variation in parental care in prairie voles. In R. S. Bridges (Ed.), *Neurobiology of the Parental Brain* (pp. 333–345). Boston, MA: Academic Press.

Olesen, K. M., & Auger, A. P. (2008). Dopaminergic activation of estrogen receptors induces fos expression with restricted regions of the neonatal female rat brain. *PloS One, 3*(5), e2177.

Omelchenko, N., Bell, R., & Sesack, S. R. (2009). Lateral habenula projections to dopamine and GABA neurons in the rat ventral tegmental area. *European Journal of Neuroscience, 30*(7), 1239–1250.

Omelchenko, N., & Sesack, S. R. (2005). Laterodorsal tegmental projections to identified cell populations in the rat ventral tegmental area. *Journal of Comparative Neurology, 483*(2), 217–235.

Ongur, D., An, X., & Price, J. L. (1998). Prefrontal cortical projections to the hypothalamus in macaque monkeys. *Journal of Comparative Neurology, 401*(4), 480–505.

Ongur, D., & Price, J. L. (2000). The organization of networks within the orbital and medial prefrontal cortex of rats, monkeys and humans. *Cerebral Cortex, 10*(3), 206–219.

Ormandy, C. J., Camus, A., Barra, J., Damotte, D., Lucas, B., Buteau, H., . . . Kelly, P. A. (1997). Null mutation of the prolactin receptor gene produces multiple reproductive defects in the mouse. *Genes and Development, 11*(2), 167–178.

Orpen, B. G., & Fleming, A. S. (1987). Experience with pups sustains maternal responding in postpartum rats. *Physiology & Behavior, 40*(1), 47–54.

Ostermeyer, M. C. (1983). Maternal aggression. In R. W. Elwood (Ed.), *Parental behaviour of rodents* (pp. 151–179). New York, NY: Wiley.

Otero-Garcia, M., Agustin-Pavon, C., Lanuza, E., & Martinez-Garcia, F. (2016). Distribution of oxytocin and co-localization with arginine vasopressin in the brain of mice. *Brain Structure and Function, 221*(7), 3445–3473.

Pace, C. S., Di Folco, S., Guerriero, V., & Muzi, S. (2019). Late-adopted children grown up: A long-term longitudinal study of attachment patterns of adolescent adoptees and their adoptive parents. *Attachment and Human Development, 21*(4), 372–388.

Padoin, M. J., Cadore, L. P., Gomes, C. M., Barros, H. M., & Lucion, A. B. (2001). Long-lasting effects of neonatal stimulation on the behavior of rats. *Behavioral Neuroscience, 115*(6), 1332–1340.

Palanza, P., & Parmigiani, S. (1991). Inhibition of infanticide in male Swiss mice: Behavioral polymorphism in response to multiple mediating factors. *Physiology & Behavior, 49*(4), 797–802.

Palombit, R. A. (1996). Pair bonds in monogamous apes: A comparison of the siamang Hylobates syndactyeus and the white-handed gibbon Hylobates lar. *Behaviour, 133*(5–6), 321–356.

Paloyelis, Y., Doyle, O. M., Zelaya, F. O., Maltezos, S., Williams, S. C., Fotopoulou, A., & Howard, M. A. (2016). A spatiotemporal profile of in vivo cerebral blood flow changes following intranasal oxytocin in humans. *Biological Psychiatry, 79*(8), 693–705.

Pan, P., Fleming, A. S., Lawson, D., Jenkins, J. M., & McGowan, P. O. (2014). Within- and between-litter maternal care alter behavior and gene regulation in female offspring. *Behavioral Neuroscience, 128*(6), 736–748.

Pan, P., Lawson, D. O., Dudin, A., Vasquez, O. E., Sokolowski, M. B., Fleming, A. S., & McGowan, P. O. (2018). Both maternal care received and genotype influence stress-related phenotype in female rats. *Developmental Psychobiology, 60*(8), 889–902.

Parada, M., King, S., Li, M., & Fleming, A. S. (2008). The roles of accumbal dopamine D1 and D2 receptors in maternal memory in rats. *Behavioral Neuroscience, 122*(2), 368–376.

Parade, S. H., Ridout, K. K., Seifer, R., Armstrong, D. A., Marsit, C. J., McWilliams, M. A., & Tyrka, A. R. (2016). Methylation of the glucocorticoid receptor gene promoter in preschoolers: Links with internalizing behavior problems. *Child Development, 87*(1), 86–97.

Parent, J., Parade, S. H., Laumann, L. E., Ridout, K. K., Yang, B. Z., Marsit, C. J., . . . Tyrka, A. R. (2017). Dynamic stress-related epigenetic regulation of the glucocorticoid receptor gene promoter during early development: The role of child maltreatment. *Development and Psychopathology, 29*(5), 1635–1648.

Parker, K. J., Hoffman, C. L., Hyde, S. A., Cummings, C. S., & Maestripieri, D. (2010). Effects of age on cerebrospinal fluid oxytocin levels in free-ranging adult female and infant rhesus macaques. *Behavioral Neuroscience, 124*(3), 428–433.

Parker, K. J., & Lee, T. M. (2001). Central vasopressin administration regulates the onset of facultative paternal behavior in *Microtus pennsylvanicus* (meadow voles). *Hormones and Behavior, 39*(4), 285–294.

Parreiras-e-Silva, L. T., Vargas-Pinilla, P., Duarte, D. A., Longo, D., Espinoza Pardo, G. V., Dulor Finkler, A., . . . Bortolini, M. C. (2017). Functional New World monkey oxytocin forms elicit an altered signaling profile and promotes parental care in rats. *Proceedings of the National Academy of Sciences USA, 114*(34), 9044–9049.

Pawluski, J. L., Li, M., & Lonstein, J. S. (2019). Serotonin and motherhood: From molecules to mood. *Frontiers in Neuroendocrinology, 53*, 100742.

Pawluski, J. L., Lonstein, J. S., & Fleming, A. S. (2017). The neurobiology of postpartum anxiety and depression. *Trends in Neurosciences, 40*(2), 106–120.

Payne, J. L., & Maguire, J. (2019). Pathophysiological mechanisms implicated in postpartum depression. *Frontiers in Neuroendocrinology, 52*, 165–180.

Pedersen, C. A. (2004). Biological aspects of social bonding and the roots of human violence. *Annals of the New York Academy of Sciences, 1036*, 106–127.

Pedersen, C. A., Ascher, J. A., Monroe, Y. L., & Prange, A. J., Jr. (1982). Oxytocin induces maternal behavior in virgin female rats. *Science, 216*(4546), 648–650.

Pedersen, C. A., & Boccia, M. L. (2003). Oxytocin antagonism alters rat dams' oral grooming and upright posturing over pups. *Physiology & Behavior, 80*(2–3), 233–241.

Pedersen, C. A., Caldwell, J. D., Walker, C., Ayers, G., & Mason, G. A. (1994). Oxytocin activates the postpartum onset of rat maternal behavior in the ventral tegmental and medial preoptic areas. *Behavioral Neuroscience, 108*(6), 1163–1171.

Pedersen, C. A., & Prange, A. J., Jr. (1979). Induction of maternal behavior in virgin rats after intracerebroventricular administration of oxytocin. *Proceedings of the National Academy of Sciences USA, 76*(12), 6661–6665.

Pedersen, C. A., Vadlamudi, S. V., Boccia, M. L., & Amico, J. A. (2006). Maternal behavior deficits in nulliparous oxytocin knockout mice. *Genes, Brain, and Behavior, 5*(3), 274–281.

Pena, C. J., & Champagne, F. A. (2015). Neonatal overexpression of estrogen receptor-alpha alters midbrain dopamine neuron development and reverses the effects of low maternal care in female offspring. *Developmental Neurobiology, 75*(10), 1114–1124.

Pena, C. J., Neugut, Y. D., Calarco, C. A., & Champagne, F. A. (2014). Effects of maternal care on the development of midbrain dopamine pathways and reward-directed behavior in female offspring. *European Journal of Neuroscience, 39*(6), 946–956.

Pena, C. J., Neugut, Y. D., & Champagne, F. A. (2013). Developmental timing of the effects of maternal care on gene expression and epigenetic regulation of hormone receptor levels in female rats. *Endocrinology, 154*(11), 4340–4351.

Pereira, M., & Morrell, J. I. (2009). The changing role of the medial preoptic area in the regulation of maternal behavior across the postpartum period: Facilitation followed by inhibition. *Behavioural Brain Research, 205*(1), 238–248.

Pereira, M., & Morrell, J. I. (2010). The medial preoptic area is necessary for motivated choice of pup- over cocaine-associated environments by early postpartum rats. *Neuroscience, 167*(2), 216–231.

Pereira, M., & Morrell, J. I. (2011). Functional mapping of the neural circuitry of rat maternal motivation: Effects of site-specific transient neural inactivation. *Journal of Neuroendocrinology, 23*(11), 1020–1035.

Pereira, M., Uriarte, N., Agrati, D., Zuluaga, M. J., & Ferreira, A. (2005). Motivational aspects of maternal anxiolysis in lactating rats. *Psychopharmacology (Berlin), 180*(2), 241–248.

Peris, J., MacFadyen, K., Smith, J. A., de Kloet, A. D., Wang, L., & Krause, E. G. (2017). Oxytocin receptors are expressed on dopamine and glutamate neurons in the mouse ventral tegmental area that project to nucleus accumbens and other mesolimbic targets. *Journal of Comparative Neurology, 525*(5), 1094–1108.

Perkeybile, A. M., Carter, C. S., Wroblewski, K. L., Puglia, M. H., Kenkel, W. M., Lillard, T. S., . . . Connelly, J. J. (2019). Early nurture epigenetically tunes the oxytocin receptor. *Psychoneuroendocrinology, 99*, 128–136.

Perrigo, G., Belvin, L., Quindry, P., Kadir, T., Becker, J., van Look, C., . . . vom Saal, F. S. (1993). Genetic mediation of infanticide and parental behavior in male and female domestic and wild stock house mice. *Behavior Genetics, 23*(6), 525–531.

Perrin, G., Meurisse, M., & Levy, F. (2007). Inactivation of the medial preoptic area or the bed nucleus of the stria terminalis differentially disrupts maternal behavior in sheep. *Hormones and Behavior, 52*(4), 461–473.

Petrides, M., & Pandya, D. N. (2007). Efferent association pathways from the rostral prefrontal cortex in the macaque monkey. *Journal of Neuroscience, 27*(43), 11573–11586.

Petrovich, G. D., Canteras, N. S., & Swanson, L. W. (2001). Combinatorial amygdalar inputs to hippocampal domains and hypothalamic behavior systems. *Brain Research Reviews, 38*(1–2), 247–289.

Petrovich, G. D., Risold, P. Y., & Swanson, L. W. (1996). Organization of projections from the basomedial nucleus of the amygdala: A PHAL study in the rat. *Journal of Comparative Neurology, 374*(3), 387–420.

Pfaff, D., & Keiner, M. (1973). Atlas of estradiol-concentrating cells in the central nervous system of the female rat. *Journal of Comparative Neurology, 151*(2), 121–158.

Pfaff, D. W. (1982). Motivational concepts: Definitions and distinctions. In D. W. Pfaff (Ed.), *The physiological mechanisms of motivation* (pp. 3–24). New York, NY: Springer-Verlag.

Phan, K. L., Sripada, C. S., Angstadt, M., & McCabe, K. (2010). Reputation for reciprocity engages the brain reward center. *Proceedings of the National Academy of Sciences USA, 107*(29), 13099–13104.

Phelps, E. A., Delgado, M. R., Nearing, K. I., & LeDoux, J. E. (2004). Extinction learning in humans: Role of the amygdala and vmPFC. *Neuron, 43*(6), 897–905.

Phelps, S. M., Campbell, P., Zheng, D. J., & Ophir, A. G. (2010). Beating the boojum: Comparative approaches to the neurobiology of social behavior. *Neuropharmacology, 58*(1), 17–28.

Pi, X. J., & Grattan, D. R. (1998a). Differential expression of the two forms of prolactin receptor mRNA within microdissected hypothalamic nuclei of the rat. *Molecular Brain Research, 59*(1), 1–12.

Pi, X. J., & Grattan, D. R. (1998b). Distribution of prolactin receptor immunoreactivity in the brain of estrogen-treated, ovariectomized rats. *Journal of Comparative Neurology, 394*(4), 462–474.

Pissonnier, D., Thiery, J. C., Fabre-Nys, C., Poindron, P., & Keverne, E. B. (1985). The importance of olfactory bulb noradrenalin for maternal recognition in sheep. *Physiology & Behavior, 35*(3), 361–363.

Plant, D. T., Barker, E. D., Waters, C. S., Pawlby, S., & Pariante, C. M. (2013). Intergenerational transmission of maltreatment and psychopathology: The role of antenatal depression. *Psychological Medicine, 43*(3), 519–528.

Plotsky, P. M., Thrivikraman, K. V., Nemeroff, C. B., Caldji, C., Sharma, S., & Meaney, M. J. (2005). Long-term consequences of neonatal rearing on central corticotropin-releasing factor systems in adult male rat offspring. *Neuropsychopharmacology, 30*(12), 2192–2204.

Poindron, P., & Le Neindre, P. (1980). Endocrine and sensory regulation of maternal behavior in the ewe. *Advances in the Study of Behavior, 11*, 75–119.

Poindron, P., Levy, F., & Keller, M. (2007). Maternal responsiveness and maternal selectivity in domestic sheep and goats: The two facets of maternal attachment. *Developmental Psychobiology, 49*(1), 54–70.

Poindron, P., Orgeur, P., Le Neindre, P., Kann, G., & Raksanyi, I. (1980). Influence of the blood concentration of prolactin on the length of the sensitive period for establishing maternal behavior in sheep at parturition. *Hormones and Behavior, 14*(2), 173–177.

Pomrenze, M. B., Tovar-Diaz, J., Blasio, A., Maiya, R., Giovanetti, S. M., Lei, K., . . . Messing, R. O. (2019). A corticotropin releasing factor network in the extended amygdala for anxiety. *Journal of Neuroscience, 39*(6), 1030–1043.

Poole, J. C., Dobson, K. S., & Pusch, D. (2018). Do adverse childhood experiences predict adult interpersonal difficulties? The role of emotion dysregulation. *Child Abuse & Neglect, 80*, 123–133.

Pornpattananangkul, N., Zhang, J., Chen, Q., Kok, B. C., & Yu, R. (2017). Generous to whom? The influence of oxytocin on social discounting. *Psychoneuroendocrinology, 79*, 93–97.

Pose, S., Zuluaga, M. J., Ferreno, M., Agrati, D., Bedo, G., & Uriarte, N. (2019). Raising overlapping litters: Differential activation of rat maternal neural circuitry after interacting with newborn and juvenile pups. *Journal of Neuroendocrinology*, e12701.

Potter, H., Ashbrook, D. G., & Hager, R. (2019). Offspring genetic effects on maternal care. *Frontiers in Neuroendocrinology, 52*, 195–205.

Post, C., & Leuner, B. (2019). The maternal reward system in postpartum depression. *Archives of Women's Mental Health, 22*(3), 417–429.

Pratt, M., Apter-Levi, Y., Vakart, A., Feldman, M., Fishman, R., Feldman, T., . . . Feldman, R. (2015). Maternal depression and child oxytocin response: Moderation by maternal oxytocin and relational behavior. *Depression and Anxiety, 32*(9), 635–646.

Preston, S. D. (2013). The origins of altruism in offspring care. *Psychological Bulletin, 139*(6), 1305–1341.

Price, J. L. (2007). Definition of the orbital cortex in relation to specific connections with limbic and visceral structures and other cortical regions. *Annals of the New York Academy of Science, 1121*, 54–71.

Pro-Sistiaga, P., Mohedano-Moriano, A., Ubeda-Banon, I., Del Mar Arroyo-Jimenez, M., Marcos, P., Artacho-Perula, E., . . . Martinez-Marcos, A. (2007). Convergence of olfactory and vomeronasal projections in the rat basal telencephalon. *Journal of Comparative Neurology, 504*(4), 346–362.

Potter, H. G., Ashbrook, D. G., & Hager, R. (2019). Offspring genetic effects on maternal care. *Frontiers in Neuroendocrinology, 52*, 195–205.

Pryce, C. R. (1993). The regulation of maternal behaviour in marmosets and tamarins. *Behavioral Processes, 30*(3), 201–224.

Pryce, C. R. (1996). Socialization, hormones, and the regulation of maternal behavior in nonhuman simian primates. *Advances in the Study of Behavior, 25*, 423–473.

Pryce, C. R., Dobeli, M., & Martin, R. D. (1993). Effects of sex steroids on maternal motivation in the common marmoset (*Callithrix jacchus*): Development and application of an operant system with maternal reinforcement. *Journal of Comparative Psychology, 107*(1), 99–115.

Puetz, V. B., Viding, E., Gerin, M. I., Pingault, J-B., Sethi, A., Knodt, A. R., . . . McCrory, E. (2019). Investigating patterns of neural response associated with childhood abuse versus childhood neglect. *Psychological Medicine.* doi:10.1017/s003329171900134x [Epub ahead of print]

Puglia, M. H., Lillard, T. S., Morris, J. P., & Connelly, J. J. (2015). Epigenetic modification of the oxytocin receptor gene influences the perception of anger and fear in the human brain. *Proceedings of the National Academy of Sciences USA, 112*(11), 3308–3313.

Putnam, K. T., Wilcox, M., Robertson-Blackmore, E., Sharkey, K., Bergink, V., Munk-Olsen, T., . . . Meltzer-Brody, S. (2017). Clinical phenotypes of perinatal depression and time of symptom onset: Analysis of data from an international consortium. *Lancet Psychiatry, 4*(6), 477–485.

Quintana, D. S., Rokicki, J., van der Meer, D. Alnaes, D., Kaufmann, T., Cordova-Palomera, A., . . . Westlye, L. T. (2019). Oxytocin pathway gene networks in the human brain. *Nature Communications, 10*, 668.

Raby, K. L., Labella, M. H., Martin, J., Carlson, E. A., & Roisman, G. I. (2017). Childhood abuse and neglect and insecure attachment states of mind in adulthood: Prospective, longitudinal evidence from a high-risk sample. *Development and Psychopathology, 29*(2), 347–363.

Rafacz, M. L., Margulis, S., & Santymire, R. M. (2012). Hormonal correlates of paternal care differences in the Hylobatidae. *America Journal of Primatology, 74*(3), 247–260.

Ragan, C. M., & Lonstein, J. S. (2014). Differential postpartum sensitivity to the anxiety-modulating effects of offspring contact is associated with innate anxiety and brainstem levels of dopamine beta-hydroxylase in female laboratory rats. *Neuroscience, 256*, 433–444.

Raggenbass, M. (2001). Vasopressin- and oxytocin-induced activity in the central nervous system: Electrophysiological studies using in-vitro systems. *Progress in Neurobiology, 64*(3), 307–326.

Ragnauth, A. K., Devidze, N., Moy, V., Finley, K., Goodwillie, A., Kow, L. M., . . . Pfaff, D. W. (2005). Female oxytocin gene-knockout mice, in a semi-natural environment, display exaggerated aggressive behavior. *Genes, Brain, and Behavior, 4*(4), 229–239.

Rainville, J., Pollard, K., & Vasudevan, N. (2015). Membrane-initiated non-genomic signaling by estrogens in the hypothalamus: Cross-talk with glucocorticoids with implications for behavior. *Frontiers in Endocrinology, 6*, 18.

Raij, T., Nummenmaa, A., Marin, M. F., Porter, D., Furtak, S., Setsompop, K., & Milad, M. R. (2018). Prefrontal cortex stimulation enhances fear extinction memory in humans. *Biological Psychiatry, 84*(2), 129–137.

Rajhans, P., Goin-Kochel, R. P., Strathearn, L, & Kim. S. (2019). It takes two! Exploring sex differences in parenting neurobiology and behavior. *Journal of Neuroendocrinology, 31*(9), e12721.

Rameson, L. T., Morelli, S. A., & Lieberman, M. D. (2012). The neural correlates of empathy: Experience, automaticity, and prosocial behavior. *Journal of Cognitive Neuroscience, 24*(1), 235–245.

Rebello, T. J., Yu, Q., Goodfellow, N. M., Caffrey Cagliostro, M. K., Teissier, A., Morelli, E., . . . Ansorge, M. S. (2014). Postnatal day 2 to 11 constitutes a 5-HT-sensitive period impacting adult mPFC function. *Journal of Neuroscience, 34*(37), 12379–12393.

Reburn, C. J., & Wynne-Edwards, K. E. (1999). Hormonal changes in males of a naturally biparental and a uniparental mammal. *Hormones and Behavior, 35*(2), 163–176.

Rees, S. L., Panesar, S., Steiner, M., & Fleming, A. S. (2004). The effects of adrenalectomy and corticosterone replacement on maternal behavior in the postpartum rat. *Hormones and Behavior, 46*(4), 411–419.

Rees, S. L., Panesar, S., Steiner, M., & Fleming, A. S. (2006). The effects of adrenalectomy and corticosterone replacement on induction of maternal behavior in virgin female rats. *Hormones and Behavior, 49*(3), 337–345.

Reisbick, S., Rosenblatt, J. S., & Mayer, A. D. (1975). Decline of maternal behavior in the virgin and lactating rat. *Journal of Comparative and Physiological Psychology, 89*(7), 722–732.

Ren, J., Friedmann, D., Xiong, J., Liu, C. D., Ferguson, B. R., Weerakkody, T., . . . Luo, L. (2018). Anatomically defined and functionally distinct dorsal raphe serotonin subsystems. *Cell, 175*(2), 472–487.

Renier, N., Adams, E. L., Kirst, C., Wu, Z., Azevedo, R., Kohl, J., . . . Tessier-Lavigne, M. (2016). Mapping of brain activity by automated volume analysis of immediate early genes. *Cell, 165*(7), 1789–1802.

Rheingold, H. L. (1963). *Maternal behavior in mammals*. New York, NY: Wiley.

Ribeiro, A. C., Musatov, S., Shteyler, A., Simanduyev, S., Arrieta-Cruz, I., Ogawa, S., & Pfaff, D. W. (2012). siRNA silencing of estrogen receptor-α expression specifically in medial preoptic area neurons abolishes maternal care in female mice. *Proceedings of the National Academy of Sciences USA, 109*(40), 16324–16329.

Richard, J. M., Ambroggi, F., Janak, P. H., & Fields, H. L. (2016). Ventral pallidum neurons encode incentive value and promote cue-elicited instrumental actions. *Neuron, 90*(6), 1165–1173.

Rickenbacher, E., Perry, R. E., Sullivan, R. M., & Moita, M. A. (2017). Freezing suppression by oxytocin in central amygdala allows alternate defensive behaviours and mother–pup interactions. *elife, 6*.

Riedman, M. L. (1982). The evolution of alloparental care and adoption in mammals and birds. *Quarterly Review of Biology, 57*(4), 405–435.

Riem, M. M., Bakermans-Kranenburg, M. J., Huffmeijer, R., & van Ijzendoorn, M. H. (2013). Does intranasal oxytocin promote prosocial behavior to an excluded fellow player? A randomized-controlled trial with Cyberball. *Psychoneuroendocrinology, 38*(8), 1418–1425.

Riem, M. M., Bakermans-Kranenburg, M. J., Pieper, S., Tops, M., Boksem, M. A., Vermeiren, R. R., . . . Rombouts, S. A. (2011). Oxytocin modulates amygdala, insula, and inferior frontal gyrus responses to infant crying: A randomized controlled trial. *Biological Psychiatry, 70*(3), 291–297.

Riem, M. M., Bakermans-Kranenburg, M. J., & van Ijzendoorn, M. H. (2016). Intranasal administration of oxytocin modulates behavioral and amygdala responses to infant crying in females with insecure attachment representations. *Attachment & Human Development, 18*(3), 213–234.

Riem, M. M., Bakermans-Kranenburg, M. J., van Ijzendoorn, M. H., Out, D., & Rombouts, S. A. (2012). Attachment in the brain: Adult attachment representations predict amygdala and behavioral responses to infant crying. *Attachment & Human Development, 14*(6), 533–551.

Rigney, A. E., Koski, J. E., & Beer, J. S. (2018). The functional role of the ventral anterior cingulate cortex in social evaluation: Disentangling valence from subjectively rewarding opportunities. *Social, Cognitive, and Affective Neuroscience, 13*(1), 14–21.

Rigo, P., Kim, P., Esposito, G., Putnick, D. L., Venuti, P., & Bornstein, M. H. (2019). Specific maternal brain responses to their own child's face: An fMRI meta-analysis. *Developmental Review, 51*, 58–69.

Rilling, J. K. (2013). The neural and hormonal bases of human parental care. *Neuropsychologia, 51*(4), 731–747.

Rilling, J. K., & Mascaro, J. S. (2017). The neurobiology of fatherhood. *Current Opinion in Psychology, 15*, 26–32.

Rincon-Cortes, M., & Sullivan, R. M. (2014). Early life trauma and attachment: Immediate and enduring effects on neurobehavioral and stress axis development. *Frontiers in Endocrinology (Lausanne), 5*, 33.

Ring, R. H., Malberg, J. E., Potestio, L., Ping, J., Boikess, S., Luo, B., . . . Rosenzweig-Lipson, S. (2006). Anxiolytic-like activity of oxytocin in male mice: Behavioral and autonomic evidence, therapeutic implications. *Psychopharmacology (Berlin), 185*(2), 218–225.

Rizvi, T. A., Ennis, M., & Shipley, M. T. (1992). Reciprocal connections between the medial preoptic area and the midbrain periaqueductal gray in rat: A WGA-HRP and PHA-L study. *Journal of Comparative Neurology, 315*(1), 1–15.

Rizvi, T. A., Murphy, A. Z., Ennis, M., Behbehani, M. M., & Shipley, M. T. (1996). Medial preoptic area afferents to periaqueductal gray medullo-output neurons: A combined Fos and tract tracing study. *Journal of Neuroscience, 16*(1), 333–344.

Roberts, R. L., Jenkins, K. T., Lawler, T., Jr., Wegner, F. H., & Newman, J. D. (2001). Bromocriptine administration lowers serum prolactin and disrupts parental responsiveness in common marmosets (*Callithrix jacchus*). *Hormones and Behavior, 39*(2), 106–112.

Roberts, W. W., & Nagel, J. (1996). First-order projections activated by stimulation of hypothalamic sites eliciting attack and flight in rats. *Behavioral Neuroscience, 110*(3), 509–527.

Robinson, D. L., Zitzman, D. L., & Williams, S. K. (2011). Mesolimbic dopamine transients in motivated behaviors: Focus on maternal behavior. *Frontiers in Psychiatry, 2*, 23.

Rocchetti, M., Radua, J., Paloyelis, Y., Xenaki, L. A., Frascarelli, M., Caverzasi, E., . . . Fusar-Poli, P. (2014). Neurofunctional maps of the "maternal brain" and the effects of oxytocin: A multimodal voxel-based meta-analysis. *Psychiatry and Clinical Neuroscience, 68*(10), 733–751.

Rodrigo, M. J., Leon, I., Gongora, D., Hernandez-Cabrera, J. A., Byrne, S., & Bobes, M. A. (2016). Inferior fronto-temporo-occipital connectivity: A missing link between maltreated girls and neglectful mothers. *Social Cognitive and Affective Neuroscience, 11*(10), 1658–1665.

Rodrigues, S. M., Saslow, L. R., Garcia, N., John, O. P., & Keltner, D. (2009). Oxytocin receptor genetic variation relates to empathy and stress reactivity in humans. *Proceedings of the National Academy of Sciences USA, 106*(50), 21437–21441.

Rogers, C. N., Ross, A. P., Sahu, S. P. Siegel, E. R. Dooyema, J. M., Cree, M. A., . . . Preuss, T. M. (2018). Oxytocin- and arginine vasopressin-containing fibers in the cortex of humans, chimpanzees, and rhesus macaques. *American Journal of Primatology, 80*(10), e22875.

Roland, A. B., & O'Connell, L. A. (2015). Poison frogs as a model system for studying the neurobiology of parental care. *Current Opinion in Behavioral Sciences, 6*, 76–81.

Romero-Fernandez, W., Borroto-Escuela, D. O., Agnati, L. F., & Fuxe, K. (2013). Evidence for the existence of dopamine D2-oxytocin receptor heteromers in the ventral and dorsal striatum with facilitatory receptor–receptor interactions. *Molecular Psychiatry, 18*(8), 849–850.

Romero-Morales, L., Cardenas, M., Martinez-Torres, M., Garcia-Saucedo, B., Carmona, A., & Luis, J. (2018). Neuronal activation associated with paternal and aversive interactions toward pups in the Mongolian gerbils (*Meriones unguiculatus*). *Hormones and Behavior, 105*, 47–57.

Romero-Morales, L., Martinez-Torres, M., Cardenas, M., Alvarez, C., Carmona, A., Cedillo, B., . . . Luis, J. (2018). An increase in estradiol facilitates the onset of paternal behavior in the dwarf hamster (*Phodopus campbelli*). *Hormones and Behavior, 99*, 35–40.

Rondini, T. A., Donato, J., Rodrigues, C., Bittencourt, J. C., & Elias, C. F. (2010). Chemical identity and connections of medial preoptic area neurons expressing melanin-concentrating hormone during lactation. *Journal of Chemical Neuroanatomy, 39*(1), 51–62.

Root, D. H., Melendez, R. I., Zaborszky, L., & Napier, T. C. (2015). The ventral pallidum: Subregion-specific functional anatomy and roles in motivated behaviors. *Progress in Neurobiology, 130*, 29–70.

Rosati, A. G., DiNicola, L. M., & Buckholtz, J. W. (2018). Chimpanzee cooperation is fast and independent from self-control. *Psychological Science, 29*(11), 1832–1845.

Rosen, G. J., de Vries, G. J., Goldman, S. L., Goldman, B. D., & Forger, N. G. (2008). Distribution of oxytocin in the brain of a eusocial rodent. *Neuroscience, 155*(3), 809–817.

Rosenblatt, J. S. (1967). Nonhormonal basis of maternal behavior in the rat. *Science, 156*(3781), 1512–1514.

Rosenblatt, J. S., & Ceus, K. (1998). Estrogen implants in the medial preoptic area stimulate maternal behavior in male rats. *Hormones and Behavior, 33*(1), 23–30.

Rosenblatt, J. S., Hazelwood, S., & Poole, J. (1996). Maternal behavior in male rats: Effects of medial preoptic area lesions and presence of maternal aggression. *Hormones and Behavior, 30*(3), 201–215.

Rosenblatt, J. S., & Lehrman, D. S. (1963). Maternal behavior in the laboratory rat. In H. L. Rheingold (Ed.), *Maternal behavior in mammals* (pp. 8–57). New York, NY: Wiley.

Rosenblatt, J. S., & Mayer, A. D. (1995). An analysis of approach/withdrawal processes in the initiation of maternal behavior in the laboratory rat. In K. E. Hood, G. Greenberg, & E. Tobach (Eds.), *Behavioral development* (pp. 117–230). New York, NY: Garland Press.

Rosenblatt, J. S., Mayer, A. D., & Siegel, H. I. (1985). Maternal behavior among the nonprimate mammals. In N. Adler, D. Pfaff, & R. W. Goy (Eds.), *Handbook of behavioral neurobiology: Vol. 7, Reproduction* (pp. 229–298). New York, NY: Plenum Press.

Rosenblatt, J. S., & Siegel, H. I. (1975). Hysterectomy-induced maternal behavior during pregnancy in the rat. *Journal of Comparative and Physiological Psychology, 89*(7), 685–700.

Ross, H. E., & Young, L. J. (2009). Oxytocin and the neural mechanisms regulating social cognition and affiliative behavior. *Frontiers in Neuroendocrinology, 30*, 534–547.

Roth, T. L., Lubin, F. D., Funk, A. J., & Sweatt, J. D. (2009). Lasting epigenetic influence of early-life adversity on the BDNF gene. *Biological Psychiatry, 65*(9), 760–769.

Rubin, B. S., Menniti, F. S., & Bridges, R. S. (1983). Intracerebroventricular administration of oxytocin and maternal behavior in rats after prolonged and acute steroid pretreatment. *Hormones and Behavior, 17*(1), 45–53.

Ruppenthal, G. C., Arling, G. L., Harlow, H. F., Sackett, G. P., & Suomi, S. J. (1976). A 10-year perspective of motherless-mother monkey behavior. *Journal of Abnormal Psychology, 85*(4), 341–349.

Russell, J. A., Leng, G., & Douglas, A. J. (2003). The magnocellular oxytocin system, the fount of maternity: Adaptations in pregnancy. *Frontiers in Neuroendocrinology, 24*(1), 27–61.

Ruthschilling, C. A., Albiero, G., Lazzari, V. M., Becker, R. O., de Moura, A. C., Lucion, A. B., . . . Giovenardi, M. (2012). Analysis of transcriptional levels of the oxytocin receptor in different areas of the central nervous system and behaviors in high and low licking rats. *Behavioural Brain Research, 228*(1), 176–184.

Sahuque, L. L., Kullberg, E. F., McGeehan, A. J., Kinder, J. R., Hicks, M. P., Blanton, M. G., . . . Olive, M. F. (2006). Anxiogenic and aversive effects of corticotropin-releasing factor (CRF) in the bed nucleus of the stria terminalis in the rat: Role of CRF receptor subtypes. *Psychopharmacology (Berlin), 186*(1), 122–132.

Sairenji, T. J., Ikezawa, J., Kaneko, R., Masuda, S., Uchida, K., Takanashi, Y., . . . Shimokawa, N. (2017). Maternal prolactin during late pregnancy is important in generating nurturing behavior in the offspring. *Proceedings of the National Academy of Sciences USA, 114*(49), 13042–13047.

Sakanaka, M., Shibasaki, T., & Lederis, K. (1986). Distribution and efferent projections of corticotropin-releasing factor-like immunoreactivity in the rat amygdaloid complex. *Brain Research, 382*(2), 213–238.

Sakurai, K., Zhao, S., Takatoh, J., Rodriguez, E., Lu, J., Leavitt, A. D., . . . Wang, F. (2016). Capturing and manipulating activated neuronal ensembles with CANE delineates a hypothalamic social-fear circuit. *Neuron, 92*(4), 739–753.

Saltzman, W., & Abbott, D. H. (2005). Diminished maternal responsiveness during pregnancy in multiparous female common marmosets. *Hormones and Behavior, 47*(2), 151–163.

Saltzman, W., & Abbott, D. H. (2009). Effects of elevated circulating cortisol concentrations on maternal behavior in common marmoset monkeys (*Callithrix jacchus*). *Psychoneuroendocrinology, 34*(8), 1222–1234.

Saltzman, W., Boettcher, C. A., Post, J. L., & Abbott, D. H. (2011). Inhibition of maternal behaviour by central infusion of corticotrophin-releasing hormone in marmoset monkeys. *Journal of Neuroendocrinology, 23*(11), 1139–1148.

Saltzman, W., Liedl, K. J., Salper, O. J., Pick, R. R., & Abbott, D. H. (2008). Post-conception reproductive competition in cooperatively breeding common marmosets. *Hormones and Behavior, 53*(1), 274–286.

Saltzman, W., & Maestripieri, D. (2011). The neuroendocrinology of primate maternal behavior. *Progress in Neuropsychopharmacology & Biological Psychiatry, 35*(5), 1192–1204.

Samuel, S., Hayton, B., Gold, I., Feeley, N., Carter, C. S., & Zelkowitz, P. (2015). Attachment security and recent stressful life events predict oxytocin levels: A pilot study of pregnant women with high levels of cumulative psychosocial adversity. *Attachment & Human Development, 17*(3), 272–287.

Sanchez, M. M., McCormack, K. M., & Howell, B. R. (2015). Social buffering of stress responses in nonhuman primates: Maternal regulation of the development of emotional regulatory brain circuits. *Social Neuroscience, 10*(5), 512–526.

Sanchez, S. M., Ziegler, T. E., & Snowdon, C. T. (2014). Both parents respond equally to infant cues in the cooperatively breeding common marmoset, *Callithrix jacchus. Animal Behaviour, 97*, 95–103.

Santiago, A. N., Lim, K. Y., Opendak, M., Sullivan, R. M., & Aoki, C. (2018). Early life trauma increases threat response of peri-weaning rats, reduction of axo-somatic synapses formed by parvalbumin cells and perineuronal net in the basolateral nucleus of amygdala. *Journal of Comparative Neurology, 526*(16), 2647–2664.

Sarkar, A., Chachra, P., & Vaidya, V. A. (2014). Postnatal fluoxetine-evoked anxiety is prevented by concomitant 5-HT2A/C receptor blockade and mimicked by postnatal 5-HT2A/C receptor stimulation. *Biological Psychiatry, 76*(11), 858–868.

Sawchenko, P. E., & Swanson, L. W. (1982). Immunohistochemical identification of neurons in the paraventricular nucleus of the hypothalamus that project to the medulla or to the spinal cord in the rat. *Journal of Comparative Neurology, 205*(3), 260–272.

Scalia, F., & Winans, S. S. (1975). The differential projections of the olfactory bulb and accessory olfactory bulb in mammals. *Journal of Comparative Neurology, 161*(1), 31–55.

Schacht, R., & Bell, A. V. (2016). The evolution of monogamy in response to partner scarcity. *Science Reports, 6*, 32472.

Scheele, D., Plota, J., Stoffel-Wagner, B., Maier, W., & Hurlemann, R. (2016). Hormonal contraceptives suppress oxytocin-induced brain reward responses to the partner's face. *Social, Cognitive, and Affective Neuroscience, 11*(5), 767–774.

Scheele, D., Striepens, N., Gunturkun, O., Duetschlander, S., Maier, W., Kendrick, K. M., & Hurlemann, R. (2012). Oxytocin modulates social distance between males and females. *Journal of Neuroscience, 32*(46), 16074–16079.

Scheele, D., Wille, A., Kendrick, K. M., Stoffel-Wagner, B., Becker, B., Gunturkun, O., . . . Hurlemann, R. (2013). Oxytocin enhances brain reward system responses in men viewing the face of their female partner. *Proceedings of the National Academy of Sciences USA, 110*(50), 20308–20318.

Schindler, S., Schonknecht, P., Schmidt, L., Anwander, A, Straub, M. Trampel, R., . . . Geyer, S. (2013). Development and evaluation of an algorithm for computer-assisted segmentation of the human hypothalamus on 7-Tesla magnetic resonance images. *PloS One, 8*(7), e66394.

Schmidt, M. V. (2011). Animal models for depression and the mismatch hypothesis of disease. *Psychoneuroendocrinology, 36*, 330–338.

Schneiderman, I., Zagoory-Sharon, O., Leckman, J. F., & Feldman, R. (2012). Oxytocin during the initial stages of romantic attachment: Relations to couples' interactive reciprocity. *Psychoneuroendocrinology, 37*(8), 1277–1285.

Schofield, T. J., Conger, R. D., & Conger, K. J. (2017). Disrupting intergenerational continuity in harsh parenting: Self-control and a supportive partner. *Development and Psychopathology, 29*(4), 1279–1287.

Schofield, T. J., Lee, R. D., & Merrick, M. T. (2013). Safe, stable, nurturing relationships as a moderator of intergenerational continuity of child maltreatment: A meta-analysis. *Journal of Adolescent Health, 53*(4 Suppl), S32–38.

Schorscher-Petcu, A., Dupre, A., & Tribollet, E. (2009). Distribution of vasopressin and oxytocin binding sites in the brain and upper spinal cord of the common marmoset. *Neuroscience Letters, 461*(3), 217–222.

Schott, B. H., Minuzzi, L., Krebs, R. M., Elmenhorst, D., Lang, M., Winz, O. H., . . . Bauer, A. (2008). Mesolimbic functional magnetic resonance imaging activations during reward anticipation correlate with reward-related ventral striatal dopamine release. *Journal of Neuroscience, 28*(52), 14311–14319.

Schradin, C., Reeder, D. M., Mendoza, S. P., & Anzenberger, G. (2003). Prolactin and paternal care: Comparison of three species of monogamous new world monkeys (*Callicebus cupreus, Callithrix jacchus*, and *Callimico goeldii*). *Journal of Comparative Psychology, 117*(2), 166–175.

Schradin, C., Vuarin, P., & Rimbach, R. (2018). The neoteny-helper hypothesis: When to expect and when not to expect endocrine mechanisms to regulate allo-parental care? *Physiology & Behavior, 193*(Pt A), 127–134.

Schulkin, J. (2011). Evolutionary conservation of glucocorticoids and corticotropin releasing hormone: Behavioral and physiological adaptations. *Brain Research, 1392*, 27–46.

Schulte, L. M., & Summers, K. (2017). Searching for hormonal facilitators: Are vasotocin and mesotocin involved in parental care behaviors in poison frogs? *Physiology & Behavior, 174*, 74–82.

Schurz, M., & Perner, J. (2015). An evaluation of neurocognitive models of theory of mind. *Frontiers in Psychology, 6*, 1610.

Schuwerk, T., Schurz, M., Muller, F., Rupprecht, R., & Sommer, M. (2017). The rTPJ's overarching cognitive function in networks for attention and theory of mind. *Social, Cognitive, and Affective Neuroscience, 12*(1), 157–168.

Scott, N., Prigge, M., Yizhar, O., & Kimchi, T. (2015). A sexually dimorphic hypothalamic circuit controls maternal care and oxytocin secretion. *Nature, 525*(7570), 519–522.

Sear, R., & Mace, R. (2008). Who keeps children alive? A review of the effects of kin on child survival. *Evolution and Human Behavior, 29*(1), 1–18.

Sears, R. M., Liu, R. J., Narayanan, N. S., Sharf, R., Yeckel, M. F., Laubach, M., . . . DiLeone, R. J. (2010). Regulation of nucleus accumbens activity by the hypothalamic neuropeptide melanin-concentrating hormone. *Journal of Neuroscience, 30*(24), 8263–8273.

Seelke, A. M. H., Bond, J. M., Simmons, T. C., Joshi, N., Settles, M. L., Stolzenberg, D., . . . Bales, K. L. (2018). Fatherhood alters gene expression within the MPOA. *Environmental Epigenetics, 4*(4), dvy026.

Segovia, S., & Guillamon, A. (1993). Sexual dimorphism in the vomeronasal pathway and sex differences in reproductive behaviors. *Brain Research Reviews, 18*(1), 51–74.

Seifritz, E., Esposito, F., Neuhoff, J. G., Luthi, A., Mustovic, H., Dammann, G., . . . Di Salle, F. (2003). Differential sex-independent amygdala response to infant crying and laughing in parents versus nonparents. *Biological Psychiatry, 54*(12), 1367–1375.

Seip, K. M., & Morrell, J. I. (2009). Transient inactivation of the ventral tegmental area selectively disrupts the expression of conditioned place preference for pup- but not cocaine-paired contexts. *Behavioral Neuroscience, 123*(6), 1325–1338.

Shahrokh, D. K., Zhang, T. Y., Diorio, J., Gratton, A., & Meaney, M. J. (2010). Oxytocin-dopamine interactions mediate variations in maternal behavior in the rat. *Endocrinology, 151*(5), 2276–2286.

Shams, S., Pawluski, J. L., Chatterjee-Chakraborty, M., Oatley, H., Mastroianni, A., & Fleming, A. S. (2012). Dendritic morphology in the striatum and hypothalamus differentially exhibits experience-dependent changes in response to maternal care and early social isolation. *Behavioural Brain Research, 233*(1), 79–89.

Sharma, K., LeBlanc, R., Haque, M., Nishimori, K., Reid, M. M., & Teruyama, R. (2019). Sexually dimorphic oxytocin receptor-expressing neurons in the preoptic area of the mouse brain. *PloS One, 14*(7), e0219784.

Sheehan, T., Cirrito, J., Numan, M. J., & Numan, M. (2000). Using c-Fos immunocytochemistry to identify forebrain regions that may inhibit maternal behavior in rats. *Behavioral Neuroscience, 114*(2), 337–352.

Sheehan, T., & Numan, M. (2000). The septal region and social behavior. In R. Numan (Ed.), *The behavioral neuroscience of the septal region* (pp. 175–209). New York, NY: Springer.

Sheehan, T., & Numan, M. (2002). Estrogen, progesterone, and pregnancy termination alter neural activity in brain regions that control maternal behavior in rats. *Neuroendocrinology, 75*(1), 12–23.

Sheehan, T., Paul, M., Amaral, E., Numan, M. J., & Numan, M. (2001). Evidence that the medial amygdala projects to the anterior/ventromedial hypothalamic nuclei to inhibit maternal behavior in rats. *Neuroscience, 106*(2), 341–356.

Sheng, M., & Greenberg, M. E. (1990). The regulation and function of c-Fos and other immediate early genes in the nervous system. *Neuron, 4*(4), 477–485.

Shepard, J. D., Barron, K. W., & Myers, D. A. (2000). Corticosterone delivery to the amygdala increases corticotropin-releasing factor mRNA in the central amygdaloid nucleus and anxiety-like behavior. *Brain Research, 861*(2), 288–295.

Sherer, M. L., Posillico, C. K., & Schwarz, J. M. (2018). The psychoneuroimmunology of pregnancy. *Frontiers in Neuroendocrinology, 51*, 25–35.

Shingo, T., Gregg, C., Enwere, E., Fujikawa, H., Hassam, R., Geary, C., . . . Weiss, S. (2003). Pregnancy-stimulated neurogenesis in the adult female forebrain mediated by prolactin. *Science, 299*(5603), 117–120.

Shnitko, T. A., Mace, K. D., Sullivan, K. M., Martin, W. K., Andersen, E. H., Williams Avram, S. K., . . . Robinson, D. L. (2017). Use of fast-scan cyclic voltammetry to assess phasic dopamine release in rat models of early postpartum maternal behavior and neglect. *Behavioral Pharmacology, 28*(8), 648–660.

Shughrue, P. J., Lane, M. V., & Merchenthaler, I. (1997). Comparative distribution of estrogen receptor-alpha and -beta mRNA in the rat central nervous system. *Journal of Comparative Neurology, 388*(4), 507–525.

Siegel, H. I., & Rosenblatt, J. S. (1975a). Estrogen-induced maternal behavior in hysterectomized-ovariectomized virgin rats. *Physiology & Behavior, 14*(4), 465–471.

Siegel, H. I., & Rosenblatt, J. S. (1975b). Hormonal basis of hysterectomy-induced maternal behavior during pregnancy in the rat. *Hormones and Behavior, 6*(3), 211–222.

Siegel, H. I., & Rosenblatt, J. S. (1978). Duration of estrogen stimulation and progesterone inhibition of maternal behavior in pregnancy-terminated rats. *Hormones and Behavior, 11*(1), 12–19.

Silk, J. B., & House, B. R. (2011). Evolutionary foundations of human prosocial sentiments. *Proceedings of the National Academy of Sciences USA, 108*(Suppl 2), 10910–10917.

Silva, B. A., Mattucci, C., Krzywkowski, P., Murana, E., Illarionova, A., Grinevich, V., . . . Gross, C. T. (2013). Independent hypothalamic circuits for social and predator fear. *Nature Neuroscience, 16*(12), 1731–1733.

Simerly, R. B. (1995). Anatomical substrates of hypothalamic integration. In G. Paxinos (Ed.), *The rat nervous system* (2nd ed.). San Diego, CA: Academic Press.

Simerly, R. B., & Swanson, L. W. (1988). Projections of the medial preoptic nucleus: A Phaseolus vulgaris leucoagglutinin anterograde tract-tracing study in the rat. *Journal of Comparative Neurology, 270*(2), 209–242.

Simerly, R. B., Zee, M. C., Pendleton, J. W., Lubahn, D. B., & Korach, K. S. (1997). Estrogen receptor-dependent sexual differentiation of dopaminergic neurons in the preoptic region of the mouse. *Proceedings of the National Academy of Sciences USA, 94*(25), 14077–14082.

Singer, L. M., Brodzinsky, D. M., Ramsay, D., Steir, M., & Waters, E. (1985). Mother–infant attachment in adoptive families. *Child Development, 56*(6), 1543–1551.

Singer, T., & Lamm, C. (2009). The social neuroscience of empathy. *Annals of the New York Academy of Sciences, 1156*, 81–96.

Skrundz, M., Bolten, M., Nast, I., Hellhammer, D. H., & Meinlschmidt, G. (2011). Plasma oxytocin concentration during pregnancy is associated with development of post-partum depression. *Neuropsychopharmacology, 36*(9), 1886–1893.

Slattery, D. A., & Hillerer, K. M. (2016). The maternal brain under stress: Consequences for adaptive peripartum plasticity and its potential functional implications. *Frontiers in Neuroendocrinology, 41*, 114–128.

Slawski, B. A., & Buntin, J. D. (1995). Preoptic area lesions disrupt prolactin-induced parental feeding behavior in ring doves. *Hormones and Behavior, 29*(2), 248–266.

Smearman, E. L., Almli, L. M., Conneely, K. N., Brody, G. H., Sales, J. M., Bradley, B., . . . Smith, A. K. (2016). Oxytocin receptor genetic and epigenetic variations: Association with child abuse and adult psychiatric symptoms. *Child Development, 87*(1), 122–134.

Smiley, K. O. (2019). Prolactin and avian parental care: New insights and unanswered questions. *Hormones and Behavior, 111*, 114–130.

Smiseth, P., Kölliker, M., & Royle, N. (2012). What is Parental Care? In N. Royle, P. Smiseth, & M. Kölliker (Eds.), *The evolution of parental care* (pp. 1–14). Oxford, England: Oxford University Press.

Smith, A. L., Freeman, S. M., Voll, R. J., Young, L. J., & Goodman, M. M. (2013). Carbon-11 N-methyl alkylation of L-368,899 and in vivo PET imaging investigations for neural oxytocin receptors. *Bioorganic and Medicinal Chemistry Letters, 23*(3), 902–906.

Smith, A. S., Agmo, A., Birnie, A. K., & French, J. A. (2010). Manipulation of the oxytocin system alters social behavior and attraction in pair-bonding primates, *Callithrix penicillata*. *Hormones and Behavior, 57*(2), 255–262.

Smith, C. J. W., DiBenedictis, B. T., & Veenema, A. H. (2019). Comparing vasopressin and oxytocin fiber and receptor density patterns in the social behavior neural network: Implications for cross-system signaling. *Frontiers in Neuroendocrinology, 53*, 100737.

Smith, K. E., Porges, E. C., Norman, G. J., Connelly, J. J., & Decety, J. (2014). Oxytocin receptor gene variation predicts empathic concern and autonomic arousal while perceiving harm to others. *Social Neuroscience, 9*(1), 1–9.

Smith, M. S., & Neill, J. D. (1977). Inhibition of gonadotropin secretion during lactation in the rat: Relative contribution of suckling and ovarian steroids. *Biology of Reproduction, 17*(2), 255–261.

Smotherman, W. P., Bell, R. W., Starzec, J., Elias, J., & Zachman, T. A. (1974). Maternal responses to infant vocalizations and olfactory cues in rats and mice. *Behavioral Biology, 12*(1), 55–66.

Soares, M. J. (2004). The prolactin and growth hormone families: Pregnancy-specific hormones/cytokines at the maternal-fetal interface. *Reproductive Biology and Endocrinology, 2*, 51.

Solomon, N. G., & French, J. A. (1997). *Cooperative breeding in mammals*. Cambridge, England: Cambridge University Press.

Somaweera, R., Brien, M., & Shine, R. (2013). The role of predation in shaping crocodilian natural history. *Herpetological Monographs, 27*(1), 23–51.

Soroker, V., & Terkel, J. (1988). Changes in incidence of infanticide and parental responses during the reproductive cycle in male and female wild *Mus musculus. Animal Behaviour, 136*, 1275–1282.

Spence-Aizenberg, A., Di Fiore, A., & Fernandez-Duque, E. (2016). Social monogamy, male-female relationships, and biparental care in wild titi monkeys (*Callicebus discolor*). *Primates, 57*(1), 103–112.

Spertus, I. L., Yehuda, R., Wong, C. M., Halligan, S., & Seremetis, S. V. (2003). Childhood emotional abuse and neglect as predictors of psychological and physical symptoms in women presenting to a primary care practice. *Child Abuse & Neglect, 27*(11), 1247–1258.

Sripada, C. S., Phan, K. L., Labuschagne, I., Welsh, R., Nathan, P. J., & Wood, A. G. (2013). Oxytocin enhances resting-state connectivity between amygdala and medial frontal cortex. *International Journal of Neuropsychopharmacology, 16*(2), 255–260.

Sroufe, L. A. (2005). Attachment and development: A prospective, longitudinal study from birth to adulthood. *Attachment & Human Development, 7*(4), 349–367.

Stack, E. C., Balakrishnan, R., Numan, M. J., & Numan, M. (2002). A functional neuroanatomical investigation of the role of the medial preoptic area in neural circuits regulating maternal behavior. *Behavioural Brain Research, 131*(1–2), 17–36.

Stack, E. C., & Numan, M. (2000). The temporal course of expression of c-Fos and Fos B within the medial preoptic area and other brain regions of postpartum female rats during prolonged mother–young interactions. *Behavioral Neuroscience, 114*(3), 609–622.

Stamatakis, A. M., & Stuber, G. D. (2012). Activation of lateral habenula inputs to the ventral midbrain promotes behavioral avoidance. *Nature Neuroscience, 15*(8), 1105–1107.

Starr, L. R., Hammen, C., Conway, C. C., Raposa, E., & Brennan, P. A. (2014). Sensitizing effect of early adversity on depressive reactions to later proximal stress: Moderation by polymorphisms in serotonin transporter and corticotropin releasing hormone receptor genes in a 20-year longitudinal study. *Development and Psychopathology, 26*(4 Pt 2), 1241–1254.

Steidl, S., O'Sullivan, S., Pilat, D., Bubula, N., Brown, J., & Vezina, P. (2017). Operant responding for optogenetic excitation of LDTg inputs to the VTA requires D1 and D2 dopamine receptor activation in the NAcc. *Behavioural Brain Research, 333*, 161–170.

Stern, J. A., Borelli, J. L., & Smiley, P. A. (2015). Assessing parental empathy: A role for em pathy in child attachment. *Attachment & Human Development, 17*(1), 1–22.

Stern, J. M. (1996). Somatosensation and maternal care in Norway rat. *Advances in the Study of Behavior, 25*, 243–294.

Stern, J. M., & Johnson, S. K. (1990). Ventral somatosensory determinants of nursing behavior in Norway rats: I. Effects of variations in the quality and quantity of pup stimuli. *Physiology & Behavior, 47*(5), 993–1011.

Stern, J. M., & Kolunie, J. M. (1993). Maternal aggression of rats is impaired by cutaneous anesthesia of the ventral trunk, but not by nipple removal. *Physiology & Behavior, 54*(5), 861–868.

Stern, J. M., & Mackinnon, D. A. (1976). Postpartum, hormonal, and nonhormonal induction of maternal behavior in rats: Effects on T-maze retrieval of pups. *Hormones and Behavior, 7*(3), 305–316.

Stevenson-Hinde, J., Chicot, R., Shouldice, A., & Hinde, C. A. (2013). Maternal anxiety, maternal sensitivity, and attachment. *Attachment & Human Development, 15*(5–6), 618–636.

Stockley, P., & Hobson, L. (2016). Paternal care and litter size coevolution in mammals. *Proceedings: Biological Sciences, 283*(1829).

Stolzenberg, D. S., Grant, P. A., & Bekiranov, S. (2011). Epigenetic methodologies for behavioral scientists. *Hormones and Behavior, 59*(3), 407–416.

Stolzenberg, D. S., & Mayer, H. S. (2019). Experience-dependent mechanisms in the regulation of parental care. *Frontiers in Neuroendocrinology, 54*, 100745.

Stolzenberg, D. S., McKenna, J. B., Keough, S., Hancock, R., Numan, M. J., & Numan, M. (2007). Dopamine D1 receptor stimulation of the nucleus accumbens or the medial preoptic area promotes the onset of maternal behavior in pregnancy-terminated rats. *Behavavioral Neuroscience, 121*(5), 907–919.

Stolzenberg, D. S., & Numan, M. (2011). Hypothalamic interaction with the mesolimbic DA system in the control of the maternal and sexual behaviors in rats. *Neuroscience and Biobehavioral Reviews, 35*(3), 826–847.

Stolzenberg, D. S., & Rissman, E. F. (2011). Oestrogen-independent, experience-induced maternal behaviour in female mice. *Journal of Neuroendocrinology, 23*(4), 345–354.

Stolzenberg, D. S., Stevens, J. S., & Rissman, E. F. (2012). Experience-facilitated improvements in pup retrieval: Evidence for an epigenetic effect. *Hormones and Behavior, 62*(2), 128–135.

Stolzenberg, D. S., Zhang, K. Y., Luskin, K., Ranker, L., Balkema, J., Bress, J., & Numan, M. (2009). A single injection of 17beta-estradiol at the time of pup presentation promotes the onset of maternal behavior in pregnancy-terminated rats. *Hormones and Behavior, 56*(1), 121–127.

Stolzenberg, D. S., Zhang, K. Y., Luskin, K., Ranker, L., Bress, J., & Numan, M. (2010). Dopamine D1 receptor activation of adenylyl cyclase, not phospholipase C, in the nucleus accumbens promotes maternal behavior onset in rats. *Hormones and Behavior, 57*(1), 96–104.

Stoop, R. (2012). Neuromodulation by oxytocin and vasopressin. *Neuron, 76*(1), 142–159.

Storey, A. E., & Ziegler, T. E. (2016). Primate paternal care: Interactions between biology and social experience. *Hormones and Behavior, 77*, 260–271.

Strathearn, L., Fonagy, P., Amico, J., & Montague, P. R. (2009). Adult attachment predicts maternal brain and oxytocin response to infant cues. *Neuropsychopharmacology, 34*(13), 2655–2666.

Stuebe, A. M., Grewen, K., & Meltzer-Brody, S. (2013). Association between maternal mood and oxytocin response to breastfeeding. *Journal Women's Health (Larchmont), 22*(4), 352–361.

Sturgis, J. D., & Bridges, R. S. (1997). *N*-methyl-DL-aspartic acid lesions of the medial preoptic area disrupt ongoing parental behavior in male rats. *Physiology & Behavior, 62*(2), 305–310.

Su, J., Leerkes, E. M., & Augustine, M. E. (2018). DRD4 interacts with adverse life events in predicting maternal sensitivity via emotion regulation. *Journal of Family Psychology, 32*(6), 783–792.

Suchecki, D., Rosenfeld, P., & Levine, S. (1993). Maternal regulation of the hypothalamic-pituitary-adrenal axis in the infant rat: The roles of feeding and stroking. *Developmental Brain Research, 75*(2), 185–192.

Sugiura, M., Kawashima, R., Nakamura, K., Sato, N., Nakamura, A., Kato, T., . . . Fukuda, H. (2001). Activation reduction in anterior temporal cortices during repeated recognition of faces of personal acquaintances. *NeuroImage, 13*, 877–890.

Sukikara, M. H., Mota-Ortiz, S. R., Baldo, M. V., Felicio, L. F., & Canteras, N. S. (2010). The periaqueductal gray and its potential role in maternal behavior inhibition in response to predatory threats. *Behavioural Brain Research, 209*(2), 226–233.

Sun, P., Smith, A. S., Lei, K., Liu, Y., & Wang, Z. (2014). Breaking bonds in male prairie vole: Long-term effects on emotional and social behavior, physiology, and neurochemistry. *Behavioural Brain Research, 265*, 22–31.

Svare, B., & Gandelman, R. (1973). Postpartum aggression in mice: Experiential and environmental factors. *Hormones and Behavior, 4*(4), 323–334.

Svare, B., & Gandelman, R. (1976). Postpartum aggression in mice: The influence of suckling stimulation. *Hormones and Behavior, 7*(4), 407–416.

Swain, J. E. (2011). The human parental brain: In vivo neuroimaging. *Progress in Neuropsychopharmacology & Biological Psychiatry, 35*(5), 1242–1254.

Swanson, L. W. (1982). The projections of the ventral tegmental area and adjacent regions: A combined fluorescent retrograde tracer and immunofluorescence study in the rat. *Brain Research Bulletin, 9*(1–6), 321–353.

Swanson, L. W. (1998). *Brain maps: Structure of the rat brain* (2nd rev. ed.). Amsterdam, The Netherlands: Elsevier.

Swanson, L. W., & Petrovich, G. D. (1998). What is the amygdala? *Trends in Neurosciences, 21*(8), 323–331.

Szabo, E. R., Cservenak, M., & Dobolyi, A. (2012). Amylin is a novel neuropeptide with potential maternal functions in the rat. *FASEB Journal, 26*(1), 272–281.

Tabatadze, N., Sato, S. M., & Woolley, C. S. (2014). Quantitative analysis of long-form aromatase mRNA in the male and female rat brain. *PLoS One, 9*(7), e100628.

Tachikawa, K. S., Yoshihara, Y., & Kuroda, K. O. (2013). Behavioral transition from attack to parenting in male mice: A crucial role of the vomeronasal system. *Journal of Neuroscience, 33*(12), 5120–5126.

Takayanagi, Y., Yoshida, M., Bielsky, I. F., Ross, H. E., Kawamata, M., Onaka, T., . . . Nishimori, K. (2005). Pervasive social deficits, but normal parturition, in oxytocin receptor-deficient mice. *Proceedings of the National Academy of Sciences USA, 102*(44), 16096–16101.

Tallot, L., Doyere, V., & Sullivan, R. M. (2016). Developmental emergence of fear/threat learning: Neurobiology, associations and timing. *Genes, Brain, and Behavior, 15*(1), 144–154.

Tang, Y., Chen, Z., Tao, H., Li, C., Zhang, X., Tang, A., & Liu, Y. (2014). Oxytocin activation of neurons in the ventral tegmental area and interfascicular nucleus of mouse midbrain. *Neuropharmacology, 77*, 277–284.

Taylor, A., Atkins, R., Kumar, R., Adams, D., & Glover, V. (2005). A new mother-to-infant bonding scale: Links with early maternal mood. *Archives of Women's Mental Health, 8*(1), 45–51.

Taylor, J. H., & French, J. A. (2015). Oxytocin and vasopressin enhance responsiveness to infant stimuli in adult marmosets. *Hormones and Behavior, 75*, 154–159.

Tecot, S. R., & Baden, A. L. (2018). Profiling caregivers: Hormonal variation underlying allomaternal care in wild red-bellied lemurs, *Eulemur rubriventer. Physiology & Behavior, 193*(Pt A), 135–148.

Tecot, S. R., Singletary, B., & Eadie, E. (2016). Why "monogamy" isn't good enough. *American Journal of Primatology, 78*(3), 340–354.

Terasawa, Y., Kurosaki, Y., Ibata, Y., Moriguchi, Y., & Umeda, S. (2015). Attenuated sensitivity to the emotions of others by insular lesion. *Frontiers in Psychology, 6*, 1314.

Terenzi, M. G., & Ingram, C. D. (2005). Oxytocin-induced excitation of neurones in the rat central and medial amygdaloid nuclei. *Neuroscience, 134*(1), 345–354.

Teresa, F., & Gonçalves-de-Freitas, E. (2011). Reproductive behavior and parental roles of the cichlid fish *Laetacara araguaiae. Neotropical Ichthyology, 9*(2), 355–362.

Terkel, J., Bridges, R. S., & Sawyer, C. H. (1979). Effects of transecting lateral neural connections of the medial preoptic area on maternal behavior in the rat: Nest building, pup retrieval and prolactin secretion. *Brain Research, 169*(2), 369–380.

Thijssen, J. H. (2005). Progesterone receptors in the human uterus and their possible role in parturition. *Journal of Steroid Biochemistry and Molecular Biology, 97*(5), 397–400.

Thijssen, S., Muetzel, R. L., Bakermans-Kranenburg, M. J., Jaddoe, V. W., Tiemeier, H., Verhulst, F. C., . . . van Ijzendoorn, M. H. (2017). Insensitive parenting may accelerate the development of the amygdala-medial prefrontal cortex circuit. *Development and Psychopathology, 29*(2), 505–518.

Thomas, S. A., & Wolff, J. O. (2004). Pair bonding and "the widow effect" in female prairie voles. *Behavioural Processes, 67*(1), 47–54.

Timonin, M. E., Cushing, B. S., & Wynne-Edwards, K. E. (2008). In three brain regions central to maternal behaviour, neither male nor female Phodopus dwarf hamsters show changes in oestrogen receptor alpha distribution with mating or parenthood. *Journal of Neuroendocrinology, 20*(12), 1301–1309.

Tobiansky, D. J., Roma, P. G., Hattori, T., Will, R. G., Nutsch, V. L., & Dominguez, J. M. (2013). The medial preoptic area modulates cocaine-induced activity in female rats. *Behavioral Neuroscience, 127*(2), 293–302.

Tobiansky, D. J., Will, R. G., Lominac, K. D., Turner, J. M., Hattori, T., Krishnan, K., . . . Dominguez, J. M. (2016). Estradiol in the preoptic area regulates the dopaminergic response to cocaine in the nucleus accumbens. *Neuropsychopharmacology, 41*(7), 1897–1906.

Toepfer, P., Heim, C., Entringer, S., Binder, E., Wadhwa, P., & Buss, C. (2017). Oxytocin pathways in the intergenerational transmission of maternal early life stress. *Neuroscience and Biobehavioral Reviews, 73*, 293–308.

Toepfer, P., O'Donnell, K. J., Entringer, S., Garg, E., Heim, C. M., Lin, D. T. S., . . . Buss, C. (2019). DNA methylation changes in the maternal oxytocin gene (OXT) during pregnancy predict postpartum maternal intrusiveness. *Psychoneuroendocrinology, 103*, 156–162.

Tomasello, M., Melis, A. P., Tennie, C., Wyman, E., & Hermann, E. (2012). Two key steps in the evolution of human cooperation: The interdependence hypothesis. *Current Anthropology, 53*(6), 673–692.

Tomasello, M., & Vaish, A. (2013). Origins of human cooperation and morality. *Annual Review of Psychology, 64*, 231–255.

Torner, L. (2008). Role of prolactin in the behavioral and neuroendocrine stress adaptations during lactation. In R. S. Bridges (Ed.), *Neurobiology of the parental brain.* (pp. 131–143). Boston, MA: Academic Press.

Tost, H., Kolachana, B., Hakimi, S., Lemaitre, H., Verchinski, B. A., Mattay, V. S., . . . Meyer-Lindenberg, A. (2010). A common allele in the oxytocin receptor gene (OXTR) impacts prosocial temperament and human hypothalamic-limbic structure and function. *Proceedings of the National Academy of Sciences USA, 107*(31), 13936–13941.

Tottenham, N. (2015). Social scaffolding of human amygdala-mPFC circuit development. *Social Neuroscience, 10*(5), 489–499.

Tottenham, N., Hare, T. A., Quinn, B. T., McCarry, T. W., Nurse, M., Gilhooly, T., . . . Casey, B. J. (2010). Prolonged institutional rearing is associated with atypically large amygdala volume and difficulties in emotion regulation. *Developmental Science, 13*(1), 46–61.

Tovote, P., Esposito, M. S., Botta, P., Chaudun, F., Fadok, J. P., Markovic, M., . . . Luthi, A. (2016). Midbrain circuits for defensive behaviour. *Nature, 534*(7606), 206–212.

Tovote, P., Fadok, J. P., & Luthi, A. (2015). Neuronal circuits for fear and anxiety. *Nature Reviews Neuroscience, 16*(6), 317–331.

Trainor, B. C., Bird, I. M., Alday, N. A., Schlinger, B. A., & Marler, C. A. (2003). Variation in aromatase activity in the medial preoptic area and plasma progesterone is associated with the onset of paternal behavior. *Neuroendocrinology, 78*(1), 36–44.

Trainor, B. C., & Marler, C. A. (2001). Testosterone, paternal behavior, and aggression in the monogamous California mouse (*Peromyscus californicus*). *Hormones and Behavior, 40*(1), 32–42.

Trainor, B. C., & Marler, C. A. (2002). Testosterone promotes paternal behaviour in a monogamous mammal via conversion to oestrogen. *Proceedings of the Royal Society London B, 269*, 823–829.

Trouillet, A. C., Keller, M., Weiss, J., Leinders-Zufall, T., Birnbaumer, L., Zufall, F., & Chamero, P. (2019). Central role of G protein Gαi2 and Gαi2+ vomeronasal neurons in balancing territorial and infant-directed aggression of male mice. *Proceedings of the National Academy of Sciences USA, 116*(11), 5135–5143.

Tsuneoka, Y., Maruyama, T., Yoshida, S., Nishimori, K., Kato, T., Numan, M., & Kuroda, K. O. (2013). Functional, anatomical, and neurochemical differentiation of medial preoptic area subregions in relation to maternal behavior in the mouse. *Journal of Comparative Neurology, 521*(7), 1633–1663.

Tsuneoka, Y., Tokita, K., Yoshihara, C., Amano, T., Esposito, G., Huang, A. J., . . . Kuroda, K. O. (2015). Distinct preoptic-BST nuclei dissociate paternal and infanticidal behavior in mice. *EMBO Journal, 34*(21), 2652–2670.

Tsuneoka, Y., Yoshida, S., Takase, K., Oda, S., Kuroda, M., & Funato, H. (2017). Neurotransmitters and neuropeptides in gonadal steroid receptor-expressing cells in medial preoptic area subregions of the male mouse. *Science Reports, 7*(1), 9809.

Tucker, H. A. (1994). Lactation and its hormonal control. In E. Knobil & J. Neill (Eds.), *The physiology of reproduction* (Vol. 2, pp. 1065–1098). New York, NY: Raven Press.

Turecki, G., & Meaney, M. J. (2016). Effects of the social environment and stress on glucocorticoid receptor gene methylation: A systematic review. *Biological Psychiatry, 79*(2), 87–96.

Turner, L. M., Young, A. R., Romper, H., Schoneberg, T., Phelps, S. M., & Hoekstra, H. E. (2010). Monogamy evolves through multiple mechanisms: Evidence from V1aR in deer mice. *Molecular Biology and Evolution, 27*, 1269–1278.

Tusche, A., Bockler, A., Kanske, P., Trautwein, F. M., & Singer, T. (2016). Decoding the charitable brain: Empathy, perspective taking, and attentional shifts differentially predict altruistic giving. *Journal of Neuroscience, 36*(17), 4719–4732.

Tzeng, S. J., & Linzer, D. I. (1997). Prolactin receptor expression in the developing mouse embryo. *Molecular Reproduction and Development, 48*(1), 45–52.

Uchida, S., Hara, K., Kobayashi, A., Funato, H., Hobara, T., Otsuki, K., . . . Watanabe, Y. (2010). Early life stress enhances behavioral vulnerability to stress through the activation of REST4-mediated gene transcription in the medial prefrontal cortex of rodents. *Journal of Neuroscience, 30*(45), 15007–15018.

Ulrich-Lai, Y. M., & Herman, J. P. (2009). Neural regulation of endocrine and autonomic stress responses. *Nature Reviews Neuroscience, 10*(6), 397–409.

Unger, E. K., Burke, K. J., Jr., Yang, C. F., Bender, K. J., Fuller, P. M., & Shah, N. M. (2015). Medial amygdalar aromatase neurons regulate aggression in both sexes. *Cell Reports, 10*(4), 453–462.

Unternaehrer, E., Meyer, A. H., Burkhardt, S. C., Dempster, E., Staehli, S., Theill, N., . . . Meinlschmidt, G. (2015). Childhood maternal care is associated with DNA methylation of the genes for brain-derived neurotrophic factor (BDNF) and oxytocin receptor (OXTR) in peripheral blood cells in adult men and women. *Stress, 18*(4), 451–461.

Uriarte, N., Breigeiron, M. K., Benetti, F., Rosa, X. F., & Lucion, A. B. (2007). Effects of maternal care on the development, emotionality, and reproductive functions in male and female rats. *Developmental Psychobiology, 49*(5), 451–462.

Valentino, R. J., Liouterman, L., & Van Bockstaele, E. J. (2001). Evidence for regional heterogeneity in corticotropin-releasing factor interactions in the dorsal raphe nucleus. *Journal of Comparative Neurology, 435*(4), 450–463.

Vandenbergh, J. G. (1973). Effects of central and peripheral anosmia on reproduction of female mice. *Physiology & Behavior, 10*(2), 257–261.

van Leengoed, E., Kerker, E., & Swanson, H. H. (1987). Inhibition of post-partum maternal behaviour in the rat by injecting an oxytocin antagonist into the cerebral ventricles. *Journal of Endocrinology, 112*(2), 275–282.

Veenema, A. H., Bredewold, R., & Neumann, I. D. (2007). Opposite effects of maternal separation on intermale and maternal aggression in C57BL/6 mice: Link to hypothalamic vasopressin and oxytocin immunoreactivity. *Psychoneuroendocrinology, 32*(5), 437–450.

Vella, E. T., Evans, C. C., Ng, M. W., & Wynne-Edwards, K. E. (2005). Ontogeny of the transition from killer to caregiver in dwarf hamsters (*Phodopus campbelli*) with biparental care. *Developmental Psychobiology, 46*(2), 75–85.

Verhage, M. L., Schuengel, C., Madigan, S., Fearon, R. M. P., Oosterman, M., Cassibba, R., . . . van, I. M. H. (2016). Narrowing the transmission gap: A synthesis of three decades of research on intergenerational transmission of attachment. *Psychological Bulletin, 142*(4), 337–366.

Vertes, R. P. (2004). Differential projections of the infralimbic and prelimbic cortex in the rat. *Synapse, 51*(1), 32–58.

Vinograd, A., Fuchs-Shlomai, Y., Stern, M., Mukherjee, D., Gao, Y., Citri, A., . . . Mizrahi, A. (2017). Functional plasticity of odor representations during motherhood. *Cell Reports, 21*(2), 351–365.

vom Saal, F. S. (1985). Time-contingent change in infanticide and parental behavior induced by ejaculation in male mice. *Physiology & Behavior, 34*(1), 7–15.

vom Saal, F. S., & Howard, L. S. (1982). The regulation of infanticide and parental behavior: Implications for reproductive success in male mice. *Science, 215*(4537), 1270–1272.

Wager, T. D., Davidson, M. L., Hughes, B. L., Lindquist, M. A., & Ochsner, K. N. (2008). Prefrontal-subcortical pathways mediating successful emotion regulation. *Neuron, 59*(6), 1037–1050.

Wallis, C. U., Cardinal, R. N., Alexander, L., Roberts, A. C., & Clarke, H. F. (2017). Opposing roles of primate areas 25 and 32 and their putative rodent homologs in the regulation of negative emotion. *Proceedings of the National Academy of Sciences USA, 114*(20), e4075–e4084.

Walsh, C. J., Fleming, A. S., Lee, A., & Magnusson, J. E. (1996). The effects of olfactory and somatosensory desensitization on Fos-like immunoreactivity in the brains of pup-exposed postpartum rats. *Behavioral Neuroscience, 110*(1), 134–153.

Walter, H., Adenzato, M., Ciaramidaro, A., Enrici, I., Pai, L., & Bara, B. G. (2004). Understanding intentions in social interaction: The role of the anterior paracingulate cortex. *Journal of Cognitive Neuroscience, 16*(10), 1854–1863.

Walum, H., & Young. L. J. (2018). The neural mechanisms and circuitry of the pair bond. *Nature Reviews Neuroscience, 19*(11), 643–654.

Wamboldt, M. Z., & Insel, T. R. (1987). The ability of oxytocin to induce short latency maternal behavior is dependent on peripheral anosmia. *Behavioral Neuroscience, 101*(3), 439–441.

Wan, X. S., Liang, F., Moret, V., Wiesendanger, M., & Rouiller, E. M. (1992). Mapping of the motor pathways in rats: c-Fos induction by intracortical microstimulation of the motor cortex correlated with efferent connectivity of the site of cortical stimulation. *Neuroscience, 49*(4), 749–761.

Wang, B., Li, Y., Wu, R., Zhang, S., & Tai, F. (2015). Behavioral responses to pups in males with different reproductive experiences are associated with changes in central OT, TH, OTR, D1R, D2R mRNA expression in mandarin voles. *Hormones and Behavior, 67*, 73–82.

Wang, B., Wang, L., Wang, K., & Tai, F. (2018). The effects of fathering experience on paternal behaviors and levels of central expression of oxytocin and dopamine-2 type receptors in mandarin voles. *Physiology & Behavior, 193*, 35–42.

Wang, H., Duclot, F., Liu, Y., Wang, Z., & Kabbaj, M. (2013). Histone deacetylase inhibitors facilitate partner preference formation in female prairie voles. *Nature Neuroscience, 16*(7), 919–924.

Wang, L., Burger, L. L., Greenwald-Yarnell, M. L., Myers, M. G., & Moenter, S. M. (2018). Glutamatergic transmission to hypothalamic kisspeptin neurons is differentially regulated by estradiol through estrogen receptor α in adult female mice. *Journal of Neuroscience, 38*(5), 1061–1072.

Wang, L., Chen, Z., & Lin, D. (2015). Collateral pathways from the ventromedial hypothalamus mediate defensive behaviors. *Neuron, 85*(6), 1344–1358.

Wang, L., DeFazio, R. A., & Moenter, S. M. (2016). Excitability and burst generation of AVPV kisspeptin neurons are regulated by the estrous cycle via multiple conductances modulated by estradiol action. *eNeuro, 3*(3).

Wang, Z., Moody, K., Newman, J. D., & Insel, T. R. (1997). Vasopressin and oxytocin immunoreactive neurons and fibers in the forebrain of male and female common marmosets (*Callithrix jacchus*). *Synapse, 27*(1), 14–25.

Wang, Z., & Storm, D. R. (2011). Maternal behavior is impaired in female mice lacking type 3 adenylyl cyclase. *Neuropsychopharmacology, 36*(4), 772–781.

Wang, Z., Toloczko, D., Young, L. J., Newman, J. D., & Insel, T. R. (1997). Vasopressin in the forebrain of common marmosets (Callithrix jacchus): Studies with in situ hybridization, immunocytochemistry and receptor autoradiography. *Brain Research, 768*(1–2), 147–156.

Wang, Z., Young, L., Liu, Y., & R. Insel, T. (1997). Species differences in vasopressin receptor binding are evident early in development: Comparative anatomic studies in prairie and montane voles. *Journal of Comparative Neurology, 378*(4), 535–546.

Waselus, M., Nazzaro, C., Valentino, R. J., & Van Bockstaele, E. J. (2009). Stress-induced redistribution of corticotropin-releasing factor receptor subtypes in the dorsal raphe nucleus. *Biological Psychiatry, 66*(1), 76–83.

Watarai, A., Arai, N., Miyawaki, S., Okano, H., Miura, K., Mogi, K., & Kikusui, T. (2018). Responses to pup vocalizations in subordinate naked mole-rats are induced by estradiol ingested through coprophagy of queen's feces. *Proceedings of the National Academy of Sciences USA, 115*(37), 9264–9269.

Weaver, I. C., Cervoni, N., Champagne, F. A., D'Alessio, A. C., Sharma, S., Seckl, J. R., . . . Meaney, M. J. (2004). Epigenetic programming by maternal behavior. *Nature Neuroscience, 7*(8), 847–854.

Wei, L., Meaney, M. J., Duman, R. S., & Kaffman, A. (2011). Affiliative behavior requires juvenile, but not adult neurogenesis. *Journal of Neuroscience, 31*(40), 14335–14345.

Wei, Y.-C., Wang, S.-R., Jiao, Z.-L., Zhang, W., Lin, J.-K., Li, X.-Y., . . . Xu, X.-H. (2018). Medial preoptic area in mice is capable of mediating sexually dimorphic behaviors regardless of gender. *Nature Communications, 9*(1), 279.

Weidt, A., Lindholm, A. K., & Konig, B. (2014). Communal nursing in wild house mice is not a by-product of group living: Females choose. *Naturwissenschaften, 101*(1), 73–76.

Weisner, T. S., & Gilmore, R. (1977). My brother's keeper: Child and sibling caretaking. *Current Anthropology, 18*(2), 169–190.

Wiesner, B. P., & Sheard, N. M. (1933). *Maternal behavior in the rat.* Edinburgh, England: Oliver & Boyd.

Wilkenfeld, S. R., Lin, C., & Frigo, D. E. (2018). Communication between genomic and non-genomic signaling events coordinate steroid hormone actions. *Steroids, 133*, 2–7.

Willett, J. A., Johnson, A. G., Vogel, A. R., Patisaul, H. B., McGraw, L. A., & Meitzen, J. (2018). Nucleus accumbens core medium spiny neuron electrophysiological properties and partner preference behavior in the adult male prairie vole, *Microtus ochrogaster. Journal of Neurophysiology, 119*(4), 1576–1588.

Wingfield, J. C., Hegner, R. E., Dufty, A. M., & Ball, G. F. (1990). The "challenge hypothesis": Theoretical implications for patterns of testosterone secretion, mating systems, and breeding strategies. *American Naturalist, 136*(6), 829–846.

Winslow, J. T., Noble, P. L., Lyons, C. K., Sterk, S. M., & Insel, T. R. (2003). Rearing effects on cerebrospinal fluid oxytocin concentration and social buffering in rhesus monkeys. *Neuropsychopharmacology, 28*(5), 910–918.

Wintermantel, T. M., Campbell, R. E., Porteous, R., Bock, D., Grone, H. J., Todman, M. G., . . . Herbison, A. E. (2006). Definition of estrogen receptor pathway critical for estrogen positive feedback to gonadotropin-releasing hormone neurons and fertility. *Neuron, 52*(2), 271–280.

Witteman, J., Van, I. M. H., Rilling, J. K., Bos, P. A., Schiller, N. O., & Bakermans-Kranenburg, M. J. (2019). Towards a neural model of infant cry perception. *Neuroscience and Biobehavioral Reviews, 99*, 23–32.

Wolfe, F. H., Deruelle, C., & Chaminade, T. (2018). Are friends really the family we choose? Local variations in hypothalamus activity when viewing personally known faces. *Social Neuroscience, 13*(3), 289–300.

Wonch, K. E., de Medeiros, C. B., Barrett, J. A., Dudin, A., Cunningham, W. A., Hall, G. B., . . . Fleming, A. S. (2016). Postpartum depression and brain response to infants: Differential amygdala response and connectivity. *Social Neuroscience, 11*(6), 600–617.

Wong, L. C., Wang, L., D'Amour, Yumita, T., Chen., Yamaguchi, T., . . . Lin, D. (2016). Effective modulation of male aggression through lateral septum to medial hypothalamus projection. *Current Biology, 26*(5), 593–604.

Wood, M., Adil, O., Wallace, T., Fourman, S., Wilson, S. P. Herman, J. P., & Myers, B. (2019). Infralimbic prefrontal cortex structural and functional connectivity with limbic forebrain: A combined viral genetic and optogenetic analysis. *Brain Structure and Function, 224*, 73–97.

Woods, R., Bedard, M. McQuaid, R. J., Matheson, K., & Anisman, H. (2018). Rejection sensitivity and multiple group memberships: The moderating role of an oxytocin receptor gene polymorphism. *Social Neuroscience, 13*(3), 268–276.

Woodworth, H. L., Perez-Bonilla, P. A., Beekly, B. G., Lewis, T. J., & Leinninger, G. M. (2018). Identification of neurotensin receptor expressing cells in the ventral tegmental area across the lifespan. *eNeuro, 5*(1).

Wright, D. B., Laurent, H. K., & Ablow, J. L. (2017). Mothers who were neglected in childhood show differences in their neural response to their infant's cry. *Child Maltreatment, 22*(2), 158–166.

Wu, Z., Autry, A. E., Bergan, J. F., Watabe-Uchida, M., & Dulac, C. G. (2014). Galanin neurons in the medial preoptic area govern parental behaviour. *Nature, 509*(7500), 325–330.

Wynne-Edwards, K. E., & Reburn, C. J. (2000). Behavioral endocrinology of mammalian fatherhood. *Trends in Ecology & Evolution, 15*(11), 464–468.

Xiao, L., Priest, M. F., Nasenbeny, J., Lu, T., & Kozorovitskiy, Y. (2017). Biased oxytocinergic modulation of midbrain dopamine systems. *Neuron, 95*(2), 368–384.

Xue, X., Shao, S., Li, M., Shao, F., & Wang, W. (2013). Maternal separation induces alterations of serotonergic system in different aged rats. *Brain Research Bulletin, 95*, 15–20.

Yamamoto, M. E., & Box, H. O. (1997). The role of non-reproductive helpers in infant care in captive Callithrix jacchus. *Ethology, 103*, 760–771.

Yamamoto, M. E., Box, H. O., Albuquerque, F. S., & de Fátima Arruda, M. (1996). Carrying behavior in captive and wild marmosets (*Callithrix jacchus*): A comparison between two colonies and a field site. *Primates, 37*, 297–304.

Yamamoto, Y., Liang, M., Munesue, S., Deguchi, K., Harashima, A., Furuhara, K., . . . Higashida, H. (2019). Vascular RAGE transports oxytocin into the brain to elicit its maternal bonding behaviour in mice. *Communications Biology, 2*(1), 76.

Yan, C. G., Rincon-Cortes, M., Raineki, C., Sarro, E., Colcombe, S. Guilfoyle, D. N., . . . Castellanos, F. X. (2017). Aberrant development of intrinsic brain activity in a rat model of caregiver maltreatment of offspring. *Translational Psychiatry, 7*(1), e1005.

Yeo, J. A., & Keverne, E. B. (1986). The importance of vaginal-cervical stimulation for maternal behaviour in the rat. *Physiology & Behavior, 37*(1), 23–26.

Yim, I. S., Tanner Stapleton, L. R., Guardino, C. M., Hahn-Holbrook, J., & Dunkel Schetter, C. (2015). Biological and psychosocial predictors of postpartum depression: Systematic review and call for integration. *Annual Review of Clinical Psychology, 11*, 99–137.

Yoshida, M., Takayanagi, Y., Inoue, K., Kimura, T., Young, L. J., Onaka, T., & Nishimori, K. (2009). Evidence that oxytocin exerts anxiolytic effects via oxytocin receptor expressed in serotonergic neurons in mice. *Journal of Neuroscience, 29*(7), 2259–2271.

Yoshihara, C., Numan, M., & Kuroda, K. O. (2018). Oxytocin and parental behaviors. *Current Topics in Behavioral Neuroscience, 35*, 119–153.

Young, K. A., Gobrogge, K. L., Liu, Y., & Wang, Z. (2011). The neurobiology of pair bonding: Insights from a socially monogamous rodent. *Frontiers in Neuroendocrinology, 32*(1), 53–69.

Young, L. J., & Wang, Z. (2004). The neurobiology of pair bonding. *Nature Neuroscience, 7*(10), 1048–1054.

Young, L. J., Wang, Z., Donaldson, R., & Rissman, E. F. (1998). Estrogen receptor alpha is essential for induction of oxytocin receptor by estrogen. *NeuroReport, 9*(5), 933–936.

Yu, G. Z., Kaba, H., Okutani, F., Takahashi, S., & Higuchi, T. (1996). The olfactory bulb: A critical site of action for oxytocin in the induction of maternal behaviour in the rat. *Neuroscience, 72*(4), 1083–1088.

Yu, G. Z., Kaba, H., Okutani, F., Takahashi, S., Higuchi, T., & Seto, K. (1996). The action of oxytocin originating in the hypothalamic paraventricular nucleus on mitral and granule cells in the rat main olfactory bulb. *Neuroscience, 72*(4), 1073–1082.

Yu, Q., Teixeira, C, M, Mahedevia, D., Huang, Y., Balsam, D., Mann, J. J., Gingrich, J. A., & Ansorge, M. S. (2014). Dopamine and serotonin signaling during sensitive developmental periods differentially impact adult aggressive and affective behaviors in mice. *Molecular Psychiatry, 19*, 688–698.

Yuan, W., He, Z., Hou, W., Wang, L., Li, L., Zhang, J., . . . Tai, F. (2019). Role of oxytocin in the medial preoptic area (MPOA) in the modulation of paternal behavior in mandarin voles. *Hormones and Behavior, 110*, 46–55.

Zahed, S. R., Prudom, S. L., Snowdon, C. T., & Ziegler, T. E. (2008). Male parenting and response to infant stimuli in the common marmoset (*Callithrix jacchus*). *American Journal of Primatology, 70*(1), 84–92.

Zahn, R., Moll, J., Paiva, M., Garrido, G., Krueger, F., Huey, E. D., Grafman, J. (2009). The neural basis of human social values: Evidence from functional MRI. *Cerebral Cortex, 19*, 276–283.

Zarrow, M. X., Gandelman, R., & Denenberg, V. H. (1971). Prolactin: Is it an essential hormone for maternal behavior in the mammal? *Hormones and Behavior, 2*(4), 343–354.

Zhang, L., Hernandez, V. S., Swinny, J. D., Verma, A. K., Giesecke, T., Emery, A. C., . . . Eiden, L. E. (2018). A GABAergic cell type in the lateral habenula links hypothalamic homeostatic and midbrain motivation circuits with sex steroid signaling. *Translational Psychiatry, 8*(1), 50.

Zhang, R., Asai, M., Mahoney, C. E., Joachim, M., Shen, Y. Gunner, G., & Majzoub, J. A. (2017). Loss of hypothalamic corticotropin-releasing hormone markedly reduces anxiety behavior in mice. *Molecular Psychiatry, 22*(5), 733–744.

Zhu, X., Li, T., Peng, S., Ma, X., Chen, X., & Zhang, X. (2010). Maternal deprivation-caused behavioral abnormalities in adult rats relate to a non-methylation-regulated D2 receptor levels in nucleus accumbens. *Behavioural Brain Research, 209*(2), 281–288.

Zhu, Y., Wang, Y., Yao, R., Hao, T., Cao, J. Huang, H., . . . Wu, Y. (2017). Enhanced neuroinflammation mediated by DNA methylation of the glucocorticoid receptor triggers cognitive dysfunction after sevoflurane anesthesia in adult rats subjected to maternal separation during the neonatal period. *Journal of Neuroinflammation, 14*, 6.

Index

Tables and figures are indicated by *t* and *f* following the page number

For the benefit of digital users, indexed terms that span two pages (e.g., 52–53) may, on occasion, appear on only one of those pages.